REED'S APPLIED MECHANICS FOR ENGINEERS

by

WILLIAM EMBLETON O.B.E.
C.Eng., F.I. Mar.E., M.I. Mech E.
Extra First Class Engineers' Certificate

Revised by
LESLIE JACKSON, B.Sc. (Lond.).
C.Eng., F.I. Mar.E., F.R.I.N.A.
Extra First Class Engineers' Certificate

THOMAS REED PUBLICATIONS

A DIVISION OF THE ABR COMPANY LIMITED

First Edition – 1962
Reprinted – 1966
Second Edition – 1970 *(in SI Units)*
Reprinted – 1972
Third Edition – 1975
Reprinted – 1979
Reprinted – 1982
Fourth Edition – 1983
Reprinted – 1989
Reprinted – 1990
Fifth Edition – 1994
Reprinted – 1999

ISBN 0 901281 05 0

THOMAS REED PUBLICATIONS
The Barn
Ford Farm
Bradford Leigh
Bradford-on-Avon
Wiltshire BA15 2RP
United Kingdom

Email: tugsrus@abreed.demon.co.uk

Produced by Omega Profiles Ltd, SP11 7RW
Printed and Bound in Great Britain

REED'S

APPLIED MECHANICS

PREFACE

This book covers the syllabuses in Applied Mechanics for all classes of the Marine Engineers' Certificates of Competency of the Department of Transport (DTp). The examinations are now administered by the Scottish Vocational Educational Council (SCOTVEC). It is a useful aid to students on Business and Technician Education Council (BTEC) and SCOTVEC engineering courses.

Basic principles are dealt with commencing at a fairly elementary stage. Each chapter has fully worked examples interwoven into the text, test examples are set at the end of each chapter for the student to work out, and finally there are some typical examination questions included. The prefix 'f' is used to indicate those parts of the text, and some test examples, of Class One standard.

The author has gone beyond the normal practice of merely supplying bare answers to the test examples and examination questions by providing fully worked step by step solutions leading to the final answers.

This latest revision is a major update in the subject so taking the material for study into the twenty-first century.

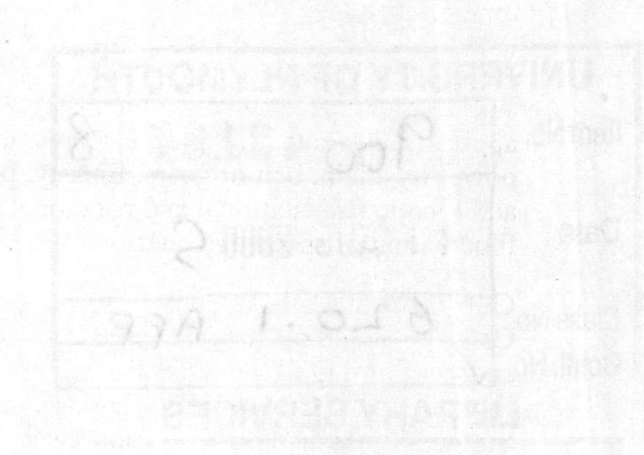

CONTENTS

CONTENTS (cont.)

CONTENTS (cont.)

CHAPTER 1

STATICS

Statics – force systems virtually at rest.

MASS, symbol m, is the quantity of matter possessed by a body; a constant proportional to its volume and density. The kilogram (kg) is the *unit* of mass; 1 Mg is 10^3 kg which is a tonne (t).

FORCE, symbol F, acts to change the state of rest or uniform rectilinear motion of a mass. The newton (N) is the *unit* of force which if applied to unit mass (1 kg) will give it unit acceleration (1m/s^2); 1 kN is 10^3N and 1 MN is 10^6N.

$$\text{Accelerating force (N)} = \text{mass(kg)} \times \text{acceleration (m/s}^2)$$
$$F = ma$$

WEIGHT, symbol W, of a body is the gravitational force on the mass of that body; gravitational acceleration, symbol g is 9.81 m/s^2.

$$\text{Weight (N)} = \text{mass (kg)} \times \text{gravitational acceleration (981 m/s}^2)$$
$$W = mg$$

To describe a force completely, its *magnitude* and also its *direction* must be specified. When two such properties are required to define a quantity it is called a *vector quantity* and it can be represented by a *vector*.

A vector is a line drawn to scale; in the case of a force the length of the line represents the magnitude of the force and the direction in which the line is drawn with an arrow on it represents the direction of the line of action of the force.

Fig 1 shows some vectors representing forces

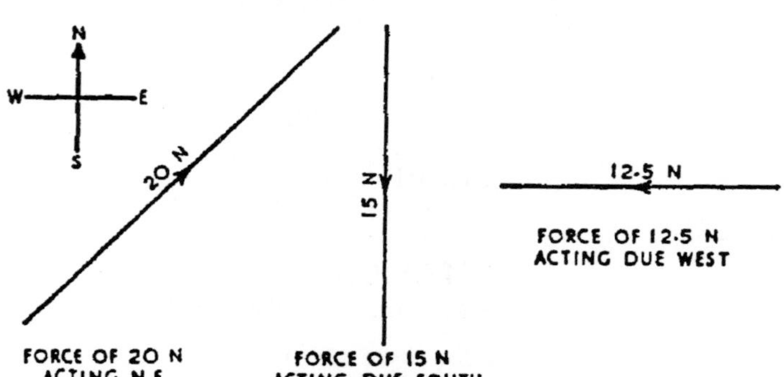

Fig. 1

RESULTANT

The resultant of a number of coplanar forces is that one force which would have the same effect if it replaced those forces.

Fig. 2 shows three forces of 8, 10 and 5 N respectively all pulling on a body in the same direction; it is obvious that the resultant of these is a single force of 23 N in the same direction. This is a simple case of parallel forces involving only the addition of the forces. The *space diagram* is an illustration of the system of forces. The *vector diagram* is a diagram drawn to scale with the vectors joined end to end.

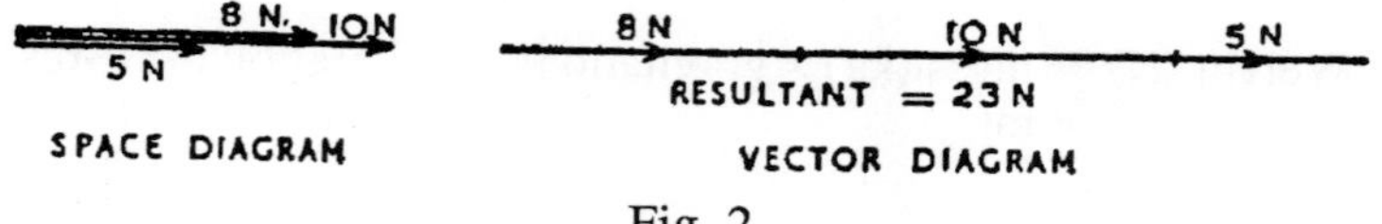

Fig. 2

When the forces are not parallel the vector diagram is 'bent' at the joints of the vectors so that each vector is drawn in the direction in which its respective force acts. Taking forces of the same magnitude as above but slightly different directions, Fig. 3 illustrates how the vector diagram is constructed.

Note how the arrows of the vectors of the given forces form a continuous path in the vector diagram. The vector diagram is drawn to scale, the resultant is the *vector addition* of the given forces, it measures 21·9 N and its direction is 23 degrees to the

8 N force. This single force can replace the three given forces to have exactly the same effect.

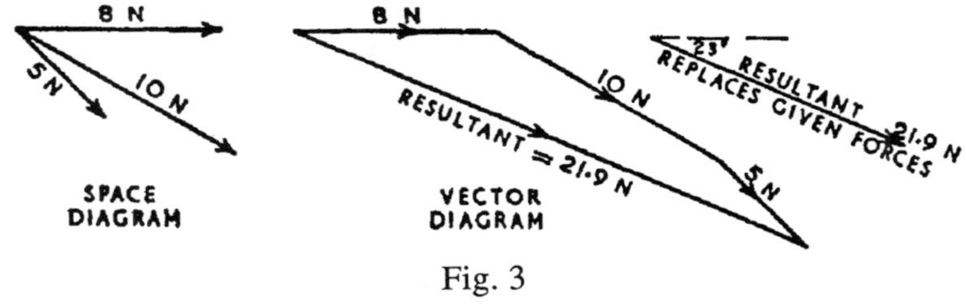

Fig. 3

EQUILIBRANT

The equilibrant is a single force which, if added to a system of forces acting on a body, would place the body in equilibrium. In other words, the equilibrant will neutralise the other forces.

Taking the last example again, if a force of the *same magnitude* but *opposite in direction* to the resultant of the three given forces was added, it would neutralise the effect of those three forces and the body would be in equilibrium. Fig 4 shows the equilibrant. Note the vector diagram. The equilibrant 'closes the vector diagram' and the direction of its arrow forms a continuous path with the others, often referred to as being 'nose to tail'.

This gives an introduction to the theorems of the triangle and polygon of forces to follow.

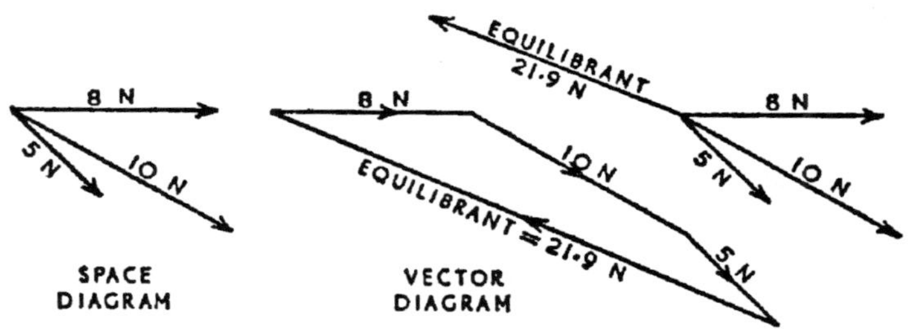

Fig. 4

TRIANGLE OF FORCES

If three forces acting at a point are in equilibrium, the vector diagram drawn to scale representing the forces in magnitude and direction, taken in order, forms a closed triangle.

POLYGON OF FORCES

If any number of forces acting at a point are in equilibrium, the vector diagram drawn to scale representing the forces in magnitude and direction, taken in order, forms a closed polygon.

The above theorems are therefore the same except that the triangle of forces refers only to three forces and the polygon of forces refers to any number greater than three.

It is now obvious that the magnitude and direction of any one of a system of forces in equilibrium could be unknowns to be solved. The given forces are taken in order and their vectors are drawn to scale, the vector required to close the diagram represents the magnitude and direction of the unknown force. Alternatively, instead of the unknown quantities being the magnitude and direction of one force, they could be the magnitudes of two forces if their directions were known, or the directions of two forces if their magnitudes were known.

CONCURRENT AND PARALLEL COPLANAR FORCES

The lines of action of three coplanar forces in equilibrium, or any number of forces in equilibrium which can be reduced to three, must either pass through a common point or be parallel to each other.

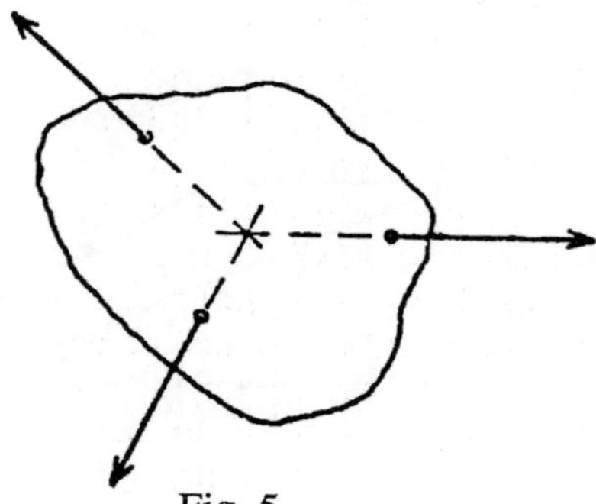

Fig. 5

Fig. 5 shows three forces pulling on a plate. If this system of forces is balanced, *i.e.* the plate is in equilibrium, the lines of action of the three forces *must pass through a common point* because they are not parallel.

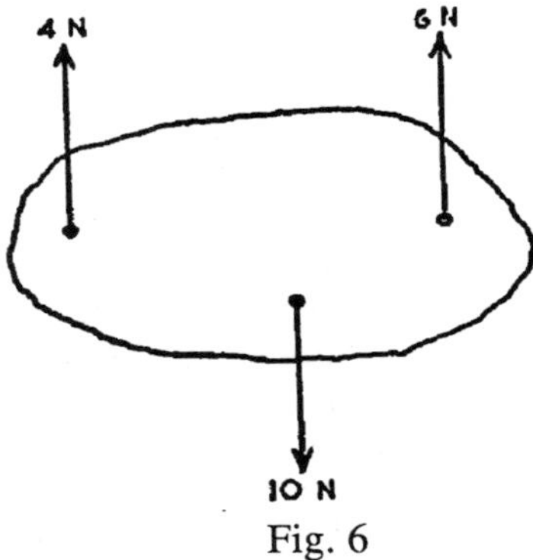

Fig. 6

Fig. 6 shows forces pulling on a plate, these forces can be in equilibrium without being concurrent because they are parallel.
Note A condition of equilibrium is that the algebraic sum of the forces (or rectangular components) perpendicular and parallel to the plane must equate to zero (see Chapter 7).

BOW'S NOTATION

This is a method of defining forces in a force system by lettering the spaces of the space diagram with capital letters, A, B, C, etc., so that each force can be referred to by the letters of the two spaces the force separates, such as force AB, force BC, and so on. The vector of each force in the vector diagram is labelled with its corresponding small letters on the two ends of the vector, *ab*, *bc*, etc., in the direction of the arrow.

Fig. 7 illustrates this method of notation. The spaces between the forces on the space diagram are lettered in a continuous (clockwise or anticlockwise) direction, preferably commencing with a vertical or horizontal force for convenience of beginning the construction of the vector diagram. The vector diagram is constructed by first drawing vector *ab* to represent the force AB, then vector *bc* to represent force BC, and so on. Although the lettering of the space diagram for one system of forces can be clockwise or anti-clockwise so long as it is continuous, it is advisable to adopt one method, say clockwise, and adhere to this for all problems. This is essential when combining vector diagrams of more than one system of forces in framed structures.

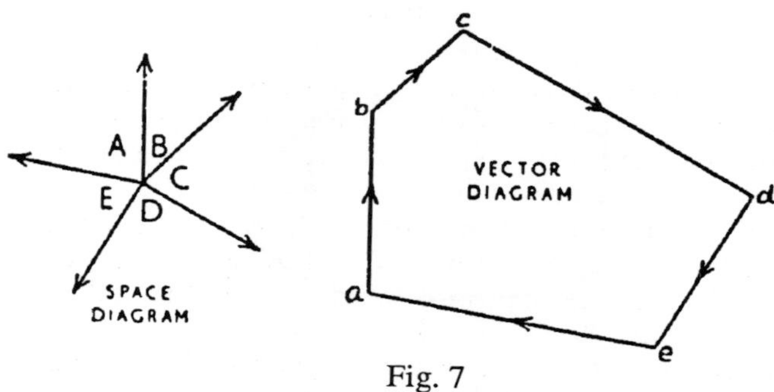

Fig. 7

COMPONENTS OF A FORCE

A force can be resolved into two components, these being two forces that could replace the given force on a body and have the same effect. It will be realised that to find the two components of a given force is the reverse process of finding the resultant of two given forces. Although components in any direction can be found, the most useful are usually the *rectangular* components, that is, those which are at right angles to each other.

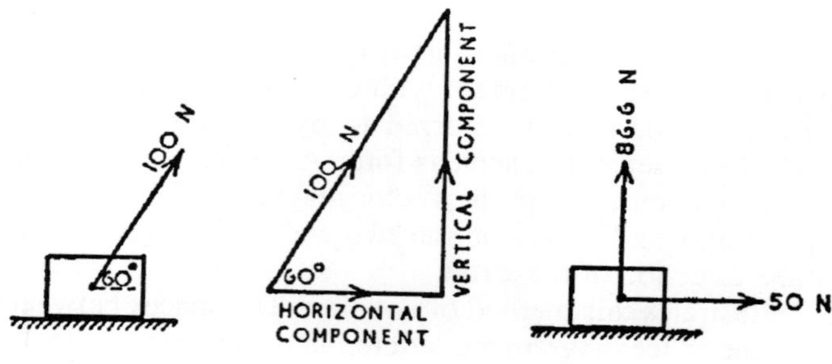

Fig.8

Fig. 8 illustrates a force of 100 N inclined at 60° to the horizontal pulling on a body. The rectangular components of this force are the horizontal and vertical pulling effects, for instance, the applied force of 100 N tends to (i) pull the body horizontally to the right, (ii) lift the body upwards. From the vector diagram the horizontal pull is $100 \times \cos 60° = 50$ N, and the vertical lift is $100 \times \sin 60° = 86.6$ N, therefore these horizontal and vertical components can take the place of the single inclined force of 100 N.

SLINGS

Example. Two ropes are slung from a beam and their lower ends are connected by a shackle from which a load of 400 N hangs. If the ropes make angles of 50° and 60° respectively to the vertical, find the pull in each rope.

Firstly, the space diagram is drawn (Fig. 9) to illustrate the connections of the ropes and load. The shackle is the 'node' where the three forces meet and arrows are inserted to indicate the directions in which the forces pull on this node.

Using Bow's notation the vector diagram is then constructed. Draw to scale the vector *ab* vertically downwards to represent the force AB which is 400 N. From *b* draw a line parallel to BC (at 50° to the vertical), as the magnitude of this force *i.e.* the length of *bc* is not known so it is drawn a little longer than it should be. Now *ca* finishes at point *a* because it is to form a closed figure, therefore draw back from point *a* in the direction of 60° to the vertical until it cuts the previous vector. This gives point *c*.

The forces in the ropes may now be found by measuring to scale the lengths of the vectors *bc* and *ca*.

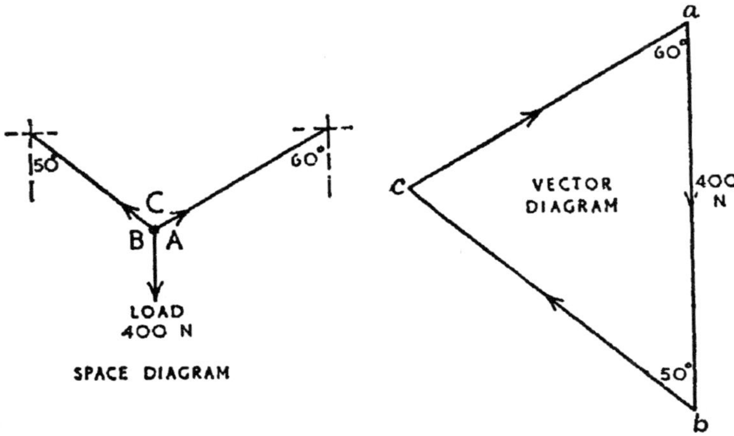

Fig. 9

To calculate the forces:
Angle *acb* (opposite 400 N vector)
$= 180 - (60 + 50) = 70°$

$$\frac{ac}{\sin 50°} = \frac{400}{\sin 70°}$$

$$ac = \frac{400 \times 0{\cdot}766}{0{\cdot}9397}$$

$$= 326 \text{ N}$$

$$\frac{bc}{\sin 60°} = \frac{400}{\sin 70°}$$

$$bc = \frac{400 \times 0{\cdot}866}{0{\cdot}9397}$$

$$= 368{\cdot}6 \text{ N}$$

∴ Force in rope AC = 326 N
„ „ BC = 368.6 N } Ans.

Example. Two rope slings, each 2 m long, are used to lift a small engine bedplate of mass 3·058 t. The attachments to the bedplate are 2·5 m apart horizontally and the top ends of the ropes are connected to a common ring on the crane hook. Find the tension in each rope sling.

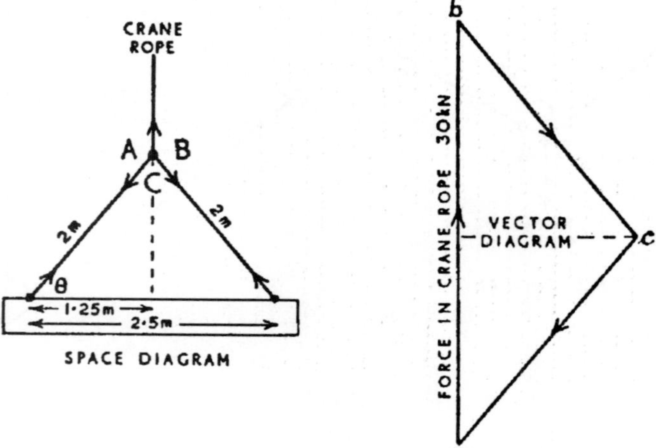

Fig. 10

When the space diagram is sketched (Fig. 10) the ring on the crane hook is a node where three forces meet. The crane rope pulls upwards on this ring, the force being equal to the total weight of the bedplate, and the top ends of each rope sling pull downwards on this ring. Note that the bottom ends of the slings pull upwards on the connections at the bedplate.

$$\text{Mass of bedplate} = 3{\cdot}058 \times 10^3 \text{ kg}$$

Force applied by crane rope to overcome gravitational force on bedplate, *i.e.* the weight of the bedplate

$$= 3{\cdot}058 \times 10^3 \times 9{\cdot}81 = 30 \times 10^3 \text{ N} = 30 \text{ kN}$$

$$\cos\theta = \frac{1{\cdot}25}{2} = 0{\cdot}625$$

$$\therefore \theta = 51^\circ\ 19'$$

Referring to the vector diagram,

$$\text{Angles } bac \text{ and } abc = 90^\circ - 51^\circ\ 19' = 38^\circ\ 41'$$

$$\frac{15}{ac} = \cos 38^\circ\ 41'$$

$$ac = \frac{15}{0{\cdot}7806}$$

$$= 19{\cdot}22 \text{ kN}$$

$\therefore$ tension in each rope sling = 19·22 kN Ans.

Example. Four forces pull on a point, the magnitudes and directions of three of them are, 12 N due North, 15 N at 30° East of North, 20 N at 40° East of South. Find the magnitude and direction of the fourth force so that the system will be in equilibrium, (*a*) by graphical means, (*b*) by calculation.

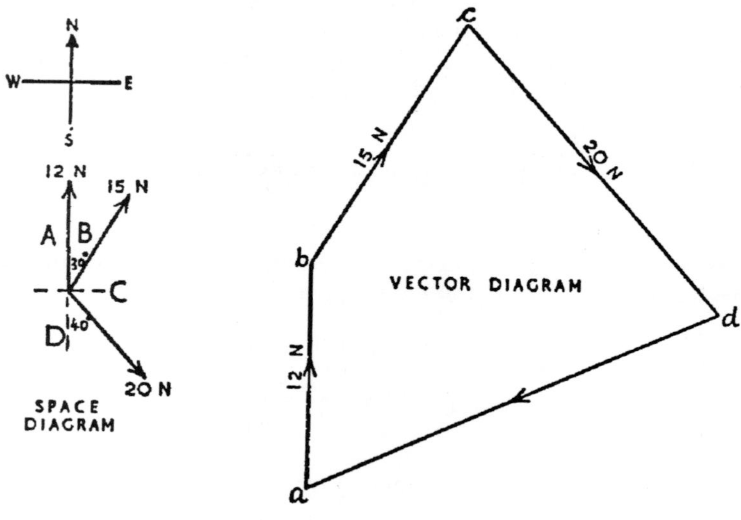

Fig. 11

(i) To construct the vector diagram (Fig. 11) choose a suitable scale and draw vertically upwards vector *ab* to represent the force of 12 N due North, from *b* draw *bc* to represent 15 N at 30° to the vertical, from *c* draw *cd* representing 20 N at 40° to the vertical. As the system is balanced the vector diagram must form a closed figure, therefore the fourth force must be represented by the vector from *d* to a which closes the diagram.

da measures $22\frac{1}{2}$ N and the angle at *a* measures $64\frac{1}{2}$°.

$\therefore$ Equilibrant = $22\frac{1}{2}$ N at S64·5°W Ans.

(ii) To calculate *da* from the vector diagram divide the figure into two triangles by drawing a line from *b* to *d*, calculating *bd* from the triangle *bcd* then calculating *da* from the triangle *bda*. However, an easier method of calculation is to resolve all forces into their North-South and East-West components and reduce the problem to a triangle of forces, and find the equilibrant of those two resultants, as follows:

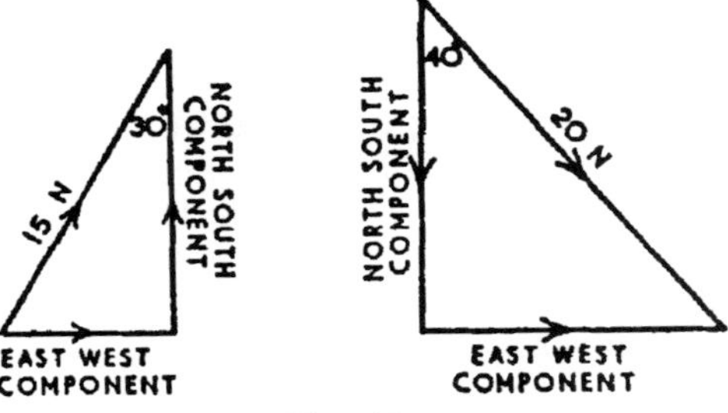

Fig. 12

N.S. component of	15 = 15 cos 30°	= 12·99 North
E.W. ,,	15 = 15 sin 30°	= 7·5 East
N.S. ,,	20 = 20 cos 40°	= 15·32 South
E.W. ,,	20 = 20 sin 40°	= 12·856 East
N.S. ,,	12	= 12 North
E.W. ,,	12	= Nil

Resultant of North-South components
= 12·99 North + 12 North – 15·32 South
= 9·67 N due North

Resultant of East-West components
= 7.5 East + 12·856 East
= 20.356 N due East

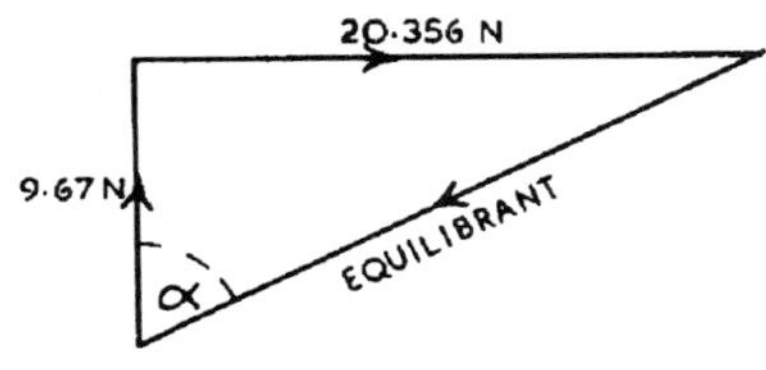

Fig. 13

$$\text{Equilibrant} = \sqrt{9{\cdot}67^2 + 20{\cdot}356^2} = 22{\cdot}54 \text{ N}$$

$$\tan \alpha = \frac{20{\cdot}356}{9{\cdot}67} = 2{\cdot}105$$

$$\therefore \alpha = 64° \ 36'$$

$\therefore$ Force = 22·54 N at S 64° 36′ W Ans.

JIB CRANES

A simple jib-crane consists of a post, a jib and a tie. The post is usually vertical, the jib is hinged at its lowest end to the bottom of the post, and the tie connects the top of the jib with the bottom of the post. The junction of the tie and jib is the crane head.

In problems on jib cranes it is sometimes taken that the load is suspended directly from a fixture at the crane head and the problem then involves a simple triangle of forces. In other cases they are described as having a pulley at the crane head, the lifting rope passing over this pulley and down to a winch behind the crane. Such cases involve more than three forces at the crane head.

Example. The angle between the jib and the vertical post of a jib crane is 42°, and between the tie and jib the angle is 36°. Find the forces in the jib and tie when a mass of $3{\cdot}822 \times 10^3$ kg is suspended from the crane head.

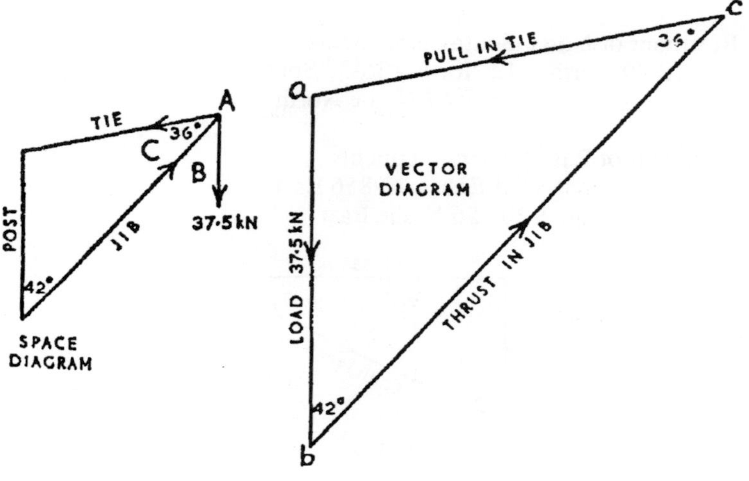

Fig. 14

Vertical downward force on crane head

$$= 3{\cdot}822 \times 10^3 \times 9{\cdot}81 \text{ N}$$
$$= 37{\cdot}5 \times 10^3 \text{ N} = 37{\cdot}5 \text{ kN}$$

At the crane head (see space diagram of Fig. 14), three forces meet which are in equilibrium. It is obvious that the jib must push upwards on the crane head to support the hanging load and the tie must pull to the left to support the top of the jib. The arrows are inserted accordingly and the vector diagram is constructed to represent these three forces at the crane head.

It will be seen that in this case the vector diagram of the forces is a similar triangle to the framework of the crane.

Referring to the vector diagram:

$$\text{Angle bac} = 180° - (42° + 36°) = 102°$$

$$\frac{\text{Force in jib}}{\sin 102°} = \frac{37{\cdot}5}{\sin 36°}$$

$$\text{Force in jib} = \frac{37{\cdot}5 \times 0{\cdot}9781}{0{\cdot}5878} = 62{\cdot}38 \text{ kN Ans. (i)}$$

$$\frac{\text{Force in tie}}{\sin 42°} = \frac{37{\cdot}5}{\sin 36°}$$

$$\text{Force in tie} = \frac{37{\cdot}5 \times 0{\cdot}6691}{0{\cdot}5878} = 42{\cdot}69 \text{ kN Ans.(ii)}$$

Example. The lengths of the vertical post and jib of a jib crane are 6·5 and 7 m respectively, and the angle between the post and jib is 40°. A mass of 2·854 t is suspended from a wire rope which passes over a pulley at the crane head and then led down at an angle of 50° to the vertical to a winch behind the post. Draw to scale the vector diagram of the forces at the crane head and measure the forces in the jib and tie.

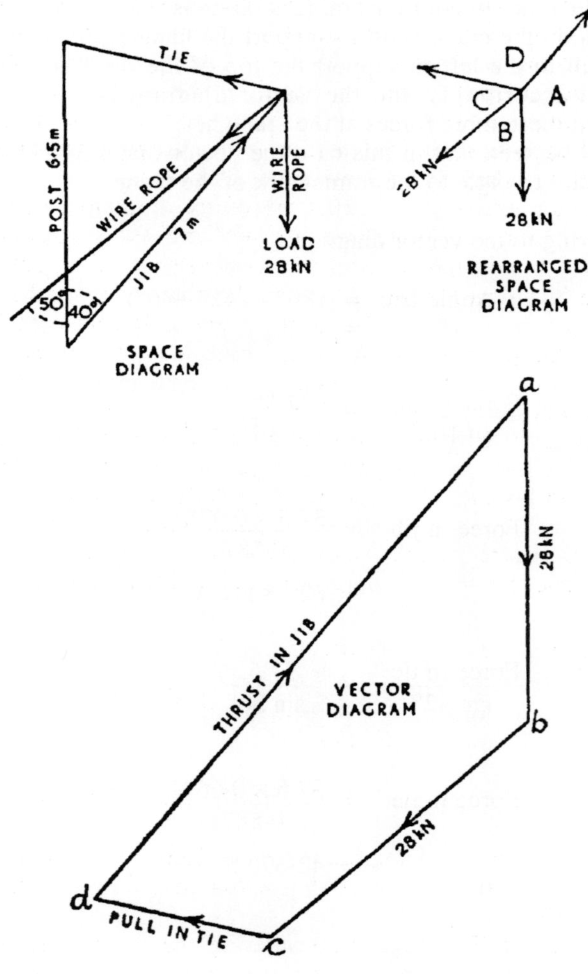

Fig. 15

$$\text{Mass} = 2{\cdot}854 \text{ t} = 2{\cdot}854 \times 10^3 \text{ kg}$$

$$\text{Load} = 2{\cdot}854 \times 10^3 \times 9{\cdot}81 \text{ N}$$

$$= 28 \times 10^3 \text{ N} = 28 \text{ kN}$$

The space diagram can be drawn to scale from the data given on the lengths of the post and jib and angle between them. The vector diagram can then be constructed by drawing the force vectors parallel to the wire ropes, jib and tie. Alternatively, the remaining angles of the space diagram could be measured and the vector diagram drawn. Note that the force in the wire rope must be the same throughout its length, that is, in the hanging part and also in the portion leading down from crane head to winch

It is much easier to construct the vector diagram when the two known forces are next to each other and not separated by an unknown. For instance, take the original space diagram as it is, with the downward pull of 28 kN on the crane head by the hanging part of the wire rope, next to this there is the upward inclined thrust of the jib, the magnitude of which is unknown, next in turn is the pull of 28 kN in the wire which leads down to the winch, then the pull in the tie which is unknown. Therefore rearrange the space diagram by extending the line of action of the force of the jib and consider it as a pulling force on the opposite side of the crane head instead of a pushing force under the head, the effect is the same and now there are the two known forces of the wire ropes together. The rearranged equivalent space diagram is drawn and lettered and the vector diagram constructed from this, as shown in Fig. 15.

Measuring the unknown vectors:

$$\left.\begin{aligned}\text{Force in jib} &= 55{\cdot}4\text{ kN}\\ \text{Force in tie} &= 14{\cdot}6\text{ kN}\end{aligned}\right\}\text{ Ans.}$$

RECIPROCATING ENGINE MECHANISM

The connecting rod and crank of a reciprocating engine converts the reciprocating motion of the piston to a rotary motion at the crank shaft.

Referring to Fig. 16 and considering the forces meeting at the crosshead, the lower end of the piston rod pushes vertically downwards on the crosshead, the thrust in the connecting rod appears as an upward resisting force at its top end inclined to the vertical, and the guide exerts a horizontal force to balance the horizontal component of the thrust in the connecting rod.

As the piston effort always acts vertically, and the guide force always horizontally, the vector diagram of the forces at the cross-head is always a right-angled triangle. Note that the angle between the centre-line of the engine and the connecting rod indicated by ϕ in the space diagram, is the same as the angle between the piston force and the force in the connecting rod in the vector diagram. Turning moment is considered in Chapter 3.

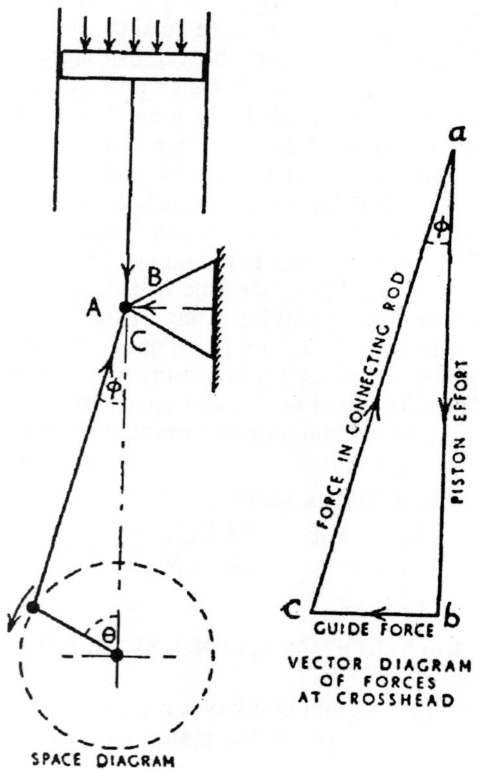

Fig. 16

Example. The piston of a reciprocating engine exerts a force of 160 kN on the crosshead when the crank is 35° past top dead centre. If the stroke of the piston is 900 mm and the length of the connecting rod is 1·65 m, find the guide force and the force in the connecting rod.

Referring to the space diagram of Fig. 16,

crank length = ½ stroke = 0·45 m
length of connecting rod = 1·65 m
crank angle from T.D.C. = θ = 35°

$$\frac{0{\cdot}45}{\sin\phi} = \frac{1{\cdot}65}{\sin 35°}$$

$$\sin\phi = \frac{0{\cdot}45 \times 0{\cdot}5736}{1{\cdot}65}$$

$$= 0{\cdot}1564$$

$$\therefore \text{Angle } \phi = 9°$$

Referring now to the vector diagram,

$$\text{Angle } \phi = 9°$$

$$\frac{\text{Guide force}}{\text{Piston force}} = \tan\phi$$

$$\text{Guide force} = 160 \times \tan 9°$$
$$= 25{\cdot}34 \text{ kN} \quad \text{Ans. (i)}$$

$$\frac{\text{Piston force}}{\text{Force in conn. rod}} = \cos\phi$$

$$\text{Force in conn. rod} = \frac{160}{\cos 9°}$$

$$= 162 \text{ kN Ans. (ii)}$$

FRAMED STRUCTURES

A framed structure is a framework of straight bars joined at their ends and, although they may be riveted or welded together, it is usually assumed in design that the end connections are pin-jointed or hinged so that the bars will be in either direct tension or direct compression.

When the external force applied to the ends of a bar tends to shorten it, the bar is in *compression.* A bar in compression is

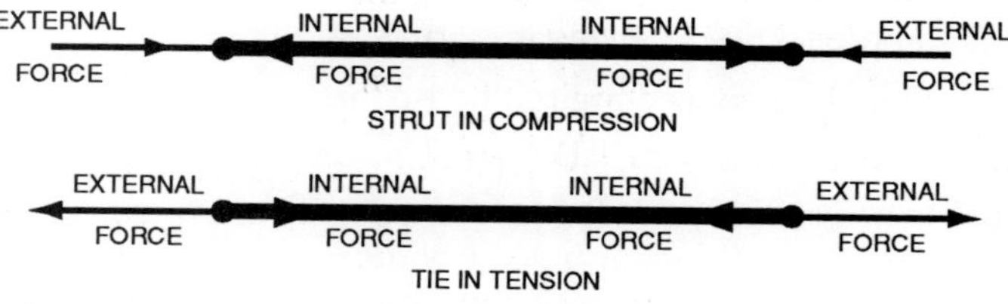

Fig. 17

referred to as a *strut* and the internal resisting force set up pushes towards its two ends.

When the external force on a bar tends to stretch it, the bar is said to be in *tension*. A bar in tension is referred to as a *tie* and the internal resisting force pulls on its two ends.

Refer back to the framework of the jib crane and insert arrows indicating the directions of the forces in the jib and tie, not only at the crane head, but at both ends of these two members, the arrows on the jib will push at its two ends indicating that it is in compression, and the arrows on the tie will pull at its ends indicating that it is in tension.

In framed structures consider not just one particular node where forces meet, as for the jib crane (at the crane head) and in the reciprocating engine mechanism (at the crosshead), but at every node where members of the structure meet.

Consider the simple common symmetrical roof structure shown in Fig 18. This consists of a horizontal bar and two sloping bars imagined to be pin-jointed at their ends. A load of W is carried at the apex. The framework rests on the two end supports and the upward reaction of each support will be equal to half the load W.

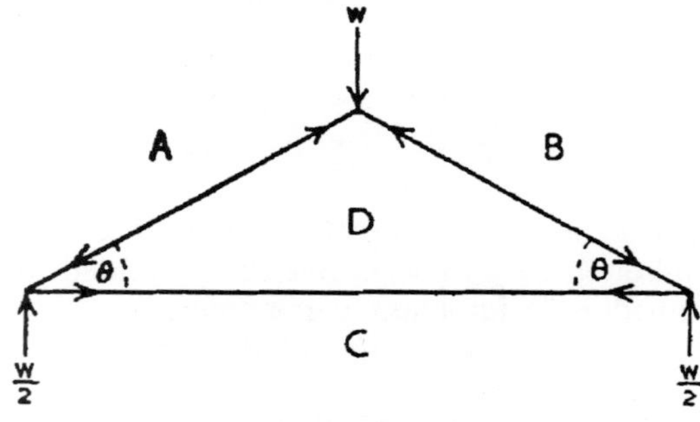

Fig.18

Using Bow's notation. To enable each member to be represented insert another letter, D, in the centre.

Now the arrows indicating the directions of the forces on each node are inserted. These are reasoned out and are fairly straightforward. At the left end node CAD, the support pushes upwards, therefore for equilibrium there must be a downward force, member DC is horizontal and cannot have a vertical component, member AD must therefore push down on this node. By reason of its slope, AD not only pushes down but also to the left, therefore DC must pull to the right to balance this horizontal thrust. Hence the arrows at that node appear as shown. The arrows indicating the directions of the forces at the other nodes are reasoned in a similar manner, also bearing in mind that the two arrows on any one member must be in opposite directions.

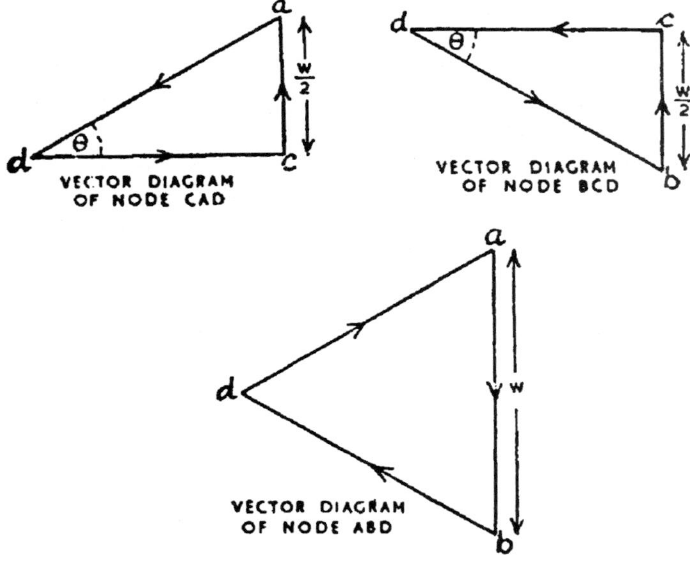

Fig. 19

A vector diagram for each separate node may be drawn as previously explained, these are illustrated in Fig. 19.

These separate diagrams could now be combined together as shown in Fig. 20 by superimposing the first two on to the third, but note particularly that the arrows in the combined vector diagram have now to be omitted because any one member of the structure applies forces in opposite directions at its two ends.

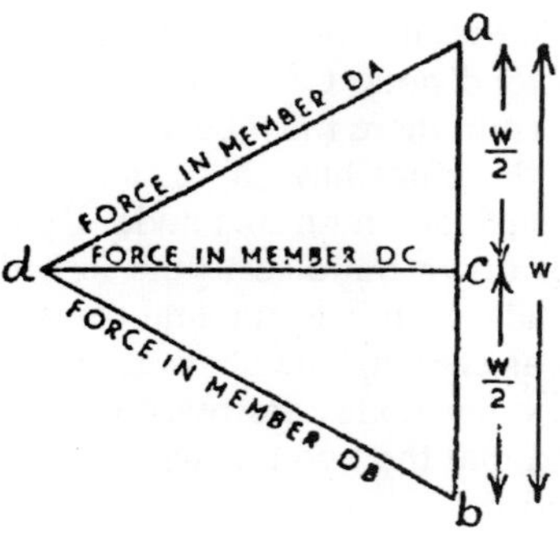

Fig. 20

After a little practice the combined vector diagram for the whole structure can be drawn without the aid of first drawing the vector diagram for each node and of course it is much quicker. To do this first draw the vector diagram of the external forces thus, beginning with the load AB draw a vertical line downwards to represent W to scale, mark this ab, from b measure $\frac{1}{2}W$ to scale vertically upwards to c to represent the right-hand reaction BC, from c vertically up to a is $\frac{1}{2}W$ which represents the left-hand reaction CA. This so far is only a straight line (this is a very simple example). Now the vector diagram for each node is quite easily added one at a time, as at least one of the vectors is now already in position from which to make a start.

Example. Fig 21 is a sketch of a roof structure, the lower inclined members are at 15° to the horizontal and the upper inclined members are at 45°. It is simply supported at each end and the structure carries a load of 50 kN on the apex. Construct the vector diagram for the whole structure, measure off the force in each member and state whether they are in compression or tension.

The structure is symmetrical therefore each reaction carries half the load, that is 25 kN at each end. The space diagram is drawn, external forces inserted, lettered by Bow's notation, directions of forces at each node reasoned out and arrows inserted, as in Fig. 22. The vector diagram is now constructed to scale, ab representing the 50 kN load, bc and ca respectively 25 kN each for the

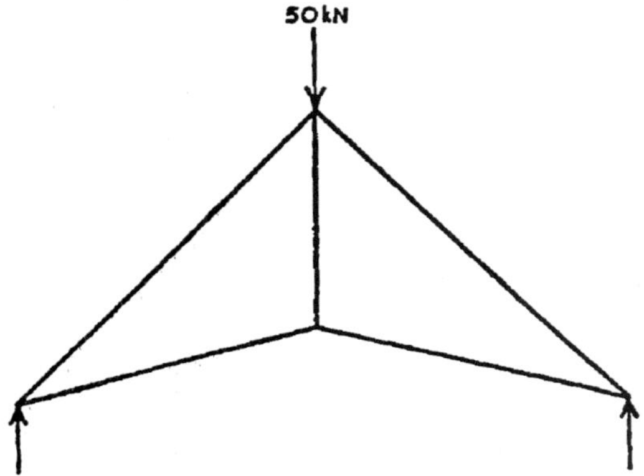

Fig. 21

reactions. The vector diagram of node CAD is added by drawing *ad* at 45° from point *a*, and *cd* at 15° to the horizontal from point *c*. The vector diagram of node BCE can be added next by drawing *be* at 45° from point *b*, and *ce* at 15° from point *c*. The vector diagram of node DEC is completed by drawing the vertical vector *de*, the other vectors *ec*, and *cd* being already in place. The vector diagram of the remaining node ABED will be found to be already complete in position.

Measuring the forces to scale on the vector diagram, and referring to the arrows on the space diagram to determine whether the members are in compression or tension, the results are tabulated thus:

MEMBER	FORCE	NATURE OF FORCE
AD	48·3 kN	Compression
BE	48·3 ,,	Compression
DC	35·4 ,,	Tension
EC	35·4 ,,	Tension
DE	18·3 ,,	Tension

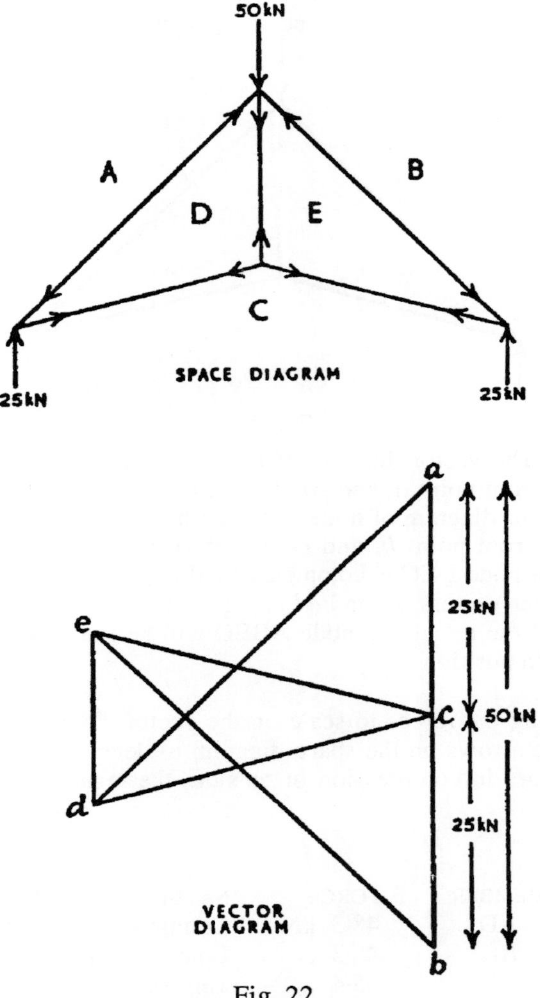

Fig. 22

NON-COPLANAR FORCES

A system of forces which are not in the same plane can be reduced to a coplanar system by substituting an imaginary member for each pair of straddled members of the structure. This is demonstrated in the following.

Example. A set of sheer legs is illustrated in Fig. 23, the front legs are each 6 m long and they are straddled 5 m apart at their bases; the back stay is 11 m long and its base fixture is 7 m horizontally from the centre of the feet of the front legs. Find the force in each member when a mass of 15·29 t hangs from the crane head, (i) by measurement of the vector diagram, (ii) by calculation.

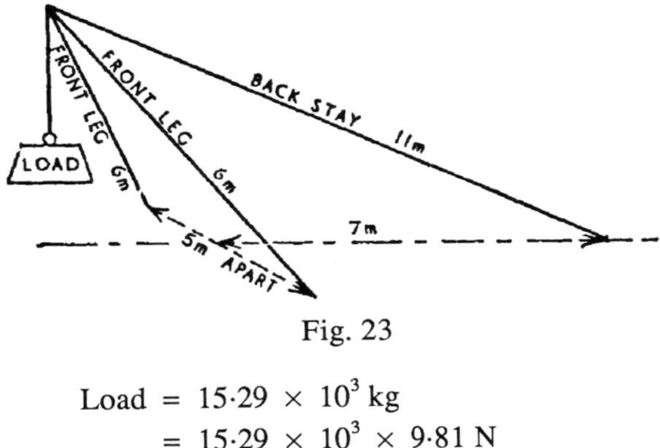

Fig. 23

$$
\begin{aligned}
\text{Load} &= 15{\cdot}29 \times 10^3 \text{ kg} \\
&= 15{\cdot}29 \times 10^3 \times 9{\cdot}81 \text{ N} \\
&= 150 \text{ kN}
\end{aligned}
$$

The two front legs can be replaced temporarily by one imaginary leg in the centre of the two and in the same plane as the back stay and the hanging load, as in Fig. 24.

$$\text{Length of imaginary leg} = \sqrt{6^2 - 2{\cdot}5^2} = 5{\cdot}455 \text{ m}$$

The space diagram is now drawn to scale with the imaginary leg in position. Being now a simple system of coplanar forces the vector diagram of the forces at the crane head can be drawn to scale by constructing the vectors parallel to the forces indicated by the arrows on the space diagram. Measuring the vector diagram, the force in the back stay scales 157 kN, and the force in the imaginary leg scales 258 kN (Fig. 24).

Now the force in the imaginary leg is really the resultant of the forces in the two actual front legs, drawing this resultant force vector diagram as in Fig. 25, the force in each front leg scales 142 kN.

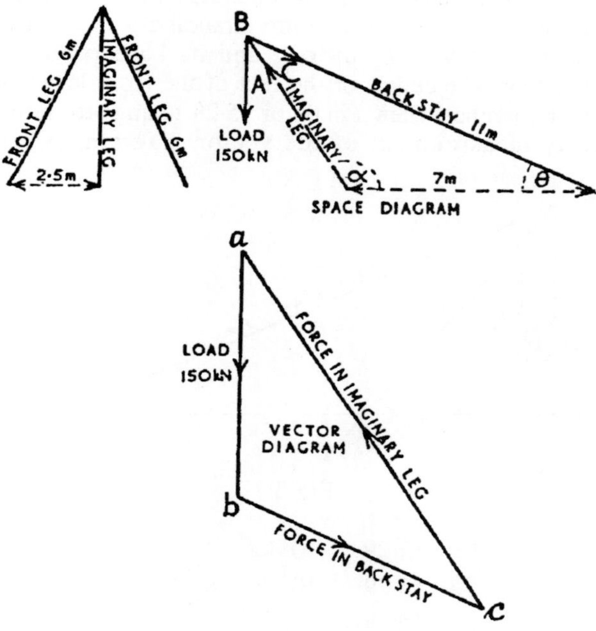

Fig. 24

Therefore by measurement,

Force in back stay = 157 kN
Force in each front leg = 142 kN } Ans. (i)

Referring to the space diagram of Fig. 24:

$$\cos C = \frac{11^2 + 5{\cdot}455^2 - 7^2}{2 \times 11 \times 5{\cdot}455}$$

$$= \frac{101{\cdot}76}{120} = 0{\cdot}8480$$

$$\therefore C = 32°$$

$$\frac{5{\cdot}455}{\sin\theta} = \frac{7}{\sin 32°}$$

$$\sin\theta = \frac{5{\cdot}455 \times 0{\cdot}5299}{7}$$

$$= 0{\cdot}4130$$

$$\therefore \theta = 24° \, 24'$$

$$\alpha = 180° - (32° + 24° \, 24') = 123° \, 36'$$

Referring to the vector diagram of Fig. 24,

$$\text{angle } c = 32°$$

$$\text{,, } a = 123° 36' - 90° = 33° 36'$$

$$\text{,, } b = 24° 24' + 90° = 114° 24'$$

$$\frac{\text{Force in back stay}}{\sin a} = \frac{\text{Load}}{\sin c}$$

$$\text{Force in back stay} = \frac{150 \times \sin 33° 36'}{\sin 32°}$$

$$= 156{\cdot}6 \text{ kN}$$

$$\frac{\text{Force in imaginary leg}}{\sin b} = \frac{\text{Load}}{\sin c}$$

$$\text{Force in imaginary leg} = \frac{150 \times \sin 114° 24'}{\sin 32°}$$

$$= 257{\cdot}8 \text{ kN}$$

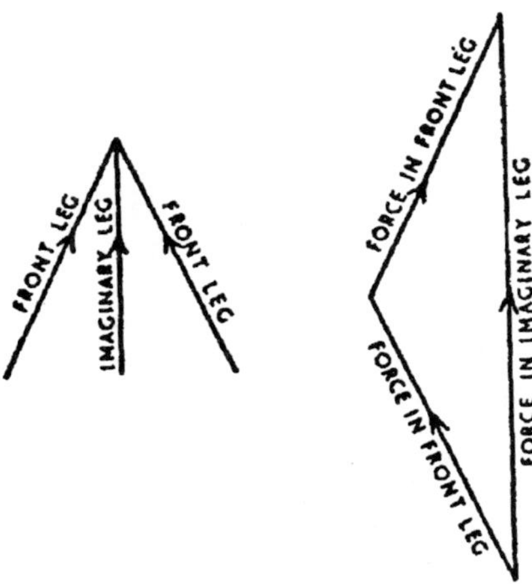

Fig. 25

Referring to Fig. 25,

$$\text{Force in each real front leg} = \frac{257{\cdot}8}{2} \times \frac{6}{5{\cdot}455}$$

$$= 141{\cdot}8 \text{ kN}$$

Therefore, by calculation,

$$\left.\begin{aligned}\text{Force in back stay} &= 156{\cdot}6 \text{ kN}\\ \text{Force in each front leg} &= 141{\cdot}8 \text{ kN}\end{aligned}\right\} \text{ Ans. (ii)}$$

TEST EXAMPLES I

1. A vertical lifting effort of 90 N is applied to a body and at the same time a force of 120 N pulls on it in a horizontal direction. Find the magnitude and direction of the resultant of these two forces.

2. Two forces act on a body, one pulls horizontally to the right and its magnitude is 20 N, the other pulls vertically downwards and its magnitude is 17 N. Find the magnitude and direction of a third force which would neutralise the effect of the other two.

3. Find the magnitude and direction of the equilibrant of two forces, one being a horizontal pull of 10 N and the other a pull of 20 N at 50° to the 10 N force.

4. Three forces pulling on a body are in equilibrium. The direction of one is due South, the direction of another is 75° East of North, and the third force acts in the direction 40° West of North. If the magnitude of the Southerly force is 35 N, find the magnitudes of the other two.

5. A block of wood is pulled along a horizontal table by a force of 25 N inclined at 20° above the horizontal. Find the vertical and horizontal components of the force.

6. Two lifting ropes are connected at their lower ends to a common shackle from which a load of 25 kN hangs. If the ropes make angles of 32° and 42° respectively to the vertical, find the tension in each rope.

7. A shaft of mass 5·097 t is lifted by two chains from a crane hook. The length of each chain is 4 m and their connections to the shaft are 4 m apart. If the centre of gravity of the shaft is 1·25 m from one of the connections, find the tension in each chain.

8. A wire rope 25·5 m long is slung between two vertical bulkheads which are 21 m apart, the end fixtures being at the same height. A freely running snatch block on the wire carries a hanging load of 30 kN. If the snatch block is pulled by a horizontal force until it is 8 m horizontally from one bulkhead, calculate the tension in the wire rope and the horizontal force applied.

9. The angle between the jib and vertical post of a jib crane is 40° and between the jib and tie the angle is 45°. Find the force in the jib and tie when a load of 15 kN hangs from the crane head.

10. A jib 6·6 m long is hinged at its foot to the base of a vertical post and connected at its top end by a 3·6 m long tie to a shackle on the post at 4·2 m up from the base. The lifting rope passes over a pulley at the jib head and is led back in the plane of the frame-

work, at an angle of 45°, to a winch. Draw to scale the vector diagram of the forces at the crane head when a load of 45 kN is being lifted at a uniform speed, and measure the forces in the jib and tie.

11. When the crank of a reciprocating engine is 60° past top dead centre, the effective piston effort on the crosshead is 180 kN. If the stroke of the piston is 600 mm and the connecting rod length is 1·25 m, find the load on the guide and the thrust in the connecting rod.

12. The roof frame shown in Fig. 26 carries a load of 50 kN at the apex. Draw the vector diagram, calculate the forces in each member and state the nature of these forces, and also the magnitudes of the two reactions.

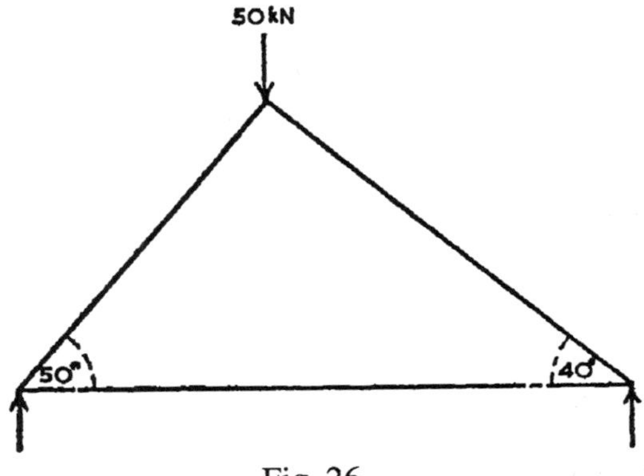

Fig. 26

13. In the framework shown in Fig. 27 all inclined members are at 45°. Draw to scale the vector diagram of the forces in the members of this structure when carrying a load of 100 kN at the centre, measure their magnitudes and tabulate results, stating also the nature of the forces.

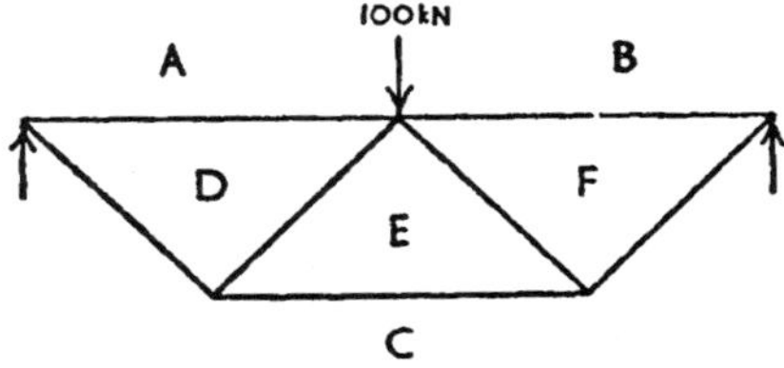

Fig. 27

14. A set of sheerlegs with front legs 10 m long straddled 10 m apart at their bases; the back stay is 18 m and the base fixture is 12 m horizontally from the centre of the feet of the front legs. Find the force in each member when a mass of 10 t is supported from the head.

*f*15. Draw the vector diagram for the structure given in Fig. 28 when carrying a load of 30 kN at the end, and calculate the wall reactions at the top and bottom connections.

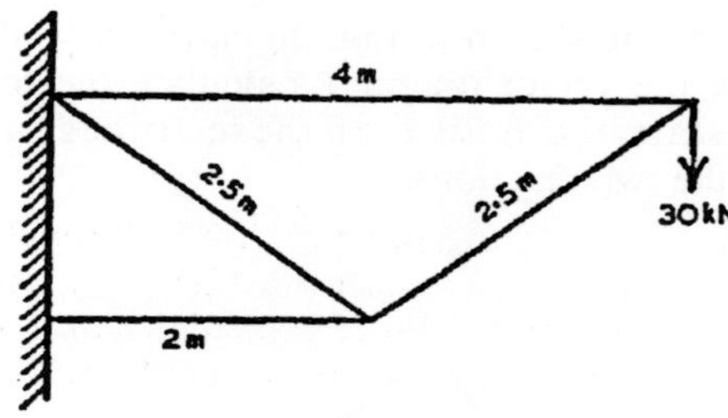

Fig. 28

CHAPTER 2

KINEMATICS

Kinematics – motion without involving mass

LINEAR MOTION

SPEED is the rate at which a body moves through space and is therefore expressed as the distance travelled in a given time. The *units* in which speed are usually expressed are m/s, km/h, etc. The speed may vary during a journey, for example, if a car travels 180 km in 3 h it is very improbable that it has been moving at an exact constant speed of 60 km/h during the 3 h, it is more likely that the speed has been well below and above that figure at times, but the *average* speed is 60 km/h.

VELOCITY indicates speed in a specified direction. Velocity therefore represents two facts about a moving body—its speed and also its direction, consequently it is a vector quantity and hence can be illustrated by drawing a vector to scale the length of which represents the speed of the body, and the direction in which it is drawn with an arrow represents its direction. See Fig. 29.

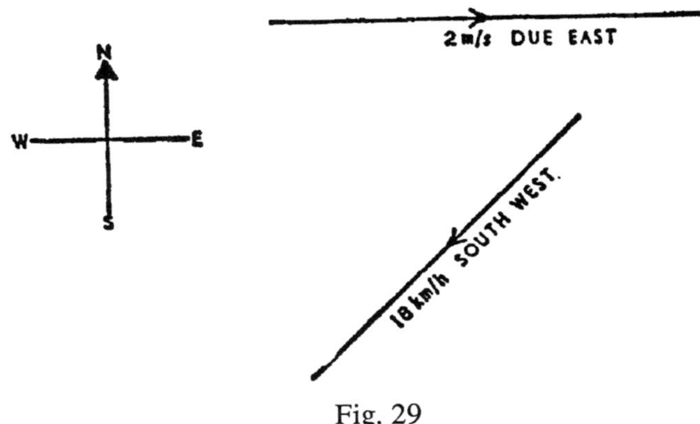

Fig. 29

RESULTANT VELOCITY is obtained from a vector diagram of velocities in the same manner as vector diagrams of forces. It is *vector addition.*

Example. A ship travelling due North at 16 knots runs into a 4 knot current moving South East. Find the resultant speed and direction of the ship.

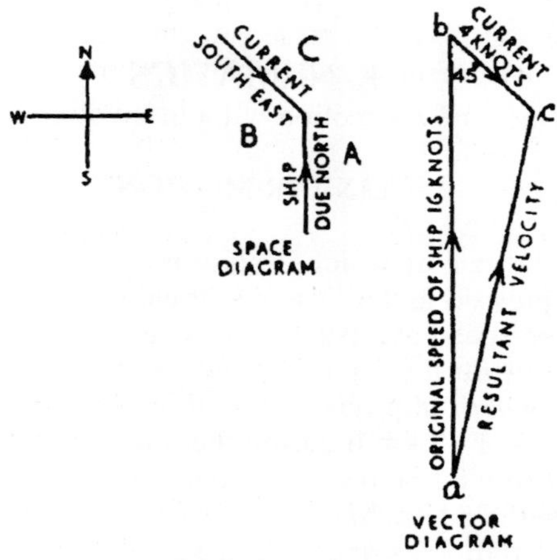

Fig. 30

$$
\begin{aligned}
(ac)^2 &= (ab)^2 + (bc)^2 - 2 \times ab \times bc \times \cos b \\
&= 16^2 + 4^2 - 2 \times 16 \times 4 \times \cos 45° \\
&= 256 + 16 - 90{\cdot}51 \\
ac &= \sqrt{181{\cdot}49} \\
&= 13{\cdot}47
\end{aligned}
$$

$$\frac{4}{\sin a} = \frac{13{\cdot}47}{\sin 45°}$$

$$
\begin{aligned}
\sin a &= \frac{4 \times 0{\cdot}7071}{13{\cdot}47} \\
&= 0{\cdot}2100 \\
\therefore \text{a} &= 12° \, 7'
\end{aligned}
$$

$\therefore$ Resultant speed = 13·47 knots
,, direction = 12° 7′ East of North } Ans.

CHANGE OF VELOCITY will take place if the speed changes, or if the direction changes, or if both speed and direction change.

In the first place, only those cases with no change of direction will be considered wherein velocity and speed can therefore be treated as being the same. Change of velocity due to change of direction and change of speed and direction will be dealt with later.

LINEAR VELOCITY is expressed in the same units as those for speed, it is usually represented by the symbols u for initial velocity and v for final velocity.

If a body travels at an average velocity of 40 m/s for a time of 5 s, the total distance travelled will be 200 m.

$$\text{Distance travelled} = \text{average velocity} \times \text{time}$$

Linear displacement (distance and direction) is the correct vector term; symbol s.

LINEAR ACCELERATION is rate of change of velocity, that is, change of velocity in a given time. If the velocity is increasing it is *accelerating*, if the velocity is decreasing it is *retarding* (or it has *negative acceleration*).

If a cyclist increases speed uniformly from 2 m/s to 12 m/s in 5 seconds, total increase in velocity is $12 - 2 = 10$ m/s. Having taken 5 s to increase velocity by 10 m/s, then increase in velocity in each one of these 5 s must have been $10 \div 5 = 2$ m/s. This acceleration is 2 m/s^2. Hence:

$$\text{Acceleration} = \frac{\text{Increase in velocity}}{\text{Time to change}}$$

Linear acceleration is usually represented by a.

As a further example, if a train, starting from rest, increases its speed uniformly for 2 min and attains a speed of 108 km/h in that time:

$$\begin{aligned}\text{Acceleration} &= 108 \text{ km/h in 2 min} \\ &= 108 \times 10^3 \text{ m/h in 2 min} \\ &= \frac{108 \times 10^3}{3600 \times 2 \times 60} \text{ m/s in 1 s} \\ &= 0{\cdot}25 \text{ m/s}^2\end{aligned}$$

One international nautical mile is 1·852 km and one international knot is 1·852 km/h.

Example. A ship's engines are stopped when travelling at a speed of 18 knots and the ship comes to rest after 20 min. Assuming uniform retardation, find the retardation (m/s^2) and the distance travelled in nautical miles in that time.

$$\text{Retardation} = 18 \text{ knots in } 20 \text{ min}$$

$$= \frac{18 \times 1{\cdot}852 \times 10^3}{3600} \text{ m/s in } 20 \text{ min}$$

$$= \frac{18 \times 1{\cdot}852 \times 10^3}{3600 \times 20 \times 60} \text{ m/s in } 1 \text{ s}$$

$$= 0{\cdot}007717 \text{ m/s}^2$$

$$\text{or } 7{\cdot}717 \times 10^{-3} \text{ m/s}^2 \text{ Ans. (i)}$$

$$\text{Distance} = \text{Average velocity} \times \text{time}$$

$$= \frac{18 + 0}{2} \times \frac{20}{60}$$

$$= 3 \text{ nautical miles Ans. (ii)}$$

Example. A motor car starting from rest attains a speed of 54 km/h over a distance of 90 m. Assuming that the rate of increase in speed is uniform, find the acceleration.

$$\text{Max. velocity} = \frac{54 \times 10^3}{3600} = 15 \text{ m/s}$$

$$\text{Average velocity} = \tfrac{1}{2}(0 + 15) = 7{\cdot}5 \text{ m/s}$$

$$\text{Distance} = \text{average velocity} \times \text{time}$$

$$90 = 7{\cdot}5 \times \text{time}$$

$$\text{time} = \frac{90}{7{\cdot}5} = 12 \text{ s}$$

$$\text{Acceleration} = 15 \text{ m/s in } 12 \text{ s}$$

$$= \frac{15}{12} \text{ m/s}^2$$

$$= 1{\cdot}25 \text{ m/s}^2 \text{ Ans.}$$

VELOCITY-TIME GRAPHS

A graph of velocity or speed on a base of time can be a very useful method of solving some problems as well as providing a picture of the facts.

The area of a velocity-time graph represents distance travelled and the slope of the curve represents acceleration. The slope of a displacement (distance)-time graph represents velocity.

Fig. 31 represents a body travelling at a constant velocity of 20 km/h for 4 h. The area enclosed by the graph is a rectangle of height 20 km/h and length 4 h, the area of a rectangle is the product of height and length, this is the product of velocity and time which gives distance travelled. Therefore the area of the graph represents distance travelled.

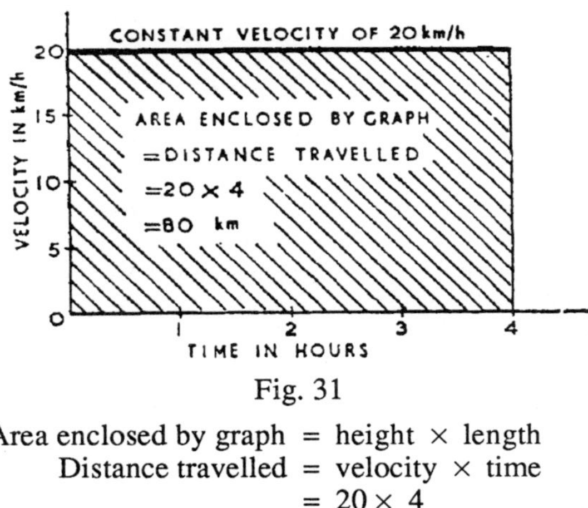

Fig. 31

$$\begin{aligned}\text{Area enclosed by graph} &= \text{height} \times \text{length}\\ \text{Distance travelled} &= \text{velocity} \times \text{time}\\ &= 20 \times 4\\ &= 80\text{ km}\end{aligned}$$

Fig. 32 represents a body starting from rest and reaching a velocity of 30 m/s in 6 s, the rate of increase of velocity (*i.e.* its acceleration) being uniform.

$$\begin{aligned}\text{Area of graph} &= \text{area of triangle}\\ &= \tfrac{1}{2} \times 30 \times 6\\ &= 90 \text{ units of distance}\\ \text{also, Distance} &= \text{average speed} \times \text{time}\\ &= \tfrac{1}{2}(0 + 30) \times 6\\ &= 90\text{ m}\end{aligned}$$

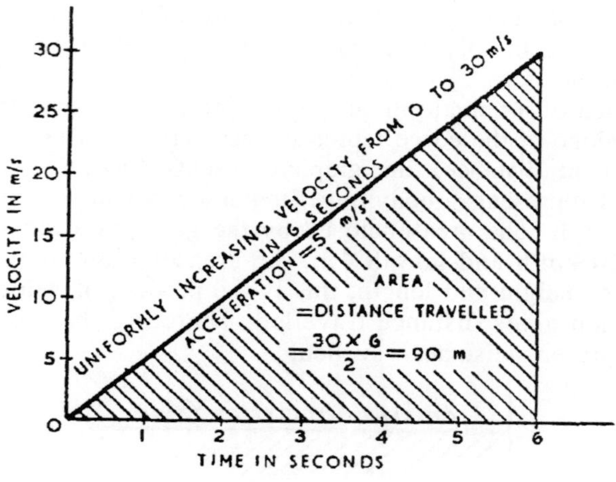

Fig. 32

Again the area of a velocity-time graph represents displacement (distance travelled). Further, in each second the increase of velocity is 5 m/s, this is the acceleration of 5 m/s^2 and is illustrated by the *slope* or *gradient* of the graph. A greater acceleration would be a steeper slope, a retardation would be a slope in the opposite direction.

Fig. 33 represents the slowing down of a ship from 16 knots to 10 knots in 12 min.

Distance travelled during this time

$$= \frac{16 + 10}{2} \times \frac{12}{60}$$

$$= 2{\cdot}6 \text{ nautical miles}$$

The ship loses 6 knots of speed in 12 min, this is equivalent to 30 knots in 60 min, so express the retardation in the same units in which the graph is plotted, as:

$$\text{Retardation} = 30 \text{ knots per h}$$

or it could be expressed in units of m/s^2 thus:

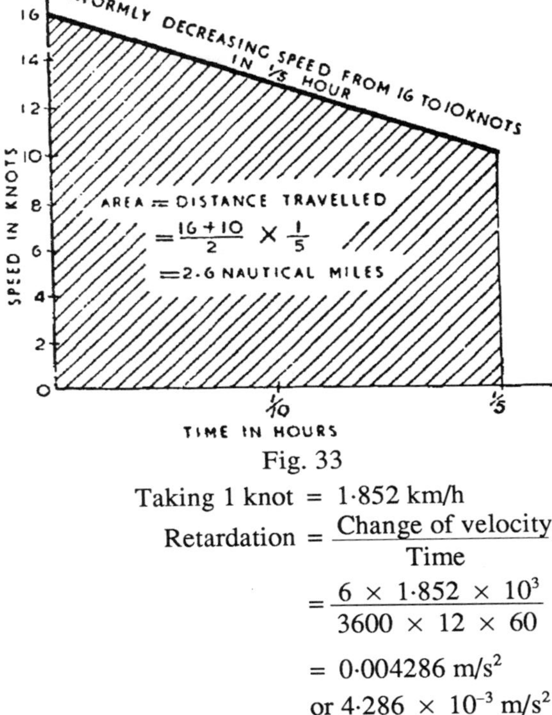

Fig. 33

$$\text{Taking 1 knot} = 1{\cdot}852 \text{ km/h}$$

$$\text{Retardation} = \frac{\text{Change of velocity}}{\text{Time}}$$

$$= \frac{6 \times 1{\cdot}852 \times 10^3}{3600 \times 12 \times 60}$$

$$= 0{\cdot}004286 \text{ m/s}^2$$

$$\text{or } 4{\cdot}286 \times 10^{-3} \text{ m/s}^2$$

Example. A locomotive starts from rest and reaches a speed of 90 km/h in 25 s, it runs at this speed for 1½ min and then reduces speed to come to rest in 20 s. Assume acceleration and retardation to be uniform, draw a speed-time graph, find the total distance travelled and express the acceleration and retardation in m/s^2.

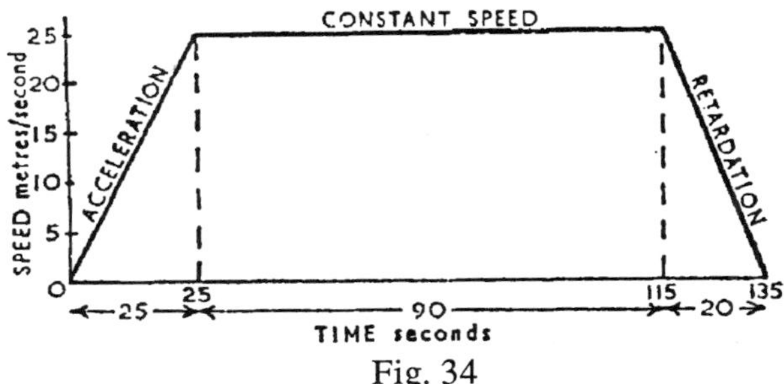

Fig. 34

$$90 \text{ km/h} = \frac{90 \times 10^3}{3600} \text{ m/s}$$

$$= 25 \text{ m/s}$$

Area under acceleration line $= \frac{1}{2} \times 25 \times 25 = 312{\cdot}5$ m

,, ,, constant speed ,, $= 25 \times 90 = 2250$,,

,, ,, retardation ,, $= \frac{1}{2} \times 25 \times 20 = 250$,,

Total $= 2812{\cdot}5$ m

Total distance travelled $= 2{\cdot}8125$ km Ans. (i)

$$\text{Acceleration} = \frac{\text{Increase in velocity}}{\text{Time}} = \frac{25}{25} = 1 \text{ m/s}^2 \text{ Ans. (ii)}$$

$$\text{Retardation} = \frac{\text{Decrease in velocity}}{\text{Time}} = \frac{25}{20} = 1{\cdot}25 \text{ m/s}^2 \text{ Ans. (iii)}$$

EQUATIONS

Although all problems can be worked out from first principles it is sometimes more expedient to solve by equations. There are four common equations connecting linear velocity, acceleration, time and displacement, and the usual symbols used in these are:

u = initial velocity in m/s

v = final velocity in m/s

a = acceleration uniform (constant) in m/s^2

t = time in s

s = displacement passed through (distance) m.

From the examples already given, the following should be readily understood.

a = increase in velocity for each second

at = ,, ,, ,, ,, t seconds,

Final velocity = initial velocity + increase in velocity

$$v = u + at \quad \ldots \ldots \ldots \ldots \ldots \ldots \ldots \ldots \text{(i)}$$

Distance travelled = average velocity × time

$$s = \left\{\frac{u + v}{2}\right\} \times t \quad \ldots \ldots \ldots \ldots \ldots \ldots \ldots \text{(ii)}$$

Substituting value of v from (i) to (ii),

$$s = \left\{\frac{u + u + at}{2}\right\} \times t$$

$$s = (u + \tfrac{1}{2}at) \times t$$

$$s = ut + \tfrac{1}{2}at^2 \quad \ldots \ldots \ldots \ldots \ldots \ldots \ldots \ldots \quad \text{(iii)}$$

Transposing (ii) to make t the subject,

$$s = \left\{\frac{u + v}{2}\right\} \times t$$

$$t = \frac{2s}{u + v}$$

Substituting this value of t into (i),

$$v = u + at$$

$$v = u + \frac{2as}{u + v}$$

Multiplying throughout by $(u + v)$

$$uv + v^2 = u^2 + uv + 2as$$

$$v^2 = u^2 + 2as \quad \ldots \ldots \ldots \ldots \ldots \ldots \ldots \ldots \quad \text{(iv)}$$

In any of the above, the acceleration a can be positive or negative; the positive sign is used when the acceleration is positive and the velocity is increasing, the negative sign is used when the acceleration is negative, that is retardation, and the velocity is decreasing.

These equations can therefore be written with the 'plus-or-minus' sign when the acceleration is included.

$$v = u \pm at$$

$$s = \tfrac{1}{2}(u + v)\, t$$

$$s = ut \pm \tfrac{1}{2}at^2$$

$$v^2 = u^2 \pm 2as$$

Example. A body with an initial velocity of 10 m/s is given uniform acceleration of 2 m/s^2 for 6 s. Find the velocity at the end of the 6 s and the distance travelled during this time.

$$u = 10 \text{ m/s}, \qquad a = 2 \text{ m/s}^2, \qquad t = 6 \text{ s.}$$

$$v = u + at$$

$$\therefore v = 10 + (2 \times 6)$$

After 6 s,

$$\text{Velocity} = 22 \text{ m/s} \qquad \text{Ans.}$$

$$s = \tfrac{1}{2}(u + v)\text{t}$$

$$\therefore s = \tfrac{1}{2}(10 + 22) \times 6$$

$$\text{Distance travelled} = 96 \text{ m} \qquad \text{Ans.}$$

Example. A ship's propellers are stopped when travelling at 25 knots and the ship travels 4 km from then until it comes to rest. Find the time taken to come to rest in min, and the average retardation in m/s^2. One knot = 1·852 km/h.

$$u = \frac{25 \times 1852}{3600} \qquad v = 0 \qquad s = 4000 \text{ m}$$

$$u = 12{\cdot}86 \text{ m/s}$$

$$s = \tfrac{1}{2}(u + v)t$$

$$4000 = \tfrac{1}{2}(12{\cdot}86 + 0) \times t$$

$$t = \frac{4000}{6{\cdot}43}$$

$$t = 622{\cdot}1 \, s$$

$$\text{Time} = 10{\cdot}37 \text{ min} \qquad \text{Ans.}$$

$$v = u + at$$

$$0 = 12{\cdot}86 + 622{\cdot}1\,a$$

$$622{\cdot}1a = -12{\cdot}86$$

$$a = \frac{-12{\cdot}86}{622{\cdot}1}$$

$$a = -0{\cdot}02067 \text{ m/s}^2$$

$$\text{Retardation} = 0{\cdot}02067 \text{ m/s}^2. \qquad \text{Ans.}$$

GRAVITATIONAL MOTION

The earth attracts all bodies towards itself so that if a body is allowed to fall freely, neglecting air resistances it will fall towards the earth with uniform acceleration. This particular acceleration, referred to as gravitational acceleration, varies slightly over different parts of the earth's surface, but it is taken as 9·81 m/s^2 and this is represented by 'g'. Thus, if a body falls from rest, its velocity increases by 9·81 m/s every second it is falling.

It also follows that if a projectile is shot vertically upwards it will lose 9·81 m/s of velocity every second it is rising.

Example. A body is allowed to fall from rest. Find the velocity after falling for 4 s and the distance fallen in that time.

$$
\begin{aligned}
v &= u + gt \\
&= 0 + 9{\cdot}81 \times 4 \\
\text{Final velocity} &= 39{\cdot}24 \text{ m/s} && \text{Ans. (i)} \\
h &= ut + \tfrac{1}{2}gt^2 \\
&= 0 + 0{\cdot}5 \times 9{\cdot}81 \times 4^2 \\
\text{Distance fallen} &= 78{\cdot}48 \text{ m} && \text{Ans. (ii)}
\end{aligned}
$$

Example. A projectile is fired vertically upwards with an initial velocity of 300 m/s. Find (i) its velocity after 20 s, (ii) the height above the ground after 20 s, (iii) the time taken to reach its maximum height, (iv) the maximum height attained, (v) the total time from leaving the ground to returning to ground.

$$
\begin{aligned}
v &= u - gt \\
&= 300 - 9{\cdot}81 \times 20 \\
\text{Velocity} &= 103{\cdot}8 \text{ m/s at 20 s} && \text{Ans. (i)} \\
h &= ut - \tfrac{1}{2}\text{gt}^2 \\
&= 300 \times 20 - 0{\cdot}5 \times 9{\cdot}81 \times 20^2 \\
\text{Height} &= 4038 \text{ m at 20 s} && \text{Ans. (ii)} \\
v &= u - gt \\
0 &= 300 - 9{\cdot}81 \times t \\
\text{Time} &= 30{\cdot}58 \text{ s To max. height} && \text{Ans. (iii)} \\
v^2 &= u^2 - 2\,gh \\
&= 300 - 2 \times 9{\cdot}81 \times h \\
\text{Max. height} &= 4587 \text{ m} && \text{Ans. (iv)} \\
\text{Total time} &= 2 \times 30{\cdot}58 \\
&= 61{\cdot}16 \text{ s} && \text{Ans. (v)}
\end{aligned}
$$

ANGULAR MOTION

ANGULAR DISPLACEMENT of a rotating body is measured in *radians*.

The symbol for angular displacement is θ.

One radian is the angle subtended at the centre of a circle by an arc of length equal to the radius of the circle.

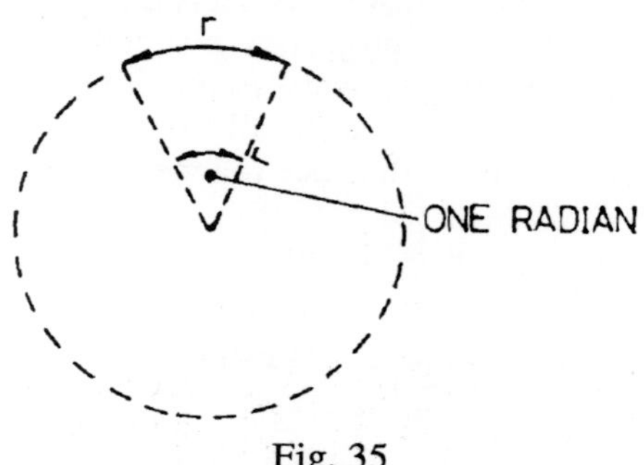

Fig. 35

$$1 \text{ revolution} = 2\pi \text{ rad}$$

ANGULAR VELOCITY is the rate of change of angular displacement and is expressed in rad/s.

The symbol for angular velocity is ω.

Since rotational speeds are usually expressed in revolutions per second or revolutions per minute, the following conversions may be necessary:

$$\omega \text{ (rad/s)} = 2\pi n \qquad \text{where } n = \text{speed in rev/s}$$

$$\omega \text{ (rad/s)} = \frac{2\pi N}{60} \qquad \text{where } N = \text{speed in rev/min}$$

ANGULAR ACCELERATION is the rate of change of angular velocity, expressed in rad/s^2.

The symbol for angular acceleration is α.

EQUATIONS

In a similar way to linear motion, four equations may be derived giving the relation, for angular motion, between displacement, velocity, acceleration and time:

ω_1 = initial angular velocity (rad/s)
ω_2 = final angular velocity (rad/s)
α = angular uniform (constant) acceleration (rad/s^2)
t = time in seconds (s)
θ = angular displacement (rad)

$$\omega_2 = \omega_1 \pm \alpha t \qquad \text{(i)}$$
$$\theta = \tfrac{1}{2}(\omega_1 + \omega_2)t \qquad \text{(ii)}$$
$$\theta = \omega_1 t \pm \tfrac{1}{2}\alpha t^2 \qquad \text{(ii)}$$
$$\omega_2^2 = \omega_1^2 \pm 2\alpha\theta \qquad \text{(iv)}$$

Example. A flywheel increases speed uniformly from 150 rev/min to 350 rev/min in 30 s. Calculate its acceleration.

$$\omega_1 = \frac{2\pi \times 150}{60} = 15{\cdot}71 \text{ rad/s}$$
$$\omega_2 = \frac{2\pi \times 350}{60} = 36{\cdot}65 \text{ rad/s}$$

From equation (i) above, $\omega_2 = \omega_1 + \alpha t$

$$36{\cdot}65 = 15{\cdot}71 + 30\alpha$$
$$\alpha = \frac{20{\cdot}94}{30}$$

Angular acceleration $= 0{\cdot}698$ rad/s^2 Ans.

Example. A shaft rotating at 40 rev/min is retarded uniformly at 0·017 rad/s^2 for 15 s.

Calculate (i) its angular velocity at the end of this time and (ii) the number of revolutions turned by the shaft during this 15 s.

Acceleration, $\alpha = -0{\cdot}017$ rad/s^2

$$\omega_1 = \frac{2\pi \times 40}{60} = 4{\cdot}19 \text{ rad/s}$$

$$\omega_2 = \omega_1 + \alpha t$$
$$\omega_2 = 4{\cdot}19 - (0{\cdot}017 \times 15)$$
$$\omega_2 = 3{\cdot}93 \text{ rad/s}$$
$$\text{Final angular velocity} = \frac{3{\cdot}93 \times 60}{2\pi} \text{ rev/min}$$
$$= 37{\cdot}53 \text{ rev/min}$$ Ans. (i)

$$\theta = \tfrac{1}{2}(\omega_1 + \omega_2)\, t$$
$$= \tfrac{1}{2}(4{\cdot}19 + 3{\cdot}93) \times 15$$
$$= 60{\cdot}9 \text{ rad}$$
$$\text{Number of revolutions} = \frac{60{\cdot}9}{2\pi}$$
$$= 9{\cdot}96$$ Ans. (ii)

RELATION BETWEEN LINEAR AND ANGULAR MOTION

Consider a point moved around on a circular path, if θ represents the angular displacement, r the radius and s the length of the arc or linear distance moved, then:

$$s = \theta r \quad \ldots \ldots \ldots \ldots \ldots \ldots \ldots \quad \text{(i)}$$

If v represents the linear velocity, ω the angular velocity and r the radius, then:

$$v = \omega r \quad \ldots \ldots \ldots \ldots \ldots \ldots \ldots \quad \text{(ii)}$$

Further, if the point is accelerating at the rate of α and the linear acceleration is represented by a, then:

$$a = \alpha r \quad \ldots \ldots \ldots \ldots \ldots \ldots \ldots \quad \text{(iii)}$$

In words, the above conversion rules are,

Linear displacement = angular displacement × radius
Linear velocity = angular velocity × radius
Linear acceleration = angular acceleration × radius

Example. A wheel 240 mm diameter is keyed to a shaft 40 mm diameter mounted in bearings which carry the shaft horizontally. A cord is wrapped around the shaft, one end of the cord being fixed to the shaft and the other end carrying a load. When the load is allowed to fall from rest, it falls a distance of 2 m in 5 s. Neglecting the thickness of the cord, find (i) the linear velocity of the load after 5 s, (ii) the angular velocity of the wheel and shaft after 5 s (iii) the linear velocity of the rim of the wheel after 5 s, (iv) the linear acceleration of the load, (v) the angular acceleration of the wheel and shaft.

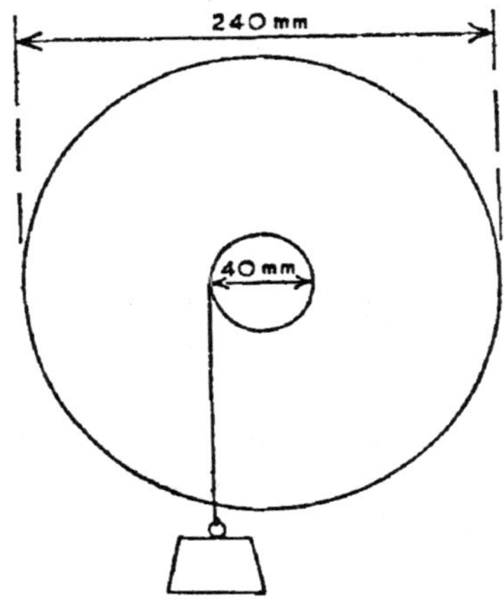

Fig. 36

The load moves through 2 m in 5 s,

$$\text{average velocity} = \frac{\text{distance}}{\text{time}}$$

$$= \frac{2}{5} = 0{\cdot}4 \text{ m/s}$$

$$\text{average velocity} = \tfrac{1}{2}(\text{initial vel.} + \text{final vel.})$$

Initial velocity is nil because it started from rest,

$$\text{Final velocity} = 2 \times \text{average velocity}$$

$$= 2 \times 0{\cdot}4$$

$$= 0{\cdot}8 \text{ m/s Ans. (i)}$$

Linear velocity = angular velocity × radius

$$\omega = \frac{v}{r} = \frac{0{\cdot}8}{0{\cdot}02}$$

$$= 40 \text{ rad/s} \quad \text{Ans. (ii)}$$

Linear velocity of rim of wheel:

$$v = \omega r$$
$$= 40 \times 0{\cdot}12$$
$$= 4{\cdot}8 \text{ m/s} \quad \text{Ans. (iii)}$$

Note the radius of the wheel rim is 6 times the radius of the shaft, they both rotate at the same angular velocity, therefore the linear velocity of the wheel rim is 6 times that of the surface of the shaft, 6 × 0·8 = 4·8 m/s.

Linear acceleration of load =

$$\frac{\text{change of velocity}}{\text{time}} = \frac{0{\cdot}8}{5}$$

$$= 0{\cdot}16 \text{ m/s}^2 \quad \text{Ans. (iv)}$$

Angular acceleration of shaft:

$$a = \alpha r$$

$$\therefore \alpha = \frac{a}{r} = \frac{0{\cdot}16}{0{\cdot}02}$$

$$= 8 \text{ rad/s}^2 \quad \text{Ans. (v)}$$

INSTANTANEOUS VALUES

For a linear moving body, at a given instant:
from the slope of the *s-t* and *v-t* curves,

$$v = \frac{ds}{dt} \qquad a = \frac{dv}{dt} = \frac{d^2s}{dt^2}$$

from the area under the *v-t* and *a-t* curves,

$$s = \int v\,dt \qquad v = \int a\,dt$$

similar expressions, different symbols, apply for angular motion.

Example. The velocity of a body at time t is given by the expression:

$$v = 3t^2 - 4t \text{ m/s}$$

Calculate the displacement, velocity and acceleration after 3 s from rest.

$$\begin{aligned}
s &= \int v\,dt \\
&= \int_0^3 (3t^2 - 4t)\,dt \\
&= [t^3 - 2t^2]_0^3 \\
&= 27 - 18 \\
\text{Displacement} &= 9 \text{ m at 3 s} \qquad \text{Ans.} \\
v &= 3t^2 - 4t \\
&= 27 - 12 \\
\text{Velocity} &= 15 \text{ m/s at 3 s} \qquad \text{Ans.} \\
a &= \frac{dv}{dt} \\
&= \frac{d}{dt}\left(3t^2 - 4t\right) \\
&= 6t - 4 \\
\text{Acceleration} &= 14 \text{ m/s}^2 \text{ at 3s} \qquad \text{Ans.}
\end{aligned}$$

Example. A body is projected vertically upwards. It's height (h) in metres after time (t) from the time of projection is given by the equation $h = 3 + 54t - 6t^2$.

By calculation, or by drawing a distance time graph, determine the time taken to reach maximum height and the maximum height attained.

$$\begin{aligned}
h &= 3 + 54t - 6t^2 \\
\frac{dh}{dt} &= 54 - 12t \\
0 &= 54 - 12t \text{ for max./min.} \\
t &= 4{\cdot}5 \\
\frac{d^2h}{dt^2} &= -12 \text{ i.e. negative (a max.)} \\
h &= 3 + (54 \times 4{\cdot}5) - (-6 \times 4{\cdot}52) \\
&= 124{\cdot}5
\end{aligned}$$

$$\left.\begin{aligned}
\text{Time to max. height} &= 4{\cdot}5 \text{ s} \\
\text{max. height} &= 124{\cdot}5 \text{ m}
\end{aligned}\right\} \text{Ans.}$$

This can be verified by (alternative), drawing h-t graph.

Note: Simple Harmonic Motion (SHM) is considered in Chapter 5.

CHANGE OF VELOCITY

Velocity is a vector quantity representing speed and direction and therefore a change of velocity takes place if the speed changes without any change of direction, or if the direction changes while the speed remains the same, or if there is a change in both speed and direction.

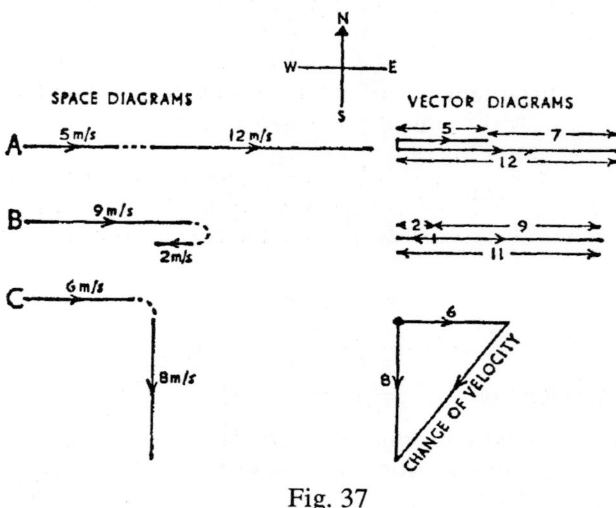

Fig. 37

Consider a few simple cases, Fig. 37 illustrates space diagrams and vector diagrams of velocities.

Case A represents a body which was moving at 5 m/s due East, having its velocity changed to 12 m/s due East, the vector of each velocity is drawn from a common point, the difference between the free ends of the vectors is the change of velocity, in this case it is 7 m/s

Case B is a body with an initial velocity of 9 m/s due East, being changed to 2 m/s due West, the vector diagram shows the vector of each velocity drawn from a common point, the difference between their free ends is the change of velocity which is 11 m/s.

Case C is that of a body with an initial velocity of 6 m/s due East changed to 8 m/s due South. The vector diagram is constructed on the same principle of the two vectors drawn from a common point. The change of velocity is, as always the difference between the free ends of the two vectors, this is $\sqrt{8^2 + 6^2} = 10$ m/s. The direction for the change of velocity is S36° 52′W. The change of velocity takes place in the direction of the applied force to bring about the change i.e. E to SW.

In all cases, the vector diagrams have been constructed by drawing the velocity vectors from a common point. This is *vector subtraction.*

Acceleration is the rate of change of velocity, therefore in all of these cases the value of the acceleration can be obtained in the usual way by dividing change of velocity by time.

Example. A body moving at 20 m/s due North is acted upon by a force for 4 s which causes the velocity to change to 20 m/s due East. Find the change of velocity and the average acceleration.

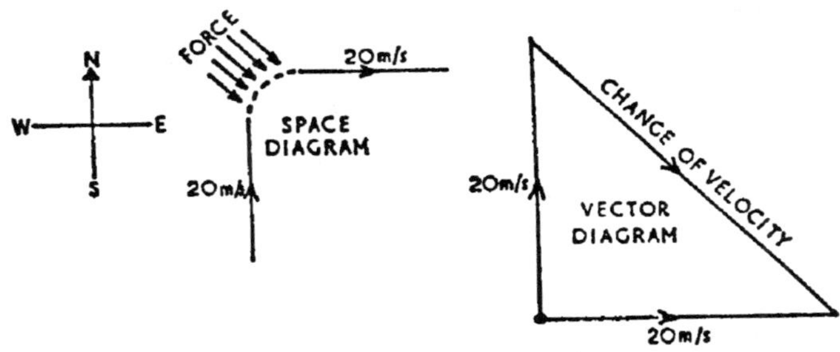

Fig. 38

$$\text{Change of velocity} = \sqrt{20^2 + 20^2}$$

$$= 28{\cdot}28 \text{ m/s} \qquad \text{Ans. (i)}$$

$$\text{Acceleration} = \frac{\text{Change of velocity}}{\text{Time to change}}$$

$$= \frac{28.28}{4}$$

$$= 7{\cdot}07 \text{ m/s}^2 \qquad \text{Ans. (ii)}$$

Example. An aircraft changes velocity from 400 km/h due West to 500 km/h North East in ½ min. Find the average acceleration in m/s².

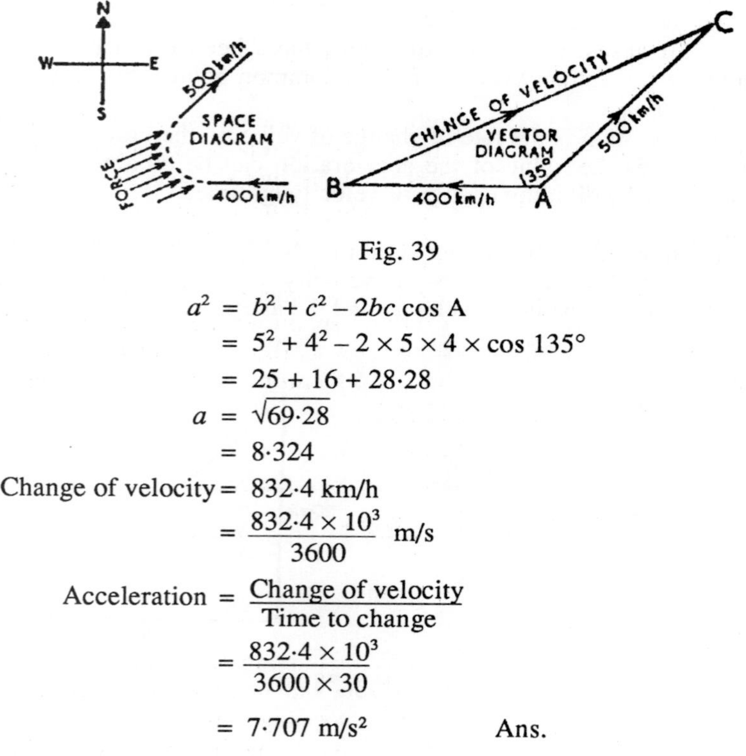

Fig. 39

$$a^2 = b^2 + c^2 - 2bc \cos A$$
$$= 5^2 + 4^2 - 2 \times 5 \times 4 \times \cos 135°$$
$$= 25 + 16 + 28{\cdot}28$$
$$a = \sqrt{69{\cdot}28}$$
$$= 8{\cdot}324$$
$$\text{Change of velocity} = 832{\cdot}4 \text{ km/h}$$
$$= \frac{832{\cdot}4 \times 10^3}{3600} \text{ m/s}$$
$$\text{Acceleration} = \frac{\text{Change of velocity}}{\text{Time to change}}$$
$$= \frac{832{\cdot}4 \times 10^3}{3600 \times 30}$$
$$= 7{\cdot}707 \text{ m/s}^2 \qquad \text{Ans.}$$

RELATIVE VELOCITY

Up to the present only velocities of moving objects as they pass fixed points on the Earth have been considered, these are termed *absolute* velocities.

When the velocity of a moving object A is expressed as the rate at which it passes another moving object B, it is termed the *relative* velocity of A with respect to B. In effect it is the velocity of A as it appears by a person moving with object B and is therefore sometimes called the *apparent* velocity.

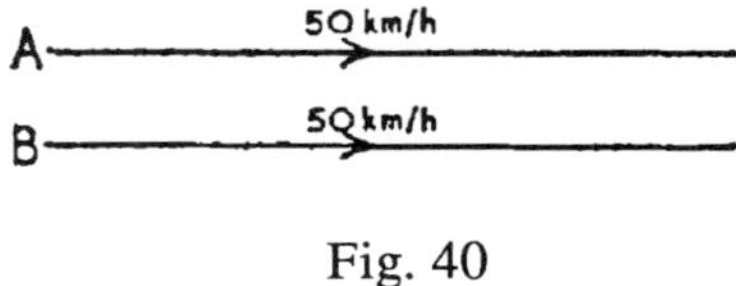

Fig. 40

If the two objects are moving on parallel courses at the same velocity such as A and B in Fig. 40, the relative velocity of one to the other is nil. A typical example is two persons sitting in the same railway carriage of a moving train, in the eyes of one the other is not moving, the apparent or relative velocity of one to the other is nil.

If, however, one object is moving exactly in the opposite direction to the other such as two trains on outward and inward parallel tracks, each travelling at 50 km/h as illustrated in Fig. 41, one appears to pass the other at 100 km/h, therefore the relative velocity of one to the other is 100 km/h.

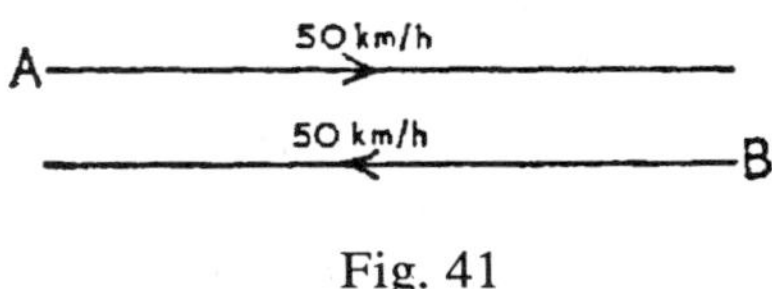

Fig. 41

The relative velocities of objects moving on parallel courses are obvious and simple to understand, but when the courses are not parallel it is necessary to draw vector diagrams.

Consider a body A moving at 30 m/s due East and another body B moving at 35 m/s 20° North of East. A space diagram can be first sketched to show the absolute velocity of each, as these velocities are 'relative to earth' they are marked A or B at the end behind the arrow, and E (for earth) at the point end. See Fig. 42.

The vector diagram is now drawn with E a common point for the two absolute velocities, the relative velocity of A to B, or B to A, is the vector connecting the two free ends. If the velocity of B relative to A is required, the arrow is put on pointing from B to A and shows how B appears to be moving in the eyes of A. If the velocity of A relative to B is required, the arrow is inserted in the direction A to B. This is *vector subtraction*.

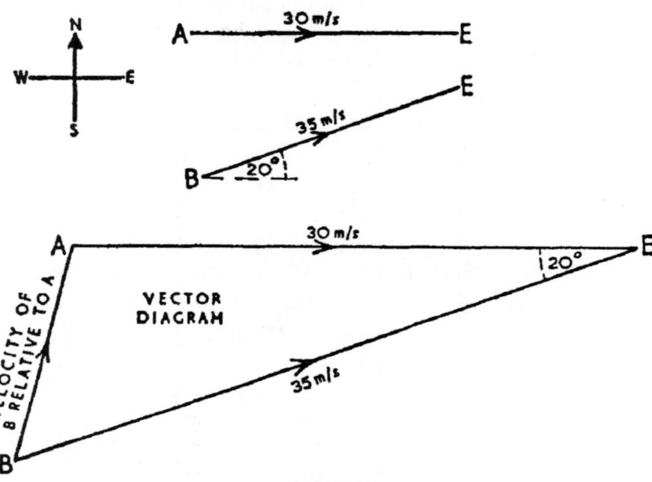

Fig. 42

Example. Rain is falling vertically at a velocity of 5m/s. Find the velocity of the rain as it appears to a cyclist moving horizontally at 4 m/s.

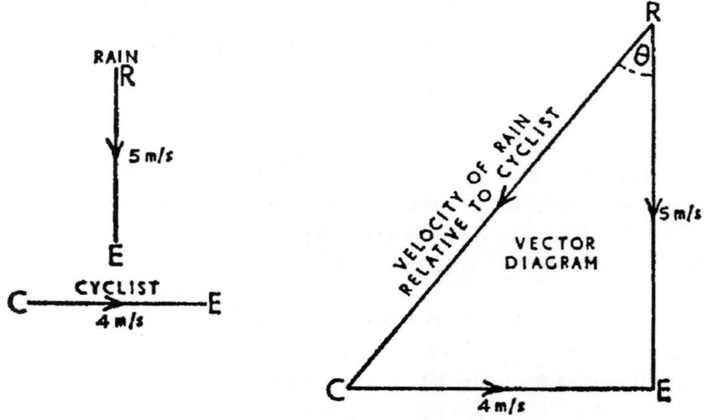

Fig. 43

$$\text{Relative velocity of rain} = \sqrt{5^2 + 4^2} = 6{\cdot}403 \text{ m/s}$$

$$\tan\theta = \frac{4}{5} = 0{\cdot}8 \quad \therefore \theta = 38^\circ\, 40'$$

Velocity of rain appears to the cyclist to be:

6·403 m/s at 38° 40′ to the vertical. Ans.

Example. If the wind is blowing from South 30° East at 15 km/h, at what speed and direction will the wind appear to blow to a cyclist travelling at 22 km/h due North?

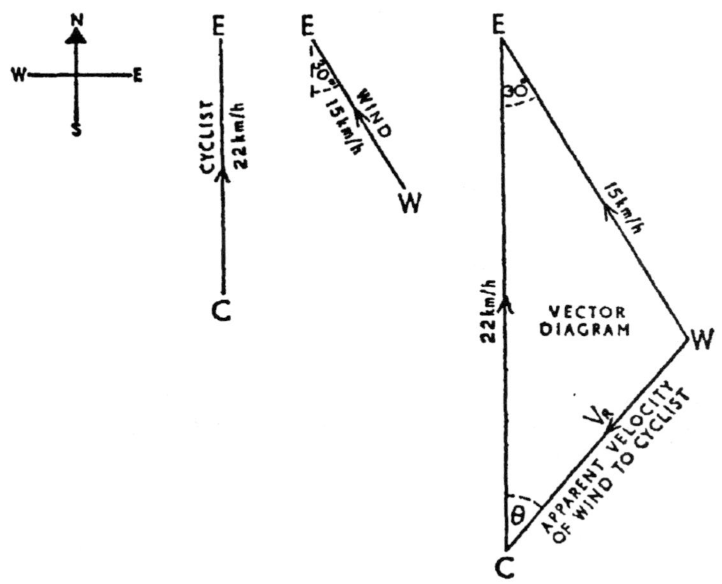

Fig. 44

$$V_R^2 = 22^2 + 15^2 - 2 \times 22 \times 15 \times \cos 30°$$
$$= 484 + 225 - 571{\cdot}6$$
$$V_R = \sqrt{137{\cdot}4}$$
$$= 11{\cdot}72 \text{ km/h}$$
$$\frac{15}{\sin \theta} = \frac{11{\cdot}72}{\sin 30}$$
$$\sin \theta = \frac{15 \times 0{\cdot}5}{11{\cdot}72} = 0{\cdot}6400$$
$$\theta = 39° \, 48'$$

Apparent velocity of wind to cyclist =

11·72 km/h from North 39° 48′ East. Ans.

Example. One ship A is going due West at 19 knots and another ship B which is 5 nautical miles South West of A is going North 30° East at 17 knots. Find the distance between the two ships when they are nearest together and the time for them to get there.

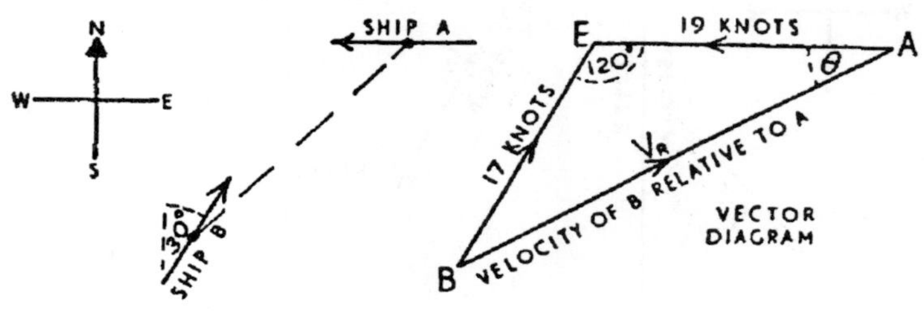

Fig. 45

$$V^2_R = 17^2 + 19^2 - 2 \times 17 \times 19 \cos 120°$$
$$= 289 + 361 + 323$$
$$V_R = \sqrt{973}$$
$$= 31{\cdot}19 \text{ knots.}$$

$$\frac{17}{\sin \theta} = \frac{31{\cdot}19}{\sin 120°}$$

$$\sin \theta = \frac{17 \times 0{\cdot}866}{31{\cdot}19} = 0{\cdot}4720$$

$$\theta = 28° \ 10'$$

Velocity of B relative to A is 31·19 knots on the course 28° 10′ North of East. Now imagine being on ship A apparently stationary and seeing ship B, which is 5 nautical miles away in a S.W. direction, moving at an apparent speed of 31·19 knots in the direction 28° 10′ North of East. A space diagram of distances is now drawn to represent these apparent conditions as in Fig. 46.

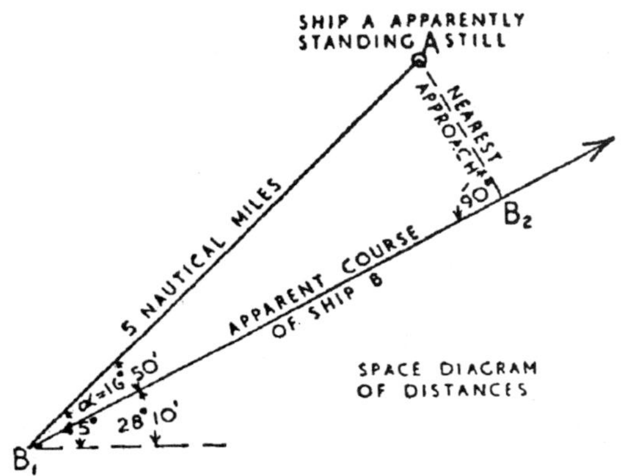

Fig. 46

$$\text{Angle } \alpha = 45° - 28°\ 10' = 16°\ 50'$$

$$AB_2 = \text{nearest approach}$$

$$= 5 \times \sin 16°\ 50' = 1{\cdot}448 \text{ naut. miles} \quad \text{Ans. (i)}$$

Apparent distance to travel by B to get to the position of nearest approach $= B_1\,B_2 = 5 \times \cos 16°\ 50'$

$$= 4{\cdot}7855 \text{ naut. miles}$$

To travel 4·7855 naut. miles at an apparent speed of 31·19 knots:

$$\frac{\text{Distance}}{\text{Speed}} = \frac{4{\cdot}7855}{31{\cdot}19}$$

$$\frac{4{\cdot}7855 \times 60}{31{\cdot}19} = 9{\cdot}2 \text{ min} \qquad \text{Ans. (ii)}$$

f INSTANTANEOUS CENTRE OF ROTATION

Consider Fig. 47, this represents a ladder resting on a horizontal ground and leaning against a vertical wall. It could also represent a link of a mechanism whose two extreme ends move in directions at right angles to each other.

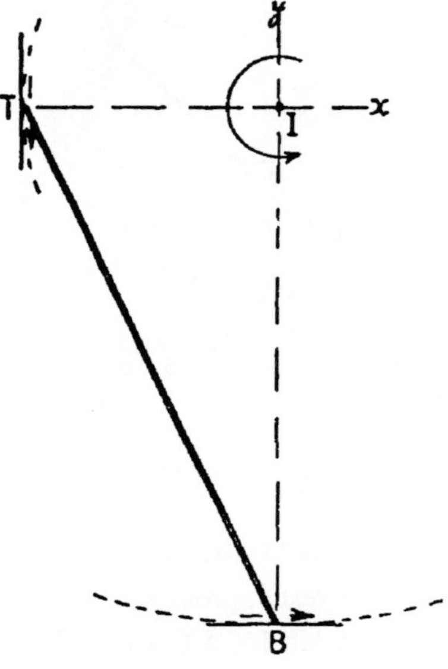

Fig. 47

If the bottom end of the ladder is moved horizontally away from the wall, the top end will slide down vertically. For a very small movement of the bottom end B imagine that the motion is along a very small length of the arc of a circle which has its centre anywhere in the direction B*y*, this line B*y* being perpendicular to the direction of motion of B. The motion of the top end T is imagined to be a very small length of the arc of a circle whose centre is anywhere in the direction T*x*, the line T*x* being perpendicular to the direction of motion of T. Hence for the instant that the ladder passes this particular position, the intersection of these two perpendiculars, denoted by I, is a common centre for the movements of both ends. This point is termed the *instantaneous centre of rotation* because it can be imagined that the whole ladder, just for an instant, is swinging about the common centre I.

Since the linear velocity of a rotating body is proportional to its radius from the centre of rotation, then the velocities of B and T are in the same ratio as their respective distances from I, thus,

$$\frac{\text{Velocity of B}}{\text{Velocity of T}} = \frac{\text{IB}}{\text{IT}}$$

This provides a useful method of determining the velocity of one point on a moving link in relation to another point.

A typical example is the motions of the crosshead (or piston) and crank pin of a reciprocating engine as illustrated in Fig. 48.

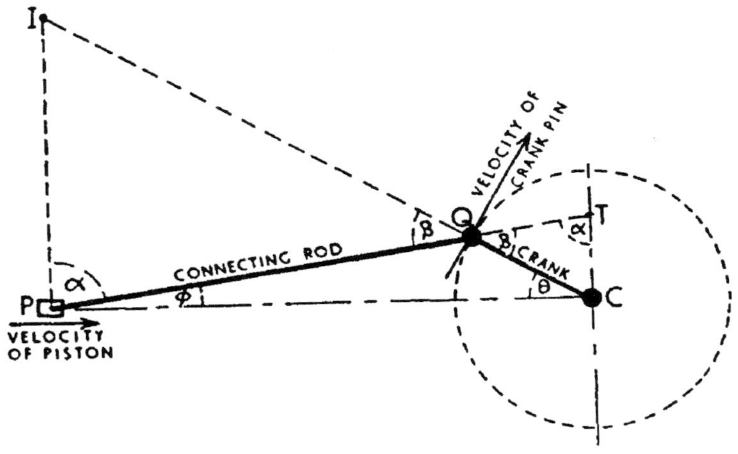

Fig. 48

The instantaneous linear velocity of the crank pin Q is tangential to the crank pin circle and perpendicular to the crank. CI is therefore drawn by extending the crank line CQ. PI is drawn perpendicular to the direction of motion of the crosshead.

$$\frac{\text{Linear velocity of P}}{\text{Linear velocity of Q}} = \frac{\text{IP}}{\text{IQ}}$$

Further, if the line of connecting rod is produced to cut the diameter of the crank pin circle at T, it can be seen that triangle QPI is similar to triangle QTC, therefore the ratio of the lengths IP and IQ can be substituted by the ratio of the lengths CT and QC, thus,

$$\frac{\text{Linear velocity of P}}{\text{Linear velocity of Q}} = \frac{\text{CT}}{r}$$

r being the length of the crank which is QC.

$$\text{Velocity of Piston} = \text{Linear velocity of crank pin} \times \frac{\text{CT}}{r}$$

Since the lengths of the sides of a triangle are proportional to the sines of their opposite angles (sine rule) then distances CT and r can be replaced by the sines of their opposite angles.

If Ω is the angular velocity of the connecting rod:

$$\Omega = \frac{\text{Linear velocity of P}}{\text{IP}} = \frac{\text{Linear velocity of Q}}{\text{IQ}}$$

Example. A reciprocating engine of 750 mm stroke runs at 120 rev/min. If the length of the connecting rod is 1500 mm, find the piston speed when the crank is 45° past top dead centre.

Referring to Fig.48,

Length of crank $= \frac{1}{2}$ stroke $= 0{\cdot}375$ m $\qquad \theta = 45°$

$$\frac{1{\cdot}5}{\sin 45°} = \frac{0{\cdot}375}{\sin \phi}$$

$$\sin \phi = \frac{0{\cdot}375 \times 0{\cdot}7071}{1{\cdot}5} = 0{\cdot}1768$$

$$\phi = 10°\ 11'$$

$$\alpha = 90° - 10°\ 11' = 79°\ 49'$$

$$\beta = \theta + \phi = 55°\ 11'$$

Linear velocity of crank pin, $v = \omega r$

$$= \frac{120 \times 2\pi}{60} \times 0{\cdot}375 \text{ m/s}$$

Velocity of piston $= \text{linear velocity of crank pin} \times \dfrac{\sin \beta}{\sin \alpha}$

$$= \frac{120 \times 2\pi \times 0{\cdot}375}{60} \times \frac{\sin 55°\ 11'}{\sin 79°\ 49'}$$

$= 3{\cdot}93$ m/s Ans.

f PROJECTILES

Consider a projectile fired horizontally at a velocity of, say, 100 m/s, from a cliff edge. Neglecting air resistances, as soon as it leaves the gun the projectile is acted upon by the force of gravity, pulling it towards the ground, giving it a downward vertical acceleration of $9{\cdot}81$ m/s^2. Since there is no horizontal force acting on it and air resistance is neglected, there is no change in the horizontal velocity. After 1 s the horizontal movement of the projectile is still 100 m/s but it also now has a downward vertical movement of 9·81 m/s, the resultant of these two is the actual velocity of the projectile after 1 s. After two seconds the horizontal component is still 100 m/s, but its vertical component is now 2 × 9·81 = 19·62 m/s. After 10 s the horizontal component remains at 100 m/s and the vertical component is now 9·81 × 10 = 98·1 m/s, and so on, this illustrated in Fig. 49.

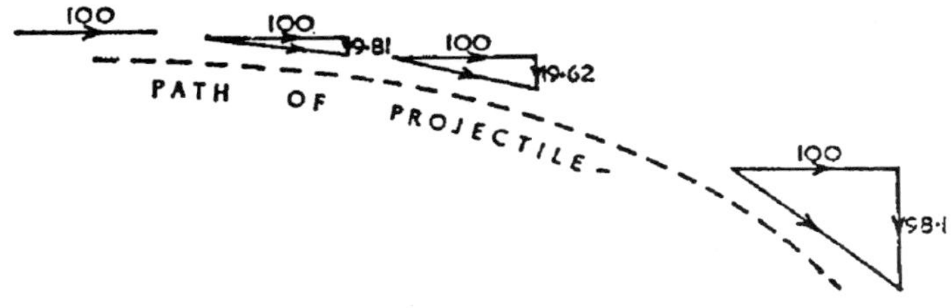

Fig. 49

As air resistances are neglected, the horizontal component of the initial velocity remains constant and the vertical component changes in the same way as a vertically rising or falling body, that is, it gains downward velocity at the rate of 9·81 m/s^2.

Example. Water flows through a hole in the vertical side of a tank. The hole is 300 mm above the ground and the jet issues horizontally at a velocity of 7·28 m/s. Find the time for the particles of water to reach the ground and the horizontal distance from the side of the tank to the point where the jet strikes the ground.

Initial vertical velocity = nil

Vertical acceleration = 9·81 m/s^2

Let time of fall = t

Velocity after falling t = 9·81 t m/s

Average vertical velocity during fall

$= \frac{1}{2}(0 + 9{\cdot}81\, t) = 4{\cdot}905\, t$ m/s

Height of fall = 300 mm = 0·3 m

Vertical distance = average vertical velocity × time

$0{\cdot}3 = 4{\cdot}905\, t \times t$

$t = \sqrt{\dfrac{0{\cdot}3}{4{\cdot}905}} = 0{\cdot}2473$ s Ans. (i)

Horizontal distance = horizontal velocity × time

= 7·28 × 0·2473

= 1·8 m Ans. (ii)

Now consider a projectile fired from a gun on a horizontal ground, let the initial velocity of the projectile be u m/s and the angle of elevation $\theta°$, referring to Fig. 50:

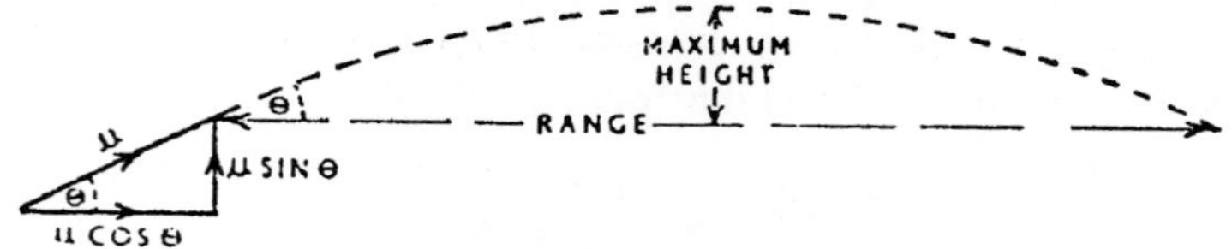

Fig. 50

$$\text{Vertical component} = u \sin \theta$$

It loses 9·81 m/s of vertical velocity every second (represented by g), therefore time to lose all its vertical velocity is

$$\frac{u \sin \theta}{g}$$

and the projectile is then at its maximum height.

$$\text{Vertical distance} = \text{average vertical velocity} \times \text{time}$$

$$\text{Maximum height} = \frac{u \sin \theta + 0}{2} \times \frac{u \sin \theta}{g}$$

$$= \frac{u^2 \sin^2 \theta}{2g} \quad \ldots \ldots \ldots \ldots \ldots \ldots \text{(i)}$$

$$\text{Total time above ground} = \text{time to go up} + \text{time to go down}$$

$$= \frac{2 \times u \sin \theta}{g}$$

$$\text{Horizontal component} = u \cos \theta$$

$$\text{Horizontal distance} = \text{horizontal velocity} \times \text{time}$$

$$= u \cos \theta \times \frac{2u \sin \theta}{g}$$

$$= \frac{2u^2 \sin \theta \cos \theta}{g}$$

$$\text{Since } \sin \theta \cos \theta = \frac{\sin 2\theta}{2}$$

$$\text{Horizontal range} = \frac{u^2 \sin 2\theta}{g} \quad \ldots \ldots \ldots \ldots \ldots \ldots \text{(ii)}$$

For a given value of the initial velocity u, the horizontal range will be greatest when sin 2θ is greatest. The maximum value of the sine of an angle is unity and this is for an angle of 90°, therefore if $2\theta = 90°$ then $\theta = 45°$.

Therefore maximum horizontal range is when the angle of elevation is 45°, and its value is,

$$\text{Maximum range} = \frac{u^2}{g}$$

Example. A projectile is fired with an initial velocity of 120 m/s at an elevation of 30° above the horizontal. Calculate (i) the range, (ii) the maximum height attained, (iii) the time of flight.

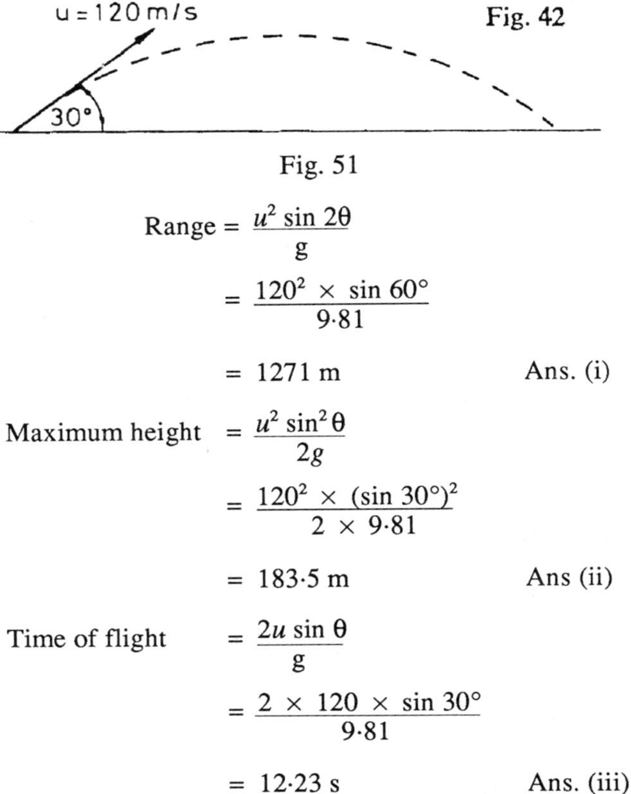

Fig. 51

$$\text{Range} = \frac{u^2 \sin 2\theta}{g}$$

$$= \frac{120^2 \times \sin 60°}{9{\cdot}81}$$

$$= 1271 \text{ m} \qquad \text{Ans. (i)}$$

$$\text{Maximum height} = \frac{u^2 \sin^2 \theta}{2g}$$

$$= \frac{120^2 \times (\sin 30°)^2}{2 \times 9{\cdot}81}$$

$$= 183{\cdot}5 \text{ m} \qquad \text{Ans (ii)}$$

$$\text{Time of flight} = \frac{2u \sin \theta}{g}$$

$$= \frac{2 \times 120 \times \sin 30°}{9{\cdot}81}$$

$$= 12{\cdot}23 \text{ s} \qquad \text{Ans. (iii)}$$

TEST EXAMPLES 2

1. A ship sailing due East at 18 knots runs into a 3 knot current moving 40° East of North. Find the resultant speed and direction of the ship.

2. A locomotive, starting from rest, is uniformly accelerated up to a maximum speed, taking 1 min and travelling a distance of 0·5 km. It then runs at maximum speed for 2 min and finally is uniformly retarded for 30 s to bring it to rest. Find the maximum speed, sketch a velocity-time graph and find the total distance travelled.

3. A motor car increases speed for 20 to 74 km/h in 12 s. Find the acceleration and the distance travelled during that time.

4. An aeroplane increases its speed from 400 km/h to 500 km/h while it travels a distance of 1½ km. Find the time taken, in seconds, over this distance, and the acceleration in m/s^2.

5. A body is projected vertically upwards with an initial velocity of 36 m/s. At what height on its upward journey will its velocity be 24 m/s and what time will it take to reach this point?

6. An object is allowed to fall from rest from a height of 70 m and at exactly the same time of release a projectile is fired upwards from the ground with an initial velocity of 35 m/s. When and where will they pass each other?

7. An electric motor comes to rest from operational speed in 1½ minutes and turns 1800 rev whilst slowing down. Find the operational speed in rev/min and the retardation in rad/s^2.

8. A wheel and axle is carried in horizontal bearings. A cord which is fixed to and wrapped around the axle carries a load on its free end. When the load is allowed to fall from rest, it falls 3 m in 8 s. If the diameter of the axle is 50 mm, find the linear acceleration of the load in m/s^2 and the angular acceleration of the wheel and axle in rad/s^2.

9. A body moves so that its distance which it travels from its starting point is given by:

$$s = 0{\cdot}2\,t^2 + 10{\cdot}4$$

Find the velocity and acceleration 5 s after the body begins to move and the average velocity in the first 10 s of motion.

10. The speed and direction of a motor launch is changed from 9 knots due North to 11 knots due West in 30 s. Find the average acceleration in m/s^2. One knot = 1·852 km/h.

11. Two trains, one 20 m long and the other 40 m long, approach each other in opposite directions on parallel tracks, the speed of the shorter train is 50 km/h and that of the longer train is 100 km/h. Find the time taken to pass each other.

*f*12. A tanker is going due South at 16 knots. At 12 noon a passenger ship is 7·5 nautical miles due West of the tanker and going South East at 18 knots. At what time will the two ships be closest together and what is then their distance apart?

*f*13. The crank CQ of a slider crank mechanism is 0·5 m and the connecting rod QP 1·5 m. The crank makes an angle of 30° past inner dead centre and is rotating at 1200 rev/min, clockwise. Determine the linear velocity of the piston P and the angular velocity of the connecting rod QP.

*f*14. A stone is thrown horizontally at 12 m/s from a 60 m high cliff. Find the time it takes for the stone to reach the ground and the horizontal distance covered.

*f*15. A projectile is fired with an initial velocity of 600 m/s at an angle of elevation of 30° to the horizontal. Find the range on horizontal ground.

CHAPTER 3

DYNAMICS

Dynamics – motion with mass (and force)

MASS is proportionally accelerated or retarded by an applied force. It is a scalar quantity.

Accelerating force (N) = mass (kg) × acceleration (m/s^2)

In symbols:

$$F = ma$$

FORCE OF GRAVITY. All bodies are attracted towards each other, the force of attraction depending upon the masses of the bodies and their distances apart. Newton's law of gravitation states that this force of attraction is proportional to the product of the masses of the bodies and inversely proportional to the square of the distance apart.

An important example of this is the huge mass of the earth which attracts all comparatively smaller earthly bodies towards it, the attractive force by which a body tends to be drawn towards the centre of the earth is the force of gravity and is called the *weight* of the body. Gravitational acceleration g = 9·81 m/s^2.

$$\text{Weight (N)} = \text{mass (kg)} \times g\ (m/s^2)$$

$$W = mg$$

The further the distance between the centre of gravity of the mass and the centre of gravity of the earth, the less is the attractive force between them. Thus, the weight of a mass measured by a spring balance (not a pair of scales which is merely a means of comparing the weight of one mass with another) will vary slightly at different parts of the earth's surface due to the earth not being a perfect sphere.

If a body is projected in a space-rocket, the attractive force of the earth on the body becomes less as its distance from the earth increases until, in complete outer-space, it becomes nil, that is, it is then weightless. The mass of the body of course remains unchanged.

INERTIA is the property possessed by matter by which it resists change of motion, and depends upon its mass. Broadly speaking, it may be regarded as a kind of sluggishness. If the mass is at rest it requires a force to give it motion, the greater the mass the greater the force required. If the mass is already moving it requires a force to change its velocity or to change its direction, again the force required being proportional to the mass.

MOMENTUM is the term given to the product of mass and velocity and therefore can be defined as the quantity of motion possessed by a moving body. This is dealt with later in greater detail.

The laws which connect motion and force are summarised in **Newton's laws of motion**:

(i) Every body continues in its state of rest, or uniform motion in a straight line, unless acted upon by an external force.

(ii) Rate of change of momentum is proportional to the force applied and takes place in the direction in which the force acts.

(iii) To every action there is a reaction, equal in magnitude and opposite in direction.

Example. Find the accelerating force required to increase the velocity of a body which has a mass of 20 kg from 30 m/s to 70 m/s in 4 s.

$$\text{Acceleration} = \frac{\text{Change of velocity}}{\text{Time to change}}$$

$$= \frac{70 - 30}{4}$$

$$= 10 \text{ m/s}^2$$

$$\text{Accelerating force (N)} = \text{mass (kg)} \times \text{acceleration (m/s}^2\text{)}$$

$$= 20 \times 10$$

$$= 200 \text{ N Ans.}$$

Example. A lift is supported by a steel wire rope, the total mass of the lift and contents is 750 kg. Find the tension in the wire rope, in newtons, when the lift is (i) moving at constant velocity, (ii) moving upwards and accelerating at 1·2 m/s^2, (iii) moving upwards and retarding at 1·2 m/s^2.

$$\text{Accelerating force} = \text{mass} \times \text{acceleration}$$

$$= 750 \times 1{\cdot}2$$

$$= 900 \text{ N}$$

Weight on wire $= 750 \times 9{\cdot}81 = 7358$ N

When the lift is at rest, or moving at constant velocity, the tension in the wire is due only to the upward force exerted to support the downward weight:

Tension in wire = supporting force = 7358 N Ans. (i)

When the lift is increasing velocity upwards an additional upward force is required to give the lift acceleration:

$$\begin{aligned}\text{Tension in wire} &= \text{supporting force} + \text{accelerating force}\\ &= 7358 + 900\\ &= 8258\text{ N} \quad \text{Ans. (ii)}\end{aligned}$$

When the lift is decreasing velocity upwards, the accelerating force is negative, that is, it is acting downwards and opposite in direction to the upward supporting force:

$$\begin{aligned}\text{Tension in wire} &= \text{supporting force} - \text{accelerating force}\\ &= 7358 - 900\\ &= 6458\text{ N} \quad \text{Ans. (iii)}\end{aligned}$$

Example. A light flexible cord is hung over a light pulley carried in frictionless bearings. A mass of 2 kg is hung from one end of the cord and another of 2·1 kg is hung from the other end, as illustrated in Fig. 52, and the system allowed to move from rest. Find (i) the acceleration of the masses, (ii) distance moved in 4 s, (iii) the tension in the cord.

$$\begin{aligned}\text{Total mass accelerated} &= 2{\cdot}1 + 2 = 4{\cdot}1\text{ kg}\\ \text{Force causing acceleration} &= (2{\cdot}1 - 2) \times 9{\cdot}81\\ &= 0{\cdot}981\text{ N}\\ \text{Accelerating force} &= \text{mass} \times \text{acceleration}\\ \text{Acceleration} &= \frac{\text{accelerating force}}{\text{mass}}\\ &= \frac{0{\cdot}981}{4{\cdot}1}\\ &= 0{\cdot}2393\text{ m/s}^2 \quad \text{Ans. (i)}\\ \text{Velocity after 4 s} &= 0{\cdot}2393 \times 4\\ &= 0{\cdot}9572\text{ m/s}\\ \text{Average velocity} &= \tfrac{1}{2}(0 + 0{\cdot}9572)\\ &= 0{\cdot}4786\text{ m/s}\end{aligned}$$

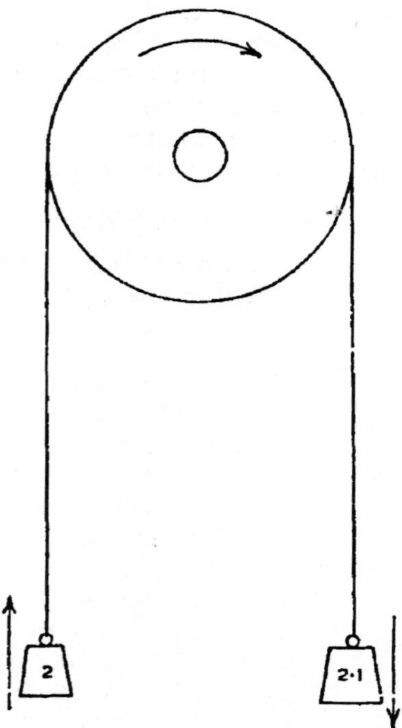

Fig. 52

Distance = average velocity × time
= 0·4786 × 4
= 1·9144 m Ans. (ii)

(This could be obtained from $s = \frac{1}{2}at^2$)

Consider the part of the cord carrying the 2 kg mass. In addition to the force to support the weight of the 2 kg mass there is the force to accelerate it.

Force to accelerate = mass × acceleration
= 2 × 0·2393
= 0·4786 N

Tension in cord = supporting force + accelerating force
= 2 × 9·81 + 0·4786
= 20·0986 N Ans. (iii)

If the other side of the cord is considered, the same result should be obtained because the tension in the cord must be uniform throughout its length (Newton's 3rd law of motion). On this side the cord applies an upward force of $2{\cdot}1 \times 9{\cdot}81$ N to support the weight of the mass, but acceleration is *downwards*, therefore the tension in the cord is the difference between the upward supporting force and the downward accelerating force.

$$\begin{aligned}\text{Tension in cord} &= \text{supporting force} - \text{accelerating force}\\ &= 2{\cdot}1 \times 9{\cdot}81 - 2{\cdot}1 \times 0{\cdot}2393\\ &= 20{\cdot}601 - 0{\cdot}5025\\ &= 20{\cdot}0985\ \text{N}\end{aligned}$$

ATWOOD'S MACHINE. The above example is based upon the principle of Atwood's machine which is used for demonstrating Newton's laws of motion. The machine consists of a grooved pulley of lightweight material so that its inertia is negligible, mounted on a spindle which runs in bearings as near frictionless as possible. This assembly is mounted on the top of a tall graduated pillar to which is attached a starting platform near the top and an adjustable stopping buffer near the bottom. A fine flexible cord hung over the pulley carries a mass on each end, these masses are usually of equal magnitude and a small rider is added to one of them to cause motion. A series of experiments can be performed by varying the masses, rider and the distance moved.

FLETCHER'S TROLLEY. This is another machine for illustrating the laws of motion and consists of a trolley on wheels which runs along a horizontal table. The pull on the trolley to accelerate it is exerted by a cord fastened to the trolley, running parallel with the table, over a guide pulley and then hanging down with a load hooked on the end, as shown in Fig. 53. The wheels of the trolley and the spindle of the guide pulley are mounted in bearings which are near frictionless. If, however, friction is not sufficiently small as to be negligible, the force required to overcome friction could be found by hanging small loads on the hook until the trolley just keeps moving without acceleration, the force causing acceleration is then due to any added loads on the hook over and above the friction force. The trolley is usually of considerable length, nearly half the length of the table, so that a smoked strip of material can be fixed along it for the needle of a vibrating arm to record oscillating waves for accurate measurement of time. Additional masses can be added to the trolley so that a series of experiments can be performed.

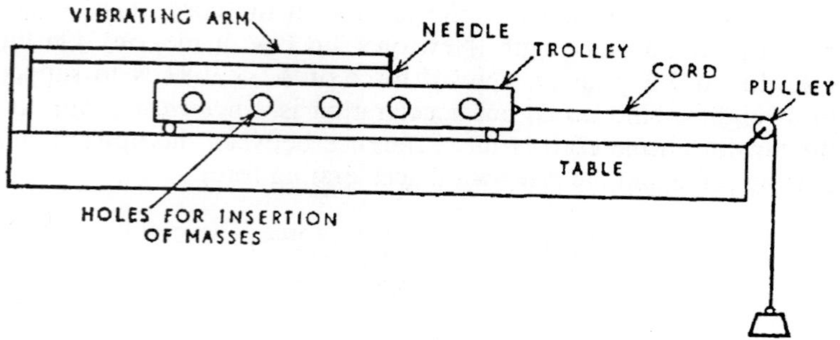

Fig. 53

Example. In an experiment on Fletcher's Trolley the total mass of the loaded trolley was 2·9 kg and the mass of the load suspended on the end of the cord was 0·1 kg. Friction was negligible and measurements taken showed that the trolley moved with uniform acceleration and, from rest, moved a distance of 654 mm in 2 s. Find the value of g from this experimental data and calculate the tension in the cord.

$$s = ut + \tfrac{1}{2}at^2$$

$$a = \frac{2s}{t^2} \text{ (because } ut = 0\text{)}$$

$$= \frac{2 \times 0{\cdot}654}{2^2}$$

$$= 0{\cdot}327 \text{ m/s}^2$$

$$\text{Total mass accelerated} = 2{\cdot}9 + 0{\cdot}1 = 3 \text{ kg}$$

$$\text{Accelerating force} = 0{\cdot}1 \times \text{N}$$

$$\text{Accelerating force} = \text{mass} \times \text{acceleration}$$

$$0{\cdot}1 \times g = 3 \times 0{\cdot}327$$

$$g = \frac{3 \times 0{\cdot}327}{0{\cdot}1}$$

$$= 9{\cdot}81 \text{ m/s}^2 \quad \text{Ans. (i)}$$

Consider horizontal portion of the cord,

$$\text{Tension in cord} = \text{Force to accelerate the trolley}$$

$$= \text{ma}$$

$$= 2{\cdot}9 \times 0{\cdot}327$$

$$= 0{\cdot}9483 \text{ N} \quad \text{Ans. (ii)}$$

As a check consider the hanging part of the cord for the same result. The hanging part of the cord supplies an upward force to support the mass of 0·1 kg but, as the mass accelerates *downwards*, the accelerating force on the 0·1 kg mass acts in a downward direction.

Tension in cord = upward supporting force – downward accel. force

$$= 0{\cdot}1 \times 9{\cdot}81 - 0{\cdot}1 \times 0{\cdot}327$$

$$= 0{\cdot}981 - 0{\cdot}0327$$

$$= 0{\cdot}9483 \text{ N}$$

LINEAR MOMENTUM

Linear momentum is the term given to the product of mass and velocity and is therefore usually defined as the quantity of motion possessed by a moving body. It is a vector quantity.

From the expression:

$$\text{Accelerating force} = \text{mass} \times \text{acceleration}$$

Substituting for acceleration, change of velocity divided by time:

$$\text{Accelerating Force} = \frac{\text{mass} \times \text{change of velocity}}{\text{time}}$$

Linear momentum being the product of mass and velocity, then change of linear momentum takes place if there is a change of velocity, therefore mass multiplied by change of velocity is change of linear momentum, and accelerating force is the force to cause that change, hence,

$$\text{Force} = \frac{\text{change of linear momentum}}{\text{time}}$$

$$\text{or Force} = \text{change of linear momentum /s.}$$

This could also be written:

$$\text{Force} \times \text{time} = \text{change of linear momentum.}$$

In calculus notation:

$$\text{Force} \propto \text{rate of change of linear momentum}$$

$$\propto \frac{d}{dt}\left(mv\right) \quad \textit{i.e.}\ m\frac{dv}{dt} \quad \textit{i.e.}\ m\frac{d^2s}{dt^2}$$

$$= ma \text{ in suitable units}$$

For a given change of linear momentum of a body free to move, the applied force varies inversely as the time taken. For instance, a force of 2 N acting for 10 s would have the same effect as 10 N acting for 2 s, or 5 N acting for 4 s, and so on. The product of force and time is referred to as the *impulse* of the force.

Example. The mass of the head of a hand hammer is 0·8 kg. When moving at 9 m/s it strikes a chisel and is brought to rest in $\frac{1}{250}$ s. What is the average force of the blow?

$$\text{Force} = \frac{\text{change of linear momentum}}{\text{time}}$$

$$= \frac{0{\cdot}8 \times 9}{\frac{1}{250}}$$

$$= 0{\cdot}8 \times 9 \times 250$$

$$= 1800 \text{ N} \quad \text{Ans.}$$

Example. A jet of fresh water, 20 mm diameter, issues from a horizontal nozzle at a velocity of 21 m/s, and strikes a stationary vertical plate. Find the mass flow of water leaving the nozzle every second and, assuming no splash back, find the force of the jet on the plate. Density of fresh water = 10^3 kg/m³.

$$\text{Volume flow (m}^3\text{/s)} = \text{area (m}^2) \times \text{velocity (m/s)}$$

$$\dot{V} = Av$$

$$= 0{\cdot}7854 \times 0{\cdot}02^2 \times 21$$

$$= 0{\cdot}006598 \text{ m}^3\text{/s}$$

$$\text{Mass flow (kg/s)} = \text{vol. flow (m}^3\text{/s)} \times \text{density (kg/m}^3)$$

$$\dot{m} = \dot{V}\rho$$

$$= 0{\cdot}006598 \times 10^3$$

$$= 6{\cdot}598 \text{ kg/s} \quad \text{Ans. (i)}$$

Assuming no splash back infers that the water has no velocity after striking the plate, therefore the change of velocity is 21 m/s.

$$\text{Force} = \text{Change of linear momentum per second}$$

$$= \text{mass} \times \text{change of velocity, per second}$$

$$= 6{\cdot}598 \times 21$$

$$= 138{\cdot}6 \text{ N} \quad \text{Ans. (ii)}$$

CONSERVATION OF LINEAR MOMENTUM. When two bodies collide, the force of one body on the other is equal in magnitude and opposite in direction and the time during which the force acts is the same, hence each body receives the same change of momentum. This means that the gain of momentum by one body is equal to the momentum lost by the other, therefore the sum of momentum of the two bodies after impact is the same as that before impact. Momentum is neither created nor destroyed by the collision. This is known as the law of the conservation of momentum.

Example. A railway truck of mass 3 t moving at 64 km/h collides with another of 2 t mass moving at 34 km/h in the same direction, and then move on together as one unit. Find the velocity of the trucks after impact.

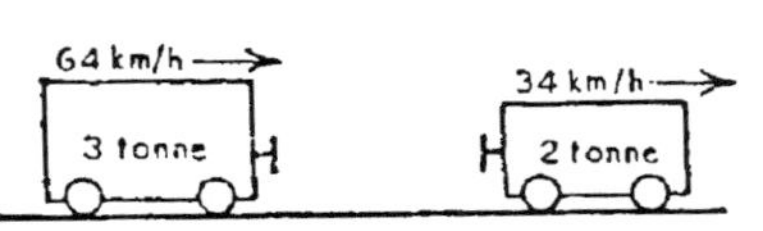

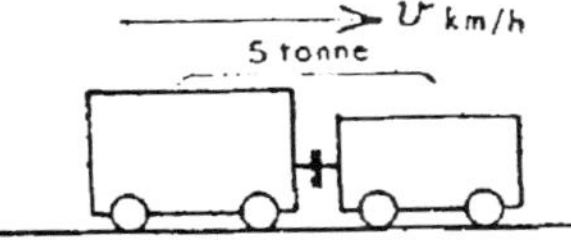

Fig. 54

Linear momentum before impact = Linear momentum after impact

$$(3 \times 64) + (2 \times 34) = (3 + 2) \times v$$

$$192 + 68 = 5v$$

$$v = 52 \text{ km/h} \quad \text{Ans.}$$

Note that in such an equation, linear momentum can be expressed in any convenient units of mass and velocity provided that the same kind are used throughout the whole equation.

Furthermore, if the smaller truck had been travelling initially in the opposite direction, its velocity would be regarded as a negative quantity in relation to the direction of the larger truck, then:

Linear momentum before impact = Linear momentum after impact

$$(3 \times 64) + (2 \times -34) = (3 + 2) \times v$$

$$192 - 68 = 5v$$

$$v = 24{\cdot}8 \text{ km/h}$$

Being a positive answer, the direction is the same as the initial direction of the larger truck which was taken as positive.

Example. A bullet of mass 0·04 kg is fired into a freely suspended and stationary block of wood whose mass is 13·6 kg and caused the wood to start moving at 1·9 m/s. Find the initial velocity of the bullet.

Let v = initial velocity of bullet.

Linear momentum before impact = Linear momentum after impact

Mom. of bullet + Mom. of wood = Mom. of (wood and bullet)

$$0{\cdot}04 \times v + 13{\cdot}6 \times 0 = (0{\cdot}04 + 13{\cdot}6) \times 1{\cdot}9$$

$$v = \frac{13{\cdot}64 \times 1{\cdot}9}{0{\cdot}04}$$

$$= 647{\cdot}9 \text{ m/s} \quad \text{Ans.}$$

ANGULAR MOMENTUM

Angular momentum is defined as the *moment* of linear momentum (see Chapter 7: first and second moments).

The mass m shown (Fig. 55) is rotating in a circular path of radius r with angular velocity ω,

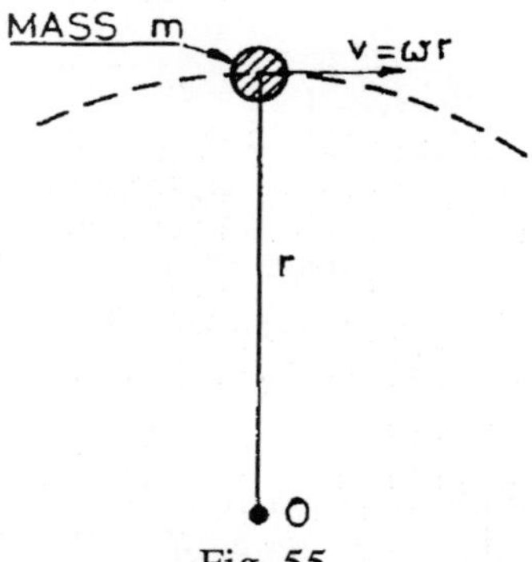

Fig. 55

$$\text{linear velocity } v = \omega r$$

$$\text{and, linear momentum} = mv$$

$$= m\omega r$$

$$\text{The moment of this momentum, about 0} = m\omega r \times r$$

$$= m\omega r^2$$

$$\text{Rearranging, angular momentum} = mr^2 \times \omega$$

But mr^2 is the *moment of inertia* (second moment of mass) of the mass about its axis of rotation, denoted by symbol I. Thus, the angular momentum of the mass = $I\omega$.

Hence, for a shaft or rotor, with *radius of gyration* (the radial distance at which mass is taken to be concentrated) k:

$$\text{Angular momentum} = I\omega$$

where $I = mk^2$

Example. A flywheel of mass 500 kg and radius of gyration 1·2 m is running at 300 rev/min. By means of a clutch, this flywheel is suddenly connected to another flywheel, mass 2000 kg and radius of gyration 0·6 m, initially at rest. Calculate their common speed of rotation after engagement.

$$\text{Initial speed} = \frac{2\pi \times 300}{60}$$

$$= 31{\cdot}42 \text{ rad/s}$$

$$\begin{array}{c}\text{Angular momentum}\\\text{before impact}\end{array} = \begin{array}{c}\text{Angular momentum}\\\text{after impact}\end{array}$$

$$I\omega_1 = I\omega_2$$

$$500 \times 1{\cdot}2^2 \times 31{\cdot}42 = (500 \times 1{\cdot}2^2 + 2000 \times 0{\cdot}6^2) \times \omega_2$$

$$\omega_2 = 15{\cdot}71 \text{ rad/s}$$

Common speed after impact = 150 rev/min Ans.

TURNING MOMENT

The moment of a force about a given point is the product of the force and the perpendicular distance from the line of action of the force to that point. The perpendicular distance is the leverage of the force.

When the moment has a tendency to twist or rotate a body, such as turning a shaft in its bearings, it is usually called a *turning moment* or *torque*.

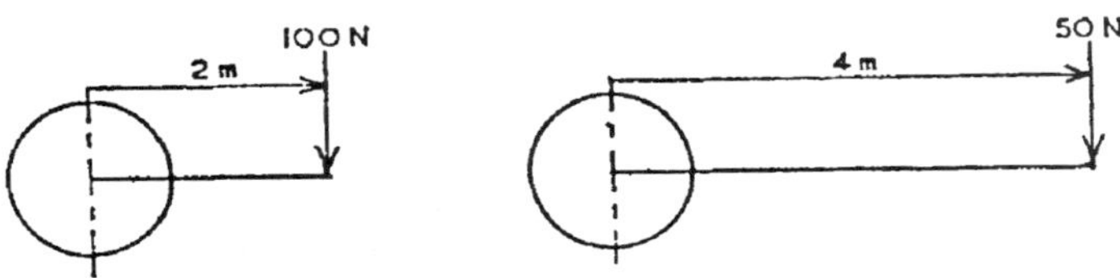

Fig. 56

Consider Fig. 56. The first sketch illustrates a turning moment of 100 N × 2 m = 200 N m applied to a shaft to turn it. The second sketch shows half as much force (50 N) with twice as much leverage (4 m), this will have the same turning or twisting effect because the turning moment of 50 N × 4 m is the same (200 N m) as before.

Consider now a thin rim of a flywheel of mean radius r. If a tangential force of F is applied to the rim, neglecting frictional resistances, the rim will accelerate and the acceleration can be found from:

$$F = ma$$

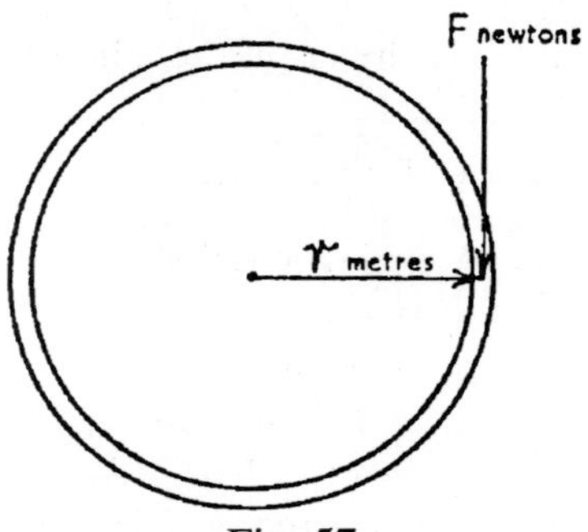

Fig. 57

This acceleration, however, is the linear acceleration of the rim. It can be converted into terms of angular acceleration.

$$\text{Linear acceleration} = \text{angular acceleration} \times \text{radius}$$

$$a = \alpha r$$

$$F = ma$$

$$F = m\alpha r$$

If, as it will be more likely, the force is not applied on the rim itself, but at a greater or less leverage, say L from the centre, the effective force on the rim causing it to accelerate will be greater or less accordingly and in the ratio of L to r, thus:

$$F \times \frac{L}{r} = m \times \alpha r$$

Multiplying both sides by r:

$$F \times L = m \times \alpha \times r^2$$

Now $F \times L$ is the torque applied, therefore,

$$\text{Accelerating torque} = m \times r^2 \times \alpha$$

For a rotating mass which is not a thin rim the whole is considered as concentrated at the radius of gyration, not to be confused with the radius of the wheel:

$$\text{Accelerating torque} = mk^2\,\alpha$$

$$T = I\alpha$$

In calculus notation:

$$\text{Torque} \propto \text{rate of change of angular momentum}$$

$$\propto \frac{\mathrm{d}}{\mathrm{dt}}\left(\mathrm{Iw}\right) \quad \textit{i.e.}\ \mathrm{I}\frac{\mathrm{dw}}{\mathrm{dt}} \quad \textit{i.e.}\ \mathrm{I}\frac{\mathrm{d}^2\theta}{\mathrm{dt}^2}$$

$$T = I\alpha \text{ in suitable units}$$

Example. The mass of a flywheel is 175 kg and its radius of gyration is 380 mm. Find the torque required to attain a speed of 500 rev/min from rest in 30 s.

$$\alpha = \frac{500 \times 2\pi}{60 \times 30}$$
$$= 1{\cdot}746 \text{ rad/s}^2$$
$$T = I\,\alpha$$
$$= mk^2\alpha$$
$$= 175 \times 0{\cdot}38^2 \times 1{\cdot}746$$
$$= 44{\cdot}12 \text{ N m} \quad \text{Ans.}$$

Example. The torque to overcome frictional and other resistances of a turbine is 317 N m and may be considered as constant for all speeds. The mass of the rotating parts is 1·59 t and the radius of gyration is 0·686 m. If the gas is cut off when the turbine is running free of load at 1920 rev/min, find the time it will take to come to rest and the number of revolutions turned during that time.

$$\text{Change in velocity} = \frac{1920 \times 2\pi}{60}$$
$$= 64\,\pi \text{ rad/s}$$

Torque to overcome friction acts as a retarding torque to bring the turbine to rest when the steam is shut off.

$$T = mk^2\alpha$$
$$\alpha = \frac{T}{mk^2}$$
$$= \frac{317}{1{\cdot}59 \times 10^3 \times 0{\cdot}686^2}$$
$$= 0{\cdot}4237 \text{ rad/s}^2$$
$$\text{Time} = \frac{\text{change of velocity}}{\text{retardation}}$$
$$= \frac{64 \times \pi}{0{\cdot}4237}$$
$$= 474{\cdot}5 \text{ s}$$
$$= 7 \text{ min } 54{\cdot}5 \text{ s} \qquad \text{Ans. (i)}$$

Average velocity while coming to rest

$$= \tfrac{1}{2}(1920 + 0) = 960 \text{ rev/min}$$
$$\text{Distance} = \text{average velocity} \times \text{time}$$
$$= 960 \times \frac{474{\cdot}5}{60}$$
$$= 7592 \text{ rev} \qquad \text{Ans. (ii)}$$

TEST EXAMPLES 3

1. Find the extra thrust, in kN, required to increase the speed of a ship of 10 000 t displacement, from 15 to 20 knots in 10 min, neglecting the increase of resistance due to increased speed. One knot = 1·852 km/h.

2. The tractive resistance of a vehicle of 2 t mass is 155 N per tonne. If a total pull of 1 kN is applied to the vehicle, find (i) the acceleration, (ii) the speed (km/h) after travelling for one min from rest.

3. Find the average acceleration force required to change the speed and direction of a motor-boat of 5 t mass from 9 knots North-East to 12 knots South-East in 2 min. One knot = 1·852 km/h.

4. The total mass of a lift and its contents amount to 1134 kg and is suspended by a single wire rope. Find the tension in the wire when the lift is going down and (i) accelerating at 2 m/s^2, (ii) moving at a steady speed, (iii) retarding at 1 m/s^2.

5 The masses on the two ends of the cord around the pulley of an Atwood's machine are each 2·5 kg. If the mass of the pulley is negligible and the bearings considered frictionless, find (i) the acceleration of the system when a rider of 0·25 kg is added to one end, (ii) the velocity of the system after 3 s from starting.

6. The mass of the loaded carriage of a Fletcher's Trolley apparatus is 5·44 kg and the mass of the load on the hook at the hanging end of the cord is 0·22 kg. If friction is negligible, find the acceleration of the carriage and the tension in the cord.

7. A 3 kg hammer head moving at a velocity of 7m/s strikes the head of a wedge and is brought to rest in $\frac{1}{100}$ s. Find the average force of the blow.

8. A rail-truck of 2 t mass moving at 10 km/h is overtaken on the same lines by a locomotive of 10 t mass moving at 16 km/h, and after colliding, the locomotive and truck move on locked together. Find the speed of the two immediately after impact.

9. Find the accelerating torque required to increase the speed of a flywheel from 470 to 700 rev/min in 10 s if the mass of the flywheel is 544 kg and its radius of gyration is 0·5 m.

10. A rotor of 68 kg mass has a radius of gyration of 229 mm. If a retarding torque of 24 Nm is applied when the rotor is running at 2800 rev/min, find the retardation and the time taken to come to rest.

11. A rotating gearwheel has a mass of 1 t and a radius of gyration of 1·5 m. The angular displacement, from rest, is:

$$\theta = 2{\cdot}1 - 3{\cdot}2\ t + 4{\cdot}8\ t^2 \text{ at any instant.}$$

Find the accelerating torque after 1·5 s.

*f*12. The total mass of the reciprocating parts of a vertical I.C. engine is 317·5 kg. At a certain position on the downward stroke the effective pressure on the piston is 6 bar and the acceleration of the piston is 21 m/s^2. If the piston diameter is 250 mm, find, neglecting friction, the thrust on the crosshead. 1 bar = 10^5 N/m^2.

*f*13. A light flexible cord is hung over a pulley of negligible mass in frictionless bearings. Masses of 1·8 kg and 1·9 kg respectively are hung on the two ends of the cord and the system allowed to move from rest. Find,

(i) the acceleration of the system,
(ii) the velocity after 4 s,
(iii) distance moved in this time,
(iv) the tension in the cord,
(v) the total load on the pulley bearings when the system is moving under the above conditions.

*f*14. A block of wood of 9 kg mass stands on a horizontal table. A cord is connected at one end to the block, is led parallel with the table, over a guide pulley, and hangs down over the side of the table. It is found that 0·9 kg hung on the end of the cord is just sufficient to overcome friction between the block and the table. If an additional 0·45 kg is hung on the end, find,

(i) the acceleration of the block,
(ii) its velocity after 2·5 s from rest,
(iii) the distance travelled in the above time,
(iv) the tension in the cord.

*f*15. The winding drum of an electric hoist is 1·5 m diameter, the mass of the rotating parts of the drive and drum is 1225 kg and the radius of gyration is 0·53 m. The wire rope from the drum carries a load of 450 kg. Find the total torque required to raise the load with an acceleration of 0·6 m/s^2 when 190 N m of torque are required to overcome friction.

CHAPTER 4

WORK, POWER AND ENERGY

WORK is done when a force applied on a body causes it to move and is measured by the product of the force and the distance through which the force moves.

The unit of work is the *joule* and is defined as the work done when the point of application of a force of 1 N moves through a distance of 1 m in the direction in which the force is applied.

When a body is being moved, the force applied may vary while it is moved from one position to another. To cover all cases the general rule for the quantity of work done is therefore:

$$\text{Work done} = \text{average force} \times \text{distance moved}$$

GRAPHICAL REPRESENTATION. If a graph is plotted to represent force and distance, the area under the graph represents work done. Fig. 58 illustrates a constant force of 5 N acting through a distance of 6 m, the work done is 30 J.

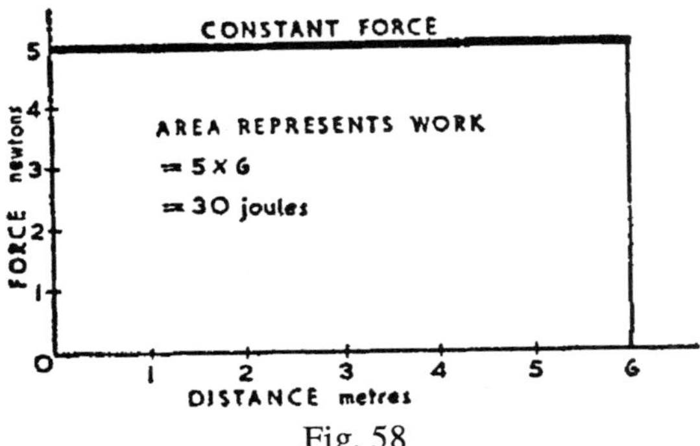

Fig. 58

The extension or compression of a spring is proportional to the force applied. For example, if the stiffness of a spring, or the spring rate, is stated to be 100 N/mm, it means that 100 N will compress the spring the first mm, another 100 N (making 200 N) is required to compress it a further mm (making 2 mm) and so on. Suppose it is required to compress this spring 40 mm, then:

$$\text{Final force} = 100 \times 40 = 4000 \text{ N}$$

Average force applied (from zero up to 4000 N)

$$= \tfrac{1}{2}(0 + 4000) = 2000 \text{ N}$$

$$\text{Work done} = 2000 \times 0{\cdot}04$$
$$= 80 \text{ J}$$

The graph shown in Fig. 59 represents the relationship between the compression from zero to 40 mm and the force applied from zero to 4000 N.

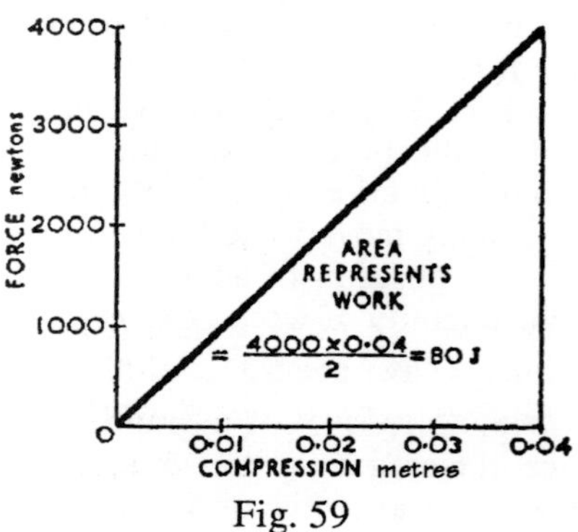

Fig. 59

LIFTING LOADS. When a mass is lifted at a steady speed, the direct force applied to raise the mass is simply the upward force to overcome the downward gravitational pull on the mass, which is its weight. The effective distance through which the mass is raised against gravity is the *vertical height* through which the centre of gravity of the mass moves, therefore:

$$\text{Work done} = \text{weight} \times \text{vert. height its c.g. is lifted}$$

In calculating the total weight of a body from its volumetric dimensions, the specific weight of the material may be expressed in force units per unit volume, such as N/m^3, the total weight (N) is simply the product of the volume (m^3) and the specific weight (N/m^3). More usually, the density of the material may be expressed in mass units per unit volume, such as kg/m^3, in which case conversion from kg to N will be necessary.

Centre of gravity (C.G.) or *Centre of Mass* is that point where all the mass is considered to be concentrated (see Chapter 7).

Example. A rectangular block of steel of 150 mm square section and 600 mm long lies lengthwise on a horizontal ground. Find the work done in lifting it about one end (i) until the base makes an angle of 60° to the horizontal, (ii) until it will begin to topple over on to its end. Take the density of steel as 7850 kg/m³.

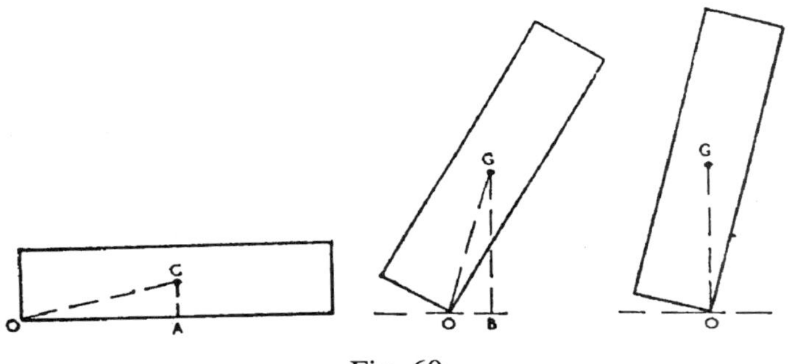

Fig. 60

Referring to Fig. 60 the procedure is (*a*) to find the weight of the block, (*b*) to find the vertical height the centre of gravity is raised, (*c*) to multiply the two to obtain work done.

$$\text{Density} = 7{\cdot}85 \times 10^3 \text{ kg/m}^3$$

$$\text{Volume of block} = 0{\cdot}15^2 \times 0{\cdot}6\text{m}^3$$

$$\text{Mass} = 0{\cdot}15^2 \times 0{\cdot}6 \times 7{\cdot}85 \times 10^3$$

$$\text{Weight} = 0{\cdot}15^2 \times 0{\cdot}6 \times 7{\cdot}85 \times 10^3 \times 9{\cdot}81$$

$$= 1040 \text{ N}$$

Centre of gravity is at the geometrical centre of the block, indicated by G.

Initial height of G above ground = GA = 75 mm

$$\text{GO} = \sqrt{(\text{GA})^2 + (\text{OA})^2}$$

$$= \sqrt{75^2 + 300^2}$$

$$= 309{\cdot}2 \text{ mm}$$

$$\text{tan of angle GOA} = \frac{\text{GA}}{\text{AO}} = \frac{75}{300} = 0{\cdot}25$$

$$\text{angle GOA} = 14^\circ\, 2'$$

$$\text{angle GOB} = 14^\circ\, 2' + 60^\circ = 74^\circ\, 2'$$

$$\text{GB} = \text{GO} \times \sin 74^\circ\, 2'$$

$$= 309{\cdot}2 \times 0{\cdot}9615$$

$$= 297{\cdot}3 \text{ mm}$$

$$\text{Vertical height G is raised} = \text{GB} - \text{GA}$$
$$= 297{\cdot}3 - 75$$
$$= 222{\cdot}3 \text{ mm} = 0{\cdot}2223 \text{ m}$$
$$\text{Work done} = \text{weight} \times \text{vertical lift of G}$$
$$= 1040 \times 0{\cdot}2223$$
$$= 231{\cdot}2 \text{ J} \qquad \text{Ans. (i)}$$

To lift until it would fall over on to its end, the block is to be raised until its centre of gravity is vertically above the point about which the block swings, that is, G is to be vertically above O, it will then fall over on its own accord.

$$\text{Vertical height G is raised} = \text{GO} - \text{GA}$$
$$= 309{\cdot}2 - 75$$
$$= 234{\cdot}2 \text{ mm} = 0{\cdot}2342 \text{ m}$$
$$\text{Work done} = \text{weight} \times \text{vertical lift of G}$$
$$= 1040 \times 0{\cdot}2342$$
$$= 243{\cdot}5 \text{ J} \qquad \text{Ans. (ii)}$$

POWER

Power is the rate of doing work, that is, the quantity of work done in a given time. The unit of power is the *watt* which is equal to the rate of 1 J of work being done every s.

$$\text{Power (W} = \text{J/s} = \text{Nm/s)} = \text{force (N)} \times \text{velocity (m/s)}$$
$$\text{Work done} = \text{F ds (for small displacement)}$$
$$\text{Power} = \text{F}\,\frac{\text{ds}}{\text{dt}} \text{ (for small time interval)}$$
$$\text{P} = \text{Fv (instantaneous velocity v)}$$

For normal powers in engineering, mechanical, electrical or hydraulic, the kW is usually more convenient.

Example. A pump lifts fresh water from one tank to another through an effective height of 12 m. If the mass flow of the water is 40 t/h, find the output power of the pump.

$$40 \text{ t} = 40 \times 10^3 \text{ kg}$$

$$\text{Weight of water lifted/s} = \frac{40 \times 10^3 \times 9{\cdot}81}{3600} \text{ N}$$

$$\text{Power} = \text{weight lifted /s} \times \text{height}$$

$$= \frac{40 \times 10^3 \times 9{\cdot}81 \times 12}{3600} \text{ W}$$

$$P = 1308 \text{ W or } 1{\cdot}308 \text{ kW} \qquad \text{Ans.}$$

MECHANICAL EFFICIENCY The ratio of the power got out of a machine to the power put in.

$$\text{Mechanical efficiency} = \frac{\text{output power}}{\text{input power}}$$

This gives a fraction less than unity. It is common practice to multiply this by 100 and express the efficiency as a percentage.

For example, if the input power of the above pump was 1·75 kW then the efficiency would be:

$$\Im = \frac{\text{output power}}{\text{input power}}$$

$$= \frac{1{\cdot}308}{1{\cdot}75} = 0{\cdot}7474 \text{ or } 74{\cdot}74\%$$

PRESSURE is force per unit area with N/m^2 as units.

POWER OF RECIPROCATING ENGINES. When a fluid such as a gas or a liquid acts on a piston in a cylinder, the pressure multiplied by the piston area on which the pressure acts, gives the total force on the piston. If the motive power is liquid, such as in a hydraulic cylinder, the pressure exerted on the piston is constant during the whole stroke but, in internal-combustion engines, the pressure of the gas varies throughout the stroke. There is also a back pressure on the other side of the piston which may or may not vary. Therefore, to obtain the work done during one stroke, the average effective pressure on the piston throughout the complete stroke must be used, this is termed the *mean effective pressure*.

Let p_m = mean effective pressure (N/m^2)

A = area of piston (m^2)

L = length of stroke (m)

N = number of power strokes per second

then,

Average force on piston

$= p_m \times A \quad N$

Work done in one power stroke

$= p_m \times A \times L$ Nm *i.e.* J

Work per second

$= p_m \times A \times L \times n$ J/s *i.e.* W

The value of n, which here represents the number of power strokes per second, depends upon the cycle of operations on which the engine works.

pV DIAGRAMS. If a graph is drawn to represent the variation of pressure on the piston of a cylinder on a base of volume, the area under the graph represents work done, thus:

$$\text{Total work done} = \int p\,dV \text{ over volume range}$$

In a hydraulic cylinder the pressure of the liquid is usually constant throughout the whole stroke of the piston and the quantity of liquid supplied per stroke is equal to the full stroke volume. The power supplied in watts is therefore the product of the supply pressure in N/m^2 and the volume flow in m^3/s.

Example. The rate of water supply to a hydraulic crane is 90 l/min at a steady pressure of 70 bar. Find the input power and its efficiency if the output is 7·5 kW (1 bar = 10^5 N/m^2).

$$\text{Flow rate} = \frac{90}{10^3 \times 60} \qquad (1 = 10^{-3}m^3)$$

$$\text{Input power} = \frac{70 \times 10^5 \times 90}{10^3 \times 60}$$

$$= 10\,500 \text{ W or } 10{\cdot}5 \text{ kW} \qquad \text{Ans.(i)}$$

$$\text{Efficiency} = \frac{\text{output}}{\text{input}}$$

$$= \frac{7{\cdot}5}{10{\cdot}5}$$

$$\eta = 0{\cdot}7143 \text{ or } 71{\cdot}43\% \qquad \text{Ans.(ii)}$$

POWER BY TORQUE

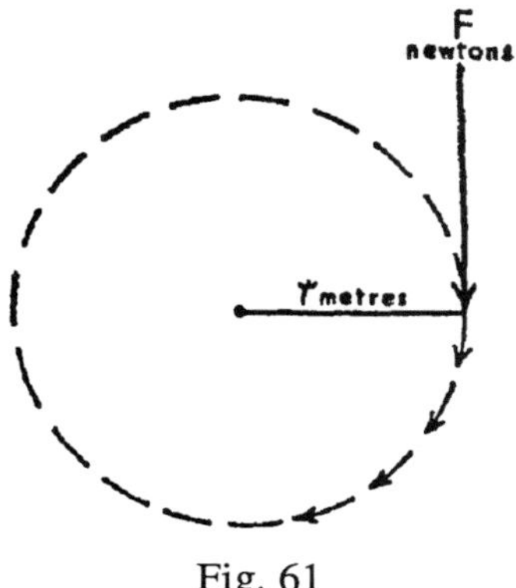

Fig. 61

Consider a force of F at a radius of r on a rotating mechanism.

$$\text{Work done in one revolution} = \text{force} \times \text{circumference}$$
$$= F \times 2\pi r \quad \text{Nm } i.e. \text{ J}$$

If it is running at n revolutions per second,

$$\text{Power} = F \times 2\pi r \times n \quad \text{J/s } i.e. \text{ W}$$

but $F \times r$ = torque (T) so,

$$P = 2\pi Tn$$
$$\omega = 2\pi n$$
$$P = T\omega$$
$$\text{Work done} = Td\theta \ \text{(for small } d\theta)$$
$$\text{Power} = \frac{Td\theta}{dt} \ \text{(for small dt)}$$
$$P = T\omega \ \text{(instantaneous } \omega)$$

Example. The mean torque in a propeller shaft is $2{\cdot}26 \times 10^5$ N m when running at 140 rev/min. Find the power transmitted.

$$\text{Rotational speed} = 140 \text{ rev/min}$$
$$= {}^{140}\!/_{60} \times 2\pi \text{ rad/s}$$
$$\text{Power} = 2{\cdot}26 \times 10^2 \times {}^{140}\!/_{60} \times 2\pi$$
$$P = 3313 \text{ kW}$$

Ans.

BRAKE POWER. The output power at the shaft of the engine is referred to as the shaft power or brake power. It is measured by coupling the shaft to some form of brake which can absorb and measure the power output. On large engines, water or electric dynamometers are used, but for small powers a rope brake is a simple and reliable arrangement for this purpose.

A rope is hung over a flanged flywheel, one end of the rope is anchored to a spring balance fixed to the bedplate and the other end carries suitable loads. The power is absorbed by the friction between the rope and wheel rim, this generates heat which is carried away by running water led into and out of a trough formed around the inside of the periphery of the wheel.

If W is the load hung on one end, and S is the reading of the spring balance, then the effective tangential braking force at the rim of the wheel is $(W - S)$. This multiplied by the radius from shaft centre to rope centre is the braking torque. The brake power is then calculated from:

$$P = T\omega$$

TRANSMISSION OF POWER BY CHAIN, BELT OR GEARS. When driving one pulley (the follower) from another pulley (the driver) by means of a chain, or a belt, if there is no slipping the linear velocity of the rim of each pulley is the same and are equal to the linear velocity of the chain or belt, therefore, rotational speed × circumference is constant for each pulley. Circumference can be represented by diameter or radius when equating because the constants π or 2π will cancel; in the case of chain drive or gear wheels in mesh, circumference can be represented by the number of sprockets or teeth in the wheel because the pitch is common.

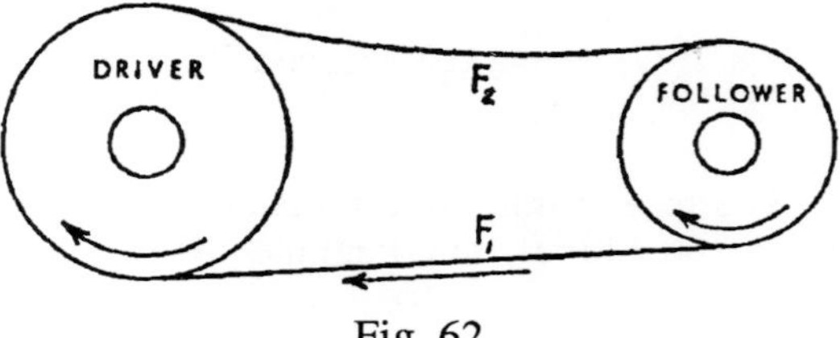

Fig. 62

Consider a belt driven pulley (Fig. 62), there is tension in all parts of the belt but more in one part than the other. To distinguish between the two parts, that which has the greatest pull is referred to as the *tight* side and the pull (tension or force) in it can be represented by F_1. The other not-so-tight part is referred to as the slack side and the pull (tension or force) in it can be represented by F_2. The effective driving force acting at the rim of the pulley is the difference between the pull on one side and the pull on the other, that is, $F_1 - F_2$ therefore the driving torque is the product of $F_1 - F_2$ and the radius.

Hence, if n represents the rotational speed (rev/s) and r the radius (m):

$$\text{Work done/rev.} = (F_1 - F_2) \times 2\pi r \quad \text{J}$$

$$\text{Work done/s} = \text{J/s } i.e. \text{ W}$$

$$\text{Power transmitted} = (\text{F}_1 - \text{F}_2) \times 2\pi r \times \text{n}$$

Since $2\pi rn$ is the linear speed of the belt then,

$$\text{Power transmitted} = (\text{F}_1 - \text{F}_2) \times \text{speed of belt}$$

Example. A pulley 270 mm diameter is driven at 300 rev/min by a belt 12 mm thick. The tensions in the tight and slack sides of the belt are 1560 and 490 N respectively. Find the power transmitted.

Effective radius, from shaft centre to mid-thickness of belt = 135 + 6 = 141 mm = 0·141 m.

$$\text{Speed of pulley} = 300 \div 60 = 5 \text{ rev/s}$$

$$\text{Speed of belt} = 2\pi \times 0{\cdot}141 \times 5 \text{ m/s}$$

$$\text{Power} = (\text{F}_1 - \text{F}_2) \times \text{speed of belt}$$

$$= (1560 - 490) \times 2\pi \times 0{\cdot}141 \times 5$$

$$\text{P} = 4741 \text{ W or } 4{\cdot}741 \text{ kW} \quad \text{Ans.}$$

Example. One gear wheel with 100 teeth of 6 mm pitch running at 250 rev/min drives another which has 50 teeth. If the power transmitted is 0·5 kW, find the driving force on the teeth and the speed of the driven wheel.

$$\text{P} = \text{T}\omega$$

$$500 = \text{T} \times \frac{250 \times 2\pi}{60}$$

$$\text{Torque} = \frac{500 \times 60}{250 \times 2\pi} \text{ N m}$$

$$\text{Circumference of wheel} = \text{no. of teeth} \times \text{pitch}$$

$$= 100 \times 6 = 600 \text{ mm} = 0{\cdot}6 \text{ m}$$

$$\text{Radius} = \frac{\text{circumference}}{2\pi} = \frac{0{\cdot}6}{2\pi} \text{ m}$$

$$\text{Torque} = \text{force} \times \text{radius}$$

$$\text{Force between teeth} = \frac{\text{torque}}{\text{radius}}$$

$$= \frac{500 \times 60 \times 2\pi}{250 \times 2\pi \times 0{\cdot}6}$$

$$= 200 \text{ N} \qquad \text{Ans. (i)}$$

$$\text{speed of driver} \times \text{teeth in driver} = \text{speed of follower} \times \text{teeth in follower}$$

$$\text{Speed of follower} = \frac{250 \times 100}{50}$$

$$= 500 \text{ rev/min} \qquad \text{Ans. (ii)}$$

ENERGY

Energy is the quantity of work stored up in a solid, liquid or gas which is capable of doing work and is consequently measured in units of work. The unit of energy is therefore the *joule*. There are different means of storing work, two kinds of mechanical energy will be dealt with later.

POTENTIAL ENERGY is the work stored up in a stationary body by virtue of its position or condition. For example, if a body of weight 50 N is lifted 20 m, the amount of work done to lift it is 50 × 20 = 1000 N m and at this height the body has stored up 1000 J of potential energy by virtue of its elevated position, it could give out this amount of work, usefully or otherwise, if allowed to fall back to its original level. A wound-up spring is another example of stored up energy. Compressed air stored in a reservoir contains potential energy by virtue of its condition and can do useful work.

KINETIC ENERGY is the energy stored up in a moving body which (linear) depends upon the mass of the body and the square of its linear velocity. One of the best examples is the flywheel, its function is to store up and give out kinetic energy. For instance as a shearing or punching machine, the speed is built up during the idle stroke and work from the driving motor is absorbed into the flywheel as kinetic energy, during the working stroke this energy is given out again in doing most of the work of shearing or punching.

KINETIC ENERGY OF TRANSLATION. Consider a body being pushed along, the force over and above that required to overcome frictional and other resistances causes acceleration of the body and is therefore increasing its velocity. Thus the kinetic energy put into the body is the whole of the work done to increase its velocity. Starting from rest, neglecting friction,

$$\text{Acceleration} = \frac{v-0}{t} = \frac{v}{t}$$

$$\text{Force} = \frac{\text{mass} \times v}{t}$$

$$\begin{aligned}\text{Displacement} &= \text{average velocity} \times \text{time}\\ &= \frac{0+v}{2} \times t\\ &= \frac{vt}{2}\end{aligned}$$

$$\begin{aligned}\text{Work done} &= \text{force} \times \text{displacement}\\ &= \frac{mv}{t} \times \frac{vt}{2}\\ &= \tfrac{1}{2}mv^2\end{aligned}$$

As the kinetic energy in the body when moving at v must be equal to the work done to attain that velocity (because no work is lost here in friction), then:

$$\text{Translational kinetic energy} = \tfrac{1}{2}mv^2$$

Also consider this from the point of view of a falling body. Let a body of mass be at rest at a height of h. Its weight is mg and it contains potential energy equal to mgh but no kinetic energy because it is not moving. If the mass is allowed to fall it gains kinetic energy as it increases velocity, the gain in kinetic energy being equal to the loss of potential energy. Just before the body reaches the ground it has fallen through h, lost all its potential energy and gained an equal amount of kinetic energy.

$$\begin{aligned}\text{kinetic energy gained} &= \text{potential energy lost}\\ &= mgh\\ v^2 &= u^2 + 2gh\\ v^2 &= 2gh\\ \text{and } h &= \frac{v^2}{2g}\end{aligned}$$

$$\text{Kinetic energy} = mg \times \frac{v^2}{2g}$$
$$= \tfrac{1}{2} mv^2$$

Example. A body of mass 30 kg is moving at a velocity of 100 km/h. Find the kinetic energy in the body and the loss of kinetic energy when the speed falls to 25 km/h.

$$\text{Kinetic energy} = \tfrac{1}{2} mv^2$$
$$= \tfrac{1}{2} \times 30 \times \left\{\frac{10^5}{3600}\right\}^2$$
$$= 11\,570 \text{ J}$$
$$= 11{\cdot}57 \text{ kJ} \qquad \text{Ans. (i)}$$

Kinetic energy is proportional to the square of the velocity,

$$K.E. \text{ at } 25 \text{ km/h} = 11\,570 \times (^{25}/_{100})^2$$
$$= 11\,570 \times (^1/_4)^2$$
$$= 723{\cdot}1 \text{ J}$$
$$\text{Loss of } K.E. = 11\,570 - 723{\cdot}1$$
$$= 10\,846{\cdot}9 \text{ J} = 10{\cdot}8469 \text{ kJ} \qquad \text{Ans. (ii)}$$

As the kinetic energy is increased by increasing the velocity of a body due to applying an accelerating force then so is the kinetic energy reduced when the velocity is diminished by the application of a retarding force, the work done being equal to the average force multiplied by the distance through which the force is applied.

If a bullet is fired from a gun into a block of wood which arrests the bullet, then the work done to stop it is the average force applied by the wood multiplied by the distance through which the force acts, that is the distance the bullet penetrates into the wood.

Example. A bullet of mass 42·5 grammes is fired at a velocity of 520 m/s into a fixed block of wood. If the bullet penetrates 230 mm into the wood, find the average resisting force of the wood on the bullet.

$$K.E. = \tfrac{1}{2}mv^2$$
$$= \tfrac{1}{2} \times 0{\cdot}0425 \times 530^2$$
$$= 5970 \text{ J}$$
$$\text{Loss of } K.E. = \text{average force} \times \text{distance}$$
$$\text{Average force} = \frac{\text{change in } K.E.}{\text{distance}}$$

$$= \frac{5970}{0{\cdot}23}$$

$$= 25\,960 \text{ N} = 25{\cdot}96 \text{ kN} \qquad \text{Ans.}$$

Note the above expression thus,

$$\text{Average force} = \frac{\text{change in kinetic energy}}{\text{distance}}$$

Compare it with the expression previously derived connecting average force and change of momentum, thus,

$$\text{Average force} = \frac{\text{change of momentum}}{\text{time}}$$

IMPACT TESTING MACHINE. This is a machine which measure the energy required to break a specimen of material by one single blow. To enable the resistance to shock offered by one material to be compared with that of another, standard sized notched test pieces are used. The machine consists of a rigid frame on a substantial base which supports a heavy swinging pendulum. The pendulum of known mass is pulled to one side until it is a definite vertical height h_1, above the bottom of its travel, and held there by a trigger-release catch. The specimen is then placed in a vice in the base of the machine set by a gauge, and firmly gripped. The pendulum is now released and it gains kinetic energy equal to the loss of potential energy as it falls in its swing, a striking edge in the bottom of the pendulum strikes the specimen and breaks it. The height to which the pendulum rises on the other side of its swing after impact, h_2, is measured, and the energy to break the specimen is derived thus:

Energy in pendulum before impact $= mgh_1$

" " after " $= mgh_2$

$\therefore$ Energy to break specimen $= mg\,(h_1 - h_2)$

A pointer, catch-operated by the pendulum suspension to move across a scale of energy may be fitted to indicate automatically the energy absorbed in breaking the specimen.

Example. The mass of the pendulum of an impact testing machine is 40 kg and the effective length of the arm is 1 m. The arm is raised so that it makes an angle of 60° to the vertical and then released to strike the test piece. Calculate (i) the kinetic energy and velocity of the mass just before impact, and (ii) the kinetic energy lost in breaking the specimen if the pendulum swings to an angle of 35° to the vertical after impact.

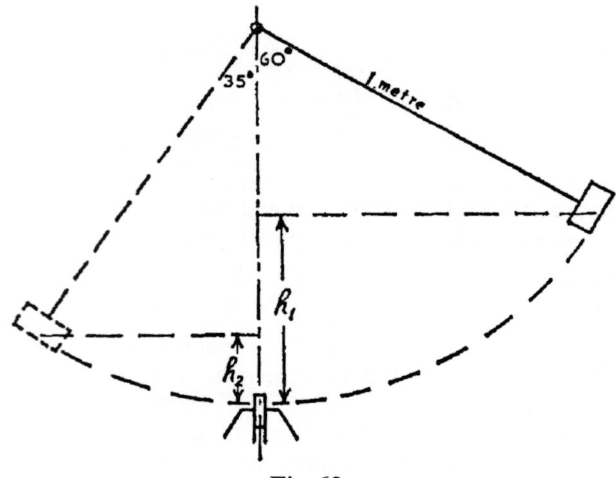

Fig.63

$$h_1 = 1 - 1 \cos 60° = 1 - 0{\cdot}5 = 0{\cdot}5 \text{ m}$$

$$\begin{aligned} K.E. \text{ just before impact} &= \text{loss of potential energy} \\ &= 40 \times 9{\cdot}81 \times 0{\cdot}5 \\ &= 196{\cdot}2 \text{ J} \qquad \text{Ans. (ia)} \end{aligned}$$

$$\begin{aligned} \text{Velocity just before impact} &= \sqrt{2gh} \\ &= \sqrt{2 \times 9{\cdot}81 \times 0{\cdot}5} \\ &= 3{\cdot}132 \text{ m/s} \qquad \text{Ans. (ib)} \end{aligned}$$

$$h_2 = 1 - 1 \cos 35° = 1 - 0{\cdot}8192 = 0{\cdot}1808 \text{ m}$$

$$\begin{aligned} K.E. \text{ after impact} &= 40 \times 9{\cdot}81 \times 0{\cdot}1808 \\ &= 70{\cdot}96 \text{ J} \end{aligned}$$

$$\begin{aligned} \text{Energy to break specimen} &= 196{\cdot}2 - 70{\cdot}96 \\ &= 125{\cdot}24 \text{ J} \qquad \text{Ans. (ii)} \end{aligned}$$

KINETIC ENERGY OF ROTATION. We have seen that translational $K.E. = \frac{1}{2}mv^2$ where v is the linear velocity of the body. For a rotating body, the effective linear velocity is that at the radius of gyration because this is the radius at which the whole mass of the rotating body can be considered as acting.

Let k = radius of gyration m

Let ω = angular velocity rad/s

$$\text{Linear velocity} = \text{angular velocity} \times \text{radius}$$
$$v = \omega \times k$$
$$\text{Substituting } v^2 = \omega^2 k^2$$
$$\text{Rotational } K.E. = \tfrac{1}{2}mv^2$$
$$= \tfrac{1}{2}mk^2\omega^2$$
$$\text{Rotational } K.E. = \tfrac{1}{2}I\omega^2$$

Example. The radius of gyration of the flywheel of a shearing machine is 0·46 m and its mass is 750 kg. Find the kinetic energy stored in it when running at 120 rev/min. If the speed falls to 100 rev/min during the cutting stroke, find the kinetic energy given out by the wheel.

At 120 rev/min,

$$\omega = \frac{120 \times 2\pi}{60} = 4\pi \text{ rad/s}$$

$$\text{Kinetic energy} = \tfrac{1}{2}\,mk^2\omega^2$$
$$= \tfrac{1}{2} \times 750 \times 0{\cdot}46^2 \times 4^2 \times \pi^2$$
$$= 12\,530 \text{ J or } 12{\cdot}53 \text{ kJ} \quad \text{Ans. (i)}$$

Kinetic energy varies as $(\text{speed})^2$

$$\therefore \text{Kinetic energy at 100 rev/min} = 12{\cdot}553 \times (100/120)^2$$
$$= 8{\cdot}704 \text{ kJ}$$
$$\text{Kinetic energy given out} = 12{\cdot}53 - 8{\cdot}704$$
$$= 3{\cdot}826 \text{ kJ} \quad \text{Ans. (ii)}$$

Example. The torque required to turn a flywheel and shaft against friction at the bearings is 34 N m. The mass of the wheel and shaft is 907 kg and the radius of gyration is 381 mm. Assuming frictional resistance to be constant at all speeds, find the number of revolutions the system will turn whilst coming to rest from a speed of 450 rev/min when the driving power is cut out, and also the time taken in coming to rest.

At 450 rev/min,

$$\omega = \frac{450 \times 2\pi}{60} = 15\pi \text{ rad/s}$$

Kinetic energy stored in flywheel

$$= \tfrac{1}{2}\,mk^2\omega^2$$
$$= \tfrac{1}{2} \times 907 \times 0{\cdot}381^2 \times 15^2 \times \pi^2$$
$$= 146\,200 \text{ J}$$

Work done against friction to turn the wheel and shaft through one revolution (*i.e.* 2π radians) is Tθ *i.e.*:

$$\text{Work done} = 34 \times 2\pi$$
$$= 213{\cdot}6 \text{ J}$$

As 213·6 J of energy is lost every revolution, then the number of revolutions it will take to lose 146 200 J (and bring the system to rest) will be:

$$\text{No. of revolutions} = \frac{\text{Loss of } K.E.}{K.E. \text{ lost per rev.}}$$
$$= \frac{146\,200}{213{\cdot}6}$$
$$= 684{\cdot}4 \qquad \text{Ans. (i)}$$
$$\text{Distance} = \text{average velocity} \times \text{time}$$
$$684{\cdot}4 = \tfrac{1}{2}(450 + 0) \times \text{time}$$
$$\text{time} = \frac{684{\cdot}4}{225}$$
$$= 3{\cdot}042 \text{ min} \qquad \text{Ans.(ii)}$$

An alternative solution, based upon the relationship between retarding torque and angular retardation, is given below:

$$\text{Accelerating or retarding torque} = I\alpha \text{ or } mk^2\alpha$$
$$\text{Angular retardation} = \frac{\text{change of angular velocity}}{\text{time}}$$
$$\alpha = \frac{450 \times 2\pi}{60 \times \text{time}} = \frac{15\pi}{\text{time}}$$
$$\text{Torque} = mk^2\alpha$$
$$34 = \frac{907 \times 0{\cdot}381^2 \times 15\pi}{\text{time}}$$
$$\text{time} = \frac{907 \times 0{\cdot}381^2 \times 15\pi}{34}$$
$$= 182{\cdot}5 \text{ s}$$
$$= 3{\cdot}042 \text{ min} \qquad \text{Ans. (ii)}$$
$$\text{Distance} = \text{average velocity} \times \text{time}$$
$$= \tfrac{1}{2}(450 + 0) \times 3{\cdot}042$$
$$= 684{\cdot}4 \text{ rev} \qquad \text{Ans. (i)}$$

f Example. The total mass of a wheel and axle is 27·2 kg. The axle is 76 mm diameter and is supported in horizontal bearings. A cord is wrapped around the axle, one end being fixed to it and the other end with a hook attached hangs freely. In an experiment to determine the radius of gyration of the wheel and axle it was found that a force of 6·8 N on the cord was just sufficient to overcome friction. When this force is removed and a mass of 3 kg is suspended from the hook, it falls through 1·25 m in 10 s from rest. Calculate the radius of gyration from this experiment.

Average velocity of falling mass

$$= \frac{\text{distance}}{\text{time}} = \frac{1{\cdot}25}{10} = 0{\cdot}125 \text{ m/s}$$

Since it started from rest, then,

$$\text{Final velocity} = 2 \times 0{\cdot}125 = 0{\cdot}25 \text{ m/s}$$

$$\text{Radius of axle} = 38 \text{ mm} = 0{\cdot}038 \text{ m}$$

Final angular velocity,

$$\omega = \frac{v}{r} = \frac{0{\cdot}25}{0{\cdot}038} = 6{\cdot}579 \text{ rad/s}$$

Total loss of potential energy by falling mass

$$= 3 \times 9{\cdot}81 \times 1{\cdot}25 \text{ N m}$$

$$= 36{\cdot}79 \text{ J}$$

Energy used up in overcoming friction

$$= 6{\cdot}8 \times 1{\cdot}25 = 8{\cdot}5 \text{ J}$$

Potential energy converted into kinetic energy

$$= 36{\cdot}79 - 8{\cdot}5 = 28{\cdot}29 \text{ J}$$

Total (translational and rotational) kinetic energy gained is the sum of that gained by the falling mass as its linear velocity is increased and the energy gained by the wheel and axle as its angular velocity is increased.

$$\text{Total } K.E. = \tfrac{1}{2} m_1 v^2 + \tfrac{1}{2} m_2 k^2 \omega^2$$

$$28{\cdot}29 = \tfrac{1}{2} \times 3 \times 0{\cdot}25^2 + \tfrac{1}{2} \times 27{\cdot}2 \times k^2 \times 6{\cdot}579^2$$

$$28{\cdot}29 \times 2 = 0{\cdot}1875 + 1178\, k^2$$

$$\text{k} = \sqrt{\frac{56{\cdot}3925}{1178}}$$

$$= 0{\cdot}2189 \text{ m} = 218{\cdot}9 \text{ mm Ans.}$$

KINETIC ENERGY OF TRANSLATION AND ROTATION. Wheels of moving vehicles and rolling bodies possess kinetic energy of translation by virtue of their linear velocity along the ground and also kinetic energy of rotation by virtue of their velocity about their own centres.

Example. A sphere of 4·5 kg mass is rolling along the ground at a velocity of 1·2 m/s. Given that for a solid sphere, $k^2 = \frac{2}{5}r^2$, find the total kinetic energy.

$$v = 1{\cdot}2 \text{ m/s}$$

$$\omega = \frac{v}{r} = \frac{1{\cdot}2}{r} \text{ rad/s}$$

$$\begin{aligned}
\text{Total } K.E. &= K.E. \text{ of translation} + K.E. \text{ of rotation}\\
&= \tfrac{1}{2}mv^2 + \tfrac{1}{2}mk^2\,\omega^2\\
&= \tfrac{1}{2} \times 4{\cdot}5 \times 1{\cdot}2^2 + \tfrac{1}{2} \times 4{\cdot}5 \times \frac{2r^2}{5} \times \frac{1{\cdot}2^2}{r^2}\\
&= \tfrac{1}{2} \times 4{\cdot}5 \times 1{\cdot}2^2 \left\{1 + \frac{2r^2}{5r^2}\right\}\\
&= \tfrac{1}{2} \times 4{\cdot}5 \times 1{\cdot}2^2 \times 1{\cdot}4\\
&= 4{\cdot}536 \text{ J Ans.}
\end{aligned}$$

*f*FLUCTUATION OF SPEED AND ENERGY

FLUCTUATION OF SPEED is the difference between the maximum and the minimum rotational speed (angular velocity) of a machine during its normal working conditions. If ω_1 represents the maximum speed and ω_2 the minimum speed, then,

$$\text{Fluctuation of speed} = \omega_1 - \omega_2$$

This can be expressed as a fraction or percentage of the mean speed. For normal practical purposes the mean speed may be taken as the arithmetical mean of ω_1 and ω_2 that is $\frac{1}{2}(\omega_1 + \omega_2)$. When expressed in fraction form it is usual to refer to it as the coefficient of fluctuation of speed, thus, if ω represents the mean speed,

$$\text{Coefficient of fluctuation of speed} = \frac{\omega_1 - \omega_2}{\omega}$$

FLUCTUATION OF ENERGY. Turning moment or torque (N m) multiplied by the angle turned (rad), is work done (J).

Therefore if a graph is drawn representing the torque transmitted to an engine shaft, on a base of angle turned, the area under the graph represents work. The energy produced in one revolution is the work done in 2π radians. The torque at the shaft of a turbine is steady and the torque graph is therefore a straight line. For a reciprocating engine, however, the torque varies considerably throughout one cycle, depending upon the piston effort and the effective leverage of the crank.

As the torque required to overcome external loading, friction and other resistances will be approximately constant, then at some parts of the cycle a reciprocating engine is supplying more torque than that required, in these brief periods the engine increases speed and the excess energy is absorbed by the flywheel; at the other parts of the cycle the engine is supplying less torque that that needed, resulting in a brief reduction of speed so that the flywheel gives out energy to the shaft.

Over one cycle the total energy supplied by the engine must be equal to the energy required to overcome the load, therefore the amount of engine energy which is momentarily surplus to requirements and is absorbed by the flywheel, is equal to the energy given out by the flywheel during the period the engine energy is deficient.

The size of the flywheel therefore decides the amount by which the speed will vary during one cycle, the greater the moment of inertia the less will be this variation.

The fluctuation of energy of a flywheel is expressed as the difference between its kinetic energy at maximum speed and its kinetic energy at minimum speed during its normal running conditions, thus,

$$
\begin{aligned}
K.E.1 - K.E.2 &= \tfrac{1}{2} I\omega_1^2 - \tfrac{1}{2} I\omega_2^2 \\
&= \tfrac{1}{2} I(\omega_1^2 - \omega_2^2) \\
&= \tfrac{1}{2} I\,(\omega_1 + \omega_2)\,(\omega_1 - \omega_2), \text{ by factorisation} \\
&= I \times \frac{(\omega_1 + \omega_2)}{2} \times (\,\omega_1 - \omega_2)
\end{aligned}
$$

but $\dfrac{\omega_1 + \omega_2}{2} = \omega$ the mean speed

$\therefore$ Fluctuation of energy $= I\omega\ (\omega_1 - \omega_2)$

The fluctuation of energy may be given as a coefficient by expressing it as a fraction or ratio of the work done by the engine in one revolution or in one cycle.

TEST EXAMPLES 4

1. The stiffness of a spring is 88 N/mm of axial compression. Find the work to increase the compression from 50 mm to 80 mm.

2. A double-bottom tank is 6 m long by 4·5 m broad and is full of fresh water. A deck tank is 3·6 m long by 3 m broad by 1·5 m deep and is empty, the height from the double-bottom tank-top to the base of the deck tank is 12 m. If the deck tank is filled by pumping water from the double-bottom tank, find (i) the quantity of water moved, in t, (ii) by what depth the level in the double-bottom tank is lowered, (iii) the effective work done. Density of fresh water = 10^3 kg/m^3.

3. A solid steel cone, 450 mm diameter at the base and of 600 mm perpendicular height, stands on its base on a level ground. Calculate the work to tilt it until it is on the point of toppling over on to its side. The density of steel is 7860 kg/m^3, the position of the centre of gravity of a cone is at one-quarter of its height from the base.

4. A mass of 544 kg is lifted by a wire rope from a winch drum of 380 mm effective diameter. If the mechanical efficiency is 70%, find the power of the driving motor when the winding drum is rotating at a steady speed of 40 rev/min.

5. Sea water is pumped through a 50 mm diameter pipe to a height of 9 m. If the velocity of water through the pipe is constant at 1·5 m/s, find the equivalent output power. Find also the input power to the pump if the efficiency of the system is 0·6. Take the density of sea water as 1024 kg/m^3.

6. A hydraulic machine is supplied with 900 l of water per hour at a pressure of 80 bar. Calculate the power supplied to the machine and the output assuming an efficiency of 75%.

7 Find the width of a belt of 8 mm thickness required to drive a pulley 500 mm diameter at 450 rev/min and transmit 4·5 kW of power, the maximum tension in the belt is not to exceed 7 N per mm of width and the tension in the tight side is to be taken as 2¼ times the tension in the slack side.

8. Calculate the translational kinetic energy in a body of 240 kg mass when it is moving at a speed of 36 km/h, and the change in kinetic energy when it slows down to 18 km/h.

9. A 28 gramme bullet is fired at 450 m/s into a fixed block of wood 100 mm thick and comes out of the other side at a velocity of 250 m/s. Find (i) the average resisting force of the wood, and (ii) the minimum thickness of similar wood required to bring the

bullet to rest.

10. The mass of a flywheel is 109 kg and its radius of gyration is 380 mm. Calculate the kinetic energy stored when rotating at 100 and 300 rev/min respectively.

11. The radius of gyration of a solid disc wheel of uniform thickness is at $r/\sqrt{2}$ from the centre, where r is the radius of the wheel. Calculate the change in kinetic energy, in kilojoules, in a solid disc flywheel of 1400 mm diameter and 1·25 t mass when its speed changes from 8π to 10π rad/s.

*f*12 The mass of the flywheel of a shearing machine is 1220 kg and its radius of gyration is 0·58 m.

(i) Find the accelerating torque required to attain a speed of 200 rev/min from rest in 60 s.

(ii) If the speed falls from 200 to 180 rev/min during the cutting stroke of 150 mm, find the average cutting force exerted assuming the whole of the work done during the cutting stroke is due to the kinetic energy given out by the flywheel.

*f*13. The total mass of a flywheel and shaft is 2·54 t, the radius of gyration is 686 mm and it is running at 150 rev/min. The torque required to overcome friction at the bearings is 27 N m. Find (i) the kinetic energy stored in the wheel and shaft, (ii) the energy lost per revolution due to friction, (iii) the number of revolutions made in coming to rest when uncoupled from the driving motor, (iv) the time to come to rest.

*f*14. A solid cylindrical roller starts from rest and rolls a distance of 2·286 m down an incline in 3 seconds. Calculate the angle of the incline given that $k = r/\sqrt{2}$.

*f*15. An engine is designed to develop 10 kW of power at a mean speed of 1000 rev/min. Find the moment of inertia, in kg m^2, of a suitable flywheel, assuming a speed variation of ± 1·5% of the mean speed and an energy fluctuation equal to 0·9 of the work done per revolution.

CHAPTER 5

CENTRIPETAL FORCES

NEWTON'S first law of motion states that every body will continue in a state of rest or of uniform motion in a straight line unless acted upon by an external force. If a stone is fastened to the end of a piece of string and whirled around in a circular path, an inward pull must be continuously exerted to keep it travelling in a circle, the stone itself is exerting an outward radial pull trying to get away. If the speed is increased the pull becomes greater until the string snaps, the stone then flies off in a *straight line* at a tangent to the circle.

The outward radial force created by a body travelling in a circular path due to its natural tendency to fly off and travel in a straight line is termed the *centrifugal force*. The inward pull applied to counteract this and keep it on its circular path is termed the *centripetal force*, it is equal in magnitude to the centrifugal force and opposite in direction (Newton's third law of motion).

Consider a body moving at a constant speed of v around a circle of radius r. Referring to Fig. 64, at the instant it is passing point a its instantaneous velocity is v in the direction tangential to the circle at a; a little further around it is passing point b and its velocity is now v tangential to the circle at b. Although the speed is constant, the velocity has changed because there is a change of direction.

Let the movement from a to b be through a small angle θ. To find the change of velocity the vector diagram of velocities is drawn. If θ is considered a small angle then the base angles of the triangle are almost 90° and the change of velocity can be taken as $v \sin\theta$, $v \tan\theta$ or $v\theta$. For small angles:

$$\sin\theta = \tan\theta = \theta \text{ rad}$$

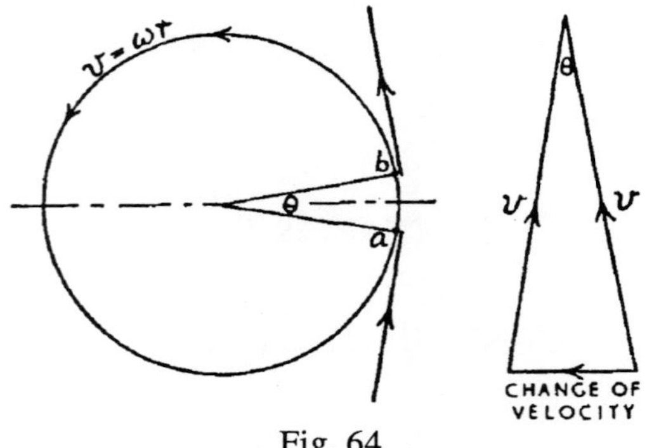

Fig. 64

Change in velocity $= v\theta$

If there is a change of velocity there is acceleration.

Time for body to move from *a* to *b*

$$= \frac{\text{distance}}{\text{speed}} = \frac{r\theta}{v}$$

$$\text{Acceleration} = \text{change of velocity} \div \text{time}$$

$$= v\theta \div \frac{r\theta}{v}$$

$$= \frac{v\theta \times v}{r\theta}$$

$$= \frac{v^2}{r}$$

This is a special form of acceleration. The body is travelling at a constant speed, the acceleration is due to the constantly changing velocity (by changing direction) as the body travels around its circular path, and the acceleration is directed towards the centre of the circle. It is therefore distinguished from the more common cases by referring to it as *centripetal acceleration.*

Being circular motion, the velocity of the body may be more conveniently expressed in angular measurement ω rad/s instead of in linear units of v. The relation between these is $v = \omega r$.

Substituting this value of v:

$$\frac{v^2}{r} = \frac{\omega^2 r^2}{r} = \omega^2 r$$

$$\therefore \text{Centripetal acceleration} = \frac{v^2}{r} \text{ or } \omega^2 r$$

Example. A ventilation fan, 0·5 m diameter rotates at 200 rev/min. Calculate the centripetal acceleration of the fan blade tips.

$$\omega = \frac{2\pi \times 200}{60}$$

$$= 20{\cdot}94 \text{ rad/s}$$

$$\text{Centripetal accln} = \omega^2 r$$

$$= 20{\cdot}94^2 \times 0{\cdot}25$$

$$= 109{\cdot}6 \text{ m/s}^2 \text{ (acting radially inward). Ans.}$$

The force required to produce the centripetal acceleration is called the centripetal force.

$$F = ma$$

$$\text{Centripetal force} = m\omega^2 r \text{ or } \frac{mv^2}{r}$$

This centripetal force is applied radially inwards to constrain a body to follow a circular or curved path. Centripetal force counter-balances the centrifugal force which acts radially outwards, equal in magnitude to the centripetal force, but in opposite in direction.

$$\text{Centrifugal force} = m\omega^2 r \text{ or } \frac{mv^2}{r}$$

Example. Calculate the pull on the root of a turbine blade due to centrifugal force when the rotor is rotating at 3000 rev/min, the mass of the blade is 0·2 kg and the radius from the axis of the rotor to the centre of gravity of the blade is 460 mm.

$$\frac{300 \text{ rev/min} \times 2\pi}{60} = 100\pi \text{ rad/s}$$

$$\text{Centrifugal force} = m\omega^2 r$$

$$= 0{\cdot}2 \times (100\pi)^2 \times 0{\cdot}46$$

$$= 9080 \text{ N} = 9{\cdot}08 \text{ kN} \qquad \text{Ans.}$$

Example. A body of 2 kg mass is attached to the end of a cord and swung around in a vertical plane of 0·75 m radius at a speed of 100 rev/min. Find (i) the centrifugal force set up, (ii) the tension in the cord when passing (*a*) top centre, (*b*) bottom centre, (*c*) a point 30° from bottom centre, (*d*) a point 60° from top centre.

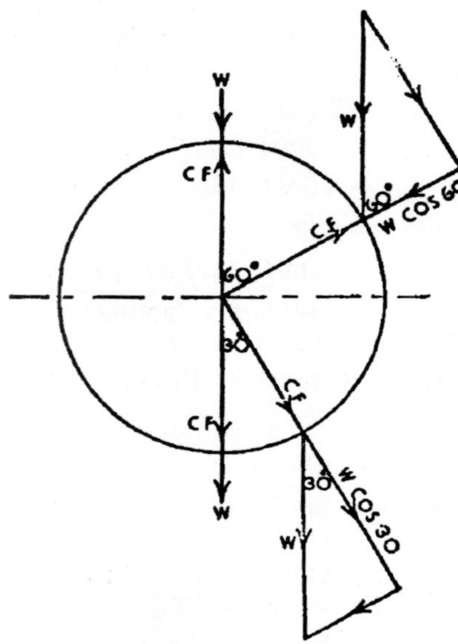

Fig 65

$$\frac{100 \text{ rev/min} \times 2\pi}{60} = 10{\cdot}47 \text{ rad/s}$$

Centrifugal force = $m\omega^2 r$

$= 2 \times 10{\cdot}47^2 \times 0{\cdot}75$

$= 164{\cdot}4$ N Ans. (i)

Force of gravity on 2 kg mass (weight)

$= 2 \times 9{\cdot}81 = 19{\cdot}62$ N

When passing top centre, centrifugal force pulls directly upwards (always radially outwards) and the force of gravity acts vertically downwards, therefore,

Tension in cord = centrifugal force − weight

$= 164{\cdot}4 - 19{\cdot}62$

$= 144{\cdot}78$ N Ans. (iia)

When passing bottom centre, centrifugal force pulls directly downwards, the force of gravity also acts downwards, therefore,

Tension in cord = centrifugal force + weight

$= 164{\cdot}4 + 19{\cdot}62$

$= 184{\cdot}02$ N Ans. (iib)

When passing the point 30° to the bottom centre, the effect of the weight of the mass on the cord is its component in the direction of the cord, see Fig. 65, this is $W \cos 30°$ and acts in the same direction as the centrifugal force, therefore,

$$\begin{aligned}\text{Tension in cord} &= \text{centrifugal force} + W \cos 30° \\ &= 164{\cdot}4 + 19{\cdot}62 \times 0{\cdot}866 \\ &= 181{\cdot}39 \text{ N} \qquad \text{Ans. (iic)}\end{aligned}$$

At 60° from top centre the component of the weight is $W \cos 60°$ in the direction of the cord, but is in the opposite direction in which the centrifugal force acts, hence,

$$\begin{aligned}\text{Tension in cord} &= \text{centrifugal force} - W \cos 60° \\ &= 164{\cdot}4 - 19{\cdot}62 \times 0{\cdot}5 \\ &= 154{\cdot}59 \text{ N} \qquad \text{Ans. (iid)}\end{aligned}$$

SIDE SKIDDING AND OVERTURNING OF VEHICLES

When a car or lorry takes a bend in the road, around a corner or road island, it moves around a circular path and centrifugal force is created which tends to cause the vehicle to skid or overturn.

SIDE SKID. Treaded rubber tyres provide a good grip on the ground and, within limits, prevent broadside skidding. To slide a vehicle sideways, the force required to overcome friction depends upon the *coefficient of friction* between the tyres and the ground, the coefficient of friction is the ratio between the force required to overcome friction and the normal force between the surfaces. Thus if a vehicle weighing W stands on a horizontal ground (Fig. 66) the force between the surfaces is simply the weight W. If the coefficient of friction is represented by μ, the force to drag it sideways is $\mu \times W$ (see Chapter 6).

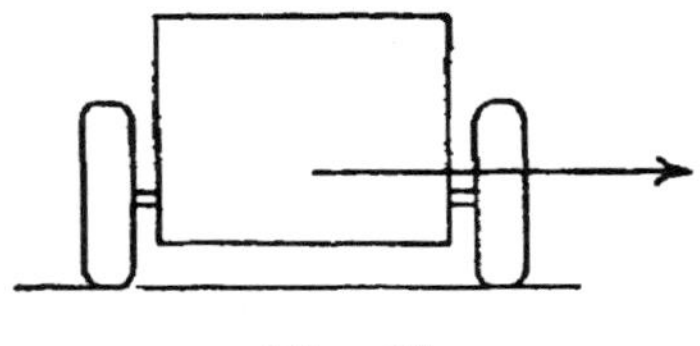

Fig. 66

For example, if the mass of a car is 1000 kg and the coefficient of friction between the tyres and ground is 0·6, the force to overcome friction is $0{\cdot}6 \times 100 \times 9{\cdot}81 = 5886$ N. This force could be reached and exceeded by the centrifugal force created when the car travels around a bend at an excessive speed.

Example. Calculate the speed at which a car will begin to skid sideways when turning a bend of 25 m radius, taking the coefficient of friction between tyres and ground as 0·7. If the coefficient of friction is halved due to worn tyres and wet road, what will now be the danger speed?

Let m = mass of car, in kg

then force between surfaces = $m \times 9{\cdot}81$ N

Force to overcome friction to push the car sideways = $0{\cdot}7 \times m \times 9{\cdot}81$ N, and when the centrifugal force reaches this value skidding will occur.

$$\frac{mv^2}{r} = 0{\cdot}7 \times m \times 9{\cdot}81$$

$$v = \sqrt{0{\cdot}7 \times 9{\cdot}81 \times 25}$$

$$= 13{\cdot}1 \text{ m/s}$$

$$13{\cdot}1 \times 10^{-3} \times 3600 = 47{\cdot}17 \text{ km/h} \qquad \text{Ans. (i)}$$

Velocity varies directly as the square root of the coefficient of friction. If the coefficient of friction is halved, then,

$$\text{New danger speed} = 47{\cdot}17 \times \sqrt{0{\cdot}5}$$

$$= 33{\cdot}35 \text{ km/h} \qquad \text{Ans. (ii)}$$

OVERTURNING. If friction provides sufficient grip to prevent side-skid, there is the other danger of overturning when turning corners at excessive speeds, especially with vehicles which have a high centre of gravity. Referring to Fig. 67, the vehicle tilts about the point of contact of the ground and those wheels which are furthest from the centre of the circular path, marked o. The centrifugal force acts horizontally through the centre of gravity and the force of gravity acts vertically downwards through this point. If the centre of gravity is h above the ground, this is the leverage or perpendicular distance from the line of action of the centrifugal force to point o, the *overturning moment* is therefore $C.F. \times h$. The moment maintaining the vehicle on its four wheels is the weight multiplied by its perpendicular distance, this distance being half the track of the wheels. When the overturning moment becomes equal to the stabilising moment, overturning will commence (see Chapter 7).

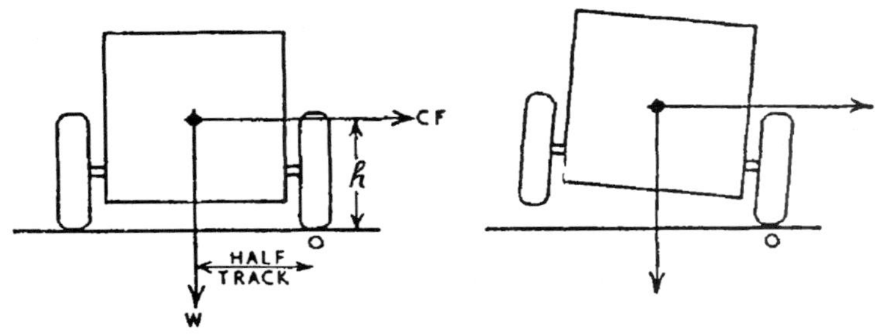

Fig. 67

Example. The track between the wheels of a lorry is 1·52 m and its centre of gravity is 0·9 m above the ground. Find the speed at which it will overturn when travelling around a bend of 25 m radius.

$$\text{Overturning moment} = \text{Stabilising moment}$$

$$\text{Centrifugal force} \times h = \text{Weight} \times \tfrac{1}{2}\,\text{track}$$

$$\frac{mv^2}{h} \times h = mg \times \tfrac{1}{2}\,\text{track}$$

$$v = \sqrt{\frac{9{\cdot}81 \times 0{\cdot}76 \times 25}{0{\cdot}9}}$$

$$= 14{\cdot}39 \text{ m/s}$$

$$14{\cdot}39 \times 10^{-3} \times 3600 = 51{\cdot}9 \text{ km/h} \qquad \text{Ans.}$$

BANKED TRACKS. Bends of roadways and railways are often banked, the incline to the horizontal being calculated on the average speed of vehicles passing around the bends. By inclining the road, the reaction of the ground which is normal (*i.e.* at right angles) to the surface, is inclined, and its two components are (i) an upward force to support the downward weight of the vehicle, and (ii) a horizontal force to counteract the centrifugal force of the vehicle. This is illustrated in Fig. 68.

The magnitude of the horizontal component to counteract the centrifugal force depends upon the tangent of the incline to which the road is banked and hence can be designed to completely balance the centrifugal force at any pre-determined speed.

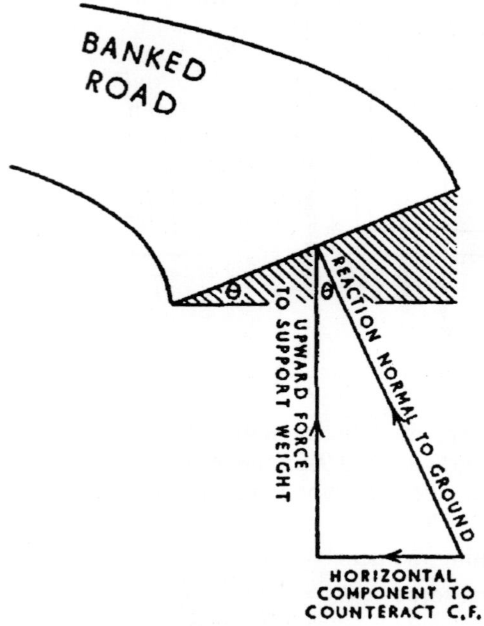

Fig. 68

Example. Find the angle to the horizontal to which a bend in the road should be banked so that there will be no tendency to skid when a vehicle is travelling at a speed of 72 km/h, the radius of the bend being 180 m. Neglect friction between tyres and ground.

$$\frac{72 \text{ km/h} \times 10^3}{3600} = 20 \text{ m/s}$$

Upward force to support weight of mass of m kg

$$= m \times 9{\cdot}81 \text{ N}$$

$$\text{tan of angle of incline} = \frac{\text{centrifugal force}}{\text{weight}}$$

$$= \frac{mv^2}{r \times 9{\cdot}81\, m}$$

$$= \frac{20^2}{180 \times 9{\cdot}81} = 0{\cdot}2266$$

$$\text{Angle of incline} = 12° \, 46' \qquad \text{Ans.}$$

In railway lines the same effect is obtained by raising the outer rail, in this case it is done to prevent an excessive side thrust on the flanges of the wheels.

Example. Calculate the height of the outer rail above the level of the inner rail around a curvature of 800 m radius of a railway of 1435 mm track so that there will be no side thrust on the wheel flanges when the train is travelling at 80 km/h.

Let θ = angle of inclination of outer rail above inner rail

h = super-elevation of outer rail

$$\frac{80 \text{ km/h} \times 10^3}{3600} = 22{\cdot}22 \text{ m/s}$$

$$\tan\theta = \frac{\text{centrifugal force}}{\text{weight}} = \frac{mv^2}{r \times mg}$$

$$= \frac{22{\cdot}22^2}{800 \times 9{\cdot}81} = 0{\cdot}06292$$

$$\theta = 3^\circ\, 36'$$

$$h = 1435 \times \sin 3^\circ\, 36'$$

$$= 1435 \times 0{\cdot}0628$$

$$= 90{\cdot}12 \text{ mm} \quad \text{Ans.}$$

BALANCING

If the centre of gravity of a rotating piece of machinery does not coincide with the centre of rotation, centrifugal force is set up which causes vibration as well as putting an extra load on bearings. To balance an eccentric load in its own plane place a counter-balance diametrically opposite, of such mass and radius from the centre so that it will create a centrifugal force of equal magnitude and opposite in direction to that of the eccentric load.

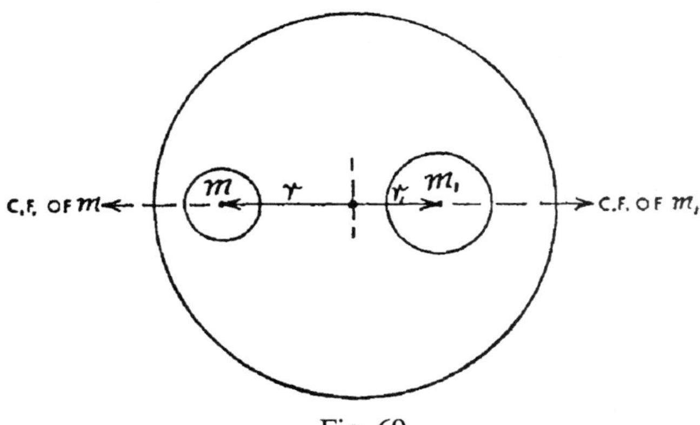

Fig. 69

Referring to Fig. 69, let m_1 be the mass to be balanced, and let r_1 be the distance from centre of rotation to the centre of gravity of the mass. Let m be the mass of the counter-balance placed diametrically opposite at a radius r. For balance, the centrifugal force of m must be equal to the centrifugal force of m_1 thus,

$$m\,\omega^2\,r = m_1\,\omega^2\,r_1$$

Since they both rotate at the same velocity, ω^2 cancels, therefore,

$$m \times r = m_1 \times r_1$$

Note that if the moments are taken about the centre of rotation, the turning moment of one mass is equal to the turning moment of the other and therefore the system is statically balanced (see Chapter 7).

The centre of gravity of the above masses must lie in the same plane (see side elevation in Fig. 70), otherwise an unbalanced couple will be set up when the system rotates, which will cause a rocking action. Taking moments about the line of action of one force, the unbalanced moment (couple) is $C.F. \times$ arm (or leverage) between the two forces (see Chapter 7).

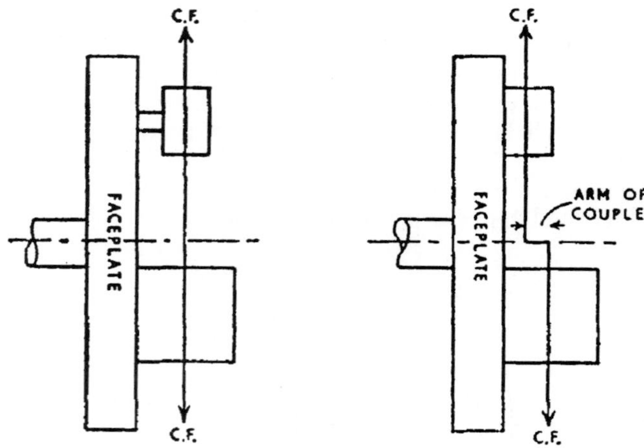

Fig. 70

Example. A valve-box casting of mass 27 kg is secured to the faceplate of a lathe in such a position that its centre of gravity is 90 mm radially from the centre of rotation and 114 mm axially from the surface of the faceplate. Calculate the radius at which to fix a counter-balance of 12·75 kg. If the centre of gravity of this counter-balance is 76 mm axially from the faceplate surface, calculate the unbalanced couple when rotating at 200 rev/min.

C.F. of counter-balance = *C.F.* of casting

$$m_1\,\omega^2\,r_1 = m_2\,\omega^2\,r_2$$

$$m_1 \times r_1 = m_2 \times r_2$$

$$12{\cdot}75 \times r_1 = 27 \times 90$$

$$r_1 = 190{\cdot}5 \text{ mm} \qquad \text{Ans.}$$

Thus, counter-balance should be fixed at a radius of 190·5 mm diametrically opposite the direction in which the centre of gravity of the casting lies.

$$\frac{200 \text{ rev/min} \times 2\pi}{60} = 20{\cdot}94 \text{ rad/s}$$

Unbalanced couple = C.F. of casting (or counter-balance) × Arm of couple

$$= m\,\omega^2\,r \times \text{arm}$$

$$= 27 \times 20{\cdot}94^2 \times 0{\cdot}09 \times (0{\cdot}114 - 0{\cdot}076)$$

$$= 40{\cdot}48 \text{ N m} \qquad \text{Ans.}$$

BALANCING A NUMBER OF ROTATING MASSES

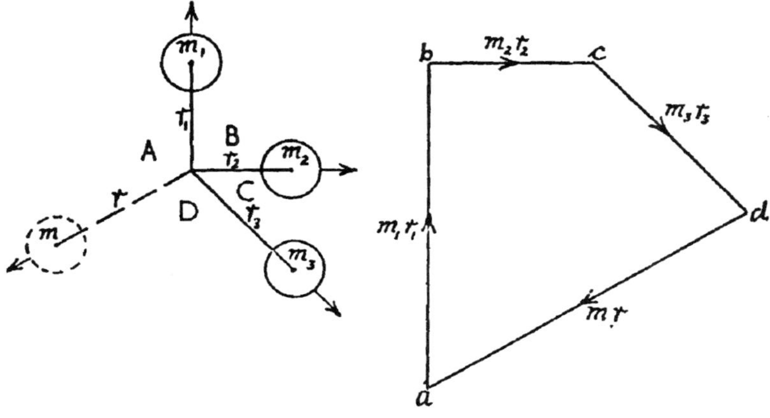

Fig. 71

If the vector diagram representing a number of coplanar forces meeting at a point forms a closed figure, the system is in equilibrium.

Centrifugal forces can be dealt with in the same manner. A number of rotating masses can be balanced if they are all in the same plane either by displacing them at such angular positions that their vector diagram forms a closed figure, or, if the positions of the existing masses cannot be arranged in this way, an additional mass can be included in such a position that the vector of its centrifugal force will close the vector diagram (see Fig. 71).

Example. Two masses are fixed at right angles to each other on a disc which is to rotate, one is 3 kg at a radius of 125 mm from the centre of rotation, and the other is 4 kg at 150 mm radius. Find the position to fix a balance mass of 7 kg to equalise the centrifugal forces.

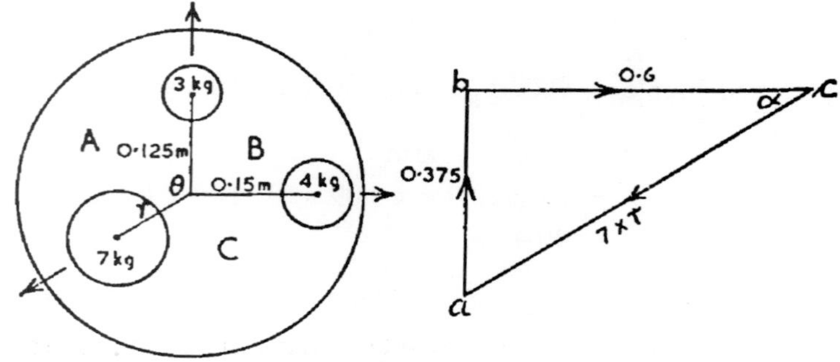

Fig. 72

Vector ab is drawn to represent force AB,
represented by $3 \times 0{\cdot}125 = 0{\cdot}375$

Vector bc is drawn to represent force BC,
represented by $4 \times 0{\cdot}15 = 0{\cdot}6$

$$ac = \sqrt{0{\cdot}375^2 + 0{\cdot}6^2}$$

$$7 \times r = 0{\cdot}7075$$

$$r = 0{\cdot}101 \text{ m} = 101 \text{ mm}$$

$$\tan \alpha = \frac{0{\cdot}375}{0{\cdot}6} = 0{\cdot}625$$

$$\alpha = 32°$$

$$\theta = 32° + 90° = 122°$$

Therefore the 7 kg counter-balance should be placed at a radius of 101 mm, at 122° to the 3 kg mass. Ans.

Again note carefully that the actual forces are centrifugal forces, each of value $m\omega^2 r$, but since all rotate at the same angular velocity, each force can be *represented* by $m \times r$.

STRESS IN FLYWHEEL RIMS DUE TO CENTRIFUGAL FORCE

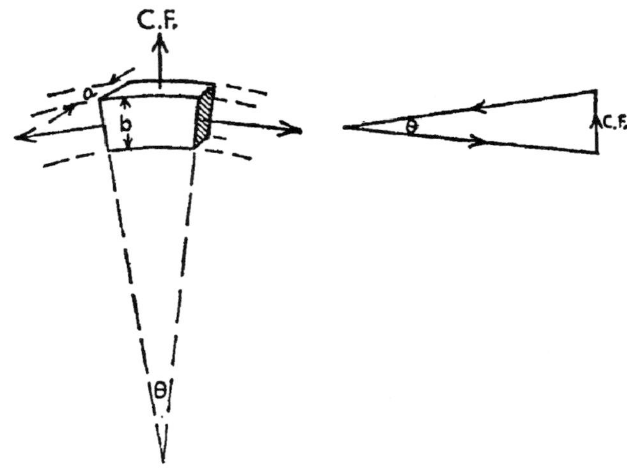

Fig. 73

Stress is the load or force carried by a material per unit of cross-section (see Chapter 9).

$$\text{Stress} = \frac{\text{total load}}{\text{area of cross section}}$$

Referring to Fig. 73, considering the equilibrium of a small piece of the flywheel rim it can be seen from the vector diagram of forces that the outward radial centrifugal force is balanced by the circumferential tensile force in the rim. This tension tends to snap the material, the stress, expressed by dividing the total tensile force by the area, is therefore termed the tensile stress or hoop stress.

length of piece $= r\theta$ (r = mean rim radius)

area of cross-section $= ab$

Mass of piece $=$ volume $\times$ density

$= r\theta \times ab \times \rho$

$C.F.$ of piece $= m\,\omega^2\,r$

$= r\theta \times ab \times \rho \times \omega^2 \times r$

From vector diagram,

$$\text{Tensile force} = \frac{C.F.}{\theta}$$

$$\text{Tensile stress} = \frac{\text{tensile force}}{\text{area of cross-section}}$$

$$= \frac{r\theta \times ab \times \rho \times \omega^2 \times r}{\theta \times ab}$$

$$= \rho\omega^2 r^2 = \rho v^2 \text{ N/m}^2$$

Example. Calculate the stress set up in a thin flywheel rim, 1 m mean diameter, made of steel of density 7860 kg/m³, when rotating at 1500 rev/min.

$$\frac{150 \text{ rev/min} \times 2\pi}{60} = 157{\cdot}1 \text{ rad/s}$$

$$\text{Stress due to centrifugal force} = \rho v^2 = \rho\omega^2 r^2$$

$$= 7860 \times 157{\cdot}1^2 \times 0{\cdot}5^2$$

$$= 4{\cdot}85 \times 107 \text{ N/m}^2$$

$$= 48{\cdot}5 \text{ MN/m}^2$$

CONICAL PENDULUM

If a mass is suspended at the end of a cord and set in motion in a circular path in a horizontal plane, the system sweeps out the shape of a cone and therefore it is referred to as a 'conical pendulum' to distinguish it from the simple pendulum which swings backwards and forwards.

Referring to Fig. 74, if the mass travels at a constant angular velocity, the angle between the cord and the vertical centre-line remains constant because the system is in equilibrium, that is, the moment of the centrifugal force which tends to increase this angle is balanced by the moment of the weight of the mass which tends to close the angle (see Chapter 7).

The moment of a force about a point is the product of the force and its effective leverage. The effective leverage is the perpendicular distance from the line of action of the force to the point about which moments are taken. The centrifugal force acts radially outwards from the centre through the centre of gravity of the moving mass, this line of action is horizontal and, if taking moments about the point of suspension o, the perpendicular distance from the line of action of the centrifugal force to o is the

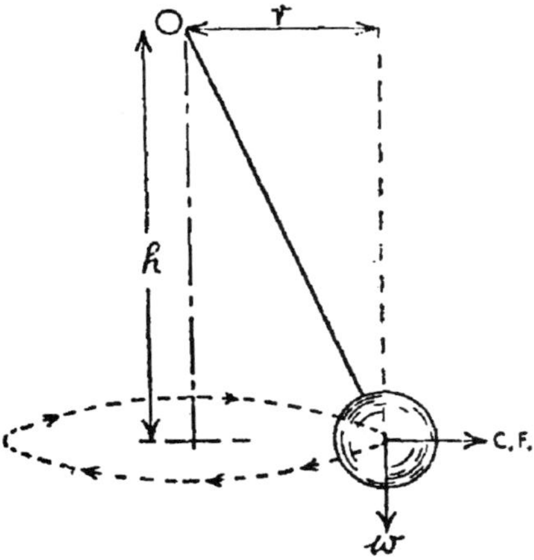

Fig. 74

vertical height h. The weight of the mass acts vertically downwards through its centre of gravity, the perpendicular distance from the line of action of this force to point o is the radius of the circular path r, therefore:

Let θ = angle between cord and vertical centre-line, taking moments about point o:

Moment tending to increase θ = moment tending to reduce θ

centrifugal force $\times$ h = weight $\times$ r

hence,

$$m\,\omega^2 r \times h = mg \times r$$

$$h = \frac{m \times g \times r}{m \times \omega^2 \times r}$$

$$h = \frac{g}{\omega^2}$$

Note that the height is inversely proportional to the square of the velocity and is independent of the mass.

SIMPLE UNLOADED GOVERNOR

The simple unloaded governor, often referred to as the Watt governor, works on the principle of the conical pendulum. This is illustrated in Fig. 75 and consists of a pair of bob masses suspended on arms from a centre spindle so that an increase in speed will cause the bobs to swing outward and upward; by connecting

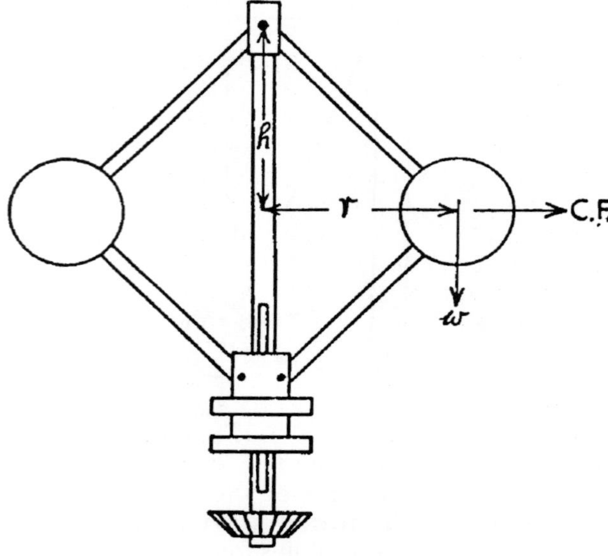

Fig. 75

through a pair of links to a sliding sleeve on the spindle, the sleeve is pulled up when the speed increases, and lowered when the speed decreases. This motion can be transmitted to an engine control valve by levers to close or open the valve, thus governing the engine speed within pre-determined limits. The height h from the plane of rotation of the bobs to the point of suspension, is referred to as 'the height of the governor'.

Example. Find the height of a Watt governor when rotating at speeds of 50 and 75 rev/min.

$$\frac{50 \text{ rev/min} \times 2\pi}{60} = 5{\cdot}236 \text{ rad/s}$$

$$h = \frac{g}{\omega^2} = \frac{9{\cdot}81}{5{\cdot}236^2}$$

$$= 0{\cdot}3579 \text{ m or } 357{\cdot}9 \text{ mm} \quad \text{Ans. (i)}$$

$$\frac{75 \text{ rev/min} \times 2\pi}{60} = 7{\cdot}854 \text{ rad/s}$$

$$h = \frac{g}{\omega^2} = \frac{9{\cdot}81}{7{\cdot}854^2}$$

$$= 0{\cdot}1591 \text{ m or } 159{\cdot}1 \text{ mm} \quad \text{Ans. (ii)}$$

Note. h is inversely proportional to ω^2

f PORTER GOVERNOR

The Porter governor is similar in its operation to the Watt governor but carries a central load on the sliding sleeve. The mass of this central load in relation to the mass of each bob determines the amount of movement of the sleeve for a given change of speed, and can be designed to suit engine control requirements.

Referring to Fig. 76 and considering the equilibrium of one bob. Let the links be all of equal length and take moments about the *instantaneous centre* (i) of the movement of the central load and the bob so that the effect of the weight of the central load is readily taken into account. h is the height from the plane of rotation of the bobs to point of suspension, and r is the radius of the circular path, therefore, since the links are of the same length, the height from the connection of the lower links and sleeve to plane of rotation is equal to h, and the distance from this connection to the instantaneous centre is $2r$ (by similar triangles).

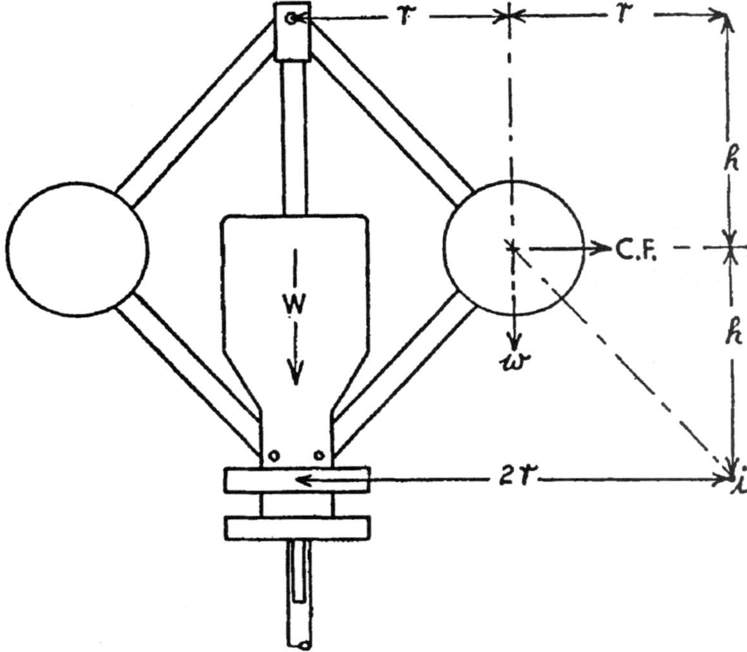

Fig. 76

Let M represent the mass of the central load and sleeve; m represents the mass of each bob.

Taking moments about i:—

$$\text{Clockwise moments} = \text{Anticlockwise moments}$$

$$C.F. \times h = mg \times r + \tfrac{1}{2}Mg \times 2r$$

Note that half the weight of the central mass is taken as being the effect on one bob,

$$C.F. \times h = gr\,(m + M)$$

$$h = \frac{gr\,(m + M)}{m\,\omega^2\,r}$$

$$h = \frac{g}{\omega^2}\left\{\frac{m + M}{m}\right\}$$

f Example. The links of a Porter governor are each 230 mm long. The mass of the central load and sleeve is 14 kg and the mass of each bob is 2 kg. When the governor rotates at a maximum speed the links make an angle of 60° to the vertical. Calculate (i) the maximum speed and, (ii) the change in height when the speed falls to 200 rev/min.

$$h = 230 \times \cos 60° = 115 \text{ mm} = 0{\cdot}115 \text{ m}$$

$$h = \frac{g}{\omega^2}\left\{\frac{m + M}{m}\right\}$$

$$0{\cdot}115 = \frac{9{\cdot}81}{\omega^2}\left\{\frac{2 + 14}{2}\right\}$$

$$\omega = \sqrt{\frac{9{\cdot}81 \times 8}{0{\cdot}115}} = 26{\cdot}12 \text{ rad/s}$$

$$\frac{26{\cdot}12 \text{ rad/s} \times 60}{2\pi} = 249{\cdot}4 \text{ rev/min} \qquad \text{Ans. (i)}$$

$$\text{height at 200 rev/min} = 115 \times \left\{\frac{249{\cdot}4}{200}\right\}^2$$

$$= 178{\cdot}8 \text{ mm}$$

$$\text{Change in height} = 178{\cdot}8 - 115$$

$$= 63{\cdot}8 \text{ mm} \qquad \text{Ans. (ii)}$$

EFFECT OF FRICTION. All frictional resistances in a governor, such as between the sleeve and spindle, at the various joints and in the operating gear can be reduced to one single friction force F acting at the sleeve. Friction always opposes motion, therefore when the speed of the governor is increasing and the sleeve rising, the friction force will act downwards, thus being equivalent to increasing the central load. When the speed of the governor is decreasing and the sleeve is descending, the friction force will act upwards and have the effect of reducing the centre load.

Hence, referring again to Fig. 76, considering the equilibrium of one bob, taking moments about i, let $+F$ represent the friction force at the central sleeve when the governor is increasing speed, and $-F$ when the governor speed is decreasing then:

$$C.F. \times h = mg \times r \; \tfrac{1}{2}(Mg \pm F) \times 2r$$

$$h = \frac{r(mg + Mg \pm F)}{m\,\omega^2\,r}$$

$$h = \frac{g}{\omega^2}\left\{\frac{m + M \pm F/g}{m}\right\}$$

f Example. The mass of each bob of a Porter governor is 2 kg and the mass of the central load and sleeve is 19·2 kg. If the friction force at the sleeve is 7·85 N find the maximum and minimum governor speeds when the height is 250 mm.

$$h = \frac{g}{\omega^2}\left\{\frac{m + M \pm F/g}{m}\right\}$$

For maximum speed,

$$0{\cdot}25 = \frac{9{\cdot}81}{\omega^2}\left\{\frac{2 + 19{\cdot}2 - 7{\cdot}85/9{\cdot}81}{2}\right\}$$

$$\omega = \sqrt{\frac{9{\cdot}81 \times 22}{0{\cdot}25 \times 2}} = 20{\cdot}78 \text{ rad/s}$$

$$\frac{20{\cdot}78 \times 60}{2\pi} = 198{\cdot}4 \text{ rev/min} \qquad \text{Ans. (i)}$$

For minimum speed,

$$0{\cdot}25 = \frac{9{\cdot}81}{\omega^2}\left\{\frac{2 + 19{\cdot}2 - 7{\cdot}85/9{\cdot}81}{2}\right\}$$

$$\omega = \sqrt{\frac{9{\cdot}81 \times 20{\cdot}4}{0{\cdot}25 \times 2}} = 20 \text{ rad/s}$$

$$\frac{20 \times 60}{2\pi} = 191 \text{ rev/min} \qquad \text{Ans. (ii)}$$

f HARTNELL GOVERNOR

The Hartnell governor is a type of spring-loaded governor which is operated by a pair of bell-crank levers, the vertical arm of each lever carries a bob at its top end, the horizontal arm engages with the sleeve, and a central spring in compression exerts a controlling force on the sleeve, as diagrammatically shown in Fig. 77.

The force of gravity on each bob, *i.e.* its weight, has no effect on the sleeve when the arm carrying the bob is purely vertical. These arms lie slightly inwards at minimum speed, move outwards due to centrifugal force until they lie slightly outward from the vertical at maximum speed.

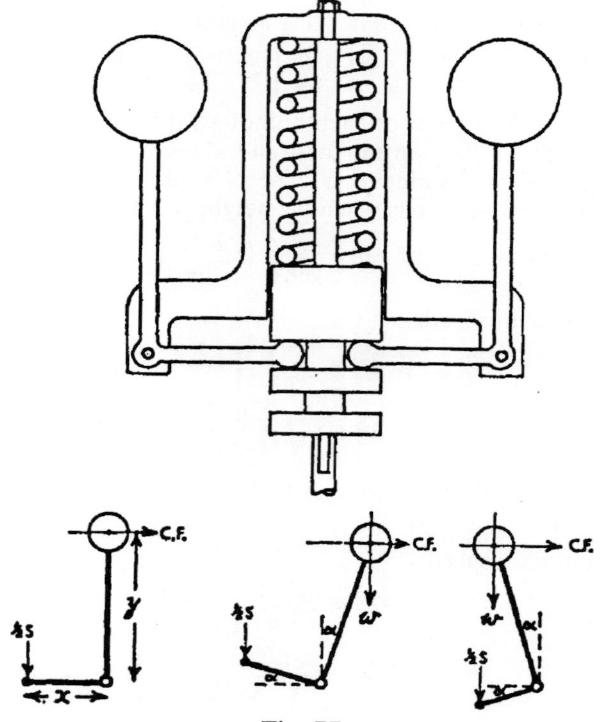

Fig. 77

Considering the equilibrium of each bob,

Let m = mass of each bob

w = weight of each bob = mg

r = radius of rotation of bobs

S = force exerted by spring on sleeve

y = length of vertical arm

x = length of horizontal arm

Taking moments about crank fulcrum,

At mean position:

$$C.F \times y = \tfrac{1}{2}S \times x$$

At high speeds:

$$C.F. \times y \cos \alpha + w \times y \sin \alpha = \tfrac{1}{2}S \times x \cos \alpha$$

At low speeds:

$$C.F. \times y \cos \alpha = w \times y \sin \alpha + \tfrac{1}{2}S \times x \cos \alpha$$

In most cases the angularity of the arms from the vertical and horizontal is small and therefore the moment of the weight of the bob about the crank fulcrum can usually be neglected.

If the weight of the sleeve is to be taken into consideration, it would be added to the spring force.

If the force to overcome sliding friction between the sleeve and the spindle is to be taken into account, it would be added to the spring force when the speed is increasing, and subtracted when the speed is decreasing, as previously shown.

f Example. In a Hartnell governor the mass of each bob is 1·5 kg. From the bell-crank lever fulcrum the length of the vertical arms to the centre of the bob is 120 mm and the length of the horizontal arms is 60 mm. When running at 300 rev/min the radius of rotation of the bobs is 80 mm, and at 320 rev.min the radius is 115 mm. Find the stiffness of the spring (N/mm) of compresssion. Neglect the effect of angularity of the arms from their respective vertical and horizontal positions.

$$\frac{300 \text{ rev/min} \times 2\pi}{60} = 31{\cdot}42 \text{ rad/s}$$

$$\frac{320 \text{ rev/min} \times 2\pi}{60} = 33{\cdot}52 \text{ rad/s}$$

Let S = force of spring on sleeve

$\tfrac{1}{2}S$ = downward force on end of each horizontal arm.

Considering equilibrium of each bob, taking moments about crank fulcrum,

At 300 rev/min:

$$C.F. \times y = \tfrac{1}{2}S_1 \times x$$

$$1{\cdot}5 \times 31{\cdot}42^2 \times 0{\cdot}08 \times 0{\cdot}12 = \tfrac{1}{2}S_1 \times 0{\cdot}06$$

$$S_1 = 473{\cdot}9 \text{ N}$$

At 320 rev/min:

$$1{\cdot}5 \times 33{\cdot}52^2 \times 0{\cdot}115 \times 0{\cdot}12 = \tfrac{1}{2}S_2 \times 0{\cdot}06$$

$$S_2 = 775{\cdot}2 \text{ N}$$

$$\text{Increase in compression} = 775{\cdot}2 - 473{\cdot}9$$

$$= 301{\cdot}3 \text{ N}$$

Horizontal movement at top of vertical arm

= difference in radius of rotation

$= 115 - 80 = 35$ mm

Ratio of lengths of vertical arm to horizontal arm

$= 120 \div 60 = 2$

Vertical movement at end of horizontal arm

$= 35 \div 2 = 17{\cdot}5$ mm this is the distance the sleeve moves up and therefore the amount the spring is compressed.

$$\text{Stiffness of spring} = \frac{301{\cdot}3}{17{\cdot}5}$$

$$= 17{\cdot}22 \text{ N/mm} \qquad \text{Ans.}$$

f SIMPLE HARMONIC MOTION

Simple harmonic motion is a particular form of reciprocating or 'to and 'fro' motion in which the acceleration and the velocity of the body varies as it moves from one end of its travel to the other, the important characteristic of this motion is that *the acceleration is proportional to the displacement from mid-travel (and directed to it).*

Note how this differs from anything which has been dealt with so far, previous to this, cases of *uniform* acceleration only have been explained.

When a body moves forwards and backwards with simple harmonic motion, at one end of its oscillation and the body is momentarily at rest, it receives maximum acceleration and causes it to move with increasing velocity. The acceleration, which is proportional to the displacement from mid-travel, decreases, and its velocity increases as the body approaches mid-travel. As it passes its mid-position the acceleration is zero and the velocity is maximum. From mid-travel onwards the acceleration is negative and increasing in magnitude while the velocity decreases until, at the other end of its travel, the velocity is momentarily nil and the acceleration is maximum to return the body in the opposite direction.

If a particle is travelling around a circular path in a vertical plane at a constant speed and its shadow could be seen on a horizontal table, the shadow would move forwards and backwards with simple harmonic motion.

From a practical point of view, neglecting the effect of the angularity of the connecting rod of a reciprocating engine, the piston moves with simple harmonic motion when the crank pin travels at a constant angular velocity.

Referring to Fig. 78, the circle represents the circular path of a particle Q moving at a constant velocity, this could be the crank pin of an engine. The projection of Q on to the plane of the diameter is point P, this could represent the relative position and motion of the piston (neglecting angularity of the connecting rod).

Q is moving at a constant angular velocity of ω around the circular path of r radius, its linear velocity is therefore ωr. The velocity vector diagram is drawn to represent the linear velocity of Q when passing the point at $\theta°$ past dead centre, the horizontal component represents the velocity of P at that instant. Thus:

$$\text{Velocity of P} = \text{Velocity of Q} \times \sin\theta$$
$$= \omega r \sin\theta$$

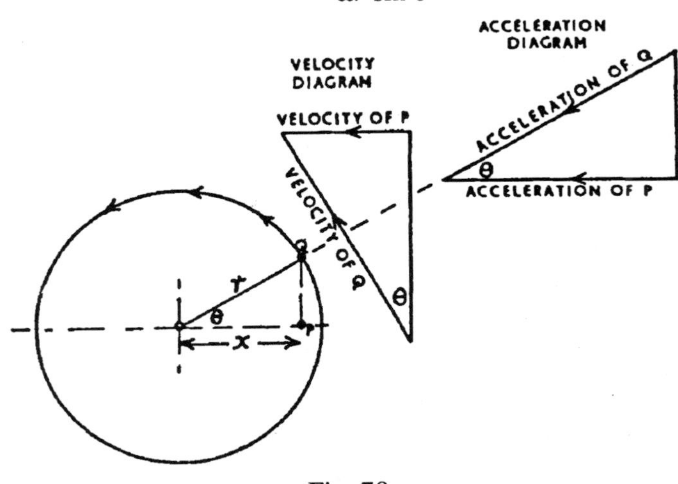

Fig. 78

If x represents the displacement of P from the middle of its travel, sin θ can be expressed in terms of r and x:

$$\text{Velocity of P} = \omega r \sin\theta$$
$$= \omega r \times \frac{\sqrt{r^2 - x^2}}{r}$$
$$= \omega\sqrt{r^2 - x^2}$$

The acceleration of a body moving around a circular path at constant velocity is centripetal acceleration, it has been shown to be $\omega^2 r$ and directed towards the centre. The acceleration of P is the horizontal component of the acceleration of Q, therefore:

$$\text{Acceleration of P} = \text{Acceleration of Q} \times \cos\theta$$
$$= \omega^2 r \cos\theta$$

or, expressing $\cos\theta$ in terms of x,

$$\text{Acceleration of P} = \omega^2 r \times \frac{x}{r}$$
$$= \omega^2 x$$
$$= \omega^2 \times \text{displacement from mid-travel}$$

The angular velocity ω is constant, therefore *the acceleration of a body moving with simple harmonic motion is proportional to its displacement from mid-travel (and directed to it).*

The AMPLITUDE is the maximum displacement to either side of its mid-travel.

The PERIODIC TIME is the time taken to make one complete oscillation, that is, to move completely across from one end to the other end and back again.

The time for P to make one complete oscillation is the same time that Q takes to move around one complete revolution, as follows,

$$\text{Time} = \frac{\text{distance}}{\text{velocity}}$$

$$\text{Time} = \frac{2\pi}{\omega}$$

The value of ω can be substituted into terms of displacement and acceleration.:–

$$\text{Acceleration} = \omega^2 \times \text{displacement}$$

$$\omega = \sqrt{\frac{\text{acceleration}}{\text{displacement}}}$$

$$\text{Periodic time} = \frac{2\pi}{\omega}$$

$$= 2\pi \div \sqrt{\frac{\text{acceleration}}{\text{displacement}}}$$

$$= 2\pi \sqrt{\frac{\text{displacement}}{\text{acceleration}}}$$

The above expression gives the periodic time for any body moving with simple harmonic motion.

The FREQUENCY is the number of oscillations made in 1 s.

If t = periodic time (s)

then, Frequency $= \frac{1}{t}$ osc/s

A MATHEMATICAL RELATION is given by:

$$x = r \cos \theta$$

$$x = r \cos wt \quad \textit{i.e.}\ \text{displacement}$$

$$\frac{dx}{dt} = -\omega r \sin wt \quad \textit{i.e.}\ \text{velocity } v$$

$$\frac{d^2x}{dt^2} = -\omega^2 r \cos wt \quad \textit{i.e.}\ \text{acceleration } a \left(= \frac{dv}{dt}\right)$$

$$\frac{d^2x}{dt^2} = -\omega^2 x$$

The minus sign (mathematically correct) is because the acceleration is in the opposite direction to that of positive x (*i.e.* increasing). SHM, such that acceleration is directly proportional to displacement from a fixed point and directed towards it. The constant of proportionality is ω^2.

*f*Example. The stroke of a reciprocating engine is 500 mm. Neglecting the effect of the angularity of the connecting rod, find the velocity and acceleration of the piston when the crank is 30° past top dead centre and the engine is running at 750 rev/min.

Angular velocity of crank

$$= \frac{750 \times 2\pi}{60} = 78{\cdot}54 \text{ rad/s}$$

Radius of crank pin circle = ½ stroke = 0·25 m

Velocity of piston = $\omega r \sin \theta$

$= 78{\cdot}54 \times 0{\cdot}25 \times \sin 30°$

$= 9{\cdot}817$ m/s Ans. (i)

Acceleration of piston $= \omega^2 r \cos\theta$

$= 78{\cdot}54^2 \times 0{\cdot}25 \times \cos 30°$

$= 1336 \text{ m/s}^2$ Ans. (ii)

*f*Example. A machine component of 2·25 kg mass moves with simple harmonic motion, the amplitude being 380 mm. If it makes 120 osc/min, find (i) the maximum accelerating force, (ii) the accelerating force when the displacement is 250 mm from mid-travel.

120 osc.min = 2 osc/s

∴ Periodic time of one oscillation = ½ s

Maximum displacement = 380 mm = 0·38 m

$$t = 2\pi\sqrt{\frac{\text{displacement}}{\text{acceleration}}}$$

$$\tfrac{1}{2} = 2\pi\sqrt{\frac{0{\cdot}38}{\text{acceleration}}}$$

$$(\tfrac{1}{2})^2 = \frac{2^2 \times \pi^2 \times 0{\cdot}38}{\text{acceleration}}$$

$$\text{Acceleration} = 2^2 \times 2^2 \times \pi^2 \times 0{\cdot}38$$

$$= 60{\cdot}01 \text{ m/s}^2$$

Accelerating force = mass × acceleration

$= 2{\cdot}25 \times 60{\cdot}01$

$= 135$ N Ans. (i)

Acceleration is proportional to displacement, therefore when the displacement is 0·25 m:

$$\text{Accelerating force} = 135 \times \frac{0{\cdot}25}{0{\cdot}38}$$

$= 88{\cdot}82$ N Ans. (ii)

f THE SIMPLE PENDULUM

A simple pendulum consists of a heavy bob swinging forwards and backwards suspended by a light cord.

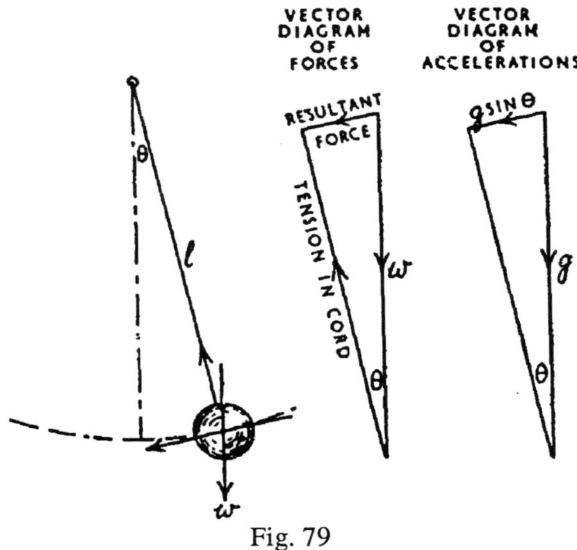

Fig. 79

Referring to Fig. 79, let m represent the mass of the bob, θ the angle made by the cord to the vertical centre line, and l the length of the cord. In the position shown, the effect of the force of gravity on the bob is to cause it to accelerate in the direction tangential to the arc of movement. The downward force of gravity on the bob is mg. From the vector diagram of the forces acting on the bob, the accelerating force causing it to move down the inclined plane is the component in that direction of the force of gravity, which is $mg \sin \theta$.

$$\text{Acceleration} = \frac{\text{accelerating force}}{\text{mass}}$$

$$= \frac{mg \sin \theta}{m} = g \sin \theta \quad \dots \dots \dots \quad \text{(i)}$$

or this can be obtained direct from the vector diagram of accelerations, the acceleration down the plane being the component in that direction of the gravitational acceleration g.

The displacement of the bob from mid-travel of its swing is the length of the arc from the centre of swing to centre of bob,

$$\text{Displacement} = l\,\theta \quad \dots \dots \dots \dots \dots \dots \dots \quad \text{(ii)}$$

From (i), acceleration is proportional to $\sin \theta$ because g is constant.

From (ii), displacement is proportional to θ radians because l is constant.

For small angles, sin θ equals θ rad, therefore for *small angles of swing*, acceleration is proportional to displacement and the motion is simple harmonic. Hence apply the expression for the periodic time of a body moving with simple harmonic motion:

$$\text{Periodic time} = 2\pi\sqrt{\frac{\text{displacement}}{\text{acceleration}}}$$

$$= 2\pi\sqrt{\frac{l\theta}{g \sin \theta}}$$

θ rad cancels with sin θ for a small angle, therefore,

$$t = 2\pi\sqrt{\frac{l}{g}}$$

Example. Find the length of a simple pendulum to make (i) one complete osc/s (*i.e.* one swing), (ii) half an osc/s.

In the first case the periodic time is 1 s,

$$t = 2\pi\sqrt{\frac{l}{g}}$$

$$l = \frac{t^2 \times g}{2^2 \times \pi^2}$$

$$= \frac{1 \times 9{\cdot}81}{2^2 \times \pi^2}$$

$$= 0{\cdot}2485 \text{ m}$$

$$= 248{\cdot}5 \text{ mm} \qquad \text{Ans. (i)}$$

In the second case, the periodic time for one complete oscillation is 2 s,

t varies as $\sqrt{l}$

$\therefore$ l varies as t^2

$\therefore l = 248{\cdot}5 \times (2/1)^2$

$= 994$ mm Ans. (ii)

f VIBRATIONS OF A SPRING

If a mass is attached to the end of a spring and the spring is allowed to extend gradually, an equilibrium condition will be reached when the spring force equals the weight of the attached mass.

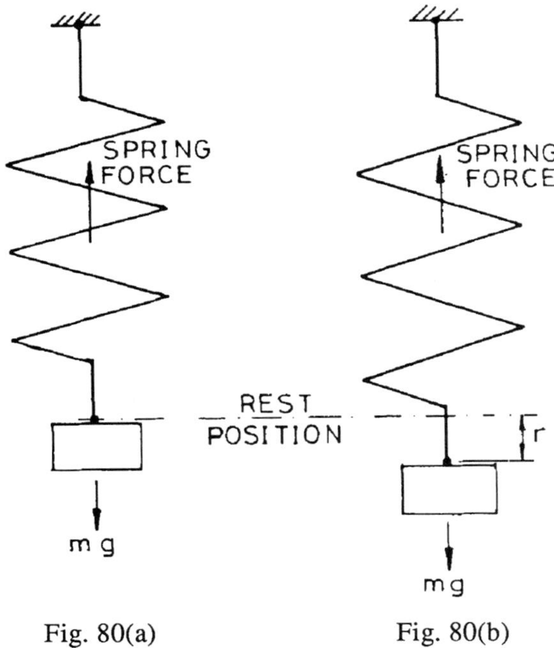

Fig. 80(a) Fig. 80(b)

Thus, referring to Fig. 80(a)

With the spring at rest,

$$\text{Spring force} = mg$$

If the mass is now pulled down some distance r below the rest position and then released, the mass will oscillate vertically through a distance r above and below the rest position.

Referring to Fig. 80(b)

Let S = spring stiffness

Thus, when the spring has been extended a distance r

$$\text{Spring force} = mg + Sr$$

At the instant of release,

$$\text{Resultant upward force} = mg + Sr - mg$$

$$= Sr$$

$$\text{Force} = \text{mass} \times \text{acceleration}$$

$$Sr = m \times a$$

$$a = \frac{Sr}{m}$$

Since stiffness S and mass m are constants for a given arrangement, acceleration must be proportional to displacement r.

Hence, the mass oscillates vertically with simple harmonic motion.

$$\text{Periodic time } t = 2\pi \sqrt{\frac{\text{displacement}}{\text{acceleration}}}$$

$$t = 2\pi \sqrt{\frac{r}{Sr/m}}$$

$$\text{Periodic time} = 2\pi \sqrt{\frac{m}{S}}$$

Example. A close coiled helical spring has a stiffness of 1400 N/m. If a 10 kg mass is attached to the end of the spring, calculate the periodic time and the frequency of oscillation when the spring vibrates vertically.

$$\text{Periodic time } t = 2\pi \sqrt{\frac{m}{S}}$$

$$= 2\pi \sqrt{\frac{10}{1400}}$$

$$\text{Time for one oscillation} = 0{\cdot}531 \text{ s} \qquad \text{Ans. (i)}$$

$$\text{Frequency} = \frac{1}{\text{periodic time}}$$

$$= \frac{1}{0{\cdot}531}$$

$$\text{Frequency} = 1{\cdot}88 \text{ osc/s} \qquad \text{Ans. (ii)}$$

EFFECT OF THE MASS OF THE SPRING. If the mass of the spring is taken into account, it can be shown that its effect is equivalent to one-third of its mass being placed at the free end

$$\text{Periodic time } t = 2\pi \sqrt{\frac{m + \frac{m_s}{3}}{S}}$$

Example. A helical spring with a mass of 6 kg has a mass of 10 kg attached to its free end. The mass is now pulled down 40 mm from the equilibrium position and then released. (i) If the spring stiffness is 532 N/m find the frequency of oscillation. (ii) Determine also the kinetic energy of the mass when it is 30 mm from the equilibrium position.

(i)

$$\text{Periodic time} = 2\pi\sqrt{\frac{m + \frac{m_s}{3}}{S}}$$

$$= 2\pi\sqrt{\frac{10 + \frac{6}{3}}{532}}$$

$$= 0{\cdot}944 \text{ s}$$

$$\text{Frequency} = \frac{1}{0{\cdot}944}$$

$$= 1{\cdot}06 \text{ osc/s} \qquad \text{Ans. (i)}$$

$$\text{Periodic time } t = \frac{2\pi}{\omega}$$

$$\omega = \frac{2\pi}{0{\cdot}944}$$

$$= 6{\cdot}66 \text{ rad/s}$$

For this spring motion,

$$\text{Amplitude } r = 0{\cdot}04 \text{ m}$$

When displacement x from mid-travel is 0·03 m:

$$\text{Instantaneous velocity} = \omega\sqrt{r^2 - x^2}$$

$$= 6{\cdot}66\sqrt{0{\cdot}04^2 - 0{\cdot}03^2}$$

$$= 0{\cdot}176 \text{ m/s}$$

$$\text{Translational K.E.} = \tfrac{1}{2}mv^2$$

$$= \tfrac{1}{2} \times 10 \times 0{\cdot}176^2$$

$$= 0{\cdot}155 \text{ J.} \qquad \text{Ans. (ii)}$$

TEST EXAMPLES 5

1. A mass of 1·2 kg is connected to the end of a cord and rotated in a vertical plane, the radius of the circular path being 600 mm. Find (i) the maximum tension in the cord when the speed of rotation is 75 rev/min, (ii) the speed of rotation when the minimum tension is nil.

2. A hole is bored in a circular disc of uniform thickness. The centre of the hole is 38 mm from the centre of the disc and the mass of material removed is 0·8 kg. Calculate the centrifugal force set up when the bored disc rotates about its geometrical centre at 240 rev/min.

3. A motor car is on the verge of skidding when travelling at 48 km/h on a level road around a curvature of 30 m radius. Find the coefficient of friction between the tyres and the ground.

4. The wheel track of a motor van is 1·68 m and its centre of gravity when fully loaded is 0·98 m above the ground. Calculate the speed at which the van will overturn when travelling around a bend of 23 m radius, assuming the road to be level.

5. Calculate the super-elevation of the outer rail of a curved railroad of 250 m radius so that there will be no side thrust when a train is travelling at 75 km/h, the track being 1·435 m.

6. Three masses, x, y and z, are to rotate in the same plane at the same angular velocity. Their masses and radii from centre of rotation are, respectively 9 kg at 400 mm radius, 10 kg at 350 mm, and 12 kg at 250 mm. Calculate the angles between them so that they will balance.

7. A thin flywheel rim, 1·2 m diameter, is made of cast iron of density 7210 kg/m^3. Find the speed in rev/min at which the stress due to centrifugal force will be 15 MN/m^2.

8. Find the change in height of a simple unloaded governor when it changes speed from 60 to 80 rev/min.

*f*9. The two balls of a Porter governor each have a mass of 2·25 kg. Assuming all links to be the same length, find the mass of the central load so that the change in height will be 25 mm when the speed changes from 240 to 270 rev/min.

*f*10 A push-rod moves with simple harmonic motion driven by an eccentric sheave running at 90 rev/min, the full travel of the rod being 50 mm. Calculate its velocity and acceleration when it is 6 mm from the beginning of its travel.

*f*11. The effective force on the piston of a vertical diesel engine when passing top dead centre is 800 kN, the mass of the reciprocating parts is 1524 kg, the length of the stroke is 1100 mm, and the engine is running at 120 rev/min. Assuming the motion of the reciprocating parts to be simple harmonic, find the effective thrust on the crosshead at the beginning of the down stroke.

*f*12. A cam rotates at 3 rev/s and imparts a vertical lift of 60 mm to a follower of mass 2 kg. The follower moves with S.H.M. and the lift of the follower is completed in one-third of a revolution of the cam. Calculate (a)the maximum acceleration of the follower and (b)the maximum force between cam and follower.

*f*13. Calculate the length of a simple pendulum to make 120 complete oscillations per minute.

*f*14. The stiffness of a helical spring is such that it stretches 1 mm for every 1·5 N of load. Find the number of vibrations it will make per minute if it is disturbed from rest when carrying a mass of 6·5 kg.

*f*15. A helical spring has a stiffness of 25 kN/m.

If a mass of 100 kg is attached to its free end, pulled down, and then released, determine the periodic time of its motion.

If the maximum deflection was 50 mm, find the velocity and acceleration of the mass when it is 300 mm from the equilibrium position.

CHAPTER 6

SLIDING FRICTION

When one body slides over another, a certain amount of resistance is set up between the surfaces in contact which tends to oppose motion. This is termed the *frictional resistance* between the surfaces of the pair of bodies and the force required to overcome this resistance and thereby cause motion is usually referred to as the *friction force*.

HORIZONTAL PLANES

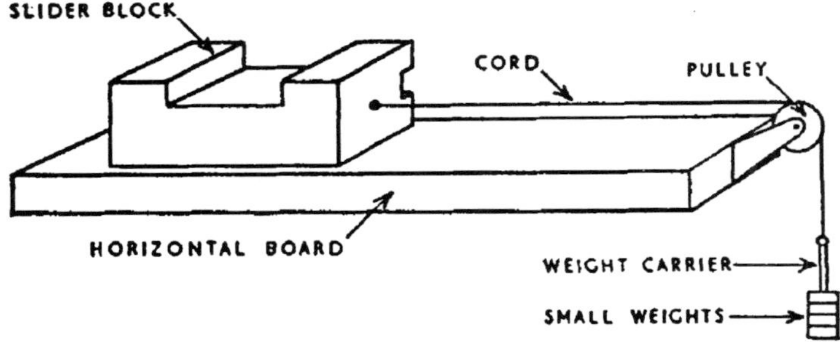

Fig. 81

Fig. 81 shows a simple piece of apparatus for investigating frictional resistance. it consists of a flat board of wood or metal which is set horizontally by a spirit level on a table; a slider block is connected to a cord running parallel to the board and passing over a guide pulley at the end of the board, a weight carrier is attached to the hanging end of the cord. The slider is usually a rectangular block with sides of different surface areas, loads can be placed on top of it to increase the pressure between the surfaces of the block and board. Small weights are added to the hanging carrier or hook until the pull in the cord is just sufficient to cause the block to slide without acceleration.

Firstly it will be found that the force to start the block moving from rest is a little greater than the force required to maintain movement at a steady speed after it has begun to move. The friction at rest is referred to as *static friction,* the friction of movement is termed *sliding friction.* It is the latter which is to be studied here, therefore in any experiment with the above apparatus, the block is tapped lightly with the fingers to overcome static friction while small weights are added to the carrier until the block moves *at steady speed* along the board.

If a series of trials are performed with various pressures between the surfaces of the block and board (by adding loads on top of the block) and adjusting the effort in each case to obtain steady movement, it will be found that the friction force is always proportional to the pressure between the surfaces. For example, if the pressure between the surfaces is doubled then the force required to slide the block along must be doubled, treble the pressure between the surfaces and the force to overcome friction must be trebled and so on. Therefore for any given pair of surface materials the friction force divided by the pressure between the surfaces is a constant, this constant is termed the *coefficient of friction* (μ).

It will also be seen that if the block is given an extra push at the beginning to start it off at a faster speed, the force applied to overcome friction will be the same, that is, the same force will keep the block moving at the greater speed as that which was required at the lower speed. it can therefore be stated that the frictional resistance is independent of the speed of sliding. This however, is only true within moderate speeds.

Now with the same loading in each case, trials can be made with the block resting on its different sides which have different surface areas. It will be noted that the force required to slide the block is the same in each case, demonstrating that friction is independent of the areas in contact.

Further experiments can be performed with smooth and rough surfaces, and with various different materials in contact, and it will be found that friction depends upon the roughness of the surfaces and the nature of the materials.

THE LAWS OF SLIDING FRICTION summarise these observations as follows:

(i) Frictional resistance is proportional to the total pressure between the surfaces (force, often called normal reaction).

(ii) It depends upon the nature and roughness of the surfaces.

(iii) It is independent of the areas in contact.

(iv) It is independent of the speed of sliding at low speeds.

(v) It opposes motion.

The total pressure between the surfaces is the total pressure pressing *normal* to the surfaces, the word normal meaning 'at right angles to'. For a straightforward case of a block resting on a horizontal plane pulled along by a horizontal force, the normal pressure between the surfaces is simply the weight of the block plus of course the loads on top of it if there are any.

THE COEFFICIENT OF FRICTION is the ratio of the force required to overcome friction to the normal pressure between the surfaces, thus,

$$\text{Coefficient of friction} = \frac{\text{friction force}}{\text{normal pressure between surfaces}}$$

$$\mu = \frac{F}{N}$$

$$\text{or, } F = \mu \times N$$

Fig. 82 shows the forces acting on a body of weight W when a force F is applied horizontally to overcome friction on a horizontal plane. Note the reaction R of the plane, when the body is at rest the plane applies a vertical upward force equal to the magnitude of W to support the weight, but when the body is moving the unseen frictional resisting force comes into action at the surface of the plane; R is the resultant of these two and swings over in the direction which opposes motion. The magnitude of the angle ϕ between R and N depends upon the magnitude of the force required to overcome friction, it is therefore termed the *friction angle.*

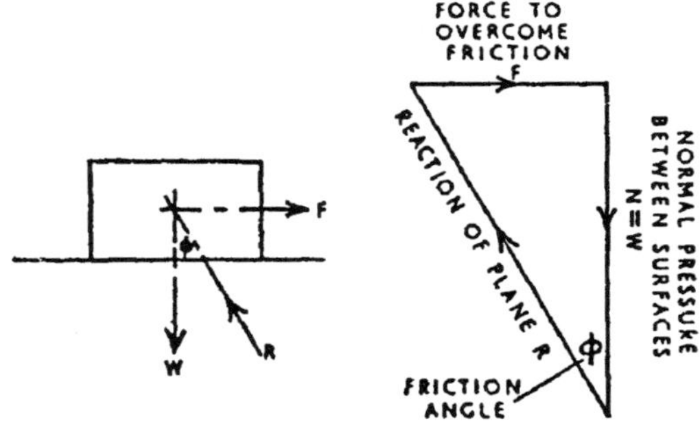

Fig. 82

From the vector diagram, $\tan \phi = \dfrac{F}{N}$

Also, as stated, $\mu = \dfrac{F}{N}$

therefore, $\tan \phi = \mu$

Stating this in words, *the tangent of the friction angle is equal to the coefficient of friction.*

Example. If a block of wood weighing 28 N requires a horizontal force of 9·8 N to pull it along a horizontal plane, what is the coefficient of friction between the block and plane? If the block is moved a distance of 5 m, what is the work done?

Normal pressure between surfaces
= weight of block = 28 N

Force to overcome friction = 9·8 N

Coefficient of friction, $\mu = \dfrac{F}{N}$

$= \dfrac{9{\cdot}8}{28} = 0{\cdot}35$ Ans. (i)

Work done = force applied × distance moved
= 9·8 × 5
= 49 J Ans. (ii)

FORCE NOT PARALLEL TO PLANE. Instead of the force applied being parallel to the horizontal plane over which the body is sliding, let it be inclined upward at θ°. This is illustrated in Fig. 83, φ is the friction angle (whose tangent is equal to the coefficient of friction), and the angle between F and R in the vector diagram of forces is $90 - \phi + \theta$.

$$\frac{F}{\sin \phi} = \frac{W}{\sin (90 - \phi + \theta)}$$

from which any one unknown can be calculated.

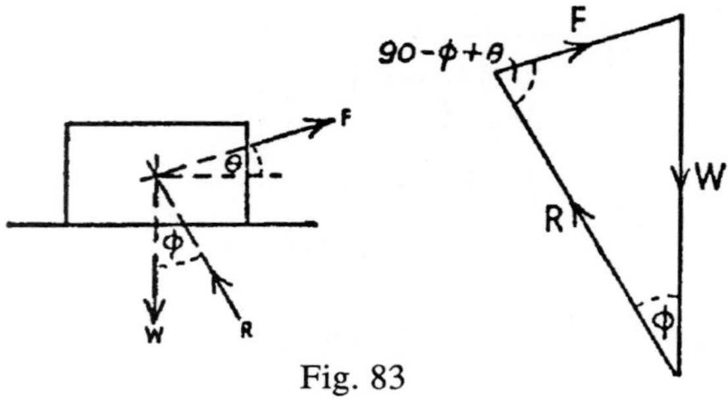

Fig. 83

Example. A body of 40 N weight is to be pulled along a horizontal plane, the coefficient of friction between the body and the plane being 0·3. If the line of action of the applied force is inclined upwards at 26° to the horizontal, find the magnitude of the force to slide the body at steady speed. Find also the magnitude and direction of the *least* force that will move the body.

Referring to Fig. 83,

$$\tan\phi = \mu = 0{\cdot}3$$

$$\text{Friction angle } \phi = 16°\,42'$$

$$\text{Angle opposite } W = 90 - \phi + \theta$$

$$= 90 - 16°\,42' + 26°$$

$$= 99°\,18'$$

$$\frac{F}{\sin 16°\,42'} = \frac{40}{\sin 99°\,18'}$$

$$F = \frac{40 \times 0{\cdot}2874}{0{\cdot}9869}$$

$$= 11{\cdot}65 \text{ N Ans. (i)}$$

The force to cause motion will be least when the vector F connecting W and R is as short as possible, that is, when the angle opposite W is 90°, and reference to Fig. 84 shows that this force will then be inclined upwards at an angle equal to the friction angle ϕ.

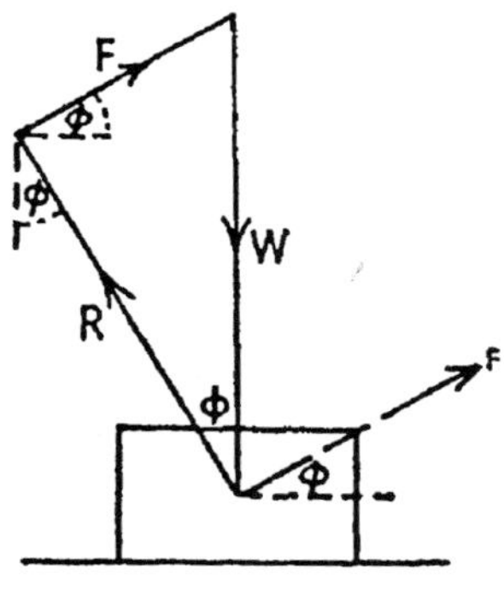

Fig. 84

The vector diagram is now a right angled triangle with W the hypotenuse, and F the opposite side to ϕ, therefore:

$$\text{Least force} = W \sin \phi$$
$$= 40 \times \sin 16° 42'$$
$$= 11{\cdot}5 \text{ N Ans. (ii}a\text{)}$$
$$\text{Angle of least force} = 16° 42' \text{ to horizontal Ans. (ii}b\text{)}$$

INCLINED PLANES

FORCES PARALLEL TO THE PLANE

Now consider sliding bodies on planes which are not horizontal. To begin with, imagine a body which has negligible friction on a plane inclined at α degrees to the horizontal, see Fig. 85.

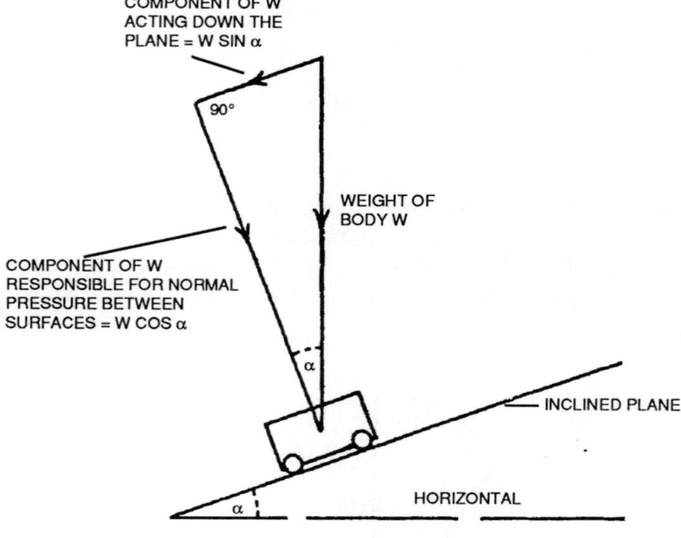

Fig. 85

The body would run down the plane on its own accord if allowed to do so, and this is due to the component in this direction of the weight W of the body. The other component of W is the pressure applied on the surface of the plane.

Thus the effect of W is seen in its two components; one which acts down the slope of the plane, its value being $W \sin \alpha$, the other which is at right angles to the surface of the plane, its value being $W \cos \alpha$.

If it is required to prevent the body from running down the plane, or to pull it up, a force must be applied equal in magnitude but opposite in direction to $W \sin \alpha$. This is often referred to as 'the force to overcome gravity' *i.e.* force of gravity.

Now take friction into account, to pull the body up the plane will require sufficient force to overcome the frictional resistance in addition to the force to overcome gravity.

On horizontal planes:

Friction force = $\mu \times$ normal pressure between surfaces.

On inclined planes the normal pressure between the surfaces is not simply the weight of the body as it is on a horizontal plane, but the component of W normal to the inclined plane, which is $W \cos \alpha$, hence the friction force is $\mu W \cos \alpha$. Therefore,

Total force to pull body up = force of gravity + friction force

$$\text{Force}_{up} = W \sin \alpha + \mu W \cos \alpha$$

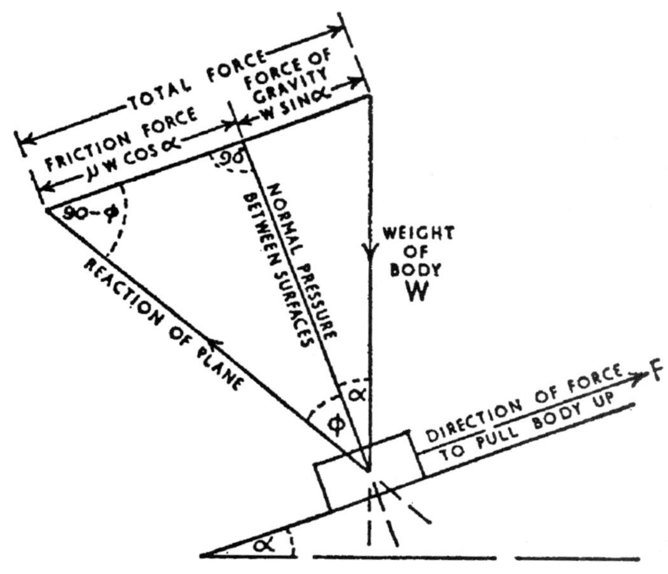

Fig. 86

The complete vector diagram is illustrated in Fig. 86, which is simply the friction vector diagram of Fig. 82 inclined to suit the inclination of the plane, added to the vector diagram of Fig.85.

Example. A body of 200 N weight is to be pulled up a slope which is inclined at 30° to the horizontal, the coefficient of friction between the body and the plane being 0·2. Find the force required assuming it to be applied parallel to the plane, and find also the work done in pulling the body a distance of 50 m up the slope.

$$\begin{aligned}
\text{Force}_{\text{up}} &= \text{force of gravity} + \text{friction force} \\
&= W \sin \alpha + \mu W \cos \alpha \\
&= 200 \times \sin 30° + 0{\cdot}2 \times 200 \times \cos 30° \\
&= 100 + 34{\cdot}64 \\
&= 134{\cdot}64 \text{ N Ans. (i)} \\
\text{Work done} &= \text{force} \times \text{distance} \\
&= 134{\cdot}64 \times 50 \\
&= 6732 \text{ J Ans. (ii)}
\end{aligned}$$

As an alternative method to the above, the force required could be found by trigonometry with reference to the vector diagram of Fig. 86 as follows:

$$\tan \phi = \mu = 0{\cdot}2$$

$$\text{Friction angle } \phi = 11° \, 19'$$

$$\text{Angle subtended by total force} = \phi + \alpha$$

$$= 11° \, 19' + 30° = 41° \, 19'$$

Angle between force and reaction

$$= 90 - \phi$$

$$= 90 - 11° \, 19' = 78° \, 41'$$

$$\frac{\text{Total force}}{\sin (\phi + \alpha)} = \frac{\text{Weight}}{\sin (90 - \phi)}$$

$$\text{Force} = \frac{\text{weight} \times \sin 41° \, 19'}{\sin 78° \, 41'}$$

$$= \frac{200 \times 0{\cdot}6602}{0{\cdot}9806}$$

$$= 134{\cdot}6 \text{ N (as before).}$$

In general most problems can be solved by the first method, but if the angle of the incline is the unknown it becomes quite a complicated solution. for such cases the latter method provides a much simpler solution. This is also true for forces not parallel to the plane, as detailed later.

Example. On a certain inclined plane, a body which weighs 50 N requires a force of 35 N to pull it up the incline. If the coefficient of friction is 0·25, find the angle of the incline.

Referring to Fig. 86,
$\tan \phi = \mu = 0{\cdot}25, \ \ \phi = 14° \, 2'$
$90 - \phi = 90 - 14° \, 2' = 75° \, 58'$

$$\frac{\text{Force}}{\sin(\phi + \alpha)} = \frac{\text{Weight}}{\sin(90 - \phi)}$$

$$\sin(\phi + \alpha) = \frac{\text{force} \times \sin 75° \, 58'}{\text{weight}}$$

$$= \frac{35 \times 0{\cdot}9702}{50}$$

$$= 0{\cdot}6791$$

$$\text{Angle } (\phi + \alpha) = 42° \, 46'$$

$$\alpha = 42° \, 46' - 14° \, 2'$$

$$= 28° \, 44' \text{ Ans.}$$

FORCE TO PULL BODY DOWN THE PLANE. When pulling a body down an inclined plane, the frictional resistance acts as a force to oppose motion, but the force of gravity acting down the plane helps the body to slide, therefore:

$$\text{Force to pull down} = \text{friction force} - \text{force of gravity}$$

$$\text{Force}_{\text{down}} = \mu \, W \cos \alpha - W \sin \alpha$$

FORCE TO HOLD. If the incline is sufficiently steep that the body would slide down the plane on its own accord, it means that the force of gravity acting down the plane is greater than the frictional resistance which tends to oppose motion, and to prevent the body from sliding a force must be applied equal to that difference, thus:

$$\text{Force to hold} = \text{force of gravity} - \text{friction force}$$

$$\text{Force}_{\text{hold}} = W \sin \alpha - \mu \, W \cos \alpha$$

ANGLE OF REPOSE. If the angle of the incline is such that the force of gravity is exactly equal to the frictional resistance, the body will be just at its critical state of being at rest but on the verge of sliding, then:

$$\text{Friction force} = \text{Force of gravity}$$

$$\mu \, W \cos \alpha = W \sin \alpha$$

$$\mu = \frac{\sin \alpha}{\cos \alpha}$$

$$\mu = \tan \alpha$$

that is, the tangent of the critical angle, or angle of repose as it is usually referred to, is equal to the coefficient of friction. The tangent of the friction angle is also equal to the coefficient of friction, therefore the angle of repose of an inclined plane is equal to the angle of friction.

Example. A casting weighs 1·12 kN and rests on an incline which rises 1 in 5. If the coefficient of friction between the casting and the plane is 0·32, find the force required to move it (i) up the plane, (ii) down the plane.

Plane rises 1 m in every 5 m

sine of angle of incline = ⅕ = 0·2

Angle of incline = 11° 32′

cosine of angle of incline = 0·9798

$$
\begin{aligned}
\text{Force}_{\text{up}} &= \text{force of gravity} + \text{friction force} \\
&= W \sin \alpha + \mu W \cos \alpha \\
&= 1120 \times 0{\cdot}2 + 0{\cdot}32 \times 1120 \times 0{\cdot}9798 \\
&= 224 + 351{\cdot}2 \\
&= 575{\cdot}2 \text{ N Ans. (i)} \\
\text{Force}_{\text{down}} &= \text{friction force} - \text{force of gravity} \\
&= 351{\cdot}2 - 224 \\
&= 127{\cdot}2 \text{ N Ans. (ii)}
\end{aligned}
$$

Example. A body on an incline which rises 1 in 4 requires a force of 1090 N to pull it up the incline, and 80 N to pull it down. Find the weight of the body and the coefficient of friction.

$$
\begin{aligned}
\text{sine of incline} &= \tfrac{1}{4} = 0{\cdot}25, \ \alpha = 14° \, 29' \\
\cos \alpha &= 0{\cdot}9682 \\
\text{Force}_{\text{up}} &= \text{friction force} + \text{force of gravity} \\
\text{Force}_{\text{down}} &= \text{friction force} - \text{force of gravity}
\end{aligned}
$$

$$
\begin{aligned}
\text{Force}_{\text{up}} &= \mu W \cos \alpha + W \sin \alpha \\
\text{Force}_{\text{down}} &= \mu W \cos \alpha - W \sin \alpha \\
1080 &= \mu W \times 0{\cdot}9682 + W \times 0{\cdot}25 \\
80 &= \mu W \times 0{\cdot}9682 - W \times 0{\cdot}25 \\
\hline
1000 &= 2 \times W \times 0{\cdot}25 \\
\therefore W &= 2000 \text{ N or 2 kN Ans. (i)}
\end{aligned}
$$

Inserting value of W into first equation,

$$1080 = \mu \times 2000 \times 0{\cdot}9682 + 2000 \times 0{\cdot}25$$
$$1080 = 1936{\cdot}4\,\mu + 500$$
$$580 = 1936{\cdot}4\,\mu$$
$$\mu = 0{\cdot}2995 \text{ Ans. (ii)}$$

ACCELERATION ON INCLINED PLANE

To accelerate a body, an accelerating force must be applied over and above any other force that may be required to move it at steady speed.

Example. A truck of mass 2 t is to be pulled up an incline of 10° and to accelerate at the rate of 0·5 m/s^2. Find the total force required if the tractive resistance on the level is 15 N per kN of weight.

Weight of truck $= 2 \times 10^3 \times 9{\cdot}81 = 19\,620$ N
Force of gravity $= W \sin\alpha$
$= 19\,620 \times \sin 10°$
$= 3407$ N ... ... (i)
Friction force $=$ 15 N per kN on the level
$= 15 \times 19{\cdot}62$ N
Friction force on incline $= 15 \times 19{\cdot}62 \times \cos 10°$
$= 289{\cdot}8$ N ... ... (ii)
Force to accelerate $=$ ma
$= 2 \times 10^3 \times 0{\cdot}5$
$= 1000$ N ... ... (iii)
Total force $= 3407 + 289{\cdot}8 + 1000$
$= 4696{\cdot}8$ N or $4{\cdot}6968$ kN Ans.

Example. A body of 20 kg mass is held at the top of an incline of 23°, the coefficient of friction between body and surface of the incline is 0·4. Find the time taken for the body to slide down the incline a distance of 6 m after it is released from rest.

Weight $= 20 \times 9{\cdot}81 = 196{\cdot}2$ N
Force of gravity $= W \sin\alpha$
$= 196{\cdot}2 \times \sin 23°$
$= 76{\cdot}67$ N
Friction force $= \mu W \cos\alpha$
$= 0{\cdot}4 \times 196{\cdot}2 \times \cos 23°$
$= 72{\cdot}24$ N

The force of gravity acts down the plane and tends to cause the body to slide.The frictional resistance tends to oppose motion. As the force of gravity is the greater of the two, the body will slide down with acceleration.

$$\text{Force causing acceleration} = 76{\cdot}67 - 72{\cdot}24$$
$$= 4{\cdot}43 \text{ N}$$
$$\text{Acceleration} = \frac{\text{accelerating force}}{\text{mass}}$$
$$= \frac{4{\cdot}43}{20} = 0{\cdot}2215 \text{ m/s}^2$$
$$s = ut + \tfrac{1}{2}at^2$$
$$\text{Initial velocity } u = 0$$
$$6 = \tfrac{1}{2} \times 0{\cdot}2215 \times t^2$$
$$t = \sqrt{\frac{6 \times 2}{0{\cdot}2215}}$$
$$= 7{\cdot}36 \text{ } s \text{ Ans.}$$

f FORCES NOT PARALLEL TO INCLINED PLANE

HORIZONTAL FORCE. Fig. 87 is a vector diagram of the forces acting on a body when it is being pulled up an inclined plane at steady speed by a force applied in a horizontal direction. As the weight acts vertically downwards and the force is horizontal, the vector diagram is a right angled triangle, hence:

Horizontal force to pull body up = weight × tan ($\phi + \alpha$)

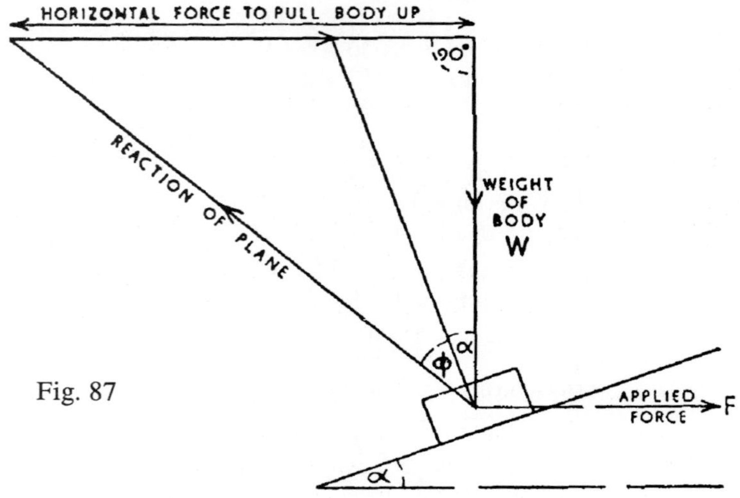

Fig. 87

If the body is pulled *down* the plane by a horizontal force, the friction angle swings over in the opposite direction because friction always opposes motion, the vector diagram is now as shown in Fig. 88. The expression for the horizontal force required to pull the body down the plane at a steady speed is therefore:

Horizontal force to pull body down = weight × tan ($\phi - \alpha$)

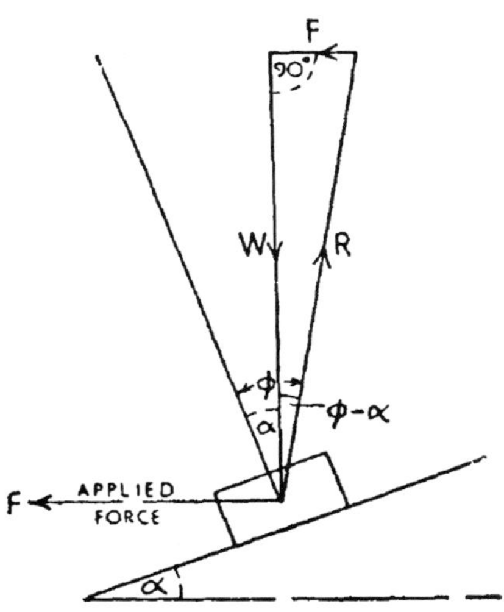

Fig. 88

One example of practical application of *horizontal forces* on inclined planes are *cotters and wedges.* Cotters and wedges are treated in the same manner, the only practical difference between them being that a cotter pulls two components together, while a wedge drives components apart. Referring to Figs. 89 and 90, if a cotter or wedge is tapered on both sides, it is regarded as two inclined planes placed back to back, the force with which the components are drawn together or forced apart is represented by W as the load on the inclined plane of the cotter or wedge. The force to drive the cotter or wedge in is the horizontal force to 'pull the body up the inclined planes' on the *two* sides, thus, if each side has the same taper, then:

$$\text{Force to drive in} = 2 \times W \tan(\phi + \alpha)$$

$$\text{Force to drive out} = 2 \times W \tan(\phi - \alpha)$$

f Example. A steel wedge is tapered equally on both sides. The length is 400 mm, thickness at butt end 55 mm, thickness at sharp end 5 mm. If it is driven between an engine bedplate and the stools with a driving force of 4·5 kN, find the lifting effect, taking the coefficient of friction as 0·15.

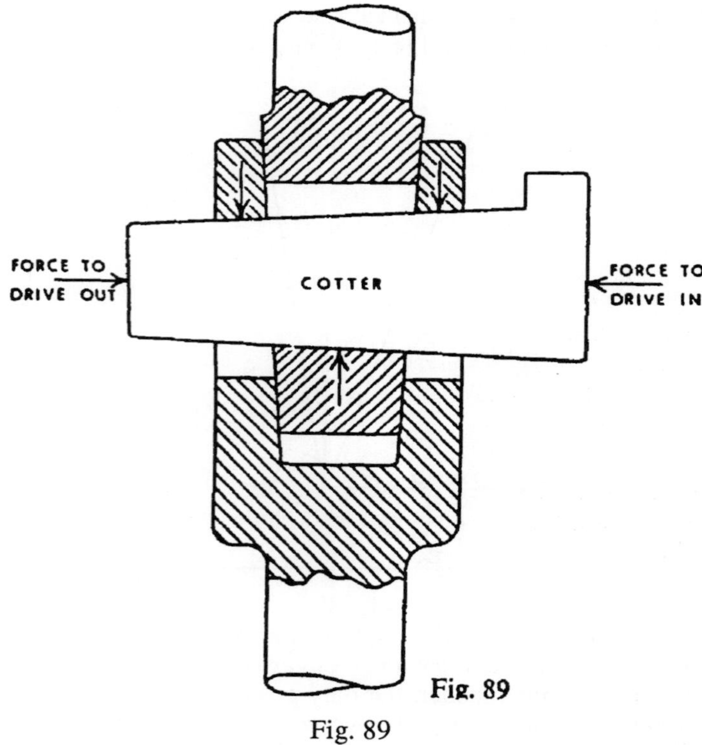

Fig. 89

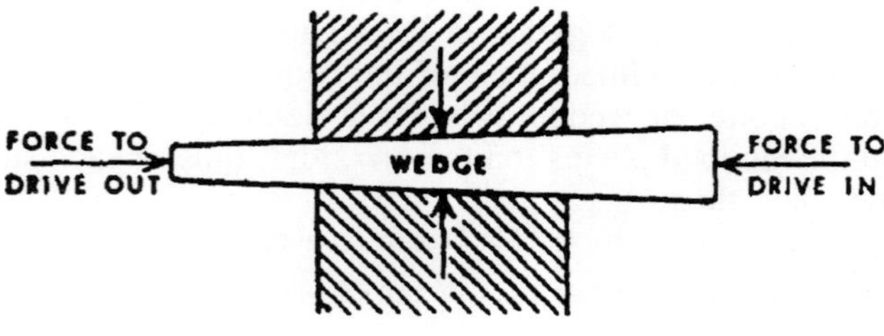

Fig. 90

Difference in end thickness = 55 − 5 = 50 mm

Taper is 25 mm over a length of 400 mm on each side,

$$\tan\alpha = \frac{25}{400} = 0{\cdot}0625$$

$$\alpha = 3^\circ\ 34'$$

$$\tan\phi = \mu = 0{\cdot}15$$

$$\phi = 8^\circ\ 32'$$

$$(\phi + \alpha) = 8^\circ\ 32' + 3^\circ\ 34' = 12^\circ\ 6'$$

$$\text{Force to drive in} = 2 \times W \tan(\phi + \alpha)$$

$$4{\cdot}5 = 2 \times W \times \tan 12^\circ\ 6'$$

$$W = \frac{4{\cdot}5}{2 \times 0{\cdot}2144}$$

$$= 10{\cdot}5 \text{ kN Ans.}$$

Many wedges and cotters have only one side tapered, the other being flat. Such cases are equivalent to an inclined plane on one side and a horizontal plane on the other side, thus:

$$\text{Force to drive in} = W \tan(\phi + \alpha) + \mu W$$

$$\text{Force to drive out} = W \tan(\phi - \alpha) + \mu W$$

LEAST FORCE. The force to pull a body up or down an inclined plane is least when the force vector of the vector diagram is shortest, this is when the vectors of the force and the reaction are at right angles.

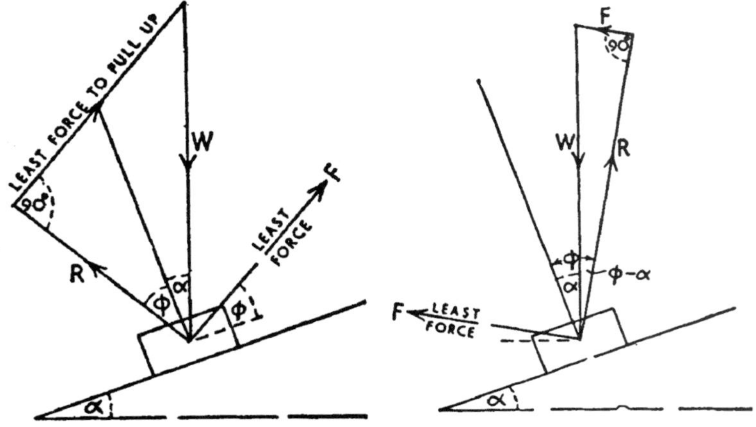

Fig. 91

Fig. 92

Fig. 91 shows the vector diagram for the least (minimum) force to pull the body up. In this right angled triangle, W is the hypotenuse, and the applied force is the opposite side to the angle $(\phi + \alpha)$, therefore:

$$\text{Least force to pull up} = W \sin(\phi + \alpha)$$

and the line of action of this force is at ϕ degrees to the plane.

Fig. 92 shows the vector diagram for the least (minimum) force to pull the body down the plane, from which:

$$\text{Least force to pull down} = W \sin(\phi - \alpha)$$

FORCE AT ANY ANGLE. The general case covering the foregoing is to let the force be applied at $\theta°$ to the plane, where θ may have any value. Referring to Fig.93 which shows the force applied at $\theta°$ to pull the body up:

$$\frac{\text{Force}}{\sin(\phi + \alpha)} = \frac{\text{Weight}}{\sin(90 - \phi + \theta)}$$

$$\therefore F = \frac{W \times \sin(\phi + \alpha)}{\sin(90 - \phi + \theta)}$$

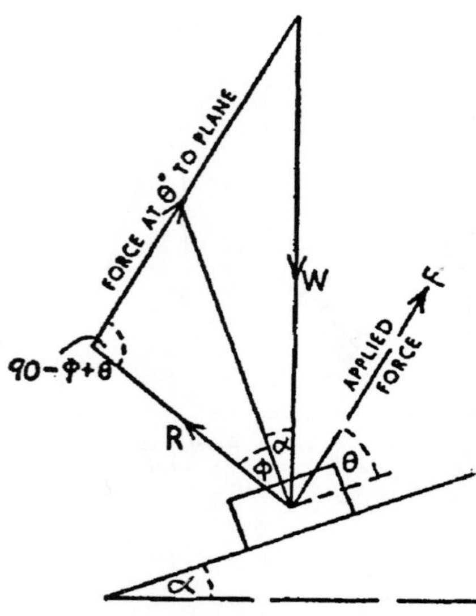

Fig. 93

To pull the body down, the vector diagram would be drawn so that the friction angle swings backwards to oppose motion as in Fig. 94.:

$$\frac{\text{Force}}{\sin(\phi - \alpha)} = \frac{\text{Weight}}{\sin(90 - \phi + \theta)}$$

$$\therefore F = \frac{W \times \sin(\phi - \alpha)}{\sin(90 - \phi + \theta)}$$

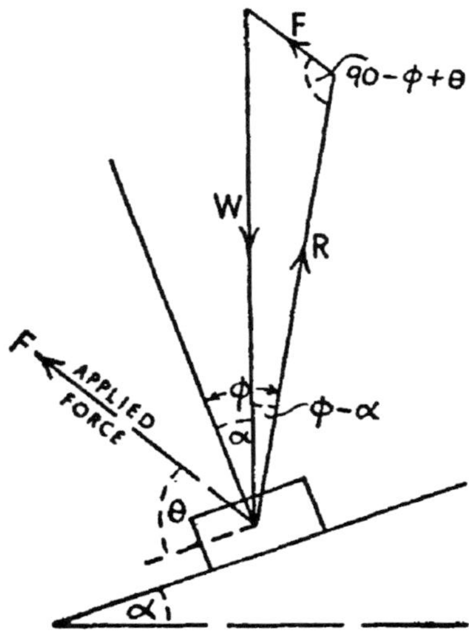

Fig. 94

f EFFICIENCY OF SQUARE THREAD

When a nut is rotated on its screwed counterpart, it can be considered from the point of view of climbing up a helical inclined plane. The force to turn the nut is taken as the equivalent to the *horizontal* force at the mean radius of the thread to move the load up the incline of the screw, the angle of the incline being the pitch angle of the thread. This, then, is another practical application.

Referring to Fig. 95, let α = pitch angle of the thread, ϕ = friction angle whose tangent is equal to the coefficient of friction, W = axial load on nut, and F = force, then,

$$F = W \tan(\phi + \alpha)$$

As the force is considered to be applied horizontally and at the mean radius of the thread, then the distance through which the force moves to turn the nut one revolution is the mean circumference of the thread, which is $2\pi \times$ radius.

In one revolution:

$$\text{Work supplied} = \text{Force} \times \text{distance moved}$$

$$= W \tan(\phi + \alpha) \times 2\pi r$$

$$\text{Work got out} = \text{Load} \times \text{distance lifted}$$

$$= W \times \text{pitch}$$

$$= W \times 2\pi \tan \alpha$$

$$\text{Efficiency} = \frac{\text{Work got out}}{\text{Work supplied}}$$

$$= \frac{W \times 2\pi r \tan \alpha}{W \tan(\phi + \alpha) \times 2\pi r}$$

$$= \frac{\tan \alpha}{\tan(\phi + \alpha)}$$

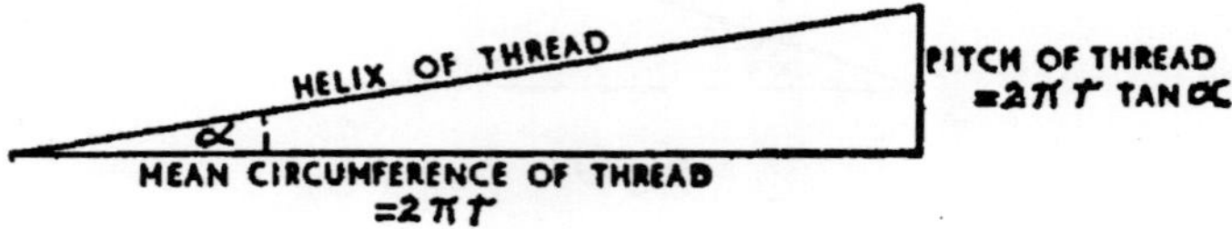

Fig. 95

TEST EXAMPLES 6

1. A casting weighing 750 N is pulled along a horizontal workshop floor for a distance of 15 m by a horizontal force. If the coefficient of friction between the casting and the floor is 0·32, calculate the force applied and the work done.

2. A shaft runs at 50 rev/s in bearings 100 mm diameter, the total load on the bearings is 25 kN and the coefficient of friction is 0·04. Calculate:
   (i) the friction force at the skin of the shaft,
   (ii) the work done to turn the shaft one revolution,
   (iii) the work lost to friction every second,
   (iv) the equivalent kW power loss.

3. An engine exerts a pull of 25 kN on a train of mass 100 t running on a horizontal track. If the tractive resistance of the train is 53 N per t, find the speed of the train after 2 min from rest, assuming uniform acceleration.

4. A body of 200 N weight rests on a horizontal plane, the coefficient of friction between the body and the plane is 0·25. Find the force inclined upwards at 30° to the horizontal which will just cause motion.

5. Find the magnitude and direction of the least force that will cause a load of 400 N weight to move along a horizontal plane when the coefficient of friction is 0·15.

6. Calculate the force necessary to pull a body of 100 N weight at a steady speed up a plane inclined at 30° to the horizontal, if the coefficient of friction between the body and the plane is 0·2 and the force is applied parallel to the plane. How much work is done in hauling it a distance of 5 m up the plane?

7. A force of 336·6 N applied parallel to the plane, is required to move a body of 500 N weight up the plane, If the coefficient of friction is 0·2, find the angle of the incline.

8. Find the force required to pull a body of 700 N weight down an incline of 1 in 4 if the coefficient of friction is 0·35 and the force is applied parallel to the plane.

9. A mass of 500 kg stands on a ramp inclined at 18° 12′ to the horizontal, the coefficient of friction between the surfaces of the mass and ramp being 0·27. If it is allowed to move from rest, find the acceleration of the mass down the plane, the velocity after 6 s and the distance moved in this time.

10. A body of 25 N weight on an inclined plane is connected by a cord to another of 50 N weight which is in a higher position on the same inclined plane, and the connecting cord is parallel to the plane. The coefficients of friction between the bodies and the plane are 0·15 and 0·3 respectively. If the inclination of the plane is such that the connected system is just on the verge of sliding, find the angle of the incline and the tension in the cord.

11. A forging requires a force of 36 kN to pull it up a ramp, and 2 kN to pull it down, the coefficient of friction between forging and ramp being 0·4. Calculate the angle the ramp is inclined to the horizontal and the weight of the forging.

*f*12. A load of 224 N rests on a plane inclined at 15° to the horizontal, the coefficient of friction between load and plane being 0·24. Find (i) the magnitude and direction of the minimum (least) force that will pull the load up the incline, (ii) the magnitude of the force required to pull the load up if it is applied horizontally.

*f*13. A cotter is driven into a plug and socket connection by a force of 500 N. The cotter has a taper of 1 in 10 equally divided between the two edges, and the coefficient of friction is 0·18. Find (i) the force holding the plug and socket together, (ii) the force required to drive the cotter out.

*f*14. A propeller wedge is tapered on one side and flat on the other, the taper being 1 in 40. Find the force driving the propeller off its shaft when the wedge is driven in with a force of 1·5 kN, the coefficient of friction is to be taken as 0·2.

*f*15. A single start square thread has a mean diameter of 50 mm and a pitch of 12·5 mm. The coefficient of friction between the screw and nut is 0·15. Find the efficiency of the thread when lifting a load of 4·5 kN and the torque required.

CHAPTER 7

MOMENTS

FIRST MOMENTS. The moment of a force is its effective turning effect and depends upon the magnitude of the force and its leverage. The effective leverage is the *perpendicular* distance from the line of action of the force to the point about which turning takes place or tends to take place. It is expressed as the moment about a given point in Nm.

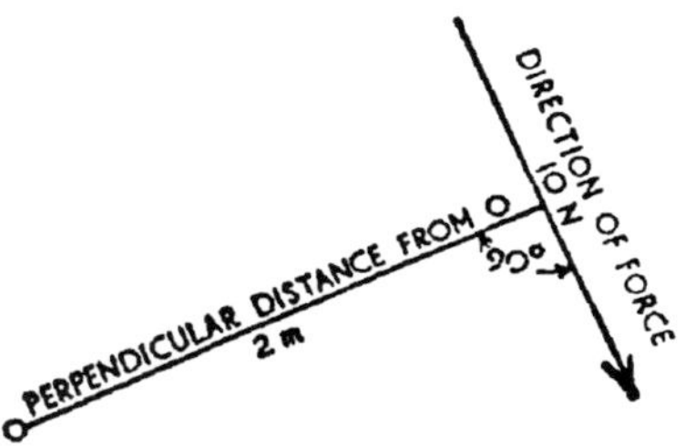

Fig. 96

Fig. 96 illustrates a force of 10 N, the perpendicular distance of the line of action of this force from the turning point *o* is 2 m, the moment is therefore expressed as 20 N m about point *o*. The direction of a moment is usually stated as being either *clockwise* or *anticlockwise*. In this case a clockwise moment around *o*.

Consider a bar of negligible weight carrying downward forces of 60 N at one end and 40 N at the other end, the distance between the forces being 0·5 m. This bar is resting on a knife-edge support at 0·2 m from the 60 N force, as illustrated in Fig. 97.

The bar is free to swing about the support as though it were hinged at one point, a single support such as this is referred to as a *fulcrum*. As the bar can swing about the fulcrum it is natural to measure the turning effects or moments of the forces about that point.

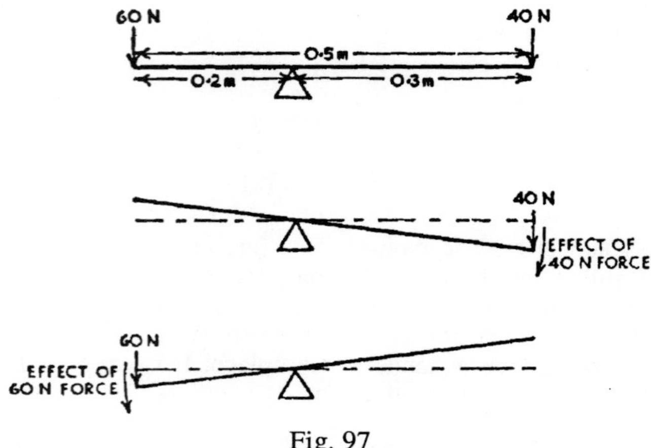

Fig. 97

The 40 N force is tending to turn the bar around in a clockwise direction, hence its moment is clockwise, and its effective leverage is the perpendicular distance from its line of action to the fulcrum, which is 0·3 m. It therefore exerts a clockwise turning moment of $40 \times 0{\cdot}3 = 12$ Nm.

At the other end there is the 60 N force at a perpendicular distance of 0·2 m from the fulcrum, this exerts an anticlockwise turning moment of $60 \times 0{\cdot}2 = 12$ Nm.

It is obvious in this example that the bar will not turn because one moment balances the other, therefore the bar is in equilibrium.

It is always true that, for equilibrium, the sum of the moments acting in a clockwise direction must be equal to the sum of the moments acting in an anticlockwise direction.

Further, the total load acting downwards is $60 + 40 = 100$ N, therefore the total of the upward supporting forces must also be 100 N, in this case there is only the one support and it must exert the whole of this force. Had there been any force on the bar pushing horizontally to the right or to the left, a force of equal magnitude and opposite direction would need to be applied by a stopper to prevent the bar sliding.

The **conditions of equilibrium** of a body at rest under the action of a number of forces, are therefore stated as:

Total upward forces = Total downward forces
Total forces pushing to the right = Total forces pushing to the left
Total clockwise moments = Total anticlockwise moments

Moments and forces may be given an algebraic sign. For example, clockwise moments could be designated positive and anticlockwise moments negative, downward forces could be positive and upward forces negative, forces to the right could be positive and forces to the left negative. By doing this, if positive and negative values of equal magnitude are added together, the answer is zero, therefore the conditions of equilibrium can be and are often stated in mathematical terms as:

The algebraic sum of all vertical and all horizontal forces must be zero.

The algebraic sum of all moments of forces about any given point must be zero.

Moments can be measured from any chosen point. If there is the equivalent of a hinge or pin-joint in the system, it is usual to first consider if that point is the most suitable when making the choice. If there is no actual swivel, a point in the system is chosen as an imaginary hinge at a position which would appear to provide the simplest calculations.

Example. A beam, 3 m long, rests on a support at each end. Neglecting the weight of the beam, find the upward force exerted by each support when the beam carries downward loads of 200, 100 and 120 N at distances of 0·75, 1·5 and 2·5 m respectively from the left end.

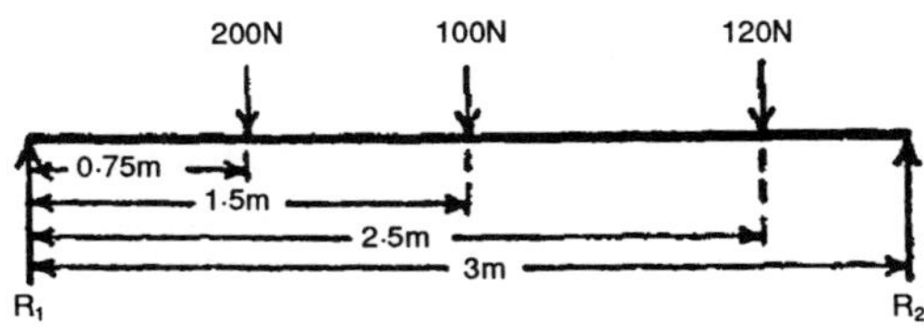

Fig. 98

The supports of a beam are usually referred to as the *reactions* and represented by R_1 and R_2 and a diagrammatic sketch such as shown in Fig. 98 is used to illustrate the problem.

Knowing that for equilibrium, clockwise moments must equal anticlockwise moments, an equation can be designed to include the required unknown. Only one unknown can be solved by one equation, therefore it must contain either R_1 or R_2 but not both. This is done by taking moments about one of them. For example, if moments are taken about R_1 this reaction will have no turning moment about itself because its distance is zero, and therefore will not appear in the equation.

To take moments about R_1 imagine that the beam is hinged at that point and can turn about this hinge in a clockwise or anticlockwise direction. The loads tend to turn the beam around clockwise, and the upward force of the reaction R_2 tends to turn it around anticlockwise.

Taking moments about R_1:

$$\text{Clockwise moments} = \text{Anticlockwise moments}$$
$$(200 \times 0{\cdot}75) + (100 \times 1{\cdot}5) + (120 \times 2{\cdot}5) = R_2 \times 3$$
$$150 + 150 + 300 = R_2 \times 3$$
$$600 = R_2 \times 3$$
$$R_2 = 200 \text{ N}$$

Now to find the value of R_1 take moments about R_2 and proceed in a similar manner to the above. However, as one of the reactions is now known it is simpler to apply the rule of vertical forces:

$$\text{Upward forces} = \text{Downward forces}$$
$$R_1 + R_1 = 200 + 100 + 120$$
$$R_1 + 200 = 420$$
$$R_1 = 420 - 200$$
$$= 220 \text{ N}$$

Supporting forces are therefore:

220 N and 200 N respectively. Ans.

When the loads are not perpendicular to the beam, either one of two methods can be used to take moments, (i) by finding the perpendicular distance from the line of action of each inclined load to the chosen point, (ii) by resolving the inclined loads into their vertical and horizontal components and treating these components as actual loads.

Example. A horizontal lever 5 m long is hinged at one end and simply supported by a vertical prop at the other end. It carries a load of 150 N inclined at 67° and another of 50 N inclined at 43° to the horizontal in the directions and at the positions shown in Fig. 99. Neglecting the weight of the lever, find the load on the prop.

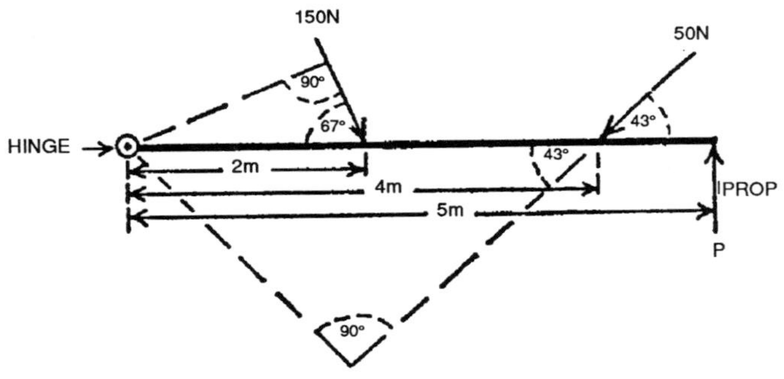

Fig. 99

Taking moments about the hinge,

Perpendicular distance from line of action of the 50 N force to hinge = $4 \times \sin 43° = 2{\cdot}728$ m.

Perpendicular distance from line of action of the 150 N force to hinge $= 2 \times \sin 67° = 1{\cdot}841$ m.

Perpendicular distance from line of action of the prop force to hinge = 5 m.

$$\begin{aligned}
\text{Clockwise moments} &= \text{Anticlockwise moments} \\
(150 \times 1{\cdot}841) + (50 \times 2{\cdot}728) &= P \times 5 \\
276{\cdot}15 + 136{\cdot}4 &= P \times 5 \\
412{\cdot}55 &= P \times 5 \\
P &= 82{\cdot}51 \text{ N} \quad \text{Ans.}
\end{aligned}$$

As stated, the alternative method is to resolve the inclined loads into their vertical and horizontal components and replace the actual loads by them. Most often it provides a simpler solution, particularly if the load carried by the hinge is also required. This is demonstrated in the following.

Example. A horizontal beam 4 m long is hinged at one end and simply supported at the other end, and carries inclined loads as shown in Fig. 100. Neglecting the weight of the beam, find the reactions at each end.

Resolving the two inclined forces into their vertical and horizontal components and replacing the actual forces by these components as in Figs. 101 and 102:

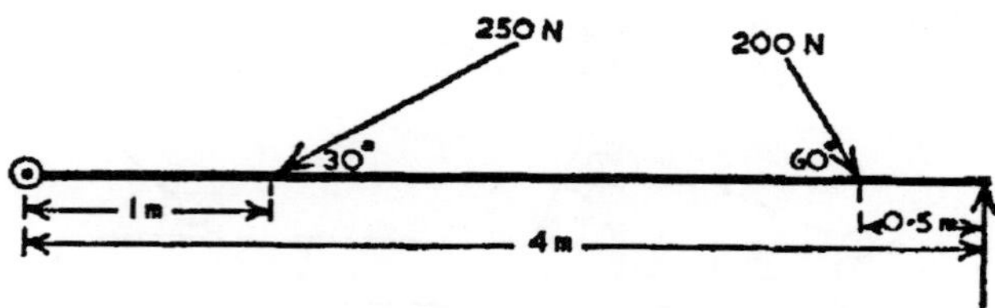

Fig. 100

Vertical component of 250 N force = 250 sin 30° = 125 N down
Horizontal " " " " = 250 cos 30°=216·5 N to left
Vertical component of 200 N force = 200 sin 60°= 173·2 N down
Horizontal " " " " = 200 cos 60°=100 N to right

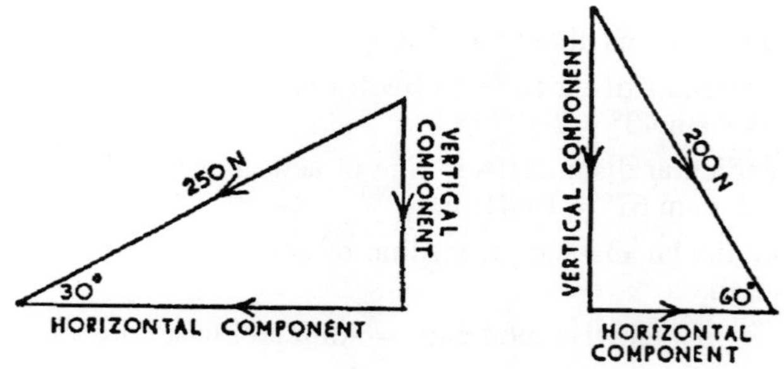

Fig. 101

Taking moments about R_1.

Note that the perpendicular distances of the vertical component forces are the horizontal distances from R_1 to the points of application of the forces. The lines of action of the horizontal component forces pass through R_1, thus their perpendicular

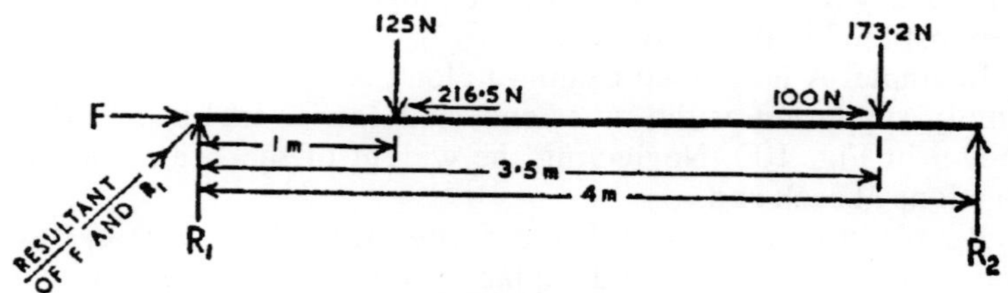

Fig. 102

distances are zero, they have no turning effect about R_1.and will not appear in the moment equation.

$$\begin{aligned} \text{Clockwise moments} &= \text{Anticlockwise moments} \\ (125 \times 1) + (173{\cdot}2 \times 3{\cdot}5) &= R_2 \times 4 \\ 125 + 606{\cdot}2 &= R_2 \times 4 \\ 731{\cdot}2 &= R_2 \times 4 \\ R_2 &= 182{\cdot}8 \text{ N} \end{aligned}$$

$$\begin{aligned} \text{Upward forces} &= \text{Downward forces} \\ R_1 + R_2 &= 125 + 173{\cdot}2 \\ R_1 &= 298{\cdot}2 - 182{\cdot}8 \\ &= 115{\cdot}4 \text{ N} \end{aligned}$$

The horizontal component forces of 216·5 N pushing to the left, and 100 N to the right, requires a force to the right to balance the difference. Let this be F.

$$\begin{aligned} \text{Forces to the right} &= \text{Forces to the left} \\ F + 100 &= 216{\cdot}5 \\ F &= 116{\cdot}5 \text{ N to the right.} \end{aligned}$$

The hinge therefore applies a force which has a vertical upward effect of 115·4 N and also a horizontal push to the right of 116·5 N, these are the rectangular components of the resultant force applied by the hinge. Referring to Fig. 103,

$$\begin{aligned} \text{Resultant} &= \sqrt{115{\cdot}4^2 + 116{\cdot}5^2} \\ &= 164 \text{ N} \\ \tan\theta &= \frac{116{\cdot}5}{115{\cdot}4} = 1{\cdot}009, \ \theta = 45° \ 15' \end{aligned}$$

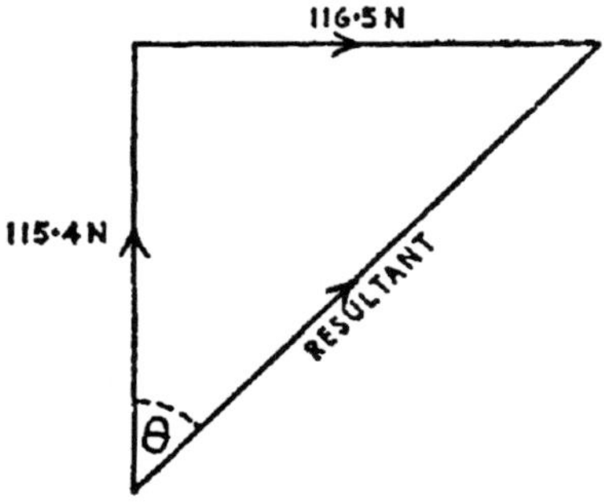

Fig. 103

$\therefore$ Right hand reaction = 182·8 N vertically upwards.
Reaction of hinge = 164 N at 45° 15′ to the vertical. } Ans.

TAKING WEIGHT OF BEAM INTO CONSIDERATION. When the weight of the beam is given, or can be calculated from its dimensions, it is treated as a single load acting through the position of its centre of gravity. For beams which are of uniform section throughout the entire length, the centre of gravity is at mid-length.

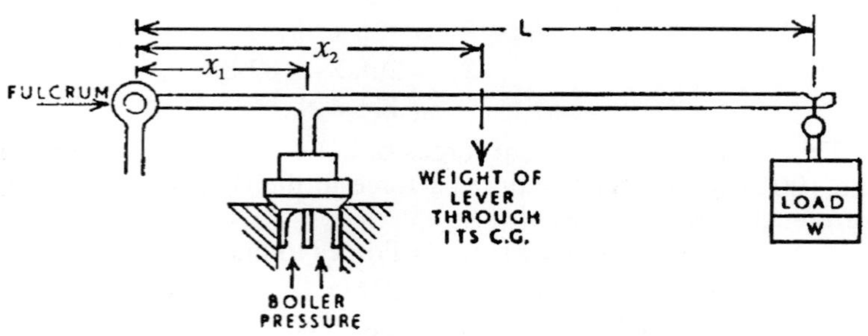

Fig. 104

Example. A lever safety valve is illustrated diagramatically in Fig. 104. The length of the lever is 0·75 m, it is fulcrumed at one end and loaded at the other, the weight of the lever is 29 N and its centre of gravity is at 0·3 m from the fulcrum. The diameter of the valve is 63·5 mm, it weighs 10 N and its centre is 0·1 m from the fulcrum. Calculate the load to be hung on the end so that the valve will lift when the boiler pressure is 8·25 bar (= $8{\cdot}25 \times 10^5$ N/m²).

Moments about the fulcrum,

$$\text{Clockwise moments} = \text{Anticlockwise moments}$$

$$(10 \times 0{\cdot}1) + (29 \times 0{\cdot}3) + (W \times 0{\cdot}75) = 8{\cdot}25 \times 10^5 \times 0{\cdot}7854 \times 0{\cdot}0635^2 \times 0{\cdot}1$$

$$1 + 8{\cdot}7 + 0{\cdot}75\,W = 261{\cdot}3$$

$$0{\cdot}75\,W = 251{\cdot}6$$

$$W = 335{\cdot}5 \text{ N Ans.}$$

COUPLE

A couple is the name given to a pair of parallel forces of equal magnitude which act in opposite directions and constitute a turning moment (see Chapter 5 and Fig. 70).

The perpendicular distance between the lines of action of the two forces is usually called the *arm* of the couple. If this is represented by L as illustrated in Fig. 105 and the magnitude of each force by F, then by taking moments about either force the turning moment is $F \times L$. If equal and opposite forces F are applied diametrically opposite on a shaft of radius r then the couple equals the torque ie. $T = 2\ F\ r$.

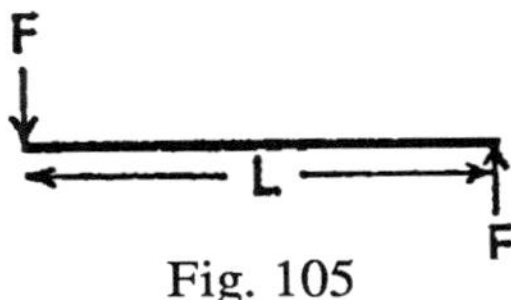

Fig. 105

CENTRE OF GRAVITY

The centre of gravity (C.G.) of a mass is that point through which the whole weight of the mass may be considered as acting. if it is imagined that a body could be compressed in volume into a tiny particle without losing mass, the position of this small heavy particle would be at the centre of gravity of the body to have the same effect. Often referred to as *centre of mass*.

In the case of a plane area which has no mass and therefore no force of gravity acting on it, the term *centroid* would be used instead of centre of gravity.

Centroid or C.G. of parallelogram = intersection of diagonals,

triangle = ⅓ height from base,

pyramid or cone = ¼ height from base,

semi-circular area = $\dfrac{4r}{3\pi}$ from diameter,

hemisphere = ⅜ r from diameter.

The centre of gravity of a system of loads can be found by taking moments. Consider a weightless bar carrying a number of loads w_1, w_2, w_3, etc., at x_1, x_2, x_3, etc., respectively from one end,as

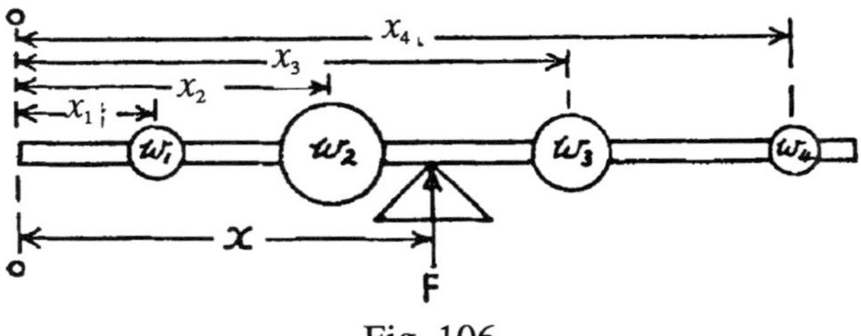

Fig. 106

illustrated in Fig. 106. If this bar is to balance on one support, that support must be placed exactly under the position of the centre of gravity of the system.

Let the required position of the support for perfect balance be at x from the end.

Taking moments about left end, oo,

$$\text{Clockwise moments} = \text{Anticlockwise moments}$$

$$(w_1 \times x_1) + (w_2 \times x_2) + (w_3 \times x_3) \text{ etc.} = F \times x$$

The magnitude of the upward supporting force F must be equal to the total downward weight, therefore,

$$(w_1 \times x_1) + (w_2 \times x_2) + (w_3 \times x_3) = (w_1 + w_2 + w_3) \times x$$

$$x = \frac{(w_1 \times x_1) + (w_2 \times x_2) + (w_3 \times x_3)}{w_1+w_2+w_3}$$

The numerator of this fraction is the 'summation of the moments of the weights', and the denominator is the 'summation of the weights'.

$$x = \frac{\Sigma \text{ moments of weights}}{\Sigma \text{ weights}}$$

Many cases arise where the system is all composed of similar material, therefore, since weight = volume × specific weight, then if the specific weight of the material is the same for all parts it can be cancelled from every term, leaving volume in the place of weight. The above expression will then be modified to:

$$x = \frac{\Sigma \text{ moments of volumes}}{\Sigma \text{ volumes}}$$

For plates and sections where the thickness is uniform throughout as well as the specific weight of the material, since volume = area × thickness, cancel thickness out of every term, leaving only area in the place of weight, thus,

$$x = \frac{\Sigma \text{ moments of areas}}{\Sigma \text{ areas}}$$

The following examples demonstrate the use of the above expressions. $\overline{x}$ horizontally and $\overline{y}$ vertically are used to locate the C.G.

Example. The pin of a fork-joint is 50 mm diameter, the shank is 100 mm long, and the head is a hemisphere 72 mm diameter. A

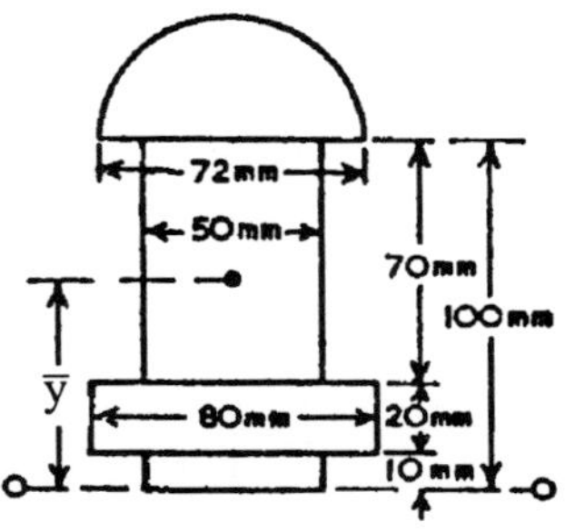

Fig. 107

circular ring-washer 80 mm diameter and 20 mm thick is fitted on the shank in such a position as to leave 70 mm between head and washer. The whole is of similar material. Calculate the position of the centre of gravity of the assembly measured from the end of the shank.

Working in cm:

$$\text{Volume of head} = \tfrac{1}{2} \times \frac{\pi}{6} \times 7{\cdot}2^3 = \pi \times 31{\cdot}1 \text{ cm}^3$$

$$\text{C.G. of head from } oo = \tfrac{3}{8} \times 3{\cdot}6 + 10 = 11{\cdot}35 \text{ cm}$$

$$\text{Volume of shank} = \frac{\pi}{4} \times 5^2 \times 10 = \pi \times 62{\cdot}5 \text{ cm}^3$$

$$\text{C.G. of shank from } oo = \tfrac{1}{2} \times 10 = 5 \text{ cm}$$

$$\text{Volume of washer} = \frac{\pi}{4}(8^2 - 5^2) \times 2 = \pi \times 19{\cdot}5 \text{ cm}^2$$

$$\text{C.G. of washer from } oo = \tfrac{1}{2} \times 2 + 1 = 2 \text{ cm}$$

$$\bar{y} = \frac{\Sigma \text{ moments of volumes}}{\Sigma \text{ volumes}}$$

$$= \frac{\pi \times 31{\cdot}1 \times 11{\cdot}35 + \pi \times 62{\cdot}5 \times 5 + \pi \times 19{\cdot}5 \times 2}{\pi \times 31{\cdot}1 + \pi \times 62{\cdot}5 + \pi \times 19{\cdot}5}$$

$$\bar{y} = \frac{353 + 312{\cdot}5 + 39}{31{\cdot}1 + 62{\cdot}5 + 19{\cdot}5}$$

$$= \frac{704{\cdot}5}{113{\cdot}1} = 6{\cdot}229 \text{ cm}$$

$$= 62{\cdot}29 \text{ m from end. Ans.}$$

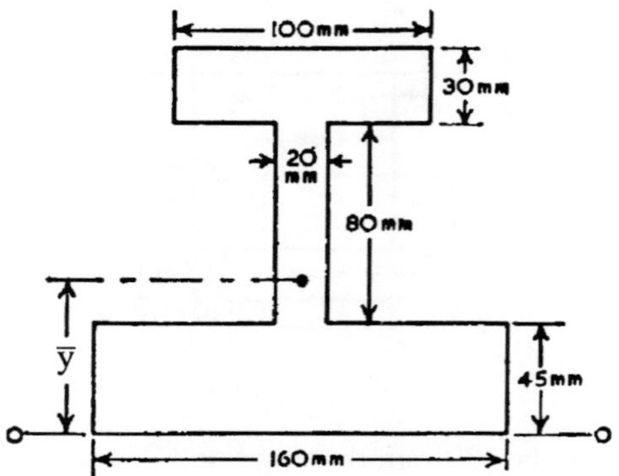

Fig. 108

Example. A steel plate of uniform thickness is composed of three rectangles and its dimensions are given in Fig. 108. Find the position of its centre of gravity from the bottom edge *oo*.

Working in centimetres:

Area of bottom rectangle $= 16 \times 4{\cdot}5 = 72$ cm^2

C.G. " " = from *oo* = 2·25 cm

Area of middle rectangle $= 8 \times 2 = 16$ cm^2

C.G. " " = from *oo* = 8·5 cm

Area of top rectangle $= 10 \times 3 = 30$ cm^2

C.G. " " = from *oo* = 14 cm

The figure is symmetrical, therefore the C.G. lies on its centre line, and it is only necessary to take moments in the one direction to express the position of its centre of gravity.

$$\bar{y} = \frac{\Sigma \text{ moments of areas}}{\Sigma \text{ areas}}$$

$$= \frac{72 \times 2{\cdot}25 + 16 \times 8{\cdot}5 + 30 \times 14}{72 + 16 + 30}$$

$$= \frac{718}{118} = 6{\cdot}085 \text{ cm}$$

$= 60{\cdot}85$ mm from base. Ans.

The position of the centre of gravity can be verified by hanging the figure from various points, say from one corner at a time, in conjunction with a plumb-line suspended from the same point. The plumb-line will pass through the centre of gravity in accordance with the principle of moments for conditions of equilibrium, because the sum of the moments on one side of the line is equal to the sum of the moments on the other side. Thus the intersection of two or more plumb-line marks struck on to the figure will give the position of the centre of gravity.

Example. A circular plate of uniform thickness is 180 mm diameter and has two circular holes cut out of it. One hole is 30 mm diameter, its centre being at 60 mm diameter from the plate centre, and the other hole is 60 mm diameter, its centre being at 50 mm from the plate centre. The centres of the two holes and the plate lie at the corners of a right-angled triangle. Find the position of the centre of gravity.

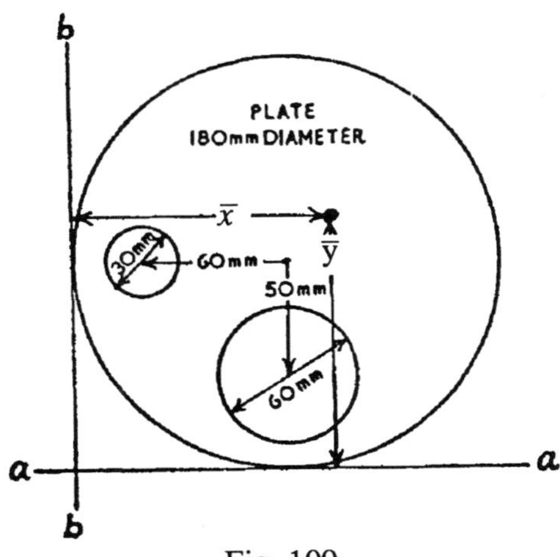

Fig. 109

As this figure is not symmetrical about a centre line it is necessary to take moments in two directions to express the position of the centre of gravity, say at $\bar{y}$ from *aa*, and at $\bar{x}$ from *bb*. Since area is lost when the holes are cut out, then moment of area is lost, these are therefore subtracted when obtaining the summation of moments of areas and summation of areas.

Firstly, taking moments about *aa*, working in cm,

$$\bar{y} = \frac{\Sigma \text{ moments of areas}}{\Sigma \text{ areas}}$$

$$= \frac{0{\cdot}7854 \times 18^2 \times 9 - 0{\cdot}7854 \times 6^2 \times 4 - 0{\cdot}7854 \times 3^2 \times 9}{0{\cdot}7854 \times 18^2 - 0{\cdot}7854 \times 6^2 - 0{\cdot}7854 \times 3^2}$$

$$= \frac{6^2 \times 9 - 2^2 \times 4 - 1 \times 9}{6^2 - 2^2 - 1}$$

$$= \frac{299}{31} = 9{\cdot}645 \text{ cm} = 96{\cdot}45 \text{ mm from } aa \text{ Ans. (i)}$$

Moments about *bb*,

$$\bar{x} = \frac{0{\cdot}7854 \times 18^2 \times 9 - 0{\cdot}7854 \times 6^2 \times 9 - 0{\cdot}7854 \times 3^2 \times 3}{0{\cdot}7854 \times 18^2 - 0{\cdot}7854 \times 6^2 - 0{\cdot}7854 \times 3^2}$$

$$= \frac{6^2 \times 9 - 2^2 \times 9 - 1 \times 3}{6^2 - 2^2 - 1}$$

$$= \frac{285}{31} = 9{\cdot}194 \text{ cm} = 91{\cdot}94 \text{ mm from } bb \text{ Ans. (ii)}$$

If required the position from the centre of the plate can be stated as 6·45 mm in the one direction and 1·94 mm in the other.

The direct distance from centre of plate could be given. Referring to Fig. 110 this is,

$$\sqrt{6{\cdot}45^2 + 1{\cdot}94^2} = 6{\cdot}736 \text{ mm}$$

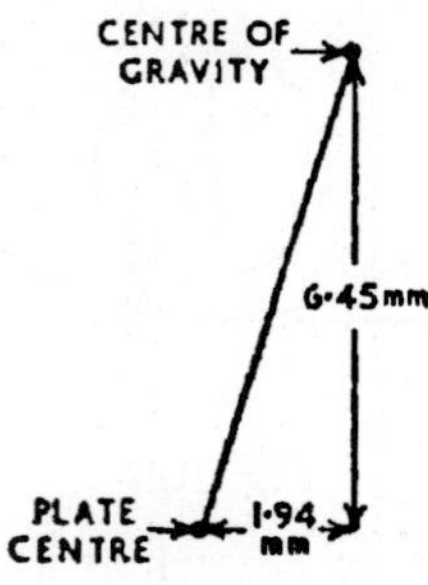

Fig. 110

CENTROID OF IRREGULAR AREA

Simpson's rule can be employed to find the moment of an irregular area in a similar manner to which it is used to finding the area.

The figure is divided into an even number of equi-distant divisions this gives an odd number of ordinates which are measured, the distance between the ordinates is termed the common interval.

TO FIND THE AREA. Add together the first ordinate, last ordinate, four times the even ordinates and twice the odd ordinates; multiply this sum by one-third of the common interval.

TO FIND THE MOMENT OF THE AREA. To express the moment about a given point,the perpendicular distance of each ordinate is measured from that point, then: Add together the moment of the first ordinate, moment of the last ordinate, four times the moments of the even ordinates and twice the moments of the odd ordinates; multiply this sum by one-third of the common interval.

TO FIND THE CENTROID. Divide the moment of the area by the area.

As an example, consider a regular triangle of base 240 mm and perpendicular height 360 mm. The area is ½ (base × perp. × height) and the centroid is at one-third of the height from the base; compare to the results produced by Simpson's rule.

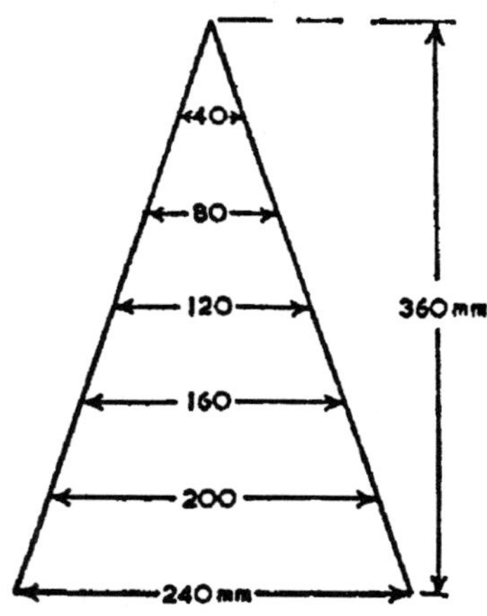

Fig. 111

Measurements from base, in cm:

i	ii	iii	iv	v
ORDINATES	SIMPSON'S MULTIPLIERS	PRODUCTS OF i & ii	DISTANCES OF ORDINATES FROM BASE	PRODUCTS OF iii & iv
24	1	24	0	0
20	4	80	6	480
16	2	32	12	384
12	4	48	18	864
8	2	16	24	384
4	4	16	30	480
0	1	0	36	0
		Sum = 216		Sum = 2592

$$\text{Common interval} = 36 \div 6 = 6 \text{ cm}$$

$$\text{Area} = 216 \times \tfrac{1}{3} \times 6 = 432 \text{ cm}^2$$

Moment of area about base

$$= 2592 \times \tfrac{1}{3} \times 6 = 5184 \text{ cm}^3$$

$$\text{Centroid from base} = \frac{\text{moment of area}}{\text{area}}$$

$$= \frac{5184}{432} = 12 \text{ cm}$$

It will be seen that the position of the centroid can be obtained by dividing the sum of column v by the sum of column iii, thus,

$$\frac{2892}{216} = 12 \text{ cm}$$

The above agrees with:

$$\text{Area} = \tfrac{1}{2} \text{ base} \times \text{perpendicular height}$$

$$= \tfrac{1}{2} \times 24 \times 36 = 432 \text{ cm}^2$$

$$\text{Centroid} = \tfrac{1}{3} \text{ of perpendicular height}$$

$$= \tfrac{1}{3} \times 36 = 12 \text{ cm}$$

SECOND MOMENTS

With *first moments* the force, volume or area is multiplied by the *first* power of the perpendicular distance from a given point or axis.

Second moments are so named because of the use of the *second* power of the distance, that is $(\text{distance})^2$. These have been considered earlier (chapter 3) as Moments of Inertia (Second Moment of Mass) involving Radius of Gyration i.e. $I = mk^2$.

Thus, referring to Fig. 112, an element of area a at a distance of h from the axis xx, the first moment of the area is ah, and the second moment is ah^2.

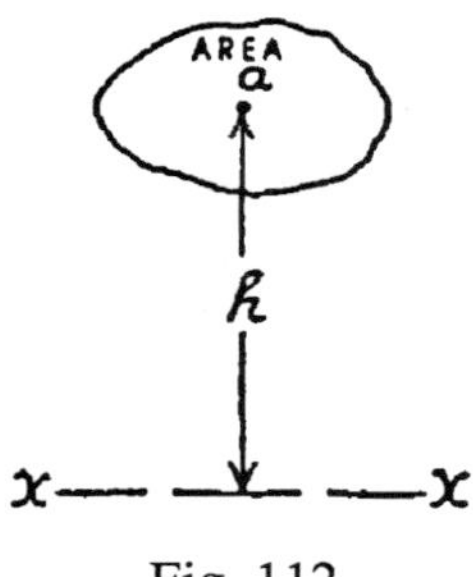

Fig. 112

Second moments of areas are used in design calculations, particularly in strength of beams and shafts. They are usually denoted by I with a suffix denoting the axis from which moments are taken. For instance in Fig. 112, $I_{xx} = ah^2$, and, if the units of area are m^2 and the distance h is in m, then the units for I are $m^2 \times m^2 = m^4$.

The second moment of a mass is termed the moment of inertia. If m represents the mass and h its distance, then the moment of inertia is $I_{xx} = mh^2$.

If an area is divided up into small elements, each element multiplied by the square of its distance from a given axis, then the summation of these products of elements of area and their $(\text{distance})^2$ gives the second moment of the whole area.

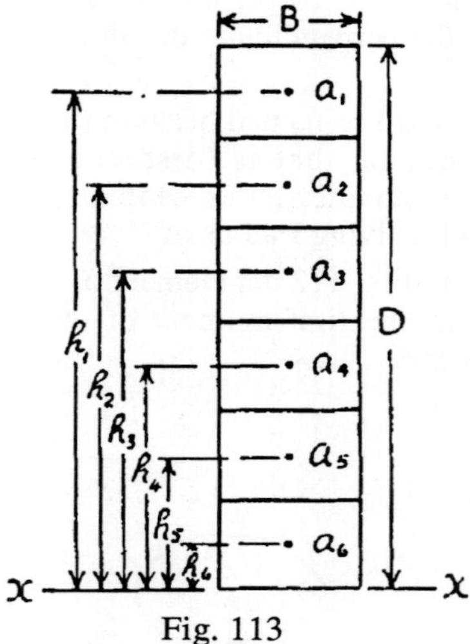

Fig. 113

Taking a rectangular area of breadth B and depth D and dividing it into strips as shown in Fig. 113,

$$Ixx = a_1h_1^2 + a_2h_2^2 + a_3h_3^2 + a_4h_4^2 + a_5h_5^2 + a_6h_6^2$$
$$= \sum ah^2$$

The smaller the strips the nearer to the exact value will the result be, this is

$$I_{base} = \frac{BD^3}{3}$$

This can be verified by integration *i.e.* consider an element of area dh deep, B wide, distance h from xx:

$$I_{xx} = \int_0^D Bh^2\,dh = \left[\frac{Bh^3}{3}\right]_0^D$$

$$I_{base} = \frac{BD^3}{3}$$

Solid rectangular section,

$$I_{CG} = \frac{BD^3}{12}$$

Hollow rectangular section or equivalent such as channel or I section as shown in Fig. 114, all of which are symmetrical about their horizontal centre-lines,

$$I_{CG} = \frac{BD^3 - bd^3}{12}$$

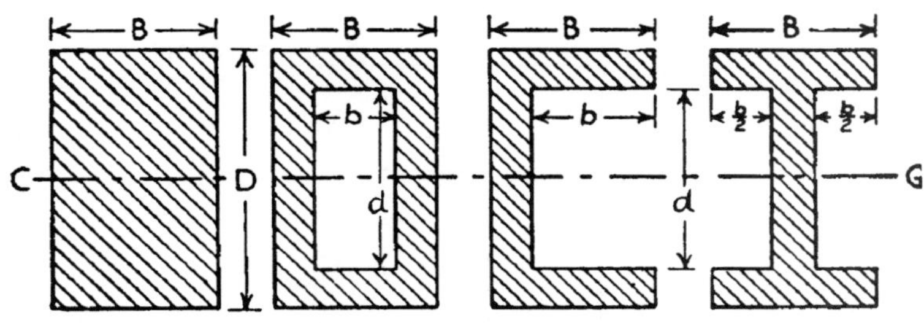

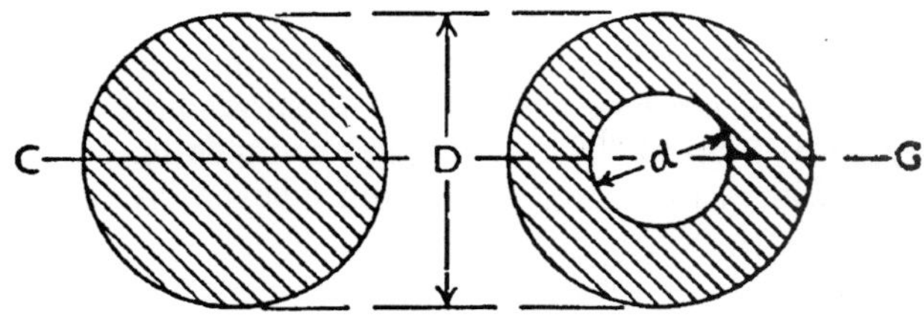

Fig. 114

Solid circular section,

$$I_{\text{dia}} = \frac{\pi}{64} D^4 \text{ or } \frac{\pi}{4} R^4$$

Hollow circular section,

$$I_{\text{dia}} = \frac{\pi}{64} (D^4 - d^4) \text{ or } \frac{\pi}{4} (R^4 - r^4)$$

These can all be verified by integration.

f THEORUM OF PARALLEL AXES

It is often required to find the second moment about an axis parallel to and at some distance from the axis passing through the centroid of the area, or vice-versa.

Let I_{CG} represent the second moment of an area (A) about an axis passing through its centroid.

Let I_{xx} represent the second moment of this same area about an axis xx parallel to that passing through the centroid and at distance h from it (see Fig. 115).

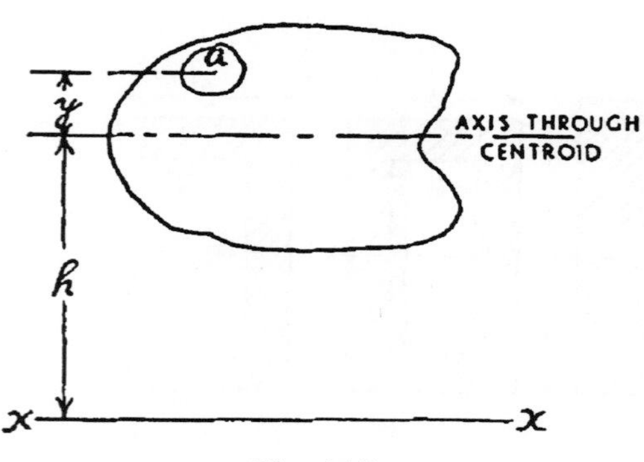

Fig. 115

Consider a small element of the area represented by a at a perpendicular distance of y from the axis through the centroid, that is, at $(y + h)$ from xx. The whole area is made up of many such elements of area, a_1, a_2, a_3, etc., at distances y_1, y_2, y_3, etc., from the centroid axis.

$$I \text{ of element about } xx = \text{area} \times \text{distance}^2$$
$$= a \times (y + h)^2$$

$$I_{xx} \text{ of whole area} = a_1 (y_1 + h)^2 + a_2 (y_2 + h)^2 + a_3 (y_3 + h)^2 + \text{etc.}$$
$$= \Sigma\, a\, (y + h)^2$$
$$= \Sigma\, a\, (y^2 + 2yh + h^2)$$
$$= \Sigma\, ay^2 + \Sigma\, 2ayh + \Sigma\, ah^2$$

Of these three terms:

$\Sigma\, ay^2$ is the summation of the products of each element of area and the square of their distances from the axis through the centroid, it is the second moment of the whole area about that axis, thus,

$$\Sigma\, ay^2 \text{ is written } I_{CG} \text{ or } I_G$$

The second term, $\Sigma 2ayh$ is equal to $2h\ \Sigma ay$ because 2 and h are constants. Σay is the summation of the first moments of area about the axis through the centroid, which, according to the principle of moments is equal to zero.

$$2h\,\Sigma ay = 2h \times 0 = 0$$

The third term, $\Sigma\, ah^2 = h^2\, \Sigma a = h^2 \times$ whole area, and can be written $A \times h^2$

$$\text{Hence, } I_{xx} = I_{CG} + Ah^2$$

*f*Example. The second moment of a triangle about an axis passing through its centroid and parallel to the base, is $\dfrac{BH^3}{36}$ where B is the base and H is the perpendicular height. Find the second moment of the triangle about its base.

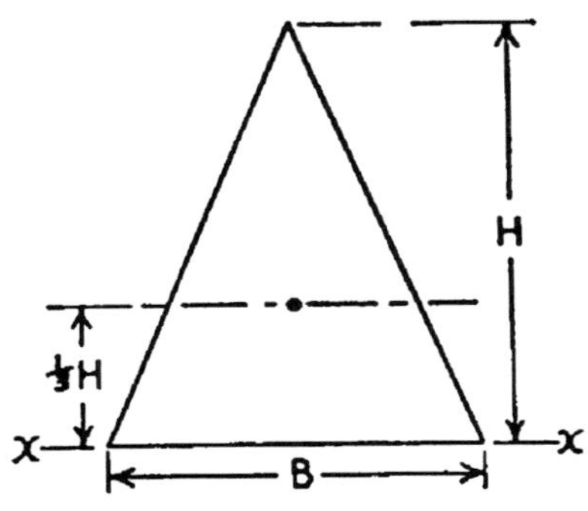

Fig. 116

$$I_{xx} = I_{base} = I_{CG} + Ah^2$$
$$= \frac{BH^3}{36} + \frac{BH}{2} \times \left\{\frac{H}{3}\right\}^2$$
$$= \frac{BH^3}{36} + \frac{BH^3}{18}$$
$$= \frac{BH^3}{12}$$

POLAR SECOND MOMENTS

When moments are taken about a given axis what is in mind is an imaginary hinge along that axis.

For instance, referring to Fig. 117, the second moment of the element of area a about the axis xx when its distance is y is, $I_{xx} = ay^2$, when taken about the axis yy (Fig. 118) its distance being x, the second moment about that axis is, $I_{yy} = ax^2$. If an axis is at right angles to both xx and yy, that is like sticking a pole or rod at right angles to the paper, this is termed the *polar axis*, and if second moments are taken about it when its distance is z from the pole, it is termed the polar second moment and written Ip or $J = az^2$.

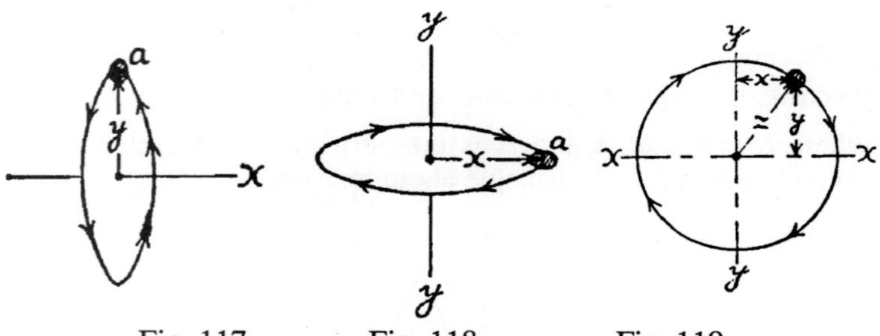

Fig. 117 Fig. 118 Fig. 119

From Fig. 119, $z^2 = x^2 + y^2$.

If each term is multiplied by a, we have,

$$az^2 = ax^2 + ay^2$$

which gives the relationship between the polar 2nd moment and the other two at right angles to it *i.e.* the *theorem of polar (perpendicular) axes* is:

$$J = I_{xx} + I_{yy}$$

As a further illustration, if a circular disc of diameter D is now considered to work like a butterfly valve or throttle across its horizontal diameter xx (Fig. 120) its second moment is

$$I_{xx} = \frac{\pi}{64} D^4$$

If it is considered to be hinged across its vertical diameter yy (Fig. 121) its second moment is

$$I_{yy} = \frac{\pi}{64} D^4$$

If it is now thought of as a spinning disc about its centre (Fig. 122) its polar second moment is

$$J = I_{xx} + I_{yy}$$

$$= \frac{\pi}{64} D^4 + \frac{\pi}{64} D^4$$

$$= \frac{\pi}{32} D^4$$

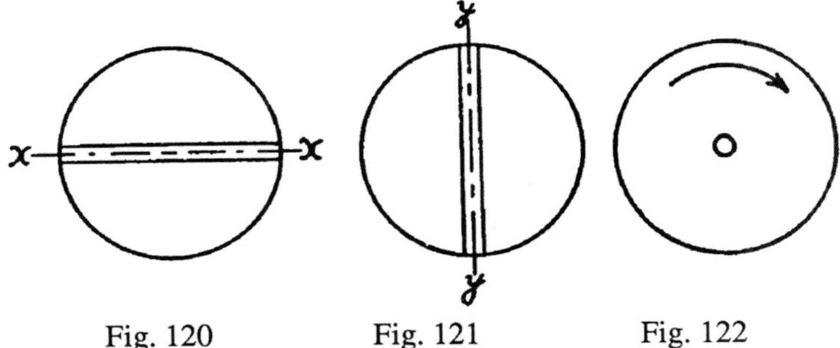

Fig. 120 Fig. 121 Fig. 122

The second moment of a circular area about its diameter is brought into the theory of bending of circular section beams. The polar second moment comes into the calculations involving the twisting of round shafts.

Example. Verify, by integration, that the polar second moment of a circular area is given by $\frac{\pi}{32} D^4$ and thus evaluate the second moment of a circular area about its diameter of 0·5 m.

Refer to Fig. 122 and consider an elemental annular ring of radial thickness δr at radius r.

Area of ring $= 2\,\pi\, r\, \delta r$

Second moment of area about polar axis (z)

$$= 2\,\pi\, r^3\, \delta r$$

$$\text{for whole, } J = \int_0^{D/2} 2\,\pi\, r^3\, \delta r$$

$$= \left[\frac{\pi\, r^4}{2}\right]_0^{D/2}$$

$$J = \frac{\pi\, D^4}{32} \quad \text{Ans.}$$

By the theorem of perpendicular axes:

$$J = I_{xx} + I_{yy}$$

$$I_{xx} = I_{yy} \text{ as symmetrical}$$

Second moment of area about diameter is ½ J

$$= \frac{\pi \times 0{\cdot}5^4}{64} = 0{\cdot}00307 \text{ m}^4 \quad \text{Ans.}$$

TEST EXAMPLES 7

1. A beam of uniform section is 15 m long and weighs 4 kN. It is simply supported at the left end and at 5 m from the right end, and carries loads of 2, 5 and 6 kN at 2, 5 and 12·5 m respectively from the left end. Find the reactions.

2. A connecting rod is 2·1 m long between its top and bottom end centres, its weight is 9·24 kN, and its centre of gravity is at 1·17 m from the top end. If it hangs freely from the top, find the least force required at the bottom end to pull it 0·7 m out of the vertical. Find also the force required if it is applied in a horizontal direction.

3. A solid hemisphere of mass 22·68 kg stands on its curved surface on a plane inclined at 20° to the horizontal, the friction of the surfaces being sufficient to prevent sliding. Calculate the downward vertical force required to apply at the periphery of its flat surface to make the hemisphere rest in a position with its flat surface horizontal.

4. An eccentric sheave of uniform thickness is 500 mm diameter, the hole for the shaft is 125 mm diameter and its centre is 100 mm from the centre of the sheave. Calculate the position of the centre of gravity of the sheave from its geometrical centre.

5. The shank of a plain steel bolt is 20 mm diameter, it has a round head 35 mm diameter and 20 mm thick, and the overall length of the bolt is 120 mm. A brass ring 30 mm diameter and 10 mm thick is forced on to the shank leaving a clearance of 70 mm between head and ring. Taking the densities of steel and brass as 7860 and 8400 kg/m^3 respectively, calculate the position of the centre of gravity from the end of the shank.

6. A segment is cut off a circular plate, the chord of the segment being 60 mm and the maximum width 20 mm. Find the diameter of the circle, draw the segment to scale, divide the segment into four equally spaced parts by lines parallel to the chord, measure the ordinates and find by Simpson's rule, (i) the area, (ii) the height of the centre of gravity from the chord.

7. A steel plate of uniform thickness is in the form of a trapezium ABCD with a circular hole cut out of it, and lies flat on a horizontal table. Sides AD and BC are parallel to each other and at right angles to AB. The dimensions of the plate are, AB = 90 mm, BC = 70 mm, and AD = 40 mm. The hole is 40 mm diameter and its centre is 30 mm from AB and 40 mm from BC. Find (i) the

position of the centre of gravity of the plate measured from the sides AB and BC, and (ii) the initial force required at the corner C to tilt the plate about the edge AB if the net weight of the plate is 46·6 N

*f*8. The 2nd moment of a triangle about its base is $\dfrac{BH^3}{12}$

Find the 2nd moment of the triangle about an axis parallel to the base and passing through (i) its centroid, (ii) its apex.

*f*9. The overall depth of an I section girder is 230 mm, the top flange is 120 mm wide and 20 mm thick, the bottom flange is 160 mm wide and 30 mm thick, and the centre web is 15 mm thick. Given that the 2nd moment of a rectangular area about an axis passing through its centroid and parallel to the base is $\dfrac{BD^3}{12}$ calculate the 2nd moment of the I section (*a*) about its base, (*b*) about an axis through the centre of gravity and parallel to the base.

*f*10. Given that the 2nd moment of a rectangle about an axis through its centroid and parallel to the base is $\dfrac{BD^3}{12}$ find the polar 2nd moment of a square section of side S.

CHAPTER 8

LIFTING MACHINES

A lifting machine is a mechanism designed to lift heavy loads by comparatively small forces. The force applied is usually referred to as the *effort* and can be represented by P, the load lifted can be represented by W.

It is obvious that no more work could be got out of a machine than that which is put into it and, as no machine is perfect, a certain amount of work is lost in overcoming friction between moving parts, hence:

Work put into machine = Work lost in friction + Useful work done.

The work put into the machine is the product of the effort applied and the distance through which it moves. The useful work done by the machine is the product of the load and the distance it is lifted. If the magnitude of the effort is to be small compared with the amount of load lifted then the distance through which the effort moves must be great compared with the distance the load moves. The ratio of the distance moved by the effort to the distance moved by the load in the same time is termed the *velocity ratio*, and this is a constant value for any one particular machine depending upon its design.

$$\text{Velocity Ratio (V.R.)} = \frac{\text{Distance moved by effort}}{\text{Distance moved by load}}$$

The velocity ratio of any machine can be found experimentally simply by moving the machine and measuring the distances moved by the points of application of the effort and the load and dividing the former by the latter, or, by calculation from the relevant dimensions of the machine which will be shown later.

The advantage of employing a lifting machine is to lift a big load by small effort, therefore the term *mechanical advantage* is used to express this ratio, thus:

$$\text{Mechanical Advantage} = \frac{\text{Load lifted}}{\text{Effort applied}}$$

or, in symbols,

$$\text{M.A.} = \frac{W}{P}$$

The efficiency of any machine or engine is expressed by the ratio of the useful work done to the work supplied, then for a lifting machine we have:

$$\mathfrak{I} \text{ (Efficiency)} = \frac{\text{Useful work done}}{\text{Work supplied}}$$

$$= \frac{W \times \text{distance } W \text{ moves}}{P \times \text{distance } P \text{ moves}}$$

Now, $\frac{W}{P}$ is the mechanical advantage = M.A.

$$\text{and } \frac{\text{distance } W \text{ moves}}{\text{distance } P \text{ moves}} = \frac{1}{\text{velocity ratio}} = \frac{1}{\text{V.R.}}$$

$$\mathfrak{I} = \frac{\text{M.A.}}{\text{V.R.}}$$

The above expression gives the efficiency in fractional form, to express it as a percentage the fraction is multiplied by 100.

The *ideal effort* is the effort that would be required to lift a given load W if there were no friction. If there is a theoretically perfect frictionless machine where the efficiency is unity or 100%, the mechanical advantage would be equal to the velocity ratio, hence,

$$\text{Ideal Effort} = \frac{W}{\text{V.R.}}$$

and therefore,

$$\text{Effort to overcome friction} = \text{Actual effort} - \text{Ideal effort}$$

$$= P - \frac{W}{\text{V.R.}}$$

The *ideal load* is the load that would be lifted by a given effort if there were no friction, thus,

$$\text{Ideal Load} = P \times V.R.$$

$$\text{and, Load lost due to friction} = \text{Ideal load} - \text{Actual load}$$

$$= P \times V.R. - W$$

ROPE PULLEY BLOCKS

Rope pulley blocks consist of two pulley blocks, one at the top and one at the bottom, each carrying a number of pulleys free to run individually on a common axle. There may be an equal number of pulleys in each block or there may be one more pulley in one than in the other. A rope is threaded over each pulley in turn from top to bottom as shown in Fig. 123, the end of the rope

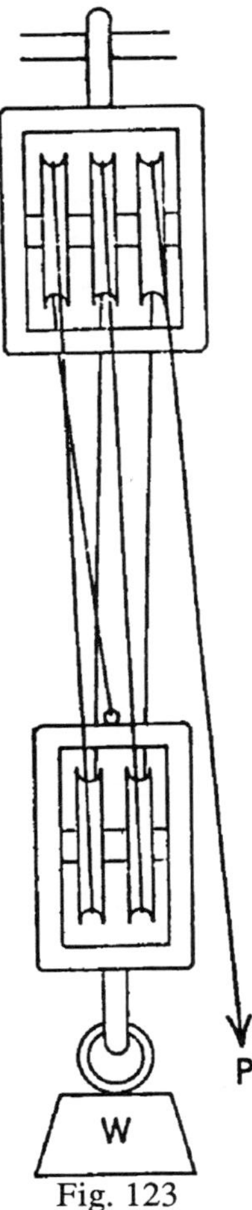

Fig. 123

being fastened to the block opposite the last pulley.

Imagine the load to be lifted 1 m, all the individual lengths of rope between top and bottom blocks must be shortened by one metre to remain taut and take their share in supporting the load, thus if there are 5 ropes connected to the load block as in Fig. 123

then 5 m of rope must be pulled away by the effort. Hence the velocity ratio is the ratio of 5 m of distance moved by the effort to the one metre of distance moved by the load, the velocity ratio of these rope pulley blocks is therefore 5. The same reasoning applies for any number of pulleys:

V.R. of rope pulley blocks = No. of ropes supporting load block

If the blocks are used in the normal way of pulling *down* to apply the effort when the load moves *up,* it will be found on examining different arrangements that the number of ropes supporting the load block is equal to the total number of pulleys in the system. For example, in a set of blocks with two pulleys in the top and one in the bottom, the velocity ratio is 3; for two pulleys in the top and two in the bottom the velocity ratio is 4, and so on.

Example. A set of rope pulley blocks has three pulleys in each block. Find the % efficiency when lifting a load of 448 N if the effort required is 90 N.

$$\text{Total number of pulleys} = 6$$

$$\text{velocity ratio} = 6$$

$$\text{Mechanical advantage} = \frac{\text{Load}}{\text{Effort}} = \frac{448}{90}$$

$$\%\mathfrak{I} = \frac{\text{M.A.}}{\text{V.R.}} \times 100 = \frac{448}{90 \times 6} \times 100 = 82{\cdot}96 \quad \text{Ans.}$$

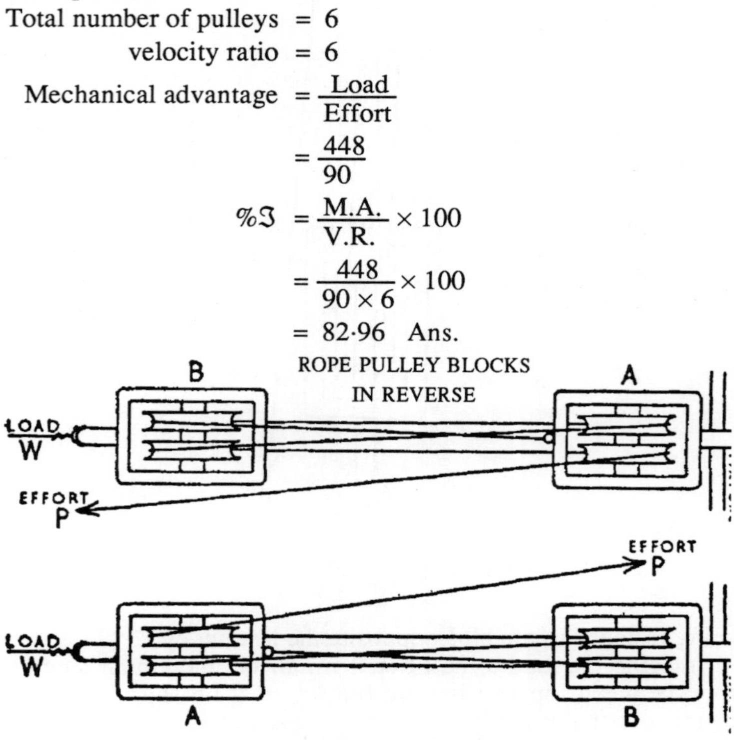

Fig. 124

Fig. 124 shows two ways in which a pair of rope pulley blocks can be used for dragging a load into position. These blocks have a total of four pulleys, the two blocks are marked A and B. In the first sketch, block A is hooked to a fixture and block B is coupled to the load, the arrangement is exactly as it would normally be used for lifting purposes, that is, the effort is applied in the *opposite* direction to that in which the load moves. There are four ropes on the block carrying the load, therefore the velocity ratio is 4.

In the second sketch the same set of blocks are reversed, block B is now hooked to the fixture and block A is coupled to the load, See now that the effort moves *in the same direction* as the movement of the load, five ropes now pull on the block which carries the load and the velocity ratio of this arrangement is 5.

Summing up, if the rope pulley blocks are used for *lifting* purposes, it is usual to take the velocity ratio as being equal to the total number of pulleys, but if they are used for pulling components into position, they can be coupled in either of the two ways shown above and the velocity ratio can be either (i) equal to the total number of pulleys, or (ii) total number of pulleys plus one. If such a problem is set which does not state how the blocks are arranged, there are two possible answers and both should be given.

Example. A set of rope pulley blocks with three pulleys in each block is used to drag a casting of 450 kg mass along the ground. If the coefficient of friction between the casting and the ground is 0·45, find the effort required assuming the efficiency of the blocks is 0·8.

$$\text{Weight of casting} = 450 \times 9{\cdot}81 \text{ newtons}$$

$$\begin{aligned}\text{Force to overcome friction} &= \mu W \\ &= 0{\cdot}45 \times 450 \times 9{\cdot}81 \\ &= 1986 \text{ N}\end{aligned}$$

This is the load on the load hook.

$$\eta = \frac{\text{M.A.}}{\text{V.R.}} = \frac{\text{load}}{\text{effort} \times \text{V.R.}}$$

$$\therefore P = \frac{\text{load}}{\text{V.R.} \times \eta}$$

If blocks are arranged so that movement of the effort is opposite in direction to movement of the load, V.R. = 6, then,

$$P = \frac{1986}{6 \times 0{\cdot}8}$$

$$= 413{\cdot}8 \text{ N}$$

If blocks are arranged so that movement of the effort is in the same direction as the movement of the load, V.R. = 7, then

$$P = \frac{1986}{7 \times 0{\cdot}8}$$

$$= 354{\cdot}6 \text{ N}$$

Therefore the effort can be 413·8 N or 354·6 N depending upon the method of coupling the blocks.

EFFICIENCY OF EACH PULLEY. The efficiency of each pulley is not important but if required it is usual to assume that all pulleys in one set of blocks have the same efficiency. The overall efficiency of any continuous system is the product of the efficiencies of each part, thus:

$$e_1 \times e_2 \times e_3 \times \text{etc.} = E$$

Therefore, if the efficiency of each pulley is the same, then,

$$e^n = E, \text{ or } e = \sqrt[n]{E}$$

Where, e = efficiency of each pulley,
n = number of pulleys,
E = overall efficiency.

WHEEL AND AXLE

This consists of a pulley wheel fixed to an axle which is supported in horizontal bearings, the rope or cord carrying the load hook is wound around and fixed to the axle, the cord to which the effort is applied is wound around and fixed to the pulley, as illustrated in Fig. 125. The effort and load cords are wound in opposite directions so that when the effort is applied and its cord winds off the pulley, the load cord is wound on to the axle and lifts the load.

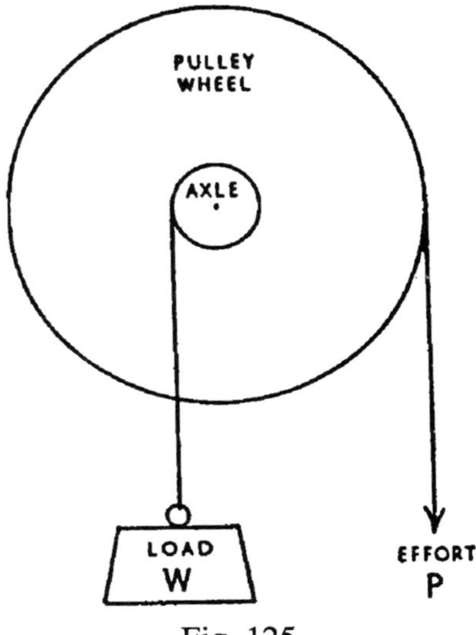

Fig. 125

Let D = diameter of pulley wheel (R = radius)
Let d = diameter of axle (r = radius)

Imagine the effort cord pulled to turn the wheel and axle one complete revolution, then:

$$\text{V.R.} = \frac{\text{Distance moved by effort}}{\text{Distance moved by load}}$$

$$= \frac{\text{Circumference of wheel}}{\text{Circumference of axle}}$$

$$= \frac{\pi \times D}{\pi \times d}$$

$$\text{V.R.} = \frac{D}{d} \text{ or } \frac{R}{r}$$

In practice a crank lever may take the place of the wheel, the length of the crank from centre to handle, L, is then the equivalent of the radius of the wheel, R, thus,

$$\text{V.R.} = \frac{L}{r}$$

In this machine and in some others, the rope may be of such a thickness as to warrant it being taken into account. In all cases where the rope size is given, the effective circumference, diameter or radius is measured to the centre of the rope as shown in Fig. 126.

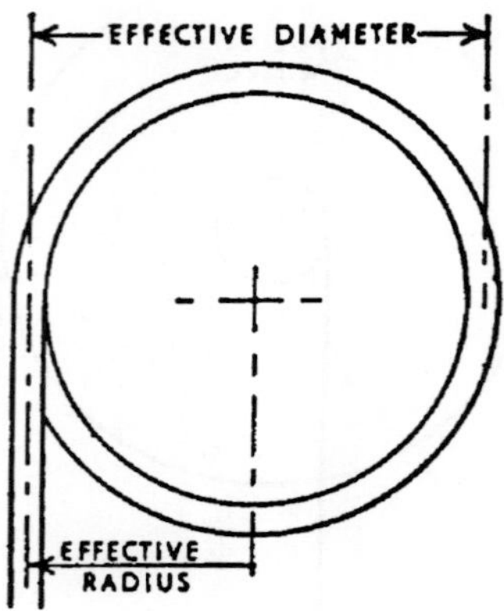

Fig. 126

Example. In a wheel and axle lifting machine, the pulley wheel is 220 mm diameter and the axle is 40 mm diameter. The load rope is 10 mm diameter and the effort rope is 5 mm diameter. Find the effort required to lift a load of 400 N if the efficiency is 0·92.

$$\text{Effective dia. of wheel} = \text{dia. of wheel} + \text{dia. of effort rope}$$
$$= 225 \text{ mm}$$

$$\text{Effective dia. of axle} = \text{dia. of axle} + \text{dia. of load rope}$$
$$= 50 \text{ mm}$$

$$\text{V.R.} = \frac{D}{d} = \frac{225}{50} = 4{\cdot}5$$

$$\text{M.A.} = \mathfrak{J} \times \text{V.R.}$$
$$= 0{\cdot}92 \times 4{\cdot}5$$
$$= 4{\cdot}14$$

$$\text{Effort } P = \frac{W}{\text{M.A.}}$$
$$= \frac{400}{4{\cdot}14}$$
$$= 96{\cdot}61 \text{ N} \quad \text{Ans.}$$

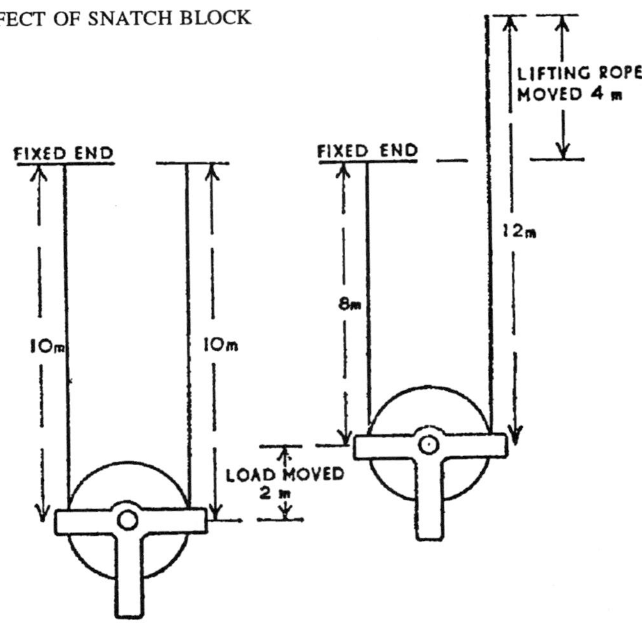

Fig. 127

In many lifting machines the load is carried on a snatch block instead of being hooked directly on the load rope. A snatch block is usually a single pulley running freely in a frame which carries the load, the rope comes down from the machine, passes around this pulley and the free end of the rope is led up to and fixed to the framework of the machine.

Fig. 127 illustrates a numerical example showing that the load is lifted only half of the distance moved by the load rope.

Alternatively, a snatch block may be considered as the lower end of a set of rope pulley blocks, there are two ropes supporting this load block, therefore the velocity ratio of it is 2. Hence the effect of fitting a snatch block to a wheel and axle or any other lifting machine, is to halve the movement of the load and therefore double the velocity of that machine.

WHEEL AND DIFFERENTIAL AXLE

This machine is a similar type to the previous one, but the axle is stepped in two diameters; the two ends of the load rope are wound in different directions around the two parts of the stepped axle and the loop between these passes around a snatch block. Fig. 128 illustrates a wheel and differential axle lifting machine, and it can be seen that as the load rope is wound up on to the larger part of the axle, it is at the same time wound off the smaller part of the axle.

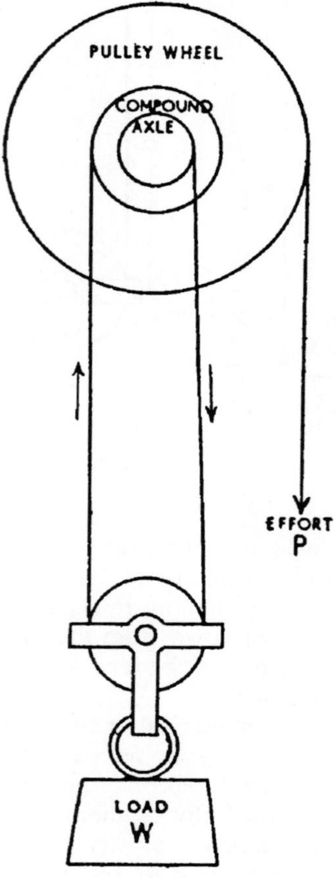

Fig. 128

Let D = diameter of pulley wheel
d_1 = diameter of large part of axle
d_2 = diameter of small part of axle

In one revolution, the lifting rope on one side of the snatch block moves up a distance of πd_1 and on the other side the rope moves down a distance of πd_2, thus the rope carrying the snatch block is shortened by $\pi d_1 - \pi d_2$. The actual distance the load is lifted is half this amount because of the effect of the snatch block as previously explained.

Also, in one revolution, the effort moves through a distance of πD.

$$\text{V.R.} = \frac{\text{Distance moved by effort}}{\text{Distance moved by load}}$$

$$= \frac{\pi D}{\frac{1}{2}(\pi d_1 - \pi d_2)} = \frac{2\pi D}{\pi d_1 - \pi d_2}$$

$$\text{V.R.} = \frac{2D}{d_1 - d_2} \quad \text{or} \quad \frac{2R}{r_1 - r_2}$$

As in the simple wheel and axle, a crank may take the place of the effort wheel, the length of the crank from centre of axle to handle being equivalent to the radius of the wheel.

DIFFERENTIAL PULLEY BLOCKS

This lifting machine is often briefly called 'Differential Pulley Blocks'. It was developed from the wheel and differential axle and has the advantage of requiring a small length of chain in comparison with the great length of rope required to operate the wheel and differential axle.

When the effort is applied to turn the compound sheave, one side of the load chain is pulled up on to the larger pulley, the other side is lowered of the smaller pulley, the movement of the load is half of this difference between lifting and lowering effects due to the chain passing around the snatch block.

Let D = diameter of larger pulley in compound sheave

d = diameter of smaller pulley in compound sheave

In one revolution of the compound sheave:

Distance moved by effort = πD

Distance moved by load = $\frac{1}{2}(\pi D - \pi d)$

$$\text{V.R.} = \frac{\text{Distance moved by effort}}{\text{Distance moved by load}}$$

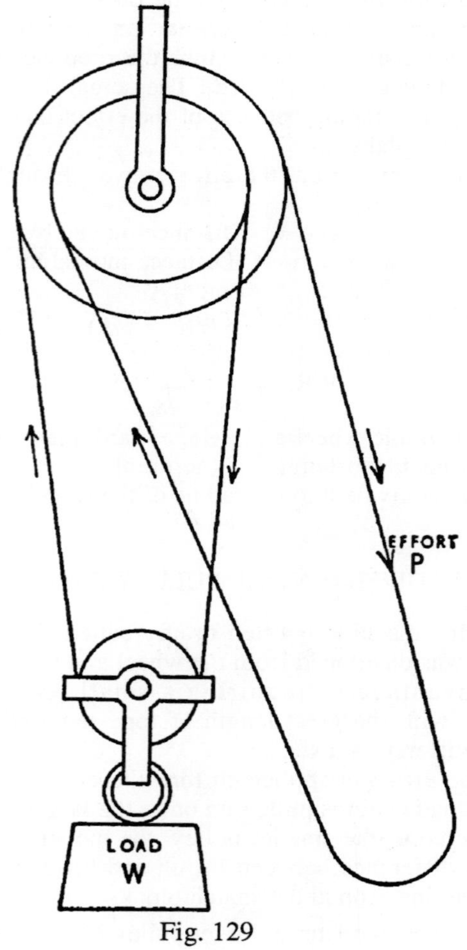

Fig. 129

$$= \frac{\pi D}{\frac{1}{2}(\pi D - \pi d)}$$

$$= \frac{2\pi D}{\pi D - \pi d}$$

$$\text{V.R.} = \frac{\text{Twice circumference of the big pulley}}{\text{Difference in circumferences of the two pulleys}}$$

As a chain is used instead of a rope (which would slip) the pulleys have sprockets or teeth to take the links of the chain.The pitch of the teeth is constant, therefore the above expression can be represented by:

$$\text{V.R.} = \frac{\text{Twice the number of teeth in the big pulley}}{\text{Difference in number of teeth in the two pulleys}}$$

$$\text{V.R.} = \frac{2D}{D-d} \quad \text{or} \quad \frac{2R}{R-r}$$

Example. The diameters of the large and small pulleys of the compound sheave in a set of differential pulley blocks are 120 mm and 110 mm respectively. Calculate the velocity ratio, mechanical advantage and efficiency when lifting a load of 2·4 kN if the effort required is 250 N. Find also how much effort is expended in friction.

$$\text{V.R.} = \frac{2D}{D-d}$$

$$= \frac{2 \times 120}{120 - 110} = 24 \text{ Ans. (i)}$$

$$\text{M.A.} = \frac{\text{Load}}{\text{Effort}}$$

$$= \frac{2400}{250} = 9{\cdot}6 \text{ Ans. (ii)}$$

$$\Im = \frac{\text{M.A.}}{\text{V.R.}}$$

$$= \frac{9{\cdot}6}{24} = 0{\cdot}4 \text{ or } 40\% \text{ Ans. (iii)}$$

$$\text{Ideal effort} = \frac{W}{\text{V.R.}}$$

$$= \frac{2400}{24} = 100 \text{ N}$$

$$\text{Effort expended} = \text{Actual effort} - \text{Ideal effort}$$

$$= 250 - 100$$

$$= 150 \text{ N Ans. (iv)}$$

WORM AND WORM WHEEL

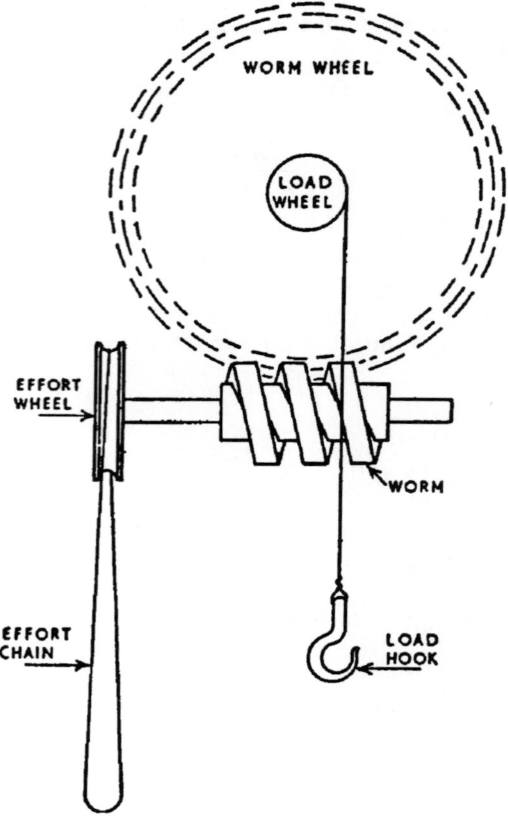

Fig. 130

In the worm and worm-wheel lifting gear the load chain passes around a hoisting pulley attached to a worm-wheel which is driven by a worm. The worm is turned by an effort wheel operated by an endless chain. The load chain may carry the load directly on a hook at its end as illustrated in Fig. 130, or it may pass around a snatch block and the end fixed to the framework of the machine.

Let D = diameter of effort wheel

d = diameter of load wheel

N = number of teeth in worm wheel.

Assuming a single-start worm (which is usual), to turn the worm-wheel one revolution the worm must revolve N revolutions and the distance moved by the effort chain is therefore $\pi D \times N$. At the same time the load wheel makes one revolution and lifts the load a distance of πd.

$$\text{V.R.} = \frac{\text{Distance moved by effort}}{\text{Distance moved by load}}$$

$$= \frac{\pi DN}{\pi d}$$

$$\text{V.R.} = \frac{DN}{d}$$

and if a snatch block is fitted, then:

$$\text{V.R.} = \frac{2DN}{d}$$

SCREW JACK

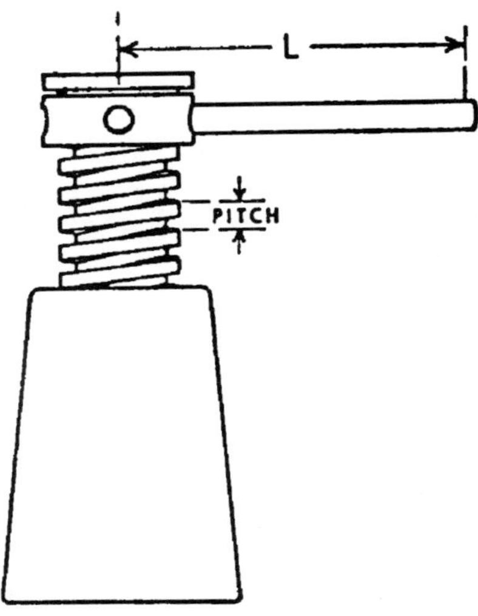

Fig. 131

The *lead* of a screw thread is the axial distance it moves when turned one complete revolution in a fixed nut.

The *pitch* of a screw thread is the axial distance from a point on one thread to the corresponding point on the next thread.

For single-start threads, pitch and lead are the same. The screw of a screw-jack is usually single-start and it is therefore common practice to use the term pitch when referring to the lead.

It can be seen by reference to Fig. 131 that when the toggle bar of effective radius L is turned one revolution, the distance moved by the effort is the circumference of the circle described by the effort, which is $2\pi L$, and the distance the load is lifted is the vertical distance the screw rises which, for a single-start thread is equal to the pitch. Therefore:

$$\text{V.R.} = \frac{\text{Distance moved by effort}}{\text{Distance moved by load}}$$

$$\text{V.R.} = \frac{2\pi L}{\text{pitch}}$$

WORM-DRIVEN SCREW JACK

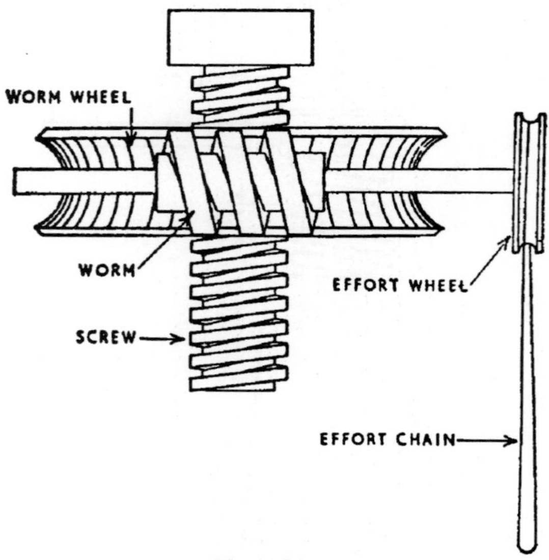

Fig. 132

A screw-jack may be operated by a worm and worm-wheel, and one common arrangement is shown diagrammatically in Fig. 132. The screw is threaded through the centre of a worm-wheel which meshes with a worm on the spindle to which the effort wheel is fitted. The screw is prevented from rotating but can move up or down, the worm-wheel rotates but is prevented from moving axially. Thus when the worm-wheel is rotated by the worm, the screw is driven up (or down).

When the worm-wheel turns one revolution, the screw, therefore the load, moves a distance equal to the pitch. If there are N teeth in the worm-wheel and the worm is single-start, the worm and effort will turn N revolutions.

In each revolution of the effort wheel the effort moves πD where D is the diameter, therefore in N revolutions it moves $\pi D \times N$.

$$\text{V.R.} = \frac{\text{Distance moved by effort}}{\text{Distance moved by load}}$$

$$\text{V.R.} = \frac{\pi DN}{\text{pitch}}$$

WARWICK SCREW

This is sometimes called a screw-bottle or turnbuckle, and is illustrated in Fig. 133. Its function is to pull together two parts of a

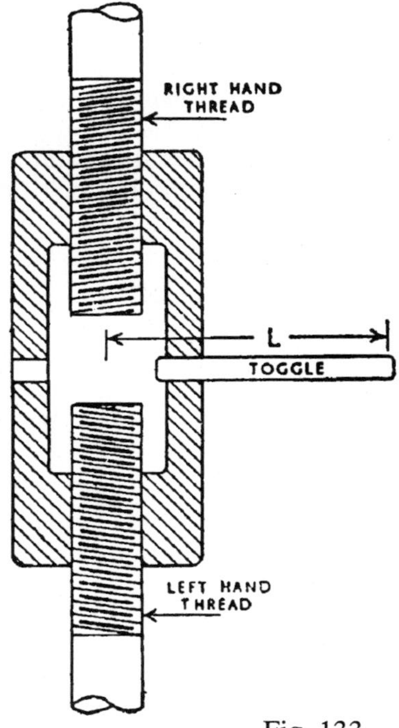

Fig. 133

stay, one end of which is screwed right-handed, the other end screwed left-handed, and an elongated nut or bottle similarly threaded connects the two.

If the toggle is turned through one revolution, the distance moved by the effort is $2\pi L$, while the stays are pulled together through a distance equal to the sum of the pitches of the right and left handed threads.

$$\text{V.R.} = \frac{\text{Distance moved by effort}}{\text{Distance moved by load}}$$

$$\text{V.R.} = \frac{2\pi L}{\text{pitch of R.H. thread} + \text{pitch of L.H. thread}}$$

In most cases the screws are of equal pitch.

CRAB WINCHES

A crab winch is a train of gear wheels in the form of speed reduction gearing to transmit the work supplied at the effort crank handle to a load drum around which the lifting rope is wound.

A single purchase winch is a single reduction gear system and has one gear wheel or pinion (the driver) meshing with one larger gear wheel (the follower). A double purchase winch is a double reduction gear system of two pairs of driver-follower wheels in series. A treble purchase winch has three pairs of wheels in series, and so on. Fig. 134 illustrates diagrammatically the double reduction gear of a double purchase winch from which all others will be readily understood.

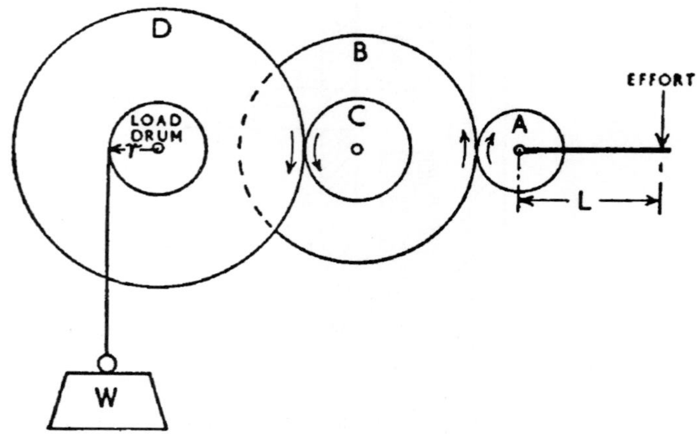

Fig. 134

Let L = length (radius) of handle

r = radius of lifting drum

N_A = number of teeth in driver A

N_B = number of teeth in follower B

N_C = number of teeth in driver C

N_D = number of teeth in follower D

If the load drum is turned one revolution:

Load is lifted $2\pi r$

Follower D turns one revolution

Driver C and therefore follower B turns $\frac{N_D}{N_C}$ revs.

Driver A and therefore crank handle turns

$$\frac{N_D}{N_C} \times \frac{N_B}{N_A} \text{ rev}$$

Effort moves a distance of $2\pi L \times \frac{N_D}{N_C} \times \frac{N_B}{N_A}$

$$\text{V.R.} = \frac{\text{Distance moved by effort}}{\text{Distance moved by load}}$$

$$= \frac{2\pi L \times N_D \times N_B}{2\pi r \times N_C \times N_A}$$

$$\text{V.R.} = \frac{L}{r} \times \frac{\text{Product of teeth in followers}}{\text{Product of teeth in drivers}}$$

If the size of the lifting rope is given, the effective radius is taken as from the centre of the drum to the centre of the rope.

HYDRAULIC JACK

A diagrammatic sketch of a hydraulic jack is given in Fig. 135. It consists of a small diameter cylinder in which the effort plunger fits and a large diameter cylinder in which the load piston works, the bottoms of these cylinders are common with a vessel containing a liquid such as oil.

As liquid is practically incompressible, any movement of the effort plunger is immediately transmitted through the liquid to move the load piston.

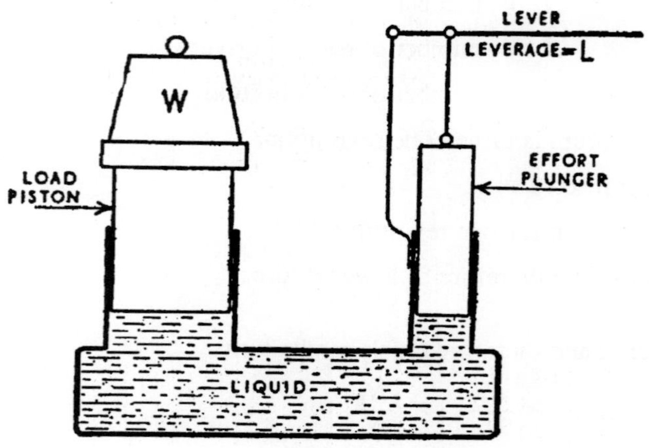

Fig. 135

Let A = area of load piston
a = area of effort plunger

If the effort plunger is pushed down a distance of x the volume of liquid displaced is $a \times x$ and the load piston must move up to allow for this volume. Therefore the distance the load piston will move up is:

$$\frac{\text{volume}}{\text{area}} = \frac{a \times x}{A}$$

$$\text{Velocity ratio} = \frac{\text{Distance moved by effort}}{\text{Distance moved by load}}$$

$$= x \div \frac{a \times x}{A}$$

$$= \frac{x \times A}{a \times x} = \frac{A}{a}$$

Usually a lever is fitted to operate the effort plunger, let this leverage be represented by L then overall velocity ratio is:

$$\text{V.R.} = \frac{A}{a} \times L$$

Example. The diameters of the effort plunger and load piston of a hydraulic jack are 25 mm and 70 mm respectively. The plunger is operated by a lever, the distance from the fulcrum to the handle is 625 mm and from fulcrum to plunger is 50 mm. Calculate the velocity ratio, mechanical advantage and efficiency if it takes a force of 80 N on the handle to lift a load of 5·88 kN.

$$\text{V.R.} = \frac{\text{Area of load piston}}{\text{Area of effort plunger}} \times \text{lever ratio}$$

$$= \frac{0{\cdot}7854 \times 70^2}{0{\cdot}7854 \times 25^2} \times \frac{625}{50}$$

$$= 98 \quad \text{Ans. (i)}$$

$$\text{M.A.} = \frac{\text{Load}}{\text{Effort}}$$

$$= \frac{5880}{80}$$

$$= 73{\cdot}5 \quad \text{Ans. (ii)}$$

$$\mathfrak{J} = \frac{\text{M.A.}}{\text{V.R.}}$$

$$= \frac{73{\cdot}5}{98}$$

$$= 0{\cdot}75 \text{ or } 75\% \quad \text{Ans. (iii)}$$

EXPERIMENTAL RESULTS

By choosing a series of suitable loads for a lifting machine and in each case measuring the effort required to just cause the machine to move and lift the load without acceleration, experimental values can be tabulated and drawn on a graph to show how the quantities are related.

Example. In an experiment on a lifting machine the following data was recorded:

Load (N)	50	100	150	200	250	300
Effort (N)	9·8	15·0	20·3	25·0	30·4	34·7

Plot a graph representing these values and find the relationship between effort and load.

The plotted graph is shown in Fig. 136 and it will be seen that the points lie on a straight line except for some slight differences. Any points which do not lie exactly on the line probably indicate irregularities in the machine or errors of observation during the experiment.

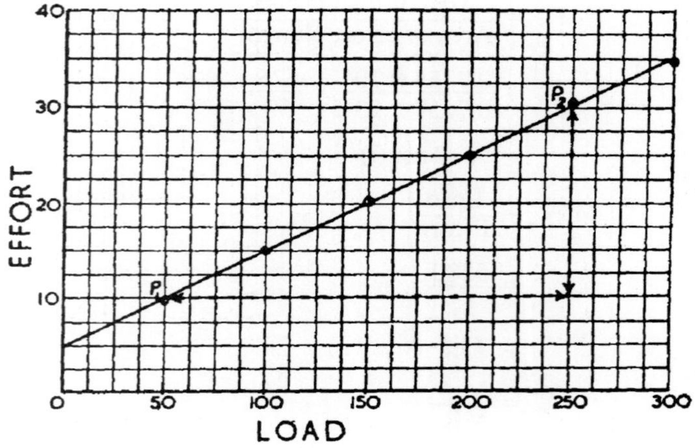

Fig. 136

The effort required to move the machine with no load on it is 5N. The slope, using points P_1 and P_2 is 0·1 N.

$$\therefore P = 5 + 0{\cdot}1\,W$$

This is called the *linear law* of this machine.

The general expression for any machine is:

$$P = a + bW$$

where P is the effort, W is the load, a and b are constants which are determined as in the previous example. The general case is illustrated in Fig. 137.

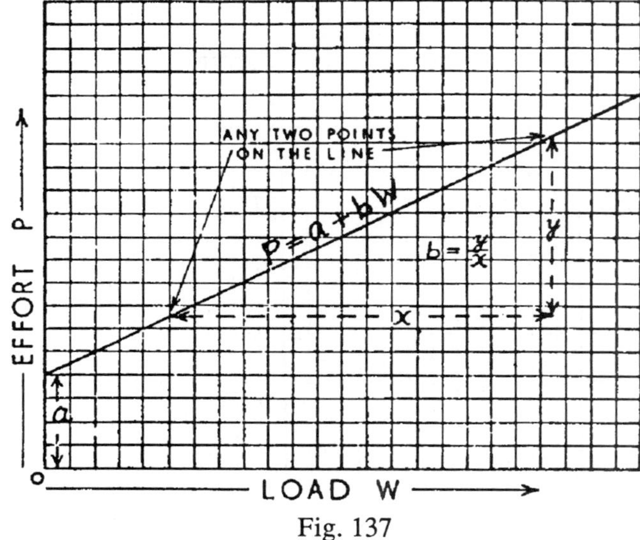

Fig. 137

Example. The diameters of the large and small pulleys of the compound sheave of a differential pulley block are 125 mm, and 112·5 mm respectively. In an experiment on this machine the following results were taken:

Load (N)	40	80	120	160	200	240
Effort (N)	7·5	11	15	18·5	22	22·5

Plot graphs of effort and percentage efficiency on a base of load, find the linear law of this machine, and use this to calculate the effort and efficiency when lifting a load of 224 N.

$$\text{V.R.} = \frac{2D}{D-d} = \frac{2 \times 125}{125 - 112{\cdot}5} = 20$$

Load W	40	80	120	160	200	240
Effort P	7·5	11	15	18·5	22	22·5
M.A. = W/P	5·33	7·27	8	8·65	9·1	9·4
%η = (M.A. / V.R.) 100	26·7	36·4	40	43·3	45·5	47

The graphs are now drawn to some suitable scale as in Fig. 138

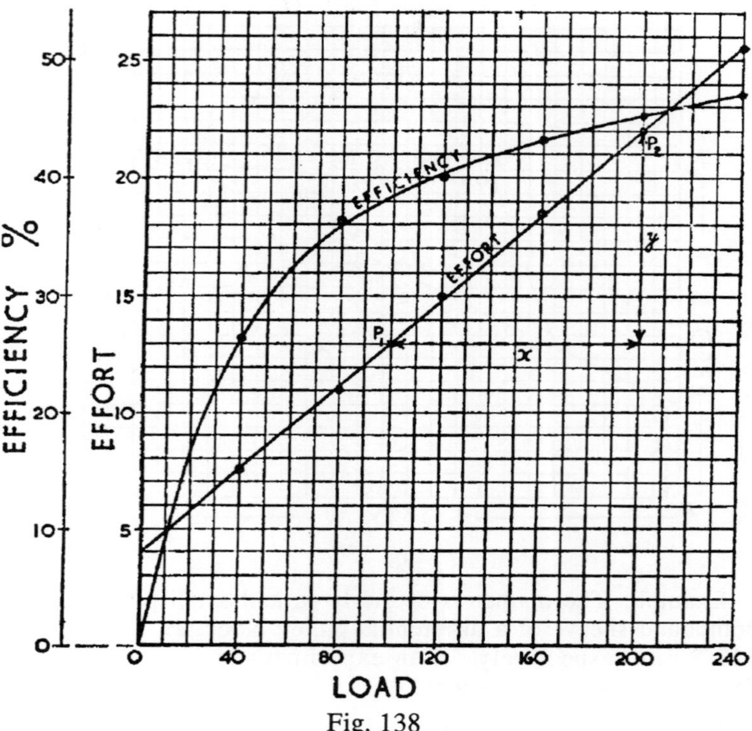

Fig. 138

The straight line effort-load graph cuts the zero load axis to give the value of the constant a as 4.

Taking two convenient points on the graph such as those shown, the slope of the line is,

$$b = \frac{y}{x} = \frac{9}{100} = 0{\cdot}09$$

Therefore the linear law for this machine is,

$$P = 4 + 0{\cdot}09\,W \quad \text{Ans. (i)}$$

When the load is 224 N, then

$$\begin{aligned} P &= 4 + 0{\cdot}09 \times 224 \\ &= 24{\cdot}16 \text{ N} \quad \text{Ans. (ii)} \end{aligned}$$

Mechanical advantage when lifting this load

$$= \frac{W}{P} = \frac{224}{24{\cdot}16} = 9{\cdot}27$$

$$\%\mathfrak{I} = \frac{\text{M.A.}}{\text{V.R.}} \times 100$$

$$= \frac{9{\cdot}27}{20} \times 100 = 46{\cdot}35 \quad \text{Ans. (iii)}$$

TEST EXAMPLES 8

1. In a wheel and differential axle type of lifting machine, a crank handle of 240 mm radius takes the place of the wheel and the diameters of the differential axle are 110 mm and 80 mm respectively. If an effort of 80 N is required at the handle to lift a load of 1·12 kN, find the velocity ratio, mechanical advantage and efficiency at this load.

2. The efficiency of a set of chain driven differential pulley blocks is 35% when lifting a load of 1·89 kN. If the large and small pulleys of the compound sheave have 27 and 24 teeth respectively, find the effort required to lift this load.

3. The diameter of the small pulley of a set of differential pulley blocks is 130 mm. When lifting a load of 560 N the effort required is 50 N and the efficiency is 40%. Find the diameter of the large pulley.

4. The effort wheel of a worm and worm-wheel chain block is 200 mm diameter, the worm is single-start and there are 40 teeth in the worm-wheel. The load wheel is 125 mm diameter and the load is carried by a snatch block on the loop of the lifting chain from the load wheel to the framework of the machine. If the effort required to lift a load of 6·72 kN is 1450 N, find (i) the efficiency when lifting this load, (ii) the ideal effort, and (iii) the effort to overcome friction.

5. Two toggle bars are used in a screw jack to raise a casting of 3 t mass. The screw thread has a pitch of 12 mm. One toggle is 500 mm long and the effort applied to its end is 220 N. The other toggle is 450 mm long, find the effort required at the end of this toggle if the efficiency when lifting this load is 35%.

6. Find the number of teeth in the worm-wheel of a worm driven screw jack to give a velocity ratio of 550 if the worm is single threaded, the screw has a pitch of 16 mm and the effort wheel is 100 mm diameter. Assuming an efficiency of 30%, find the effort required to lift a load of 50 kN.

7. A Warwick screw is used to tighten a guy rope, the right and left hand threads each have a pitch of 4 mm and the effective length of the toggle bar is 250 mm. Assuming an efficiency of 25% find the pull in the guy rope when an effort of 90 N is applied.

8. In a hand driven double purchase winch, the radius of the crank handle is 300 mm, the load drum is 200 mm diameter and

the lifting rope is 25 mm diameter. The driving wheels have 25 and 30 teeth and the followers have 90 and 125 teeth respectively. Find (i) the velocity ratio, and (ii) the efficiency when lifting a mass of 1000 kg if the effort required is 350 N.

9. The diameter of the load ram of a hydraulic jack is 50 mm and the diameter of the effort plunger is 20 mm. The plunger is operated by a handle whose effective leverage is 18 to 1. Assuming an efficiency of 80%, find the load that can be lifted by an effort of 120 N applied to the handle.

10. The following data were taken during an experiment on a model lifting machine which had a velocity ratio of 12,

Load (N)	7	14	21	28	35	42
Effort (N)	3·5	4·6	5·7	6·7	7·7	8·8

On a base of load plot graphs of effort and efficiency. Find the linear law of this machine and read from the graph the efficiency when the effort applied is 7 N.

CHAPTER 9

STRESS AND STRAIN

STRESS is the internal resistance set up in a material when an external force is applied.

When the applied force tends to shorten the material, or crush it, the material is said to be *in compression*, and the stress is referred to as a *compressive stress*. When the force tends to lengthen the material, or tear it apart, the material is said to be *in tension* and the stress is referred to as a *tensile stress*. When the force tends to cause the particles of the material to slide over each other, the material is said to be *in shear* and the stress is referred to as a *shear stress*.

Stress is always expressed as stress intensity and is therefore the force per unit area of the material.

$$\text{Stress} = \frac{\text{Total force}}{\text{Area}} \quad \text{N/m}^2$$

$$\sigma = \frac{F}{A} \quad \text{tens. or comp.}$$

$$\tau = \frac{F}{A} \quad \text{shear}$$

In the case of direct tensile or compressive forces, the area carrying the force is the cross-section in the plane of the material normal (i.e. perpendicular) to the direction of the force. In the case of a shear force, the area carrying the force is the area to be sheared through in the direction of the line of action of the force.

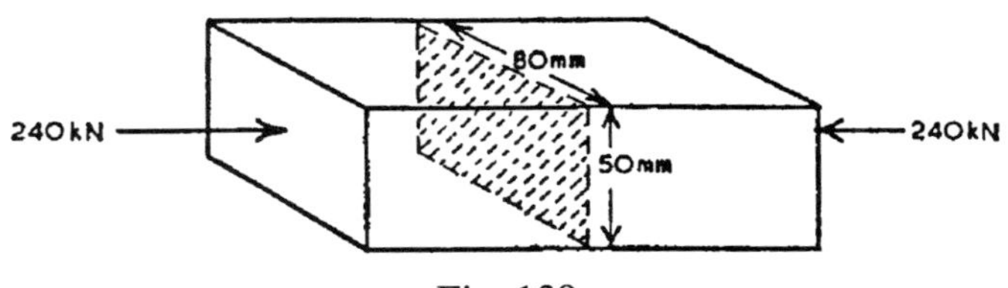

Fig. 139

Fig. 139 shows a direct compressive force of 240 kN applied to a solid piece of material of rectangular cross-section, 80 mm by 50 mm. The cross-sectional area supporting the load is 80 × 50 mm = 4000 mm² *i.e.* 4000 × 10^{-6} m².

$$\sigma = \frac{F}{A}$$

$$= \frac{240 \times 10^3}{80 \times 50 \times 10^{-6}}$$

$$= 6 \times 10^7 \text{ N/m}^2$$

$$= 60 \text{ MN/m}^2$$

The material is in compression because the applied force tends to shorten it, and the stress of 60 MN/m² is compressive.

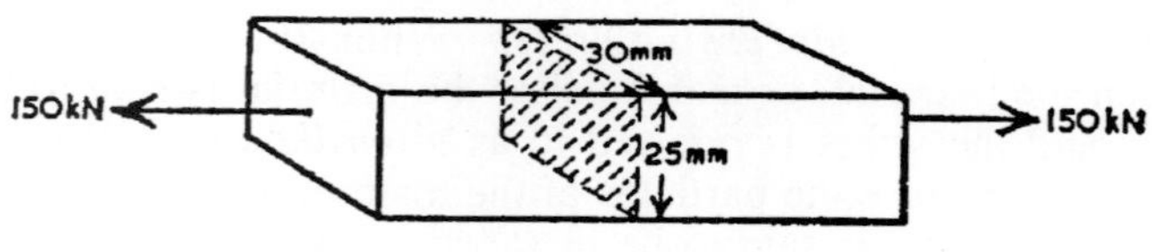

Fig. 140

A direct tensile force of 150 kN carried by a material of cross-section 30 mm by 25 mm is illustrated in Fig. 140.

$$\sigma = \frac{F}{A}$$

$$= \frac{150 \times 10^2}{30 \times 25 \times 10^{-6}} = 2 \times 10^8 \text{ N/m}^2$$

$$= 200 \text{ MN/m}^2$$

The material is in tension because the force tends to stretch it, and the stress of 200 MN/m² is tensile.

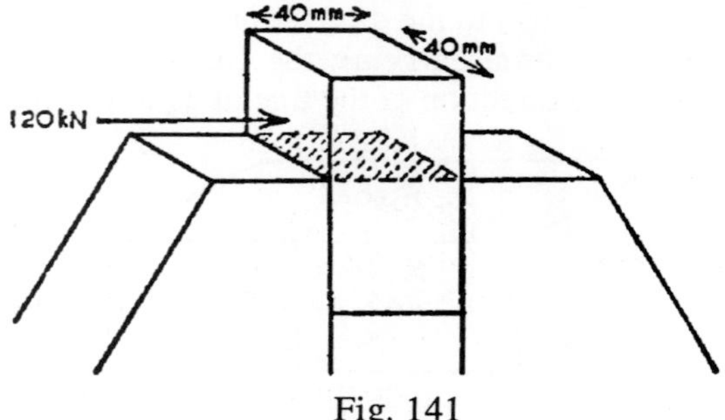

Fig. 141

The next case (Fig. 141) shows a material of rectangular section 40 mm by 40 mm gripped in a vice and a transverse force of 120 kN applied which tends to shear the material.

$$\tau = \frac{F}{A}$$

$$= \frac{120 \times 10^3}{40 \times 40 \times 10^{-6}} = 7{\cdot}5 \times 10^7 \text{ N/m}^2$$

$$= 75 \text{ MN/m}^2$$

Many cases arise where the material is in 'double shear', that is, two planes of area resist the shearing force. Fig. 142 illustrates a cotter joint and it can be seen that the area of material carrying the force is twice the cross-sectional area of the cotter, therefore the shear stress is Force ÷ (2 × cross-sectional area). It is common practice in certain kinds of joints where other factors affect the strength, to assume that the effective area is a little less than twice the cross-section, such as 1·8 or 1·9 times the area.

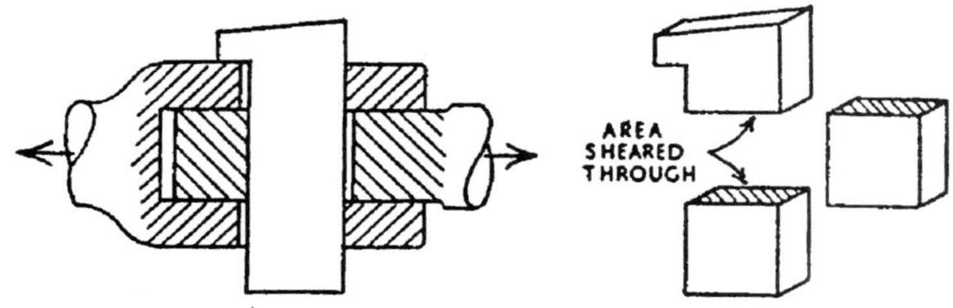

Fig. 142

ULTIMATE TENSILE STRENGTH

The strength of a material is expressed as the stress required to cause fracture. If it is to express the strength in tension, the maximum force required to break the material is divided by the original cross-sectional area at the point of fracture, and is termed the *ultimate tensile strength* (U.T.S), or the *tenacity* of the material. The original area of cross-section is taken as distinct from the actual area at fracture, the latter being much smaller due to the waist or neck which forms just before fracture occurs.

$$\text{Ultimate tensile strength} = \frac{\text{maximum breaking force}}{\text{original area of cross section}}$$

Example. A specimen of mild steel, 20 mm diameter, is tested in a testing machine and snapped when the maximum pull on the specimen was 154 kN. Calculate the tensile strength of this material.

$$\text{U.T.S} = \frac{\text{Maximum breaking force}}{\text{original area of cross section}}$$

$$= \frac{154 \times 10^3}{0{\cdot}7854 \times 0{\cdot}02^2}$$

$$= 4{\cdot}901 \times 10^8 \text{ N/m}^2$$

$$= 490{\cdot}1 \text{ MN/m}^2 \quad \text{Ans.}$$

Example. A hole 12 mm diameter is to be punched through a plate 18 mm thick. If the shear strength of the material is 300 N/mm^2, find the load required on the punch.

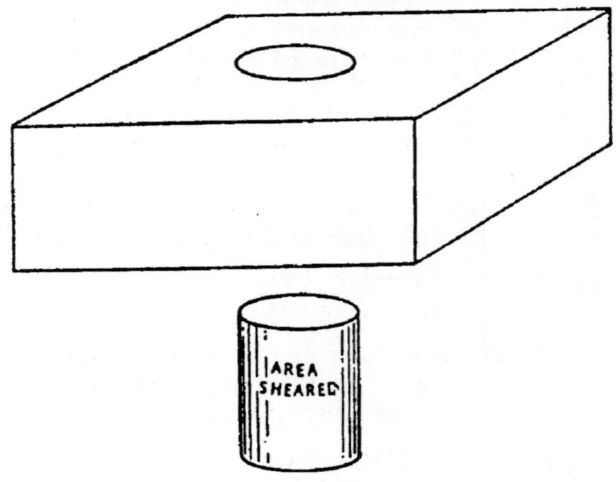

Fig. 143

By reference to Fig. 143 it can be seen that the area of material to be sheared is the circumferential curved surface of the plug to be punched out of the plate. The curved surface area of a cylinder is circumference × height, therefore:

$$\text{Area resisting load} = \text{circumference of plug} \times \text{thickness}$$

$$= \pi \times 12 \times 18 \text{ mm}^2$$

$$\text{Shear strength} = \frac{\text{maximum shear load}}{\text{area resisting load}}$$

$$\therefore \text{ Maximum load} = \text{shear strength} \times \text{area}$$

$$= 300 \times \pi \times 12 \times 18$$
$$= 2{\cdot}036 \times 10^5 \text{ N}$$
$$= 203{\cdot}6 \text{ kN} \quad \text{Ans.}$$

WORKING STRESS & FACTOR OF SAFETY

It is obvious that the stress allowed in any piece of machinery under working conditions must be much less than that which would cause the metal to fail. A safe working stress is therefore chosen with regard to the conditions under which the material is to work, taking into account whether the stress is due to a dead load, whether it varies or is subject to reversal of stress, and the rapidity of change of stress. For instance, a pressure vessel such as a boiler shell can be regarded as a dead load because the stress in the shell is due to internal pressure which changes very little and therefore a fairly high working stress can be allowed, whereas the piston rod of a double acting reciprocating engine is subjected to rapid varying and reversal stresses and consequently the working stress allowed must be very low.

The ratio of the stress which would cause fracture, to the working stress allowed in the material, is termed the *factor of safety*.

$$\text{Factor of safety} = \frac{\text{breaking stress}}{\text{working stress}}$$

Example. Find the minimum diameter of a mild steel piston rod to carry a maximum load of 560 kN, allowing a factor of safety of 14 and assuming a tensile strength of 420 MN/m^2.

$$\text{Working stress} = \frac{\text{breaking stress}}{\text{factor of safety}}$$

$$= \frac{420}{14} = 30 \text{ MN/m}^2$$

$$\text{Working stress} = \frac{\text{working load}}{\text{area of cross-section}}$$

$$30 \times 10^6 = \frac{560 \times 10^3}{0{\cdot}7854 \times d^2}$$

$$d = \sqrt{\frac{560 \times 10^3}{30 \times 106 \times 0{\cdot}7854}}$$

$$= 0{\cdot}1542 \text{ m} = 154{\cdot}2 \text{ mm} \qquad \text{Ans.}$$

STRAIN

Strain is the change of shape that takes place in a material due to it being stressed

LINEAR STRAIN is measured as the change of length per unit length which occurs when a tensile or compressive force is applied thus:

$$\text{Linear strain} = \frac{\text{change of length}}{\text{original length}}$$

$$\varepsilon = \frac{x}{l}$$

SHEAR STRAIN is measured by the angular distortion caused by a shear stress. Referring to Fig. 144, which illustrates a material strained by a shear force, the strain is expressed by the angle of distortion which is obtained by dividing x by l. This gives the tangent of the angle ϕ, or, being a small angle, it is also the angle in radians.

$$\text{Shear strain} = \frac{\text{deformation}}{\text{original dimension}}$$

$$\gamma = \frac{x}{y}$$

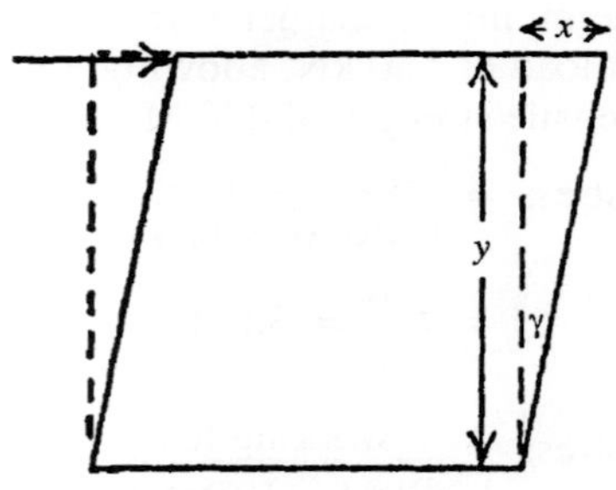

Fig. 144

Change of length and original length are in the same units, therefore in dividing one by the other the units cancel. Hence, strain is a pure number.

ELASTICITY

All metals are elastic to a certain extent. A material will stretch if a tensile force is applied to it, and return to its original length on removal of the force. Similarly, if a compressive force is removed the material will shorten, and when the force is removed the material will return to its original length. Metals are elastic under any form of straining.

There is a limit to this elastic property in every material, known as the *elastic limit*, and if this is exceeded by applying an undue force, the material, after being stretched or compressed, will not return to its original length, but remain permanently set in its deformed condition.

Provided the material is strained within its elastic limit, the amount of strain is directly proportional to the stress. This is known as HOOKE'S LAW. For example, if a piece of metal is subjected to a tensile stress of 75 MN/m^2 which causes a tensile strain of 0·00074, then for a stress of 150 MN/m^2 the strain will be 0·00148, that is, double the stress produces double the strain, and if the stress were trebled the strain would be trebled—always, the strain is proportional to the stress within the limits of elasticity of the metal.

Now consider the following graph plotted during a tensile test on a steel wire 2500 mm long and 0·8 mm diameter (cross-sectional area = 0·5027 mm^2).

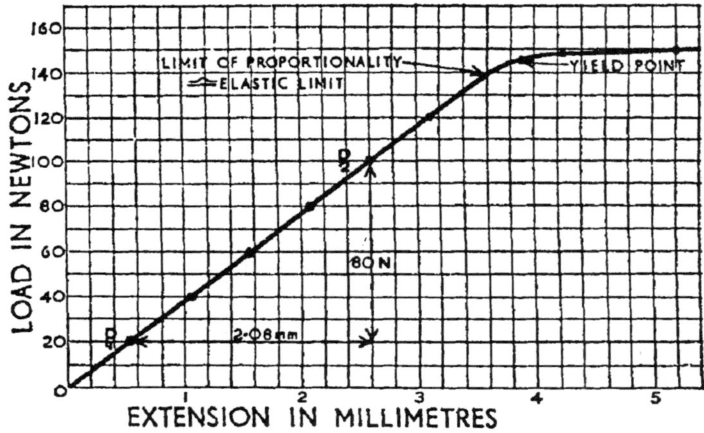

Fig. 145

Part of this graph is a straight line which means that, over this range, extension is proportional to load. As strain depends upon the extension, and stress depends upon the load, it also proves that strain is proportional to stress. If the load were removed at any point over this range, the wire would return to its original length, demonstrating that the material is elastic.

It will be seen that the limit of proportionality, *i.e.* the end of the straight part of the line, is at about the load of 140 N, a little beyond this the material begins to yield and the extension is greatly increased with very little increase of load. The elastic limit is approximately at the limit of proportionality, or in some cases a little higher but before the yield point.

MODULUS OF ELASTICITY

Within the elastic limit, stress is proportional to strain, therefore at any point during the elastic range the value of the stress divided by the corresponding value of the strain will always produce the same result for any one metal. Stress divided by strain is therefore a constant, its value depending upon the material.

When direct (tensile or compressive) stress is divided by direct strain, the constant is termed the *modulus of elasticity* (sometimes Young's modulus), and is represented by E.

$$\text{Modulus of elasticity N/m}^2 = \frac{\text{direct stress}}{\text{direct strain}}$$

$$E = \frac{\sigma}{\varepsilon}$$

When shear stress is divided by shear strain the constant obtained is termed the *modulus of rigidity* and represented by G.

$$G = \frac{\tau}{\gamma}$$

Strain being a pure number, then the units of the moduli of elasticity and rigidity are the same as for stress.

The modulus of elasticity of the wire in the previous example can be obtained from the experimental results. The straight line drawn as near as possible through the plotted points eliminates slight errors of observation during the experiment. Choose any point on this line, divide the value of the stress by the corresponding value of strain, and thus obtain the value of E.

Some graphs however, may not start from zero origin for the strain or extension of the vernier or extensometer is not set to a zero reading at the beginning of the experiment. It is then necessary, and in fact it is always better, to choose two points on the line, such as P_1 and P_2 (see Fig. 145) and divide the increase of stress by the increase of strain, this gives the slope of the graph which is the modulus of elasticity.

Measuring the values from the graph:

$$\text{Increase of load} = 80 \text{ N}$$

$$\text{Increase of stress} = \frac{\text{increase of load}}{\text{area}}$$

$$= \frac{80}{0{\cdot}5027} = 159{\cdot}2 \text{ N/mm}^2$$

$$\text{Increase of extension} = 2{\cdot}08 \text{ mm}$$

$$\text{Increase of strain} = \frac{\text{increase of extension}}{\text{original length}}$$

$$= \frac{2{\cdot}08}{2500} = 0{\cdot}000832$$

$$\text{Modulus of elasticity} = \frac{\text{stress}}{\text{strain}}$$

$$= \frac{\text{increase of stress}}{\text{increase of strain}}$$

$$= \frac{159{\cdot}2}{0{\cdot}000832}$$

$$= 1{\cdot}914 \times 10^5 \text{ N/mm}^2$$

$$E = 191{\cdot}4 \text{ kN/mm}^2$$

$$\text{or } 191{\cdot}4 \text{ GN/m}^2$$

TENSILE TEST TO DESTRUCTION

To observe the behaviour of a material subjected to tension up to breaking point, a specimen of the material is machined to a standard size, the ends are gripped in the jaws of a testing machine and a tensile load is slowly applied.

The modulus of elasticity can be obtained as previously described in the wire test, the extensions being read during the elastic range by an extensometer. Before yield point is anticipated the extensometer is removed to prevent damage. The yield point can be distinctly observed and the extensions from here to breaking point can be measured with little magnification.

The results are plotted on a stress-strain graph and, for mild steel will appear as shown in Fig. 146. When the specimen nears breaking point a waist begins to form in the material and the cross-sectional area becomes smaller. As the waist is forming some of the load can be taken off the machine but the load is removed at a slower rate than that at which the cross-sectional area is reducing, thus the actual stress in the material is still increasing up to breaking point. It is usual, however, to plot the graph of *nominal* stress values which is load divided by the *original* cross-sectional area of the specimen, this accounts for the fall in the graph towards the end near fracture. If the *actual* stress had been plotted it would show the graph continuing to rise. As previously stated, the ultimate tensile strength is taken as the maximum load divided by the original cross-sectional area.

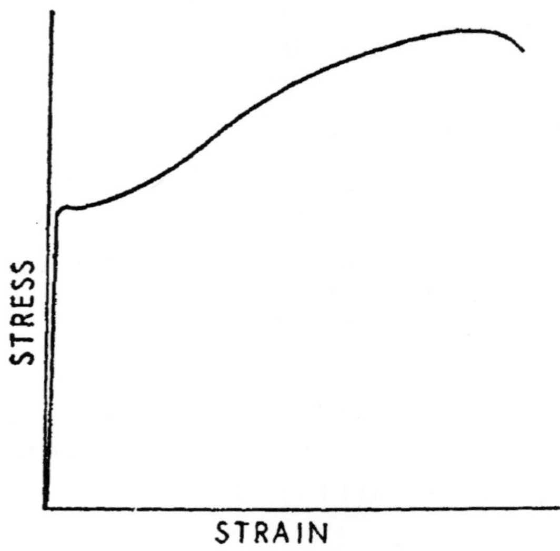

Fig. 146

Further observations can now be made to assist in assessing the quality of the material. The two broken pieces are placed together (see Fig. 147) and the distance between gauge points measured; by subtracting the original gauge length from this measurement, the total elongation is obtained. This is expressed as a percentage of the original length:

$$\% \text{ Elongation} = \frac{\text{total elongation}}{\text{original length}} \times 100$$

$$= \frac{l_2 - l_1}{l_1} \times 100$$

where l_1 = original gauge length
where l_2 = distance between gauge points after fracture.

By measuring the smallest diameter at the neck of the waist where fracture took place, the final cross-sectional area is obtained and the reduction in cross-sectional area is expressed as a percentage of the original, thus:

$$\% \text{ Reduction of area} = \frac{\text{reduction of cross-sect. area}}{\text{original cross-sect. area}} \times 100$$

$$= \frac{A_1 - A_2}{A_1} \times 100$$

where A_1 = original cross-sectional area
A_2 = minimum cross-sectional area after fracture.

The percentage elongation and reduction of cross-sectional area are indications of the ductility of the material – that property by which it can be drawn out.

Example. The test piece illustrated in Fig. 147, of diameter 10 mm and gauge length 50 mm, was made from the deposited metal of a welded joint. A tensile test on this specimen produced the following results:

Load at yield point = 25·3 kN
Maximum breaking point = 37·95 kN
Length between gauge points after fracture = 63·5 mm
Diameter at neck of waist after fracture = 7·5 mm

Calculate the stress at yield point, ultimate tensile strength, percentage elongation, percentage contraction of area.

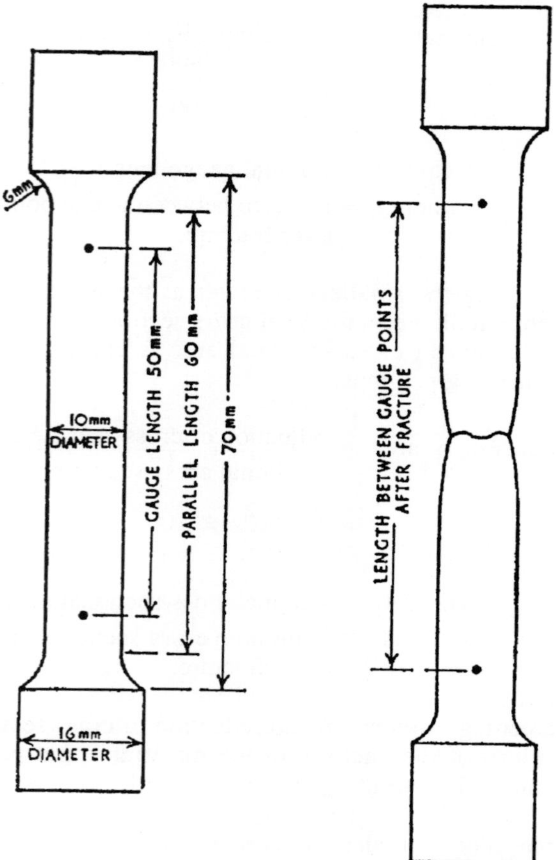

Fig. 147

$$\text{Original cross-sectional area} = 0{\cdot}7854 \times 10^2$$

$$= 78{\cdot}54 \text{ mm}^2 = 78{\cdot}54 \times 10^{-6}\text{m}^2$$

$$\text{Yield stress} = \frac{\text{yield load}}{\text{original cross-sect. area}}$$

$$= \frac{25{\cdot}3 \times 10^3}{78{\cdot}54 \times 10^{-6}}$$

$$= 3{\cdot}221 \times 10^{-6} \text{ N/m}^2$$

$$= 322{\cdot}1 \text{ MN/m}^2 \qquad \text{Ans. (i)}$$

$$\text{U.T.S.} = \frac{\text{maximum breaking load}}{\text{original cross-sect. area}}$$

$$= \frac{37{\cdot}95 \times 10^3}{78{\cdot}54 \times 10^{-6}} = 4{\cdot}832 \times 10^8 \text{ N/m}^2$$

$$= 483{\cdot}2 \text{ MN/m}^2 \qquad \text{Ans. (ii)}$$

$$\%\ \text{Elongation} = \frac{\text{elongation}}{\text{original length}} \times 100$$

$$= \frac{63{\cdot}5 - 50}{50} \times 100 = 27 \qquad \text{Ans. (iii)}$$

$$\%\ \text{Reduction of area} = \frac{\text{reduction of cross-sect. area}}{\text{original cross-sect. area}} \times 100$$

$$= \frac{0{\cdot}7854 \times 10^2 - 0{\cdot}7854 \times 7{\cdot}5^2}{0{\cdot}7854 \times 10^2} \times 100$$

$$= \frac{10^2 - 7{\cdot}5^2}{10^2} \times 100 = 43{\cdot}75 \quad \text{Ans. (iv)}$$

STRENGTH OF PRESSURE VESSELS

The total force or load due to a fluid pressure is the product of the intensity of the internal pressure and the *projected* area on which the fluid acts. For example, the total load on the piston of an internal combustion engine is the gas pressure multiplied by the projected area of the piston. The projected area of the piston is the same as the cross-sectional area of the bore of the cylinder, this is the effective piston area no matter whether the piston is flat, dished, or has a hump on the top.

Similarly in a pressure vessel, the load causing stress on a section of the shell is the product of the internal pressure and the projected area of the shell perpendicular to that section.

THIN CYLINDERS. A cylindrical vessel whose thickness of shell is small in comparison with the diameter. Boiler drums, air reservoirs and most pipes come into this category.

Consider a thin cylinder without seams, such as a solid drawn pipe or seamless boiler drum, and imagine that the tendency to rupture is by splitting along the longitudinal axis as shown in Fig. 148. The projected area of the curved semi-cylindrical shell is the area of its projected rectangular plan view which is diameter × length, and the load tending to cause longitudinal rupture is therefore the intensity of the internal pressure multiplied by this projected area.

Let p = internal fluid pressure,
d = internal diameter of cylinder,
l = length of cylinder,
t = thickness of cylinder,

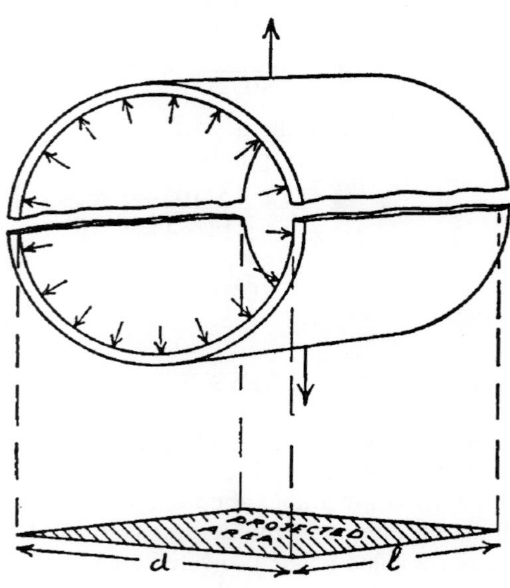

Fig. 148

Load tending to cause longitudinal fracture

= internal pressure × projected area

$= p \times d \times l$

Area of metal resisting this load

= 2 × thickness × length

$= 2 \times t \times l$

Letting the stress along the longitudinal section of the cylinder be represented by σ_1 then:

$$\text{Stress} = \frac{\text{load}}{\text{area}}$$

$$\sigma_1 = \frac{p \times d \times l}{2 \times t \times l}$$

$$\sigma_1 = \frac{pd}{2t} \qquad \dots \dots \dots \dots \dots \dots \text{(i)}$$

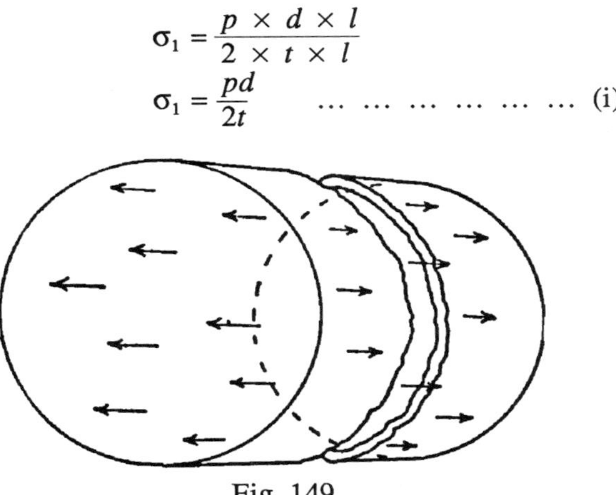

Fig. 149

Now imagine that the tendency is for the cylinder to fracture around its circumference as illustrated in Fig. 149. The cylinder may have flat ends, dished, hemispherical or otherwise, the effective area on which the internal pressure acts is the projected circular area. The total load acting on the circumferential section is therefore the product of the intensity of the internal pressure and the circular area of the end.

The area of metal carrying this load is the area of the annular section = $\frac{\pi}{4}(D^2 - d^2)$. As the thickness is small compared with the diameter, a slight approximation can be made here to simplify the final expression without affecting the calculated result very much by taking the cross-sectional area as circumference multiplied by thickness.

Load tending to cause circumferential fracture

= internal pressure × projected area

$= \pi \times \frac{\pi}{4}d^2$

Area of metal resisting this load.

= circumference × thickness

$= \pi d \times t$

Letting the stress on the circumferential section be represented by σ_2 then:

$$\text{Stress} = \frac{\text{load}}{\text{area}}$$

$$\sigma_2 = \frac{p \times \frac{\pi}{4}d^2}{\pi d \times t}$$

$$\sigma_2 = \frac{pd}{4t} \quad \ldots\ldots\ldots\ldots\ldots\ldots\ldots \text{(ii)}$$

From (i) and (ii) the stress in the material is twice as much along the longitudinal section as it is on the circumferential section.

If an elliptical manhole is to be cut in the shell of a cylinder the smallest amount of metal should be cut out of the section which is subjected to the greatest stress, therefore the minor axis should lie in the longitudinal direction and the major axis circumferentially.

Any section of a spherical shell is 'circumferential' therefore expression (ii) only will apply for spherical vessels or spherical ends of cylindrical vessels.

Example. Calculate the stresses on the longitudinal and circumferential sections of a boiler steam drum 2·15 m diameter and 45 mm thick, when working pressure is 30 bar (30×10^5 N/m^2). If the tensile strength of the material is 490 MN/m^2, what factor of safety is allowed?

$$\sigma_1 = \frac{pd}{2t} = \frac{30 \times 10^5 \times 2{\cdot}15}{2 \times 0{\cdot}045}$$

$$= 7{\cdot}166 \times 107 \text{ N/m}^2$$

$$= 71{\cdot}66 \text{ MN/m}^2 \qquad \text{Ans. (i)}$$

$$\sigma_2 = \frac{pd}{4t} = \tfrac{1}{2}\sigma_1$$

$$= 35{\cdot}83 \text{ MN/m}^2 \qquad \text{Ans. (ii)}$$

$$\text{Factor of safety} = \frac{\text{breaking stress}}{\text{working stress}}$$

$$= \frac{490}{71{\cdot}66} = 6{\cdot}837 \qquad \text{Ans. (iii)}$$

WORKING PRESSURE. The maximum internal working pressure in a pressure vessel depends upon what is considered to be the maximum allowable stress in the material within the limits of safety. If the vessel is seamless the internal working pressure can be calculated from the previous expressions, when σ represents the safe working stress, p will represent the safe working pressure.

For a cylindrical vessel the first expression would be used because it involves the greatest stress in the material which is on the longitudinal section, thus:

$$\text{Working pressure} = \frac{2t \times \text{working stress}}{d}$$

For a spherical vessel where all sections are 'circumferential' the working pressure is obtained from:

$$\text{Working pressure} = \frac{4t \times \text{working stress}}{d}$$

SEAM STRENGTH. Cylindrical shells of pressure vessels are often built up of two (sometimes more) curved plates, the seams where these plates are joined together are not as strong as the solid plate and therefore the seam strength must be taken into consideration when determining the safe working pressure.

$$\text{Working pressure} = \frac{2t \times \text{working stress}}{d} \times F_{\text{long}} \qquad \text{(iii)}$$

$$\text{Working pressure} = \frac{4t \times \text{working stress}}{d} \times F_{\text{circ}} \qquad \text{(iv)}$$

F represents seam strength fraction.

Consideration of (iii) and (iv) will show that the longitudinal seam should be as near to the strength of the solid plate as practicable, whereas the circumferential seam strength will not affect the safe working pressure unless its strength is less than half the longitudinal seam strength. Longitudinal seams are therefore designed to have a high strength.

Example. Allowing a factor of safety of 5·5, find the safe working pressure in a cylindrical boiler 3·5 mm diameter. The shell plates are mild steel 42 mm thick of tensile strength 460 MN/m^2, the fractional strength of the longitudinal seams is 0·86 and the fractional strength of the circumferential seams is 0·76.

$$\text{Working stress} = \frac{\text{tensile strength}}{\text{factor of safety}}$$

$$= \frac{460 \times 10^6}{5{\cdot}5} \text{ N/m}^2$$

$$\text{Working pressure} = \frac{2t \times \text{working stress}}{d} \times F_{\text{long}}$$

$$= \frac{2 \times 0{\cdot}042 \times 460 \times 10^6 \times 0{\cdot}86}{3{\cdot}5 \times 5{\cdot}5}$$

$$= 1{\cdot}726 \times 10^6 \text{ N/m}^2$$

$$= 1{\cdot}726 \text{ MN/m}^2 \text{ or } 17{\cdot}26 \text{ bar}$$

With regard to expression (iv) involving the fractional strength of the circumferential seam, as the circumferential seam strength is more than half the longitudinal seam strength, expression (iv) would therefore give a higher working pressure because 4 × 0·76 is more than 2 × 0·86 (other quantities being equal) and there is no need to work this out as the lesser value must be taken, hence,

Working pressure = 17·26 bar Ans.

STRESSES IN COMPOUND BARS

When two or more bars of equal length carry a load between them the stress in each can be calculated from equations formed, (i) on the principle that, as each is compressed (or stretched) by the same amount, their strains must be equal, and (ii) that the sum of the loads carried by each bar must be equal to the total load carried.

Example. A steel bar of 2580 mm² cross-sectional area and a brass bar of 3225 mm² cross-sectional area, jointly support a compressive load of 420 kN. Calculate the stress in each bar, and also the strain, taking the moduli of elasticity as:

$$E \text{ for steel} = 207 \text{ GN/m}^2$$
$$E \text{ for brass} = 92 \text{ GN/m}^2$$

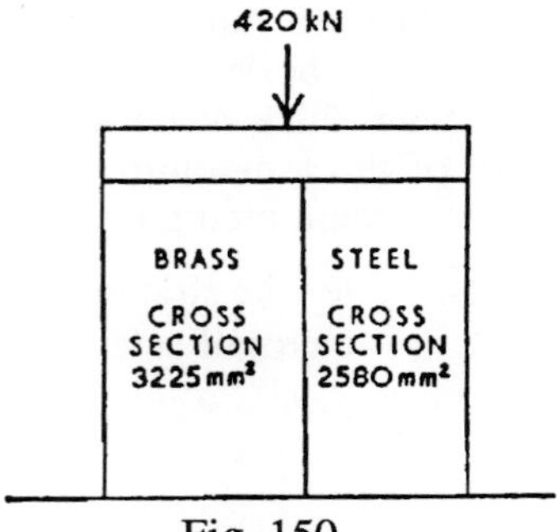

Fig. 150

From $E = \dfrac{\text{Stress}}{\text{Strain}}$ we have Strain $= \dfrac{\text{Stress}}{E}$

and from Stress $= \dfrac{\text{Load}}{\text{Area}}$ Load = Stress × Area

Letting suffix S represent steel and suffix B represent brass:

$$\text{Strain of steel} = \text{Strain of brass}$$

$$\frac{\text{Stress}_S}{E_S} = \frac{\text{Stress}_B}{E_B}$$

$$\text{Stress}_S = \frac{\text{Stress}_B \times E_S}{E_B}$$

$$= \frac{\text{Stress}_B \times 207 \times 10^9}{92 \times 10^9}$$

$$\text{Stress}_S = 2{\cdot}25 \times \text{Stress}_B \quad \dots \dots \dots \dots \text{(i)}$$

Load carried by steel + Load carried by brass = Total load

$$\text{Stress}_S \times \text{Area}_S + \text{Stress}_B \times \text{Area}_B = 420 \times 10^3$$

$$\text{Stress}_S \times 2580 \times 10^{-6} + \text{Stress}_B \times 3225 \times 10^{-6} = 420 \times 10^3 \quad \text{(ii)}$$

Substituting value of Stress_S from (i) into (ii),

$$2{\cdot}25 \times \text{Stress}_B \times 2580 \times 10^{-6} + \text{Stress}_B \times 3225 \times 10^{-6} = 420 \times 10^3$$

$$5805\ \text{Stress}_B + 3225\ \text{Stress}_B = 420 \times 10^9$$

$$9030\ \text{Stress}_B = 420 \times 10^9$$

$$\text{Stress}_B = 4{\cdot}65 \times 10^7$$

$$\sigma_B = 46{\cdot}5\ \text{MN/m}^2 \quad \text{Ans. (i)}$$

From (i),

$$\text{Stress}_S = 2{\cdot}25 \times \text{Stress}_B$$

$$\sigma_S = 104{\cdot}6\ \text{MN.m}^2 \quad \text{Ans. (ii)}$$

The strain is calculated by dividing the stress in either material by its respective modulus of elasticity:

$$\varepsilon = \frac{\sigma}{E}$$

$$= \frac{46{\cdot}5 \times 10^6}{92 \times 10^9} \text{ or } \frac{104{\cdot}6 \times 10^6}{207 \times 10^9}$$

$$= 5{\cdot}054 \times 10^{-4} \quad \text{Ans. (ii)}$$

EQUIVALENT MODULUS OF ELASTICITY OF COMPOUND BARS

The equivalent modulus of elasticity of compound bars is the value of the modulus of an equivalent single bar which would have the same total area and behave in the same manner as the compound bars under all conditions of loading within the elastic limit.

Let there be two bars, A and B, such as those shown in Fig. 150, of same length, but different cross-sectional areas and different materials.

Let σ = stress, A = area, l = length, x = change of length, E = modulus of elasticity

$$\text{Total load} = \text{load}_A + \text{load}_B = \sigma_A A_A + \sigma_B A_B$$

$$\text{Total area} = A_A + A_B$$

$$E = \frac{\text{stress}}{\text{strain}} = \frac{\text{load}}{\text{area} \times \text{strain}}$$

Strain $= \dfrac{\text{stress}}{E}$ and the strain being the same for both bars

substitute $\dfrac{\sigma_A}{E_A}$ for strain:

$$\text{Equivalent } E = \frac{\text{total load}}{\text{total area} \times \text{strain}}$$

$$= \frac{(\sigma_A A_A + \sigma_B A_B) \times E_A}{(A_A + A_B) \times \sigma_A} \quad \dots \dots \text{(i)}$$

$$\text{strain}_A = \text{strain}_B \qquad \therefore \frac{\sigma_A}{E_A} = \frac{\sigma_B}{E_B}$$

$\sigma_B = \dfrac{\sigma_A \times E_B}{E_A}$ substituting this into (i):

$$\text{Equiv. } E = \frac{\left\{\sigma_A A_A + \dfrac{\sigma_A E_A}{E_A} \times A_B\right\} \times E_A}{(A_A + A_B) \times \sigma_A}$$

$$= \frac{A_A E_A + A_B E_A}{A_A + A_{B]}}$$

Taking the compound bar in the previous example,

$$\text{E} = \frac{\text{stress}}{\text{strain}} = \frac{\text{load}}{\text{area} \times \text{strain}}$$

Equivalent E of the compound bar *from first principles*,

$$= \frac{\text{total load}}{\text{total area}} \times \frac{1}{\text{strain}}$$

$$= \frac{420 \times 10^3}{(3225 + 2580) \times 10^{-6}} \times \frac{1}{5{\cdot}054 \times 10^{-4}}$$

$$= \frac{420 \times 10^{13}}{5805 \times 5{\cdot}054}$$

$$= 1{\cdot}431 \times 10^{11} \text{ N/m}^2 \text{ or } 143{\cdot}1 \text{ GN/m}^2$$

Alternatively, the expression derived above *could* be used:

$$\text{Equiv. } E = \frac{A_S E_S + A_B E_B}{A_S + A_B}$$

$$= \frac{2580 \times 10^{-6} \times 207 \times 10^9 + 3225 \times 10^{-6} \times 92 \times 10^9}{2580 \times 10^{-6} + 3225 \times 10^{-6}}$$

$$\frac{2580 \times 207 + 3225 \times 92}{2580 + 3225} \times 10^9$$

$$= \frac{830760 \times 10^9}{5805}$$

$$= 1{\cdot}431 \times 10^{11} \text{ N/m}^2 \text{ as above.}$$

STRESS DUE TO RESTRICTED THERMAL EXPANSION

Generally speaking, metals expand when heated and contract when cooled, the amount of expansion varies with different metals and depends upon its dimensions and temperature change. The *coefficient of linear expansion* of a metal is the expansion lengthwise for each unit of length for each degree increase in temperature. For instance, the coefficient of linear expansion of copper is about 0·000018 per °C, this means that 1 mm length of copper will expand by 0·000018 mm if heated through 1°C. A length of 10 mm would expand 10 times this amount. If heated through 50°C the increase in length would be 50 times that for 1°C.

Thus, if α = coefficient of linear expansion

l = original length

θ = change of temperature.

then:

$$\text{Change of length} = \alpha l \theta$$

If the natural thermal expansion is restricted in any way, the metal will be strained from its natural length.

Example. A bar of copper 40 mm diameter and 500 mm long is firmly fixed at each end so that it cannot expand. If it is now heated through 50°C of temperature, find the strain in the bar, the stress and the equivalent load. Take the values for copper as, α = 0·000018 per °C, E = 125 GN/m^2.

Thermal expansion if free $= \alpha l \theta$

$= 0{\cdot}000018 \times 500 \times 50$

$= 0{\cdot}45$ mm

Free lengths should then be $500 + 0{\cdot}45$

$= 500{\cdot}45$ mm

If prevented from expanding the bar is then 0·45 mm shorter than it should be, therefore it is compressed 0·45 mm from its natural length of 500·45 mm,

$$\varepsilon = \frac{x}{l}$$

$$= \frac{0{\cdot}45}{500{\cdot}45}$$

It is usual however to make a slight approximation by taking the original cold length (500 mm) instead of the heated length (500·45). The difference in the final answer is negligible and it is much more convenient for the solution of more complicated problems to follow. It will also be seen that by making this approximation the length of the material is not required to be known because the strain can be obtained direct from:

$$\varepsilon = \frac{x}{l}$$

$$= \frac{\alpha l \theta}{l}$$

$$= \alpha \theta$$

Hence,

$$\varepsilon = \frac{0{\cdot}45}{500} \text{ or } 0{\cdot}000018 \times 50$$

$$= 0{\cdot}0009 \qquad \text{Ans. (i)}$$

$$\sigma = \varepsilon \times E$$

$$= 0{\cdot}0009 \times 125 \times 10^9$$

$$= 1{\cdot}125 \times 10^8 \text{ N/m}^2$$

$$= 112{\cdot}5 \text{ MN/m}^2 \qquad \text{Ans. (ii)}$$

$$\text{Cross-sectional area of bar} = 0{\cdot}7854 \times 40^2 \times 10^{-6}\text{m}^2.$$

$$\sigma = \frac{F}{A}$$

$$\begin{aligned}\text{Equivalent load} &= \text{stress} \times \text{area} \\ &= 112{\cdot}5 \times 10^6 \times 0{\cdot}7854 \times 40^2 \times 10^{-6} \\ &= 1{\cdot}414 \times 10^5\ \text{N} \\ &= 141{\cdot}4\ \text{kN} \qquad \text{Ans. (iii)}\end{aligned}$$

f STRESS DUE TO THERMAL EXPANSION OF COMPOUND BARS

If two bars of metal of different coefficients of linear expansion are firmly fastened together and then heated, the bar with the larger coefficient will be prevented from expanding its full amount by the bar with the smallest coefficient, while the latter will be pulled further out from its natural expansion by the former. The bars will therefore be strained.

Consider two bars, A and B. Bar A has a large coefficient of linear expansion (α_A) and bar B has a small coefficient (α_B). If allowed to expand quite freely when heated, the expansion of A will be $\alpha_A l\theta$ and the expansion of B will be $\alpha_B l\theta$. If these two bars were joined together when cold so that one is prevented from expanding more than the other when they are heated, neglecting any bending tendency, A will exert an outward pull in trying to expand its full amount, and B will exert an inward pull as a resistance to being pulled out more than its natural heated length. The resulting increase of length of the combined bars will be somewhere between the two amounts of free expansion. If the actual increase of length of the bars is represented by y, then:

Bar A is pulled back a distance of $\alpha_A l\theta - y$

Bar B is pulled out by a distance of $y - \alpha_B l\theta$

It can be seen by reference to Fig. 151, which is a magnified illustration, that the sum of the above two quantities is equal to the difference between their free expansions, thus:

$$(\alpha_A l\theta - y) + (y - \alpha_B l\theta) = \alpha_A l\theta - \alpha_B l\theta$$

As strain is change of length divided by original length, divide every term by l to obtain an expression in terms of the strain in each bar:

$$\frac{\alpha_A l\theta - y}{l} + \frac{y - \alpha_B l\theta}{l} = \frac{\alpha_A l\theta}{l} - \frac{\alpha_B l\theta}{l}$$

which is:

Strain in bar A + Strain in bar B = $\alpha_A\theta - \alpha_B\theta$

In words this is:

Sum of the strains = Difference in free expansion per unit length

This equation, coupled with the fundamental fact that for equilibrium:

Outward pull of one bar = Inward pull of the other provides the basis of solution for problems on this subject.

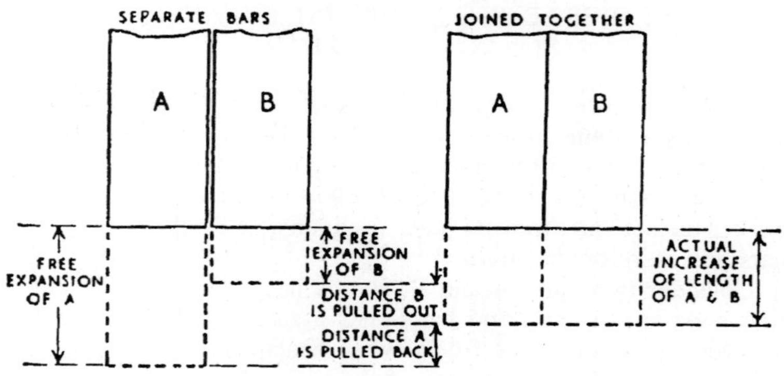

Fig. 151

Example. A solid bar of steel 80 mm diameter is placed inside a brass tube 80 mm inside diameter, 10 mm thick and of equal length. The two ends of the bar and tube are firmly fixed together and the whole is heated through 100°C. Find the stress in the steel and brass taking the values:

Coeff. of linear exp. for steel = 11×10^{-6} per °C

Coeff. of linear exp. for brass = 19×10^{-6} per °C

Modulus of elasticity for steel = 206 GN/m^2

Modulus of elasticity for brass = 103 GN/m^2

The brass tends to expand more than the steel because of its higher coefficient of linear expansion.

Outward pull of the brass = Inward pull of the steel

Stress $_{BRASS}$ × Area $_{BRASS}$ = Stress $_{STEEL}$ × Area $_{STEEL}$

$$\text{Stress}_B \times 0{\cdot}7854\,(100^2 - 80^2) = \text{Stress}_S \times 0{\cdot}7854 \times 80^2$$

$$\text{Stress}_B \times 3600 = \text{Stress}_S \times 6400$$

$$\text{Stress}_B = \text{Stress}_S \times {}^{16}\!/_9 \quad \ldots \; \ldots \quad \text{(i)}$$

Sum of the strains = Difference in expansion per unit length.

$$\frac{\text{Stress}_B}{E_B} + \frac{\text{Stress}_S}{E_S} = \alpha_B\theta - \alpha_S\theta$$

Substituting equivalent of Stress $_B$ from (i) and inserting values of E, α and θ,

$$\frac{\text{Stress}_S \times 16}{103 \times 10^9 \times 9} + \frac{\text{Stress}_S}{206 \times 10^9} = 100 \times 10^{-6}\,(19 - 11)$$

Multiplying throughout by $206 \times 10^9 \times 9$ and simplifying:

$$32\ \text{Stress}_S + 9\ \text{Stress}_S = 100 \times 10^{-6} \times 8 \times 206 \times 10^9 \times 9$$

$$41\ \text{Stress}_S = 8 \times 206 \times 9 \times 10^5$$

$$\sigma_S = 3{\cdot}618 \times 10^7\ \text{N/m}^2$$

$$= 36{\cdot}18\ \text{MN/m}^2 \qquad \text{Ans. (i)}$$

From (i),

$$\sigma_B = 36{\cdot}18 \times {}^{16}\!/_9$$

$$= 64{\cdot}32\ \text{MN/m}^2 \qquad \text{Ans. (ii)}$$

f ELASTIC STRAIN ENERGY

The work done by a force is the product of the average force applied and the distance through which the force moves.

Consider compressing a helical spring, let this spring have a stiffness of say, 100 N per mm, that is, it will compress (or stretch) 1 mm if an axial load of 100N is placed on it, or 2 mm if it carries 200 N, etc. Imagine a load of 400 N being lowered by a crane gradually on to the spring. Whilst lowering, at the moment when the spring is compressed 1 mm, the spring is carrying 100 N of the load and the crane is still holding the remaining 300 N. At the moment when the spring is compressed 2 mm, the spring is then carrying 200 N of the load and the crane is carrying the other 200 N. At 3 mm of compression the spring now carries 300 N of the load and the crane only the remaining 100 N. At 4 mm of compression, the whole load of 400 N rests on the spring and the crane carries none, the crane hook can now be uncoupled. Thus the load on the spring was gradually increased from zero to the full 400 N and the *average* load applied to the spring during its compression up to 4 m is:

$$\text{Average load} = \tfrac{1}{2}(0 + 400) = 200 \text{ N}$$

The work done in compressing the spring is the average load multiplied by the distance it moves, thus:

$$\begin{aligned}\text{Work done} &= \text{Average load} \times \text{compression}\\ &= 200 \times 4\\ &= 800 \text{ N mm} = 0{\cdot}8 \text{ N m} = 0{\cdot}8 \text{ J}\end{aligned}$$

This is one method of storing energy, by straining the material within its elastic limit. If a spring is compressed and locked in that position by a trigger, it will contain potential energy by virtue of its strained condition, on releasing the trigger the spring can do work as it returns to its original length, the work given out equal to the work which was put into the spring to compress it.

The above is a magnification of what happens when a load is gradually applied on a bar of metal (all loads are taken as being gradually applied unless stated to be otherwise). In metals the amount of compression (or extension) is very small in comparison with a spring but the principle is the same providing the metal is strained within the elastic limit.

The energy stored in a material strained within the elastic limit is termed the *elastic strain energy* or *resilience* and this quantity is equal to the work done to strain the material.

Let W = final load on the material

,, x = compression or extension

$$\begin{aligned}\text{Work done to strain material} &= \text{Average load} \times \text{distance}\\ &= \tfrac{1}{2}(0 + W) \times x\\ &= \tfrac{1}{2}Wx\end{aligned}$$

Fig. 152 is a graph showing the relationship between load and extension or compression when loaded within the elastic range, and it can be seen that the area under the graph represents work done (see also Chapter 4).

For convenience the load can be expressed in terms of the stress in the material, its dimensions, and its modulus of elasticity, this is more useful because it expresses the elastic strain energy with regard to its strength properties:

$$\begin{aligned}\text{Load} &= \text{stress} \times \text{cross-sect. area}\\ x &= \text{strain} \times \text{length}\\ &= \frac{\text{stress}}{E} \times \text{length}\end{aligned}$$

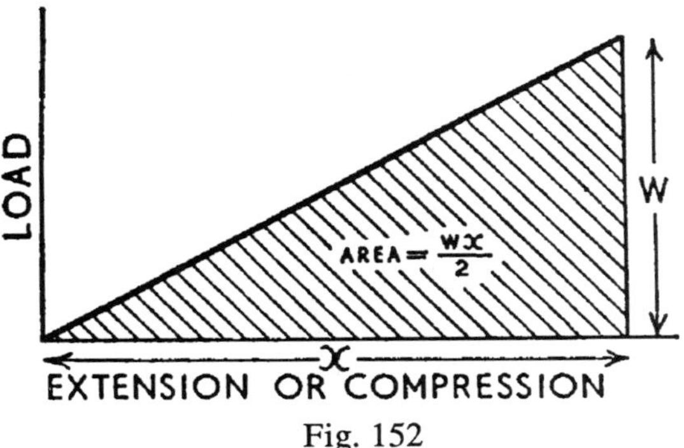

Fig. 152

Substituting these values of load (W) and extension (x)

$$\text{Elastic strain energy} = \text{work done}$$
$$= \tfrac{1}{2} Wx$$
$$= \frac{\text{stress} \times \text{area} \times \text{stress} \times \text{length}}{2 \times E}$$

$$\text{Cross-sectional area} \times \text{length} = \text{volume, therefore,}$$

$$\text{Elastic strain energy} = \frac{\text{stress}^2 \times \text{volume}}{2E} = \frac{\sigma^2 V}{2E}$$

Example. Calculate the elastic strain stored in a solid steel bar 75 mm diameter and 2 m long when subjected to a direct tensile stress of 200 MN/m^2, taking the modulus of elasticity of this material as 207 GN/m^2.

$$\text{Elastic strain energy} = \frac{\sigma^2 V}{2E}$$
$$= \frac{(200 \times 10^6) \times 0{\cdot}7854 \times 0{\cdot}075^2 \times 2}{2 \times 207 \times 10^9} \text{ Nm}$$
$$= 853{\cdot}7 \text{ J} \qquad \text{Ans.}$$

Example. Compare the resilience of two solid round bars, A and B, of similar material, equal length, and carrying an equal direct compressive load; bar A has a constant diameter over its entire length, bar B has the same diameter as A for three-quarters of its length and reduced to half diameter for its remaining quarter-length.

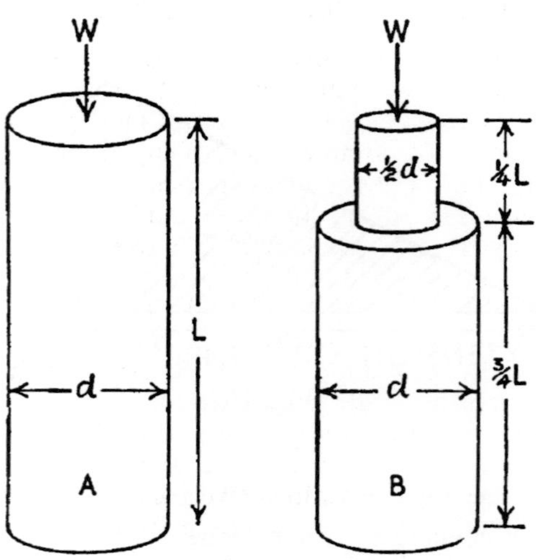

Fig. 153

Referring to Fig. 153.

Resilience of A : Resilience of B

$$\frac{\text{stress}^2 \times \text{vol.}}{2E} : \frac{\text{stress}^2 \times \text{vol.}}{2E} \text{ of large part } +$$

$$\frac{\text{stress}^2 \times \text{vol.}}{2E} \text{ of small part}$$

$$\text{volume} = \text{area} \times \text{length}$$

$$\text{and, stress} = W \div \text{area}$$

$$\frac{\text{Stress}^2 \times \text{volume}}{2E} \text{ can be written:}$$

$$\frac{W^2 \times \text{area} \times \text{length}}{\text{area}^2 \times 2E} = \frac{W^2 \times \text{length}}{\text{area} \times 2E}$$

$$\text{Also, area of diameter } d = 0{\cdot}7854d^2$$

$$\text{,,} \quad \text{,,} \quad \tfrac{1}{2}d = 0{\cdot}7854\,(\tfrac{1}{2}d)^2$$

$$= 0{\cdot}7854 \times \tfrac{1}{4}d^2$$

And, $2E$ cancels from every term, therefore Resilience ratio A to B becomes,

$$\frac{W^2 \times L}{0{\cdot}7854\,d^2} : \frac{W^2 \times \frac{3}{4}L}{0{\cdot}7854\,d^2} + \frac{W^2 \times \frac{1}{4}L}{0{\cdot}7854 \times \frac{1}{4}d^2}$$

W^2, L, 0·7854, and d^2 cancels from every term:

$$1 : \tfrac{3}{4} + 1$$

$$1 : 1\tfrac{3}{4} \text{ Ans.}$$

f SUDDENLY APPLIED AND SHOCK LOADS

Reverting back to the consideration of a load applied on a spring, this load was gradually applied. If, instead, the same load was held just about touching the top coil and suddenly released, the effect would be to compress the spring initially to twice the normal amount of compression, that is 8 mm instead of 4 mm, the spring and load would then vibrate up and down until, dampened by friction, the vibrations would die out and the spring settle down with its normal compression of 4 mm.

The important point to note is that the *initial* compression due to this *suddenly applied load* is twice as much as it would be had the same load been gradually applied, the strain in the material is doubled and therefore the stress is doubled.

Further, if the same load were allowed to fall on to the spring from a height h above the top coil, the impact of this shock load would cause a much greater initial strain and stress in the material, the amount depending upon the height h from which the load was allowed to fall before striking the spring.

Now consider a solid bar of metal, let a load W fall through a distance h before before striking the bar, and let the initial amount of compression be x. Provided the initial stress is within the elastic limit, then:

Potential Energy given up by falling weight = Elastic Strain Energy taken up by bar

$$W \times (h + x) = \frac{\sigma^2 V}{2E}$$

from which the stress can be calculated if the other quantities are known.

If the value of h is nil, it becomes the case of a 'suddenly applied load' without impact, thus:

$$W \times x = \frac{\sigma^2 V}{2E}$$

$$\text{Substituting, } x = \text{strain} \times \text{length}$$

$$= \frac{\text{stress}}{E} \times \text{length}$$

$$\text{and volume} = \text{area} \times \text{length:}$$

$$\text{W} \times \frac{\text{stress}}{E} \times \text{length} = \frac{\text{stress}^2 \times \text{area} \times \text{length}}{2E}$$

$$W = \frac{\text{stress} \times \text{area}}{2}$$

$$\sigma = \frac{2W}{A}$$

This shows that the initial stress caused by a suddenly applied load is twice the stress that would be produced if the same load had been gradually applied.

f Example. A load of 10 kN falls 6 mm on to the end of a vertical bar 200 mm long and 2500 mm^2 cross-sectional area. Calculate the instantaneous compression and stress, taking the modulus of elasticity of the material as 195 GN/m^2. If the same load had been (*b*) gradually applied, (*c*) suddenly applied without impact, what would be the stress in the bar?

Potential Energy lost by falling weight = Elastic Strain energy taken up by bar

$$W \times (h + x) = \frac{\sigma^2 V}{2E}$$

There are two unknowns in this case, therefore one must be expressed in terms of the other.

Stress in terms of x:

$$\text{Stress} = \text{strain} \times E$$

$$= \frac{x}{\text{length}} \times E$$

Substituting for stress and also area × length for volume:

$$W(h + x) = \frac{x^2 \times E^2 \times \text{area} \times \text{length}}{\text{length}^2 \times 2 \times E}$$

$$h + x = \frac{x^2 \times E \times \text{area}}{\text{length} \times 2 \times W}$$

$$0{\cdot}006 + x = \frac{x^2 \times 195 \times 10^9 \times 2500 \times 10^{-6}}{0{\cdot}2 \times 2 \times 10 \times 10^3}$$

$$0{\cdot}006 + x = 1{\cdot}281 \times 10^5 \times x^2$$

$$1{\cdot}218 \times 10^5 x^2 - x - 0{\cdot}006 = 0$$

Solving this quadratic,

$$x = \text{instantaneous compression}$$

$$= 0{\cdot}2261 \times 10^{-3}\ \text{m}$$

$$= 0{\cdot}2261\ \text{mm} \qquad \text{Ans. a(i)}$$

$$\text{Stress} = \text{strain} \times E$$

$$= \frac{x}{\text{length}} \times E$$

$$= \frac{0{\cdot}2261 \times 195 \times 10^9}{200}$$

$$= 2{\cdot}205 \times 10^8\ \text{N/m}^2$$

$$= 220{\cdot}5\ \text{MN/m}^2 \qquad \text{Ans. a(ii)}$$

Note that x is a small quantity and makes only a small difference when added to h. An approximate value of the stress can be obtained by neglecting x thus:

$$W \times (h + x) = \frac{\sigma^2 V}{2E}$$

Neglecting x,

$$W \times h = \frac{\sigma^2 V}{2E}$$

$$\text{Stress} = \sqrt{\frac{2EWh}{V}}$$

If the same load of 10 kN had been gradually applied:

$$\sigma = \frac{W}{A}$$

$$= \frac{10 \times 10^3}{2500 \times 10^{-6}}$$

$$= 4 \times 10^6 \text{ N/m}^2$$

$$= 4 \text{ MN/m}^2 \qquad \text{Ans. (b)}$$

If the same load had been suddenly applied:

$$\sigma = \frac{2W}{A}$$

$$= 8 \text{ MN/m}^2 \qquad \text{Ans. (c)}$$

f STRESSES ON OBLIQUE PLANES

Fig. 154 shows a tensile load P_1 applied to a material, if the area of the plane normal (at right angles or perpendicular) to the direction of the load is represented by a_1 and the stress on this plane by σ_1 then the total pull, being stress × area, can be represented by $\sigma_1 \times a_1$.

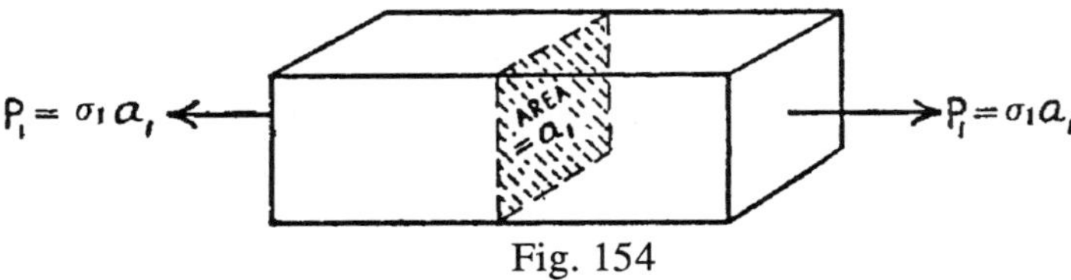

Fig. 154

Consider now a plane in the material inclined at θ° to the normal cross-section (Fig. 155), the effect of the pull P_1 in the direction of the axis is to produce two loads on the oblique plane by virtue of the rectangular components of P_1, one of these being perpendicular to the oblique plane and the other tangential to it.

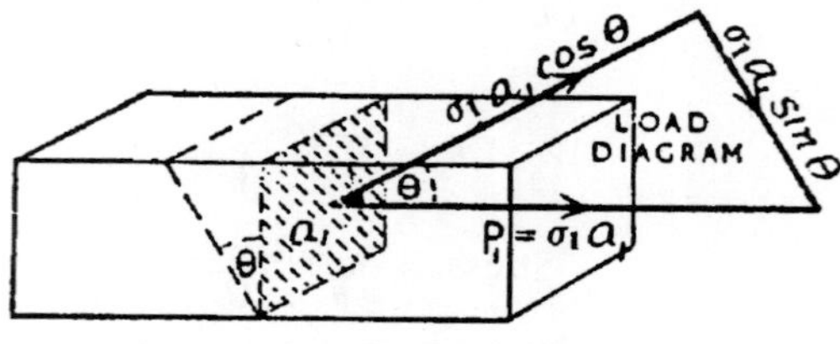

Fig. 155

Referring to the load diagram. the load perpendicular to the oblique plane is $\sigma_1 a_1 \cos\theta$ and the tangential load is $\sigma_1 a_1 \sin\theta$.

The area of the oblique plane is $a_1 \div \cos\theta$ and,stress being load ÷ area, the normal tensile stress σ_n on this oblique plane is,

$$\sigma_n = \sigma_1 a_1 \cos\theta \div \frac{a_1}{\cos\theta}$$

$$\sigma_n = \sigma_1 \cos^2\theta \quad \dots\ \dots\ \dots\ \dots \quad \text{(i)}$$

The tangential load produces a shear stress τ thus:

$$\tau = \sigma_1 a_1 \sin\theta \div \frac{a_1}{\cos\theta}$$

$$\tau = \sigma_1 \sin\theta \cos\theta$$

or, since $\sin\theta\cos\theta = \frac{1}{2}\sin 2\theta$, this may be written:

$$\tau = \tfrac{1}{2}\sigma_1 \sin 2\theta \quad \dots\ \dots\ \dots \quad \text{(ii)}$$

The maximum value of τ will be when the value of $\sin 2\theta$ is a maximum. The maximum sine is unity which is for an angle of 90°. If 2θ is 90° then $\theta = 45°$, therefore the shear stress will be maximum across a plane inclined at 45°.

$$\tau_{max} = \tfrac{1}{2}\sigma_1 \sin(2 \times 45°)$$
$$= \tfrac{1}{2}\sigma_1$$

Example. A bar 25 mm diameter is subjected to a direct tensile force of 60 kN. Calculate the normal and shear stresses on a plane at 35° to the flat end of the bar. Calculate also the maximum shear stress in the material.

$$\sigma_1 = \frac{W}{A} = \frac{60 \times 10^3}{0{\cdot}7854 \times 25^2 \times 10^{-6}}$$
$$= 1{\cdot}222 \times 10^8 \text{ N/m}^2 = 122{\cdot}2 \text{ MN/m}^2$$

$$\sigma_n = \sigma_1 \cos^2 \theta$$
$$= 122{\cdot}2 \times 0{\cdot}8192^2$$
$$= 82{\cdot}04 \text{ MN/m}^2 \qquad \text{Ans. (i)}$$
$$\tau = \tfrac{1}{2} \sigma_1 \sin 2\theta$$
$$= \tfrac{1}{2} \times 122{\cdot}2 \times 0{\cdot}9397$$
$$= 57{\cdot}41 \text{ MN/m}^2 \qquad \text{Ans. (ii)}$$
$$\tau_{max} = \tfrac{1}{2} \sigma_1$$
$$= 61{\cdot}1 \text{ MN/m}^2 \qquad \text{Ans. (iii)}$$

If the bar in Fig. 155 is subjected to a tensile load P_2 only, in a direction at right angles to the axis, as shown in Fig. 156 then, considering the equilibrium of a piece of the material such as the wedge shaped portion, let a_2 be the area normal to the load P_2 and the stress in this direction be represented by σ_2, the load can therefore be represented by $\sigma_2 a_2$.

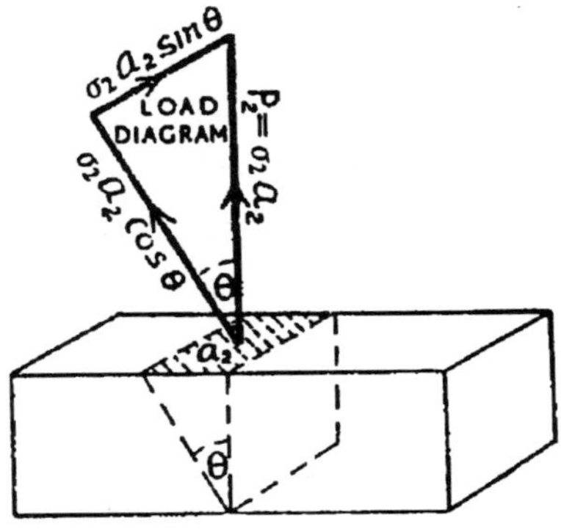

Fig. 156

The rectangular components of this load, normal and tangential to the oblique plane are $\sigma_2 a_2 \sin \theta$ and $\sigma_2 a_2 \cos \theta$. The area of the oblique plane is $a_2 \div \sin \theta$.

The normal tensile stress on the oblique plane σ_n due to P_2 is therefore:

$$\sigma_n = \sigma_2 a_2 \sin \theta \div \frac{a_2}{\sin \theta}$$

$$\sigma_n = \sigma_2 \sin^2 \theta \quad \ldots \ldots \ldots \ldots \ldots \ldots \ldots \ldots \ldots \quad \text{(iii)}$$

The shear stress on the oblique plane due to P_2 is:

$$\tau = \sigma_2 a_2 \cos\theta \div \frac{a_2}{\sin\theta}$$

$$\tau = \sigma_2 \sin\theta \cos\theta$$

$$\tau = \tfrac{1}{2}\sigma_2 \sin 2\theta \qquad \dots \dots \dots \dots \dots \dots \dots \dots \qquad \text{(iv)}$$

Combining the above results when tensile stresses of σ_1 and σ_2 at right angles to each other act at the same time on a material, the normal stress σ_n and the shear stress τ on the oblique plane are the algebraic sums of the effects of σ_1 and σ_2 as follows, and illustrated in Fig. 157.

From (i) and (iii),

$$\sigma_n = \sigma_1 \cos^2\theta + \sigma_2 \sin^2\theta \quad \dots \dots \dots \dots \dots \dots \qquad \text{(v)}$$

From (ii) and (iv),

$$\tau = \tfrac{1}{2}\sigma_1 \sin 2\theta - \tfrac{1}{2}\sigma_2 \sin 2\theta$$

$$\tau = \tfrac{1}{2}\sin 2\theta\,(\sigma_1 - \sigma_2) \qquad \dots \dots \dots \dots \dots \dots \qquad \text{(vi)}$$

Note that (iv) was subtracted from (ii) to obtain (vi) because the shear stresses are opposite in direction.

The above expressions have been derived on the assumption that σ_1 and σ_2 are *tensile* stresses. If one of these or both were compression, they may be regarded as negative tensile stresses and the above will hold good if the appropriate signs are used.

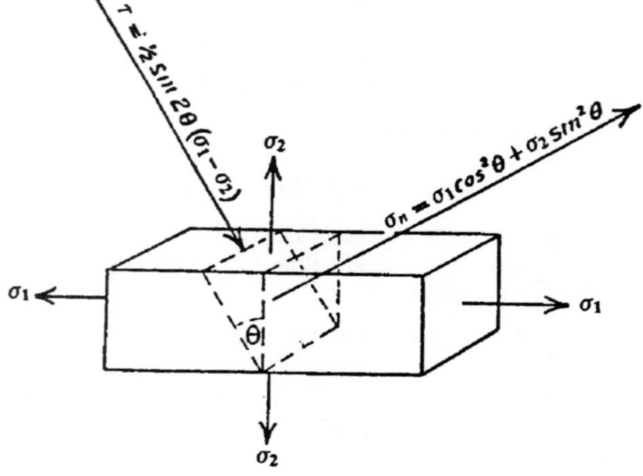

Fig. 157

Example. The circumferential tensile stress in the material of the shell plates of a starting air reservoir, due to the internal pressure, is 68 MN/m², and the longitudinal tensile stress is 34 MN/m². Calculate the normal and shear stresses on a plane of the material at 45° to the axis.

$$\begin{aligned}\sigma_n &= \sigma_1 \cos^2 45° + \sigma_2 \sin^2 45° \\ &= 68 \times 0{\cdot}7071^2 + 34 \times 0{\cdot}7071^2 \\ &= 68 \times 0{\cdot}5 + 34 \times 0{\cdot}5 \\ &= 51 \text{ MN/m}^2 \qquad \text{Ans. (i)}\end{aligned}$$

$$\begin{aligned}\tau &= \tfrac{1}{2} \sin 2\theta\,(\sigma_1 - \sigma_2) \\ &= \tfrac{1}{2} \sin 90°\,(68 - 34) \\ &= \tfrac{1}{2} \times 1 \times 34 \\ &= 17 \text{ MN/m}^2 \qquad \text{Ans. (ii)}\end{aligned}$$

COMPLEMENTARY SHEAR STRESS. If a rectangular element shown in Fig. 158, of thickness t, is subject to an applied shear stress τ there is a couple $\tau(xt)y$ acting on the element. For equilibrium this must be balanced by a couple τ^1.

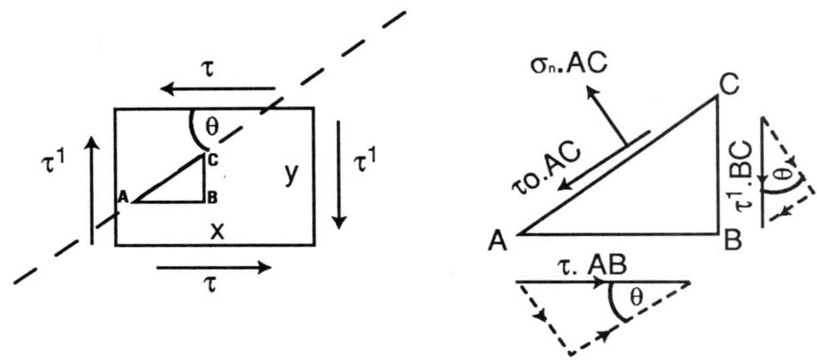

Fig. 158

For equilibrium:

$$\tau(xt)y = \tau^1(yt)x$$

$$\tau = \tau^1$$

Complementary shear stress τ^1 is equal in magnitude to the applied shear stress τ (both stresses act either towards or away from a corner).

Consider the triangular prism of material ABC and resolving forces perpendicular and parallel to the plane AC:

$$\sigma_n ACt = \tau^1 BCt \cos\theta + \tau ABt \sin\theta$$
$$\sigma_n = \tau^1 \sin\theta \cos\theta + \tau.\cos\theta \sin\theta$$
$$\sigma_n = 2\tau \sin\theta \cos\theta$$
$$\sigma_n = \tau \sin 2\theta \qquad \text{normal, tensile}$$
$$\tau_\theta ACt = \tau ABt \cos\theta - \tau^1 BCt \sin\theta$$
$$\tau_\theta = \tau \cos^2\theta - \tau^1 \sin^2\theta$$
$$\tau_\theta = \tau \cos 2\theta \qquad \text{shear}$$

TEST EXAMPLES 9

1. A tie bar made of mild steel of tensile strength 462 MN/m^2 is to carry a tensile load of 11·12 kN, find its diameter allowing a factor of safety of 12.

2. Calculate the safe load in kN that can be carried by a stud of 580·2 mm^2 cross-sectional area at the bottom of the thread allowing a safe working stress of 35 MN/m^2. Calculate also the number of these studs required to hold the cylinder cover of a diesel engine where the maximum pressure in the cylinder is 42 bar (42×10^5 N/m^2) and the diameter of the cover inside the joint is 380 mm.

3. In a cotter joint such as is shown in Fig. 142, the cotter is 100 mm wide and 25 mm thick. The tensile strength of the stay is 440 MN/m^2 and the shear strength of the cotter is 385 MN/m^2. Allowing 1·8 times the cross-sectional area of the cotter as being effective instead of twice the area for double shear, calculate the diameter of the stay and the diameter of its swelled plug end which is slotted to take the cotter, so that all parts are equal strength.

4. A tensile test is performed on a length of wire of 2 mm diameter and 4 m long, and the following data observed:

Load (N)	100	200	300	400	500	600
Extension (mm)	0·6	1·2	1·8	2·4	3·0	3·6

Plot a load-extension graph and find the modulus of elasticity of the material.

5. The following data were taken during a tensile test on a specimen of mild steel 20 mm diameter and 100 mm long between gauge points:

Yield load = 85·4 kN
Breaking load = 143·6 kN
Diameter at waist after fracture = 14·8 mm
Length between gauge points after fracture = 127 mm

Calculate (i) stress at yield point, (ii) ultimate tensile strength, (iii) percentage elongation, (iv) percentage reduction of area.

6. A cylindrical boiler, working at a pressure of 10 bar, is to have an internal diameter of 3 m. The maximum permissible tensile stress in the shell plate is 80 MN/m^2.

Calculate the minimum plate thickness (a) if joints are riveted, having a joint efficiency of 70%, (b) if the joints are welded.

The welded joints can be assumed to have the same strength as the parent metal of the shell plate.

7. Three vertical wires, each 4·5 m long, hang in the same vertical plane from a horizontal ceiling and carry a mass of 203·9 kg between them at their lower ends. The two outside wires are of steel and are each 4 mm diameter, the middle wire is brass 5 mm diameter. Find (i) the stress in each wire, (ii) the load carried by each wire, (iii) the stretch. E for steel = 200 GN/m^2, E for brass = 100 GN/m^2.

8. A straight steel steam pipe is to be fitted between two rigid bulkheads 1·5 m apart. When heated due to steam passing through, the compressive stress set up in the pipe must not exceed 177 MN/m^2. If the free expansion of the pipe in the heated condition would be 2 mm and the modulus of elasticity of the steel is 210 GN/m^2, calculate the tensile stress to induce in the pipe on assembly when cold.

9. A bronze bar 50 mm diameter and 1·25 m long is subjected to a tensile stress of 75 MN/m^2. Find the extension and the work done to stretch the bar. Take E for bronze = 85 GN/m^2.

*f*10. A bar of steel is covered by a copper sheath over its entire length and the sheath is firmly fixed to the bar so that one cannot expand more than the other. The cross-sectional area of the copper sheath is half the cross-sectional area of the steel bar. Find the stresses in the steel and the copper when this compound bar is heated through 100°C, taking the values:

Coeff. of linear expansion for steel = 12×10^{-6} per °C
Coeff. of linear expansion for copper = 17×10^{-6} per °C
E for steel = 206 GN/m^2
E for copper = 103 GN/m^2

*f*11. A bar of hexagonal cross-section measures 50 mm across the hexagonal face from one corner to its opposite corner. Find the percentage increase in resilience that would result by drilling a hole 30 mm diameter axially through the centre of the bar from end to end.

*f*12. A mass of 102 kg falls through 12 mm directly on to the end of a bar 25 mm diameter and 1·5 m long. Taking the modulus of elasticity of the material as 207 GN/m^2, find the instantaneous initial stress produced.

*f*13. A piston of 2 t mass is suspended from an engine room crane at the end of a steel cable 15 mm dia. (E = 210 GN/m^2). Calculate the stress in the cable. When the piston is being lowered at a steady speed of 0·2 m/s and is 10 m below the crane, the crane drum jams and stops suddenly. Calculate:

(a) the instantaneous extension of the cable caused by the stoppage.

(b) the momentary additional stress in the cable due to the sudden stop.

*f*14. A bar 20 mm diameter is subjected to a direct pull of 50 kN. Find the normal and shear stresses on a plane at 45° to the axis.

*f*15. Explain the meaning of the term complementary shear stress. A horizontal beam of solid rectangular section is subject at one point to a pure shear stress of 15 MN/m^2 acting in a vertical direction. Determine the normal and tangential stresses on a plane in the beam at 60° to the horizontal.

CHAPTER 10

BENDING OF BEAMS

When a material is bent, tensile and compressive stresses are set up in the material. Calculations can be made to determine these stresses and to design beams to carry loads within safe limits. Loads also tend to shear the beam.

The term *beam* is given to a single rigid length of material, usually supported horizontally to carry vertical loads. The types of beams are distinguished by the method by which they are supported.

Referring to Fig. 159, the first illustration shows a beam resting on a support at each end, this is a *simply supported beam*, the two supports on which the beam rests apply upward forces called the *reactions* to support the weight of the beam and the loads on it. The two supports may be at any two points along the beam and not necessarily at the ends. The second illustration shows a beam supported only by being built into a wall at one end, this is called *cantilever*. These two are the fundamental cases which will be dealt with in this Chapter.

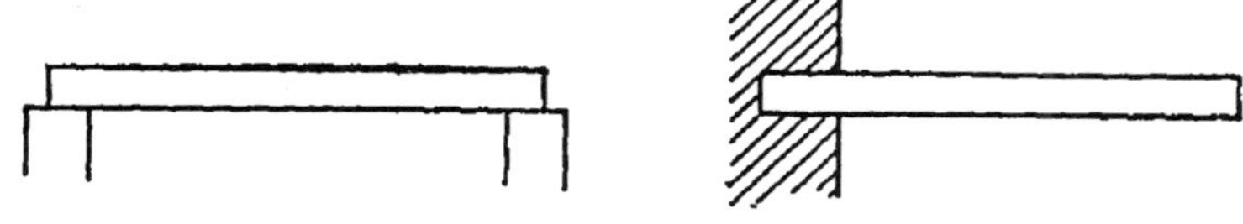

Fig. 159

The two kinds of *loading* on beams are (i) *concentrated loads* which are assumed to be loads concentrated at definite points along the beam, and (ii) *distributed loads* which are spread or distributed over the whole or part of the length of the beam. Most distributed loads will be uniformly spread, that is, every metre length of the beam will carry an equal amount of loading, but some cases may arise where the distribution is not uniform. When the weight of the beam is to be taken into account, it is treated as an additional uniformly distributed load (u.d.l.) over the entire length.

A diagrammatic sketch is always necessary to illustrate the problem, an example of such a sketch is given in Fig. 160. The supports on which the beam rests, being the reactions which apply the upward supporting forces, are denoted by R_1 and R_2. A concentrated load is denoted by an arrow on the point of the beam where the load acts and in the direction of the line of action of the load, this is usually vertically downwards but there may be some cases where the load is inclined to the vertical. Uniformly distributed loading may be denoted by a wavy line along that part of the beam over which this load is spread.

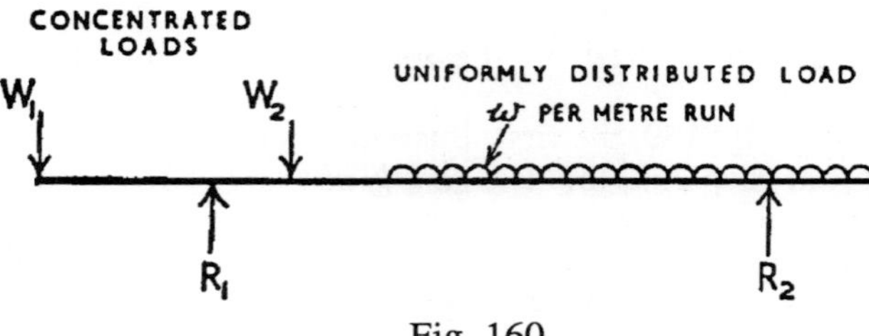

Fig. 160

CONDITIONS OF EQUILIBRIUM. A beam is 'in equilibrium' if it is at rest and obeys the simple rules of equilibrium (see also Chapter 7).

(i) The total of all upward forces must be equal to the total of all downward forces and, the total of all the forces pushing horizontally to the right must be equal to the total of all the forces pushing the the left.

(ii) The total of all moments of forces tending to turn the beam around in a clockwise direction must be equal to the total of all the moments of forces tending to turn it anticlockwise.

It is sometimes convenient in calculations to apply positive and negative signs to indicate directions. For instance, upward forces can be called positive and downward forces negative; forces pushing to the right could be positive and forces pushing to the left negative; clockwise moments can be positive and anticlockwise moments negative. For equilibrium the magnitude of all the quantities acting in one direction is equal to the magnitude of all the quantities in the opposite direction , therefore, if one direction is positive and the opposite direction negative, then equal values of positive and negative quantities added together will equal zero. Hence, the conditions of equilibrium are often stated thus:-

(i) The algebraic sum of all vertical forces must equal zero and the algebraic sum of all horizontal forces must equal zero.

(ii) The algebraic sum of all moments of forces about any given point must equal zero.

Example. A beam 10 m long rests on a support at each end and carries concentrated loads of 20 kN at 2 m from the left end and 30 kN at 4m from the right end. Calculate the two reactions.

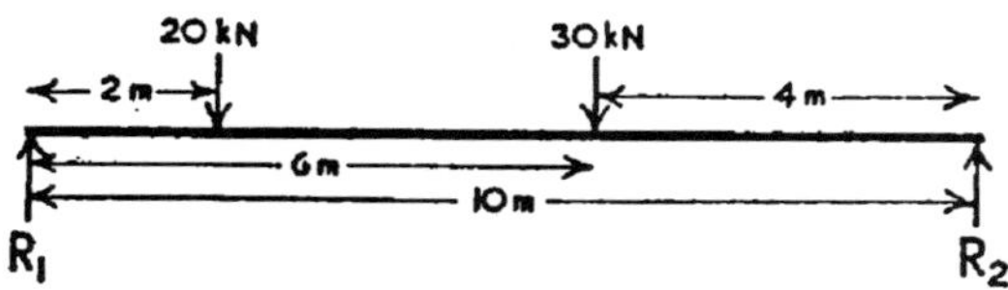

Fig. 161

If the weight of the beam is not given it infers that it is to be neglected. As the beam is at rest it obeys the conditions of equilibrium, however, the first condition cannot be applied yet because the two reactions are two unknowns. There are no horizontal forces. Therefore the second condition must be used to begin with, but take moments about one of the reactions so that this one will be eliminated in the equation, that is, if moments are taken about R_1 the moment of the force R_1 is nil.

Taking moments about R_1,

$$\begin{aligned} \text{Clockwise moments} &= \text{Anticlockwise moments} \\ (20 \times 2) + (30 \times 6) &= R_2 \times 10 \\ 40 + 180 &= R_2 \times 10 \\ 220 &= R_2 \times 10 \\ R_2 &= 22 \text{ kN} \end{aligned}$$

To find the value of R_1 take moments about R_2 and proceed, but as the value of one of the reactions is known apply the first condition which is easier to do.

$$\begin{aligned} \text{Upward forces} &= \text{Downward forces} \\ R_1 + R_2 &= 20 + 30 \\ R_1 &= 20 + 30 - 22 \\ R_1 &= 28 \text{ kN} \end{aligned}$$

The reactions are 28 and 22 kN Ans.

Example. A rolled steel joist is 20 m long, its mass is 122·3 kg/m length and is simply supported at each end. It carries concentrated loads of 20, 40 and 30 kN at points 6, 12 and 15 m respectively from one end. Calculate the two reactions.

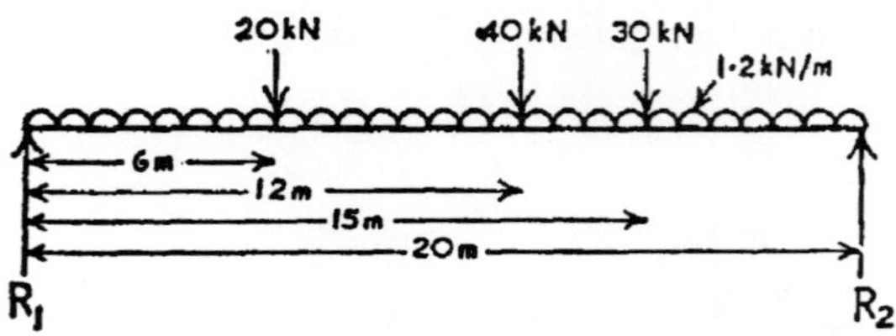

Fig. 162

$$\begin{aligned} \text{Total weight of joist} &= 122{\cdot}3 \times 20 \times 9{\cdot}81 \\ &= 2{\cdot}4 \times 10^4 \text{ N} = 24 \text{ kN} \end{aligned}$$

As the beam is uniform in section, its centre of gravity is at mid-length, that is at 10 m from one end, and the whole weight of the beam may be considered as acting through this point.

Moments about R_1 in kN m,

$$\begin{aligned} \text{Clockwise moments} &= \text{Anticlockwise moments} \\ (20 \times 6) + (40 \times 12) + (30 \times 15) + (24 \times 10) &= R_2 \times 20 \\ 120 + 480 + 450 + 240 &= R_2 \times 20 \\ 1290 &= R_2 \times 20 \\ R_2 &= 64.5 \text{ kN} \end{aligned}$$

$$\begin{aligned} \text{Upward forces} &= \text{Downward forces} \\ R_1 + R_2 &= 20 + 40 + 30 + 24 \\ R_1 &= 114 - 64{\cdot}5 \\ R_1 &= 49{\cdot}5 \text{ kN} \end{aligned}$$

The reactions are 49·5 and 64·5 kN Ans.

SHEARING FORCES & BENDING MOMENTS

The loads on a beam tend to shear the beam and also bend it.

The SHEARING FORCE (F) at any section of a beam is the algebraic sum of all the external forces perpendicular to the beam on *one* side of that section.

The BENDING MOMENT (M) at any section of a beam is the algebraic sum of all the moments of the forces on *one* side of that section. Bending moments are simply moments of forces but here they are termed 'bending moments' because they tend to bend the beam.

In each case, the algebraic sum of forces or moments of forces may be obtained from *either* side of the section, that is, to the right or to the left, as one is equal to the other.

Example. Find the shearing force and bending moment at the centre of length of a 4 m long beam which is simply supported at each end and carrying concentrated loads of 20 kN at 1·5 m from the left end and 40 kN at 1·25 m from the right end.

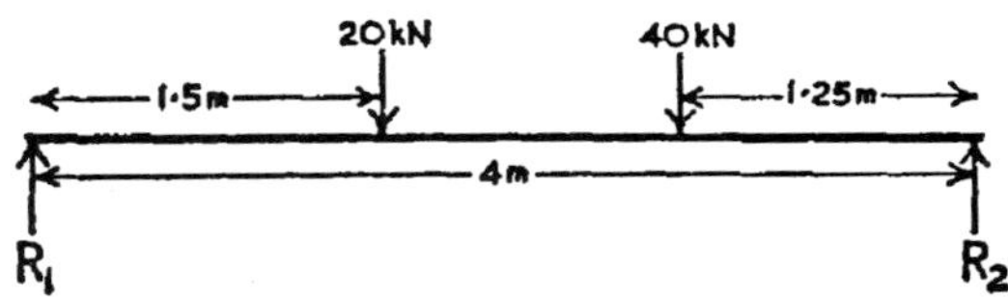

Fig. 163

Taking moments about R_1,

$$\begin{aligned} \text{Clockwise moments} &= \text{Anticlockwise moments} \\ (20 \times 1{\cdot}5) + (40 \times 2{\cdot}75) &= R_2 \times 4 \\ 30 + 110 &= R_2 \times 4 \\ 140 &= R_2 \times 4 \\ R_2 &= 35 \text{ kN} \\ \text{Upward forces} &= \text{Downward forces} \\ R_1 + R_2 &= 20 + 40 \\ R_1 &= 60 - 35 \\ R_1 &= 25 \text{ kN} \end{aligned}$$

The shearing force at the centre is the *algebraic* sum of the forces on *one* side of that section, therefore downward forces can be called negative and upward forces positive (or vice versa if preferred). Taking forces to the right of the centre, 40 kN of load acts downwards, call this negative, and 35 kN (the reaction R_2) acts upwards which will be taken as positive.

$$\begin{aligned} \text{Shearing force at centre} &= 35 - 40 \\ &= -5 \text{ kN} \quad \text{Ans. (i)} \end{aligned}$$

If the algebraic sum of the forces to the left of the centre were taken, the same result would be obtained.

The bending moment at the centre is the *algebraic* sum of the moments of forces on *one* side of that section. Therefore one direction of moment will be taken as positive and the other as negative. Consider moments to the right of the centre and examine their effects.

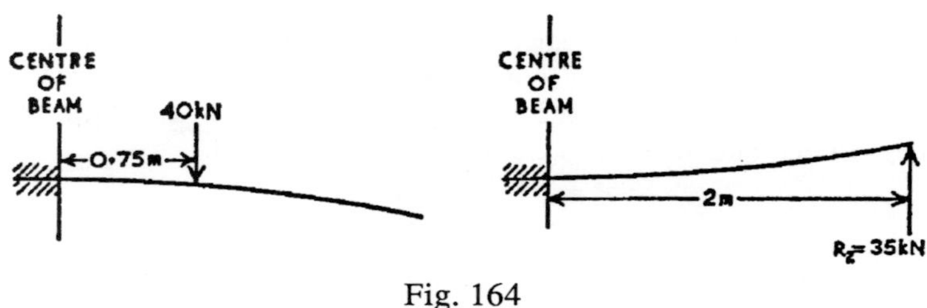

Fig. 164

Imagine the beam gripped at its centre as in Fig. 164, the effect of the 40 kN load acting alone is to cause the beam to bend in a hogging manner, and the effect of the upward force of the reaction R_2 acting alone would be to cause the beam to bend in a sagging manner. The sign of the answer to the bending moment at the centre will depend upon which is the greater and will therefore indicate whether the beam is hogging or sagging. Thus, if the moment of R_2 is taken as negative and the moment of the load positive, the answer would be a negative value because, in this case, the beam is obviously sagging.

The *sign convention* adopted is:

Positive shear force up to the right of the section considered (and, similarly down to the left).

Positive bending moment caused by clockwise moment to the right of the section considered (and, similarly anticlockwise to the left). This results in convex upwards i.e. hogging of the beam.

$$\begin{aligned}\text{Bending moment at centre} &= 40 \times 0{\cdot}75 - 35 \times 2 \\ &= 30 - 70 \\ &= -40 \text{ kN m} \quad \text{Ans. (ii)}\end{aligned}$$

Example. A cantilever 6 m long carries a concentrated load of 45 kN at its free end and also a uniformly distributed load of 20 kN/m over its entire length. Find the shearing force and bending moment at the wall.

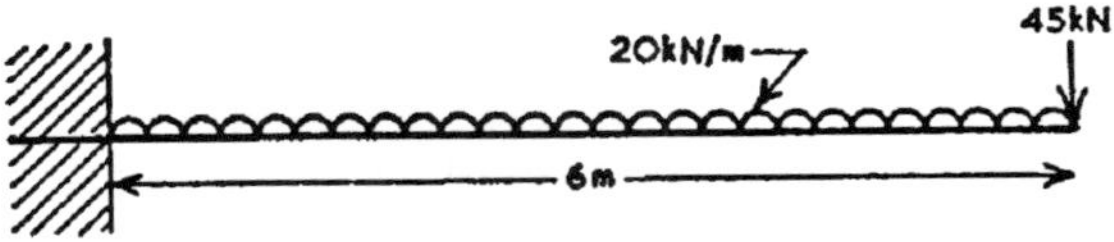

Fig. 165

The only reaction here is at the wall which supports the total load, there are no other upward forces between the wall and the free end of the beam.

$$\begin{aligned}\text{Total distributed load} &= 20 \text{ kN/m} \times 6 \text{ m}\\ &= 120 \text{ kN}\\ \text{Shearing force at the wall} &= 120 + 45\\ &= 165 \text{ kN} \quad \text{Ans. (i)}\end{aligned}$$

The centre of gravity of the distributed load is at mid-length (3 m from the wall) because this loading is uniformly spread.

$$\begin{aligned}\text{Bending moment at the wall} &= 120 \times 3 + 45 \times 6\\ &= 360 + 270\\ &= 630 \text{ kN m} \quad \text{Ans. (ii)}\end{aligned}$$

SHEARING FORCE & BENDING MOMENT DIAGRAMS

Graphs are drawn to represent the shearing force and bending moment variations over the length of the beam, these graphs are termed shearing force diagrams and bending moment diagrams. In drawing these diagrams the graphs may be plotted above or below the base line and they are usually drawn to some scale such as 1 cm to x m of beam length, 1 cm to y kN of shearing force, and 1 cm to z kN m of bending moment, x,y and z being chosen as the most convenient quantities.

In the first few examples the bending moments will be calculated for many points along the beam, say at every metre of length, so that the diagram can be plotted. If the shapes of the diagrams are carefully observed they follow standard patterns depending upon the type of loading and it will only be necessary to find the values at certain points along the beam and to join up these points by either straight lines or curves.

Example. A cantilever 4 m long carries a concentrated load of

50 kN at the free end, neglecting the weight of the beam, draw the shearing force and bending moment diagrams.

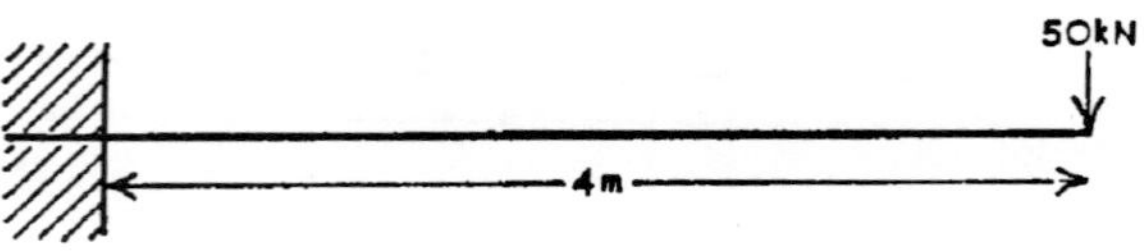

Fig. 166

The shearing force diagram is simply a diagram of up and down forces. First draw the base line to represent the length of the beam. Starting at the free end, there is a vertical downward force of 50 kN, therefore draw vertically downwards a line to represent 50 kN to scale. From the free end, moving towards the left, there is no up or down force on the beam until the wall, therefore there is no up or down movement of the graph till there, hence the graph is a straight horizontal line over this length. At the wall there is an upward force of 50 kN (this being the reaction of equal magnitude and opposite direction to the load), therefore an upward vertical line is now drawn to represent 50 kN to scale. This 'closes the diagram' to appear as in Fig. 167.

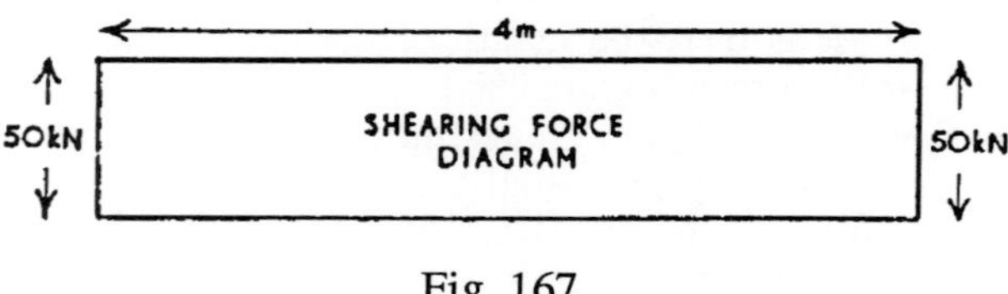

Fig. 167

For the bending moment diagram, take moments at every metre along the beam starting from the free end. Letting M represent the bending moment:

M at 1 m $= 50 \times 1 = 50$ kN m
M at 2 m $= 50 \times 2 = 100$ kN m
M at 3 m $= 50 \times 3 = 150$ kN m
M at wall $= 50 \times 4 = 200$ kN m

Draw the baseline to represent the length of the beam and measure upwards the above values of bending moments. Join the plotted points as in Fig. 168 and note that the diagram is a straight line sloping from zero at the free end to the maximum value at the wall. A hogged condition (positive *M*).

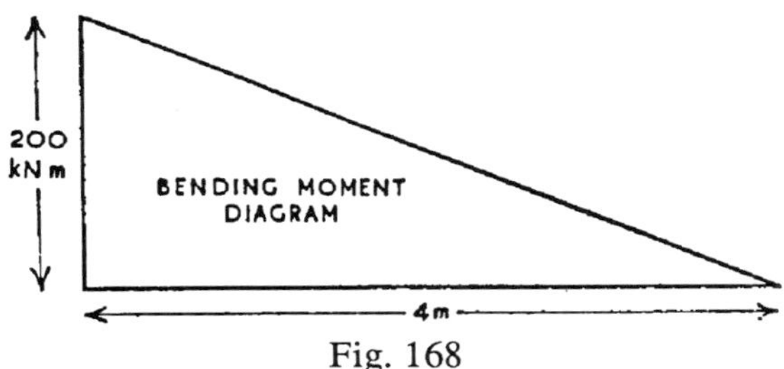

Fig. 168

Example. A cantilever 4 m long carries a uniformly distributed load over its entire length which, together with the weight of the beam, amounts to 20 kN/m run. Draw the shearing force and bending moment diagrams.

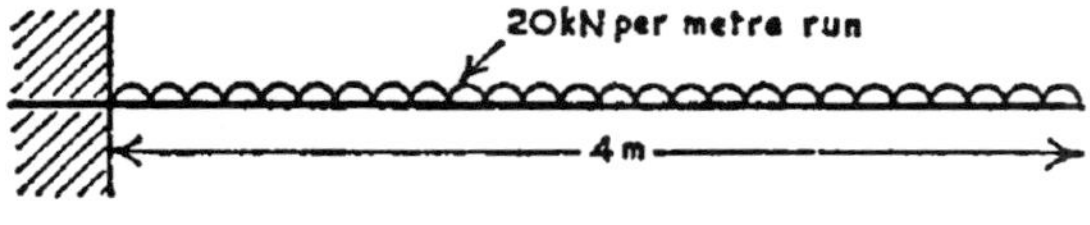

Fig. 169

At the free end the shearing force is zero. At 1 m from the free end, the load on one side of this section (to the right), is 20 kN, therefore the shearing force is 20 kN. At 2 m from the free end, the load to the right of this section is 40 kN because there are 20 kN on each metre over a distance of 2 m, therefore the shearing force there is 40 kN. At 3 m from the free end the shearing force is 60 kN. At 4 m from the free end, *i.e.* at the wall, the shearing force is 80 kN. Thus the shearing force varies from zero at the free end to a maximum of 80 kN at the wall, therefore the shearing force diagram is drawn accordingly, dropping uniformly from the base line at the free end, at at the rate of 20 kN/m of length, until it is 80 kN at the wall, that is, by a straight sloping line as in Fig. 170. At the wall there is an upward force (the reaction) of 80 kN, the upward line to represent this joins the end of the sloping line with the end of the base line to close the diagram.

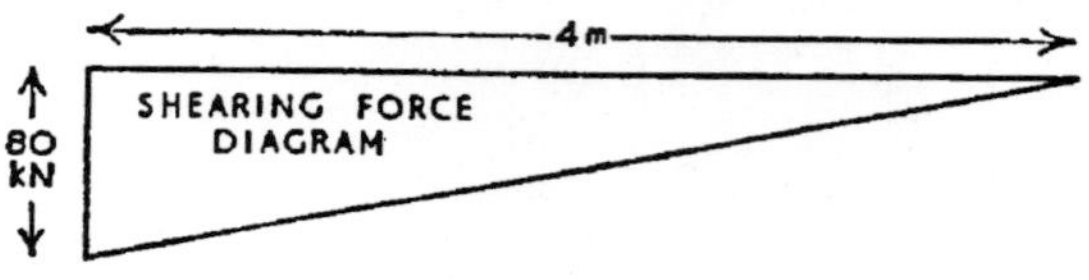

Fig. 170

The bending moment at the free end is zero. Considering the section at 1 m from the free end, to the right of this section there is a load of 20 kN spread over 1 m, its centre of gravity is mid-way between this point and the free end, *i.e.* 0·5 m, therefore the bending moment at 1 m from the free end is 20 kN × 0·5 m = 10 kN m. At 2 m from the free end there is 40 kN of load between this point and the free end and, being uniformly spread, the centre of gravity of this 40 kN is at its own mid-length which is 1 m from this point, therefore the bending moment here is 40 kN × 1 m = 40 kN m. At 3 m from the free end, there are 60 kN of load to the right of this section and the centre of gravity of this 60 kN is at the mid-point of its 3 m, which is 1·5 m from this section, the bending moment here is therefore 60 × 1·5 = 90 kN m. At 4 m from the free end, which is at the wall, 80 kN of load to the right of this section, the centre of gravity of which is at 2 m from the wall, hence the bending moment at the wall is 80 × 2 = 160 kN m:-

M at 1 m from free end = 20 × 0·5 = 10 kN m
M at 2 m from free end = 40 × 1 = 40 kN m
M at 3 m from free end = 60 × 1·5 = 90 kN m
M at 4 m from free end = 80 × 2 = 160 kN m

Plotting these values from the baseline and drawing through the plotted points the bending moment diagram is a curve, in fact it is a parabola.

If the uniformly distributed load was w per metre run, consider a section of the beam at AA which is x distance from the free end as in Fig. 172 then:-

F at AA = total load to the right
= $-wx$ (negative as down to right)

M at AA = load to the right × distance of its c.g. from AA
= $wx \times \frac{1}{2}x$
= $\frac{1}{2} wx^2$ (positive as hogging)

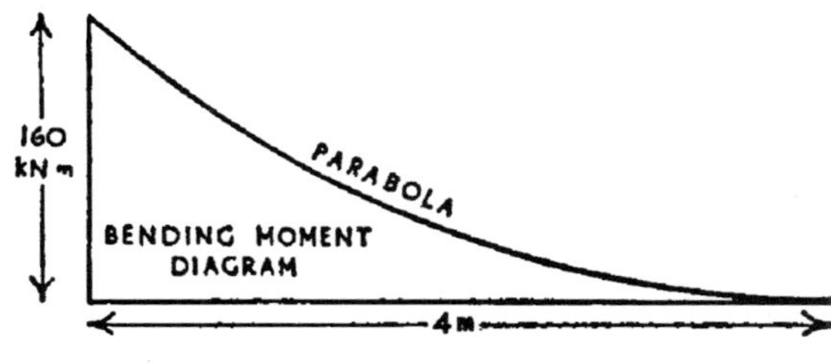

Fig. 171

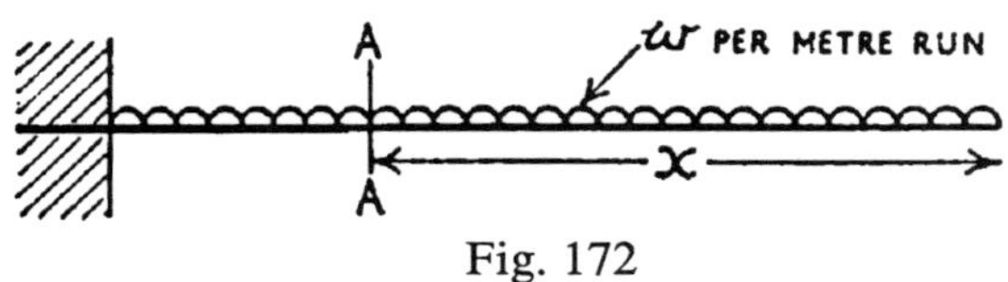

Fig. 172

Fig. 172 shows that the shearing force F varies as x and therefore the shearing force diagram is a straight sloping line; the bending moment M varies as x^2 and therefore the bending moment diagram is a parabolic curve.

Example. A beam 10 m long is simply supported at each end and carries concentrated loads of 20, 40 and 50 kN at 3, 6 and 8 m respectively from one end. Draw the shearing force and bending moment diagrams.

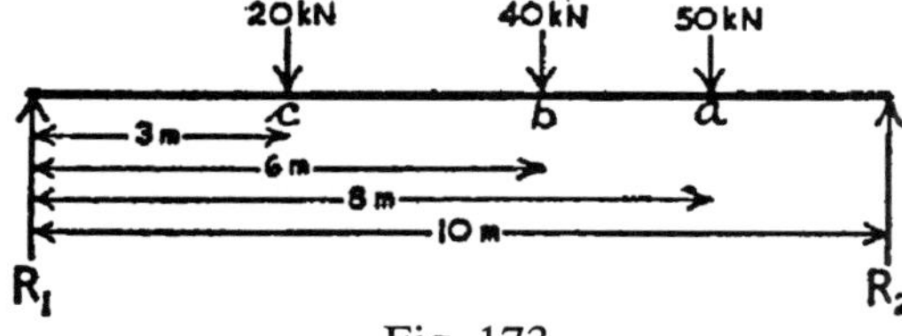

Fig. 173

Taking moments about R_1,

$$
\begin{aligned}
\text{Clockwise moments} &= \text{Anticlockwise moments} \\
20 \times 3 + 40 \times 6 + 50 \times 8 &= R_2 \times 10 \\
60 + 240 + 400 &= R_2 \times 10 \\
700 &= R_2 \times 10 \\
R_2 &= 70 \text{ kN}
\end{aligned}
$$

$$\begin{aligned} \text{Upward forces} &= \text{Downward forces} \\ R_1 + R_2 &= 20 + 40 + 50 \\ R_1 &= 110 - 70 \\ R_1 &= 40 \text{ kN} \end{aligned}$$

The shearing force diagram can now be drawn (see Fig. 174).

Note that although this shearing force diagram is drawn by commencing at the right hand end, *either* end could be taken as the starting point, the only difference in these diagrams is that one would be a mirror reflection of the other.

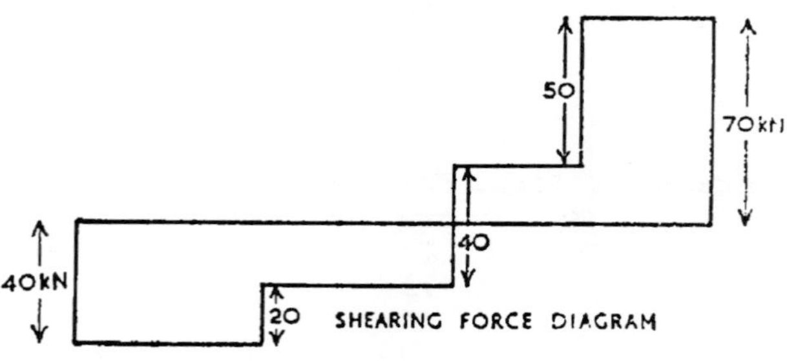

Fig. 174

As the bending moment at any section is the algebraic sum of all the moments of the forces to *either* side of that section, take the easier direction by choosing that which contains the least number of forces.

M at a (taking moments to the right of a)

$$= -70 \times 2 = -140 \text{ kN m}$$

M at b (taking moments to the right of b)

$$= -70 \times 4 + 50 \times 2$$
$$= -280 + 100 = -180 \text{ kN m}$$

M at c (taking moments to the left of c)

$$= -40 \times 3 = -120 \text{ kN m}$$

The bending moment at each end of the beam is obviously zero. The sagging beam gives a negative bending moment diagram. Fig. 175 shows the completed bending moment diagram.

Note the two diagrams. From the bending moment diagram the maximum bending moment in the beam occurs at 4 m from the right hand end. From the shearing force diagram the diagram crosses the baseline at this point. This is important because it is always true that the maximum bending moment in the beam occurs at the section where the shearing force diagram crosses the baseline, *i.e.* zero shear force.

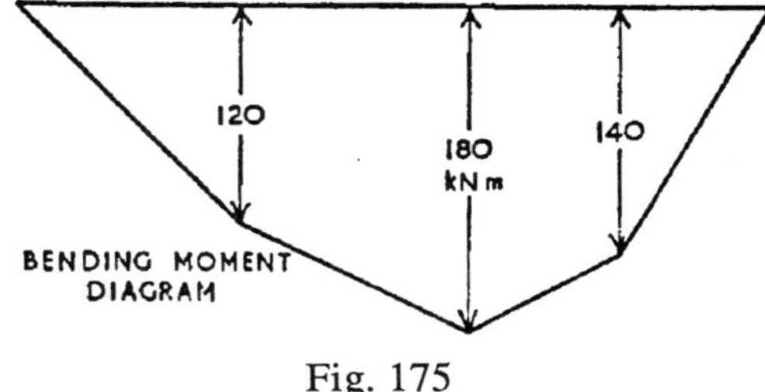

Fig. 175

Example. A beam 8 m long is simply supported at each end and carries a uniformly distributed load including the weight of the beam of 30 kN/m length. Draw the shearing force and bending moment diagrams.

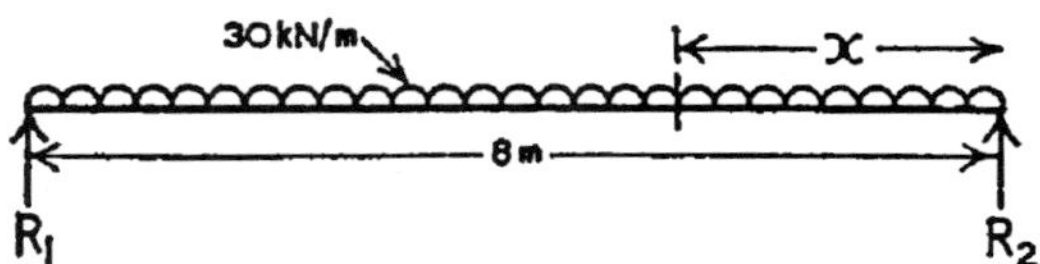

Fig. 176

$$\text{Total load on beam} = 30 \times 8 = 240 \text{ kN}$$

As it is uniformly distributed, each reaction must carry half the total load,

$$R_1 = R_2 = 120 \text{ kN}$$

The shearing force diagram can now be drawn. Starting from the baseline at the right hand end, an upward line is drawn to represent $R_2 = 120$ kN. Coming along the beam from here, the line is a straight sloping line representing a downward slope of 30 kN for each m, without interruption until the other end of the beam is reached where an upward force of $R_2 = 120$ kN closes the diagram, as in Fig. 177.

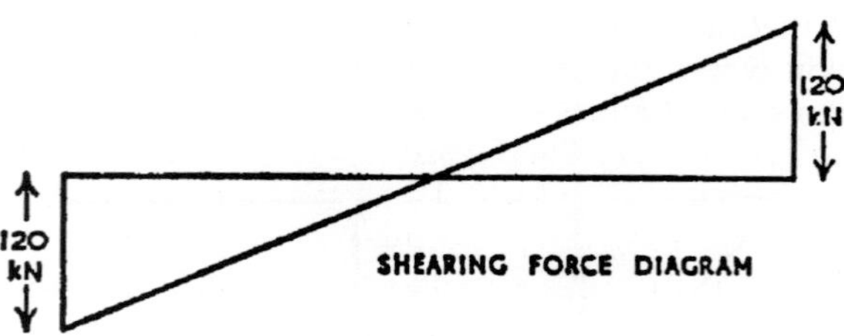

Fig. 177

Consider a section of the beam at x m from the right hand end. Taking moments about this point, to the right $R_2 = 120$ kN pushing *upwards* with a leverage of x m, and $30 \times x$ kN of load acting *downwards* whose centre of gravity is at its mid-length of $\frac{1}{2}x$. Therefore,

$$M \text{ at } x = -120 \times x + 30x \times \tfrac{1}{2}x$$
$$= -120x + 15x^2$$

From this the bending moment diagram is a parabolic curve and as a guide to the plotting of this curve the bending moments at say three points can be calculated, thus,

$$\text{When } x = 2 \text{ m}, M = -120 \times 2 + 15 \times 2^2$$
$$= -180 \text{ kN m}$$
$$\text{When } x = 4 \text{ m}, M = -120 \times 4 + 15 \times 4^2$$
$$= -240 \text{ kN m}$$
$$\text{When } x = 6 \text{ m}, M = -120 \times 6 + 15 \times 6^2$$
$$= -180 \text{ kN m}$$

Therefore the bending moment diagram appears as in Fig. 178.

The maximum bending moment is at mid-length of the beam, note again that the shearing force diagram crosses its baseline at the point of maximum bending moment and the value of the shearing force here is zero.

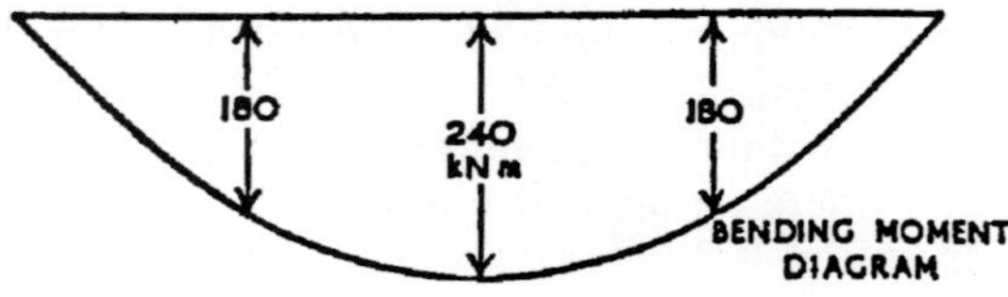

Fig. 178

Draw all diagrams under each other, with the load diagram at the top followed by the shearing force and bending moment diagrams.

Example. A beam 16 m long is simply supported at each end. It carries a uniformly distributed load of 5 kN/m, and concentrated loads of 30 and 50 kN at 2 and 10 m respectively from one end. Draw the shearing force and bending moment diagrams, state the position of maximum bending moment and calculate its value.

Total distributed load $= 5 \times 16 = 80$ kN

Its centre of gravity is at mid-length $= 8$ m

Moments about R_1

$$\begin{aligned} \text{Clockwise moments} &= \text{Anticlockwise moments} \\ (30 \times 2) + (50 \times 10) + (80 \times 8) &= R_2 \times 16 \\ 60 + 500 + 640 &= R_2 \times 16 \\ 1200 &= R_2 \times 16 \\ R_2 &= 75 \text{ kN} \\ \text{Upward forces} &= \text{Downward forces} \\ R_1 + R_2 &= 30 + 50 + 80 \\ R_1 &= 160 - 75 \\ R_1 &= 85 \text{ kN} \end{aligned}$$

The shearing force diagram is drawn by starting at R_2 (see Fig. 179).

The bending moment diagram could be plotted by finding the bending moments at every metre along the beam but a neater method is to consider (i) the distributed load acting alone, and (ii) the concentrated loads only, finding sufficient plotting points for each and drawing separate diagrams, one arbitrarily *above* the base line and the other *below* the baseline. The bending moment at any point along the beam is then the value measured right down across the whole combined diagram. *Strictly* the whole of the diagram should be plotted below the line (negative, sagging beam). Note that if the beam carried the distributed load only, each reaction would be half of the total load of 80 kN, which is 40 kN each; if the beam carried the concentrated loads only, each reaction would be 40 kN less than their values for the combined loading, that is, $R_1 = 85 - 40 = 45$ kN, and $R_2 = 75 - 40 = 35$ kN.

The shearing force diagram crosses its baseline under the 50 kN load, *i.e.* at 6 m from the right-hand end, therefore the maximum bending moment is at this section.

$$\begin{aligned} M \text{ at } 6 \text{ m} &= -75 \times 6 + 5 \times 6 \times 3 \\ &= -450 + 90 \end{aligned}$$

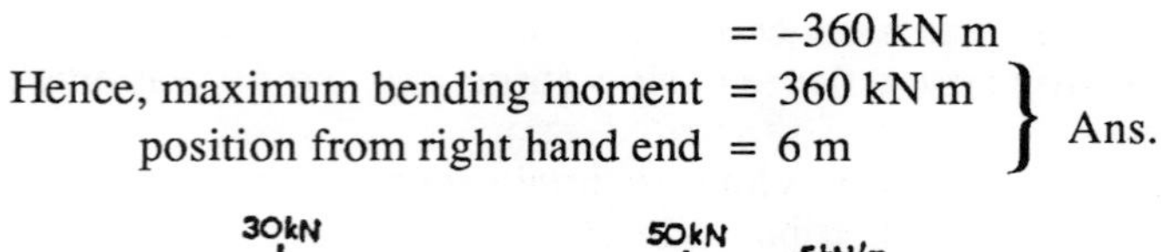

$= -360$ kN m

Hence, maximum bending moment $= 360$ kN m
position from right hand end $= 6$ m } Ans.

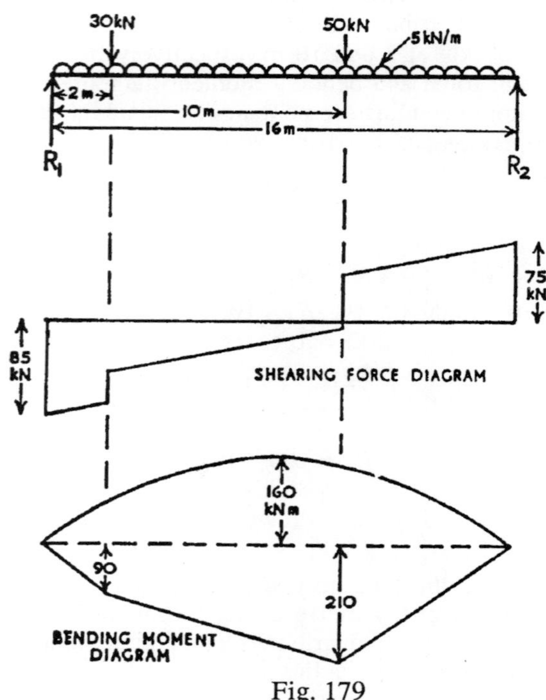

Fig. 179

Example. A beam 20 m long is simply supported at 4 m from each end and carries a uniformly distributed load of 2·5 kN/m run. Calculate the bending moments at mid-span and at the reactions, sketch the shearing force and bending moment diagrams and find the position along the beam where the bending moment is zero.

$$\text{Total load} = 2{\cdot}5 \times 20 = 50 \text{ kN}$$

$$\text{Each reaction} = \tfrac{1}{2} \text{ of } 50 = 25 \text{ kN}$$

The shearing force diagram is drawn as in Fig. 180.

$$M \text{ at centre} = -25 \times 6 + 2{\cdot}5 \times 10 \times 5$$
$$= -150 + 125$$
$$= -25 \text{ kN m} \quad \text{Ans. (i)}$$

$$M \text{ at reactions} = 2{\cdot}5 \times 4 \times 2$$
$$= 20 \text{ kNm} \quad \text{Ans. (ii)}$$

The bending moment (hogging) at each reaction is opposite in direction to the bending moment (sagging) at mid-span, this is illustrated in the bending moment diagram.

Let x = distance from end where the bending moment is zero,

$$M_x = -25 \times (x - 4) + 2{\cdot}5x \times \tfrac{1}{2}x$$
$$= -25x + 100 + 1{\cdot}25x^2$$

and this to be equal to zero:

$$1{\cdot}25x^2 - 25x + 100 = 0$$
$$x^2 - 20x + 80 = 0$$

Solving this quadratic,

$$x = 14{\cdot}472 \text{ or } 5{\cdot}528 \text{ m}$$

Note that 14·472 is 20 – 5·528

Hence, bending moment is zero at the points:

5·528 m from each end of the beam. Ans. (iii)

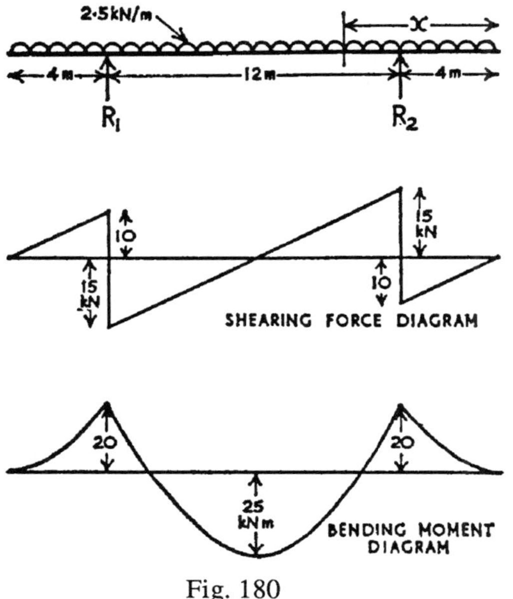

Fig. 180

These points, where the bending moment is zero (i.e. where the diagram crosses the baseline), are known as *points of contraflexure* where the slope of the beam is changing direction (sign).

STANDARD EXAMPLES

It is good practice to calculate the maximum bending moment in terms of symbols such as W for load and L for length of beam for some simple standard cases and thus prove the following.

(i) Cantilever with concentrated load W at the free end, $M_{max} = WL$ at the wall.

(ii) Cantilever with uniformly distributed load of w per unit length over the entire length, total load on beam being $W = wL$

$$M_{max} = \frac{WL}{2} \text{ at the wall } = \frac{wL^2}{2}$$

(iii) Beam supported simply at each end, carrying a concentrated load of W at mid-length.

$$M_{max} = \frac{WL}{4} \text{ at mid-length.}$$

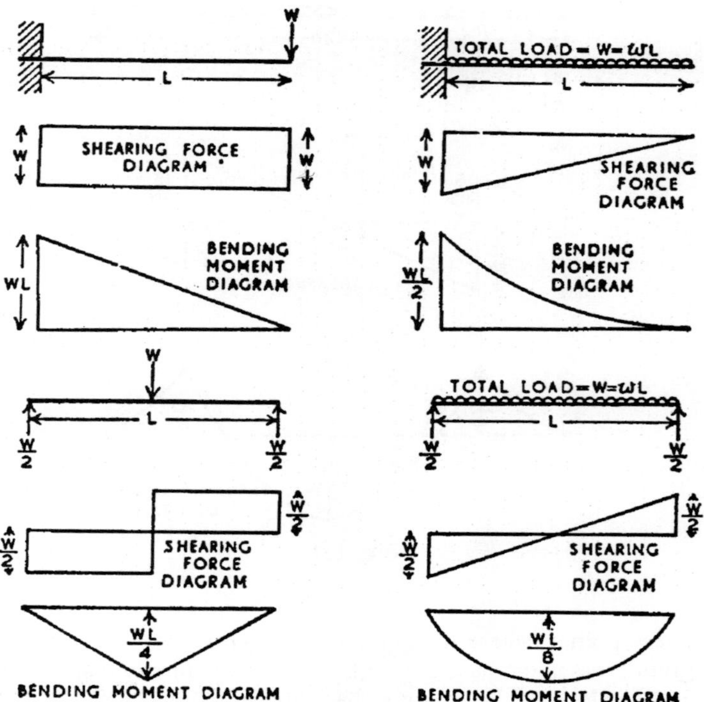

Fig. 181

(iv) Beam supported simply at each end, carrying a uniformly distributed load of w per unit length over the entire length, the total load on the beam being $W = wL$,

$$M_{max} = \frac{WL}{8} \text{ at mid-length } = \frac{wL^2}{8}$$

The shearing force and bending moment diagrams are show in Fig. 181.

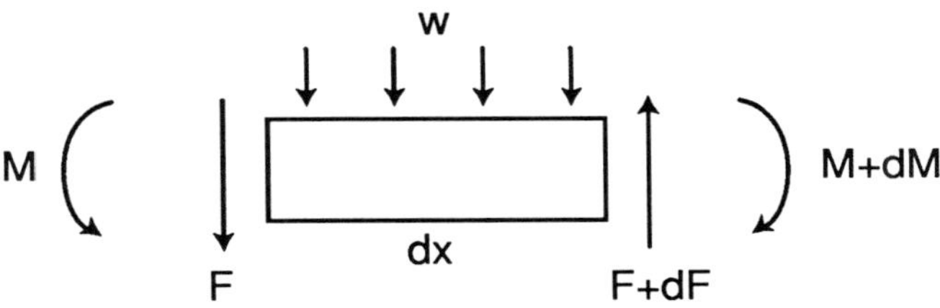

Fig. 182

MATHEMATICAL RELATIONSHIPS. It is now useful to note some calculus applications:

Equating vertical forces,

$$dF = wdx$$

Moments (about left side),

$$dM = Fdx$$

(ignoring multiples of small quantities)

At a given point:

F value equals area of load diagram to that point

$$F = \int wdx$$

Slope of F diagram equals load value

$$\frac{dF}{dx} = w$$

M value equals area of F diagram to that point

$$M = \int Fdx$$

Slope of M diagram equals F value

$$\frac{dM}{dx} = F$$

STRESSES IN BEAMS

In a beam which sags (Fig. 183) tensile stresses are set up in the lower half of the section, and compressive stresses in the upper half. In a beam which hogs, such as a cantilever, tensile stresses are in the top layers and compressive stresses in the bottom. The greatest stresses occur at the outer skin or 'outer fibres' of the material, the stress diminishes until there is no stress in the centre of the section.

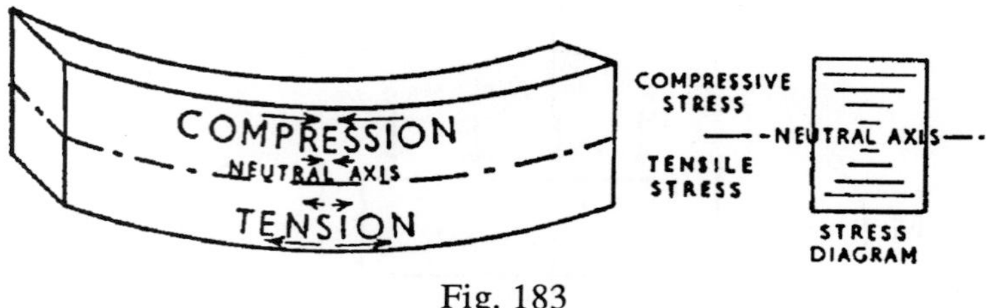

Fig. 183

In Fig. 183 the tensile stress is maximum at the bottom fibres of the section of the beam, this stress becomes uniformly less towards the centre of the section where it is zero, then changes to compressive, becoming uniformly greater until at the top fibres the compressive stress is maximum. Through the centre of gravity (centroid) of the beam section there is no stress, hence the axis through the centre of gravity is referred to as the *neutral axis*. In all cases of simple bending, the neutral axis runs through the centre of gravity of the section. In a rectangular section this is at mid-depth.

It will be seen that the material in a beam of rectangular section is not disposed to full advantage because most of the material near the centre is carrying very little stress. A beam of circular section like a round bar, is most uneconomical because, at the top and bottom, the minimum of material carries the maximum stress, and the maximum width is across the centre where there is no stress.

Hence, the economical advantage of the rolled steel joist of I section, which has the greatest width in the form of top and bottom flanges to carry the maximum stress, reduced to a thin web across the centre. Such beams as the I section are usually designed so that the flanges carry the tensile and compressive stresses due to bending, and the web carries the shearing forces.

Referring to Fig. 184, consider a part of the length of a bent beam, let the radius of curvature to which this part is bent be represented by R, and let the length under consideration span an angle of θ radians.

As the length along the neutral axis does not change because there is no stress nor strain along that axis, then:

Original length of part $= R\theta$

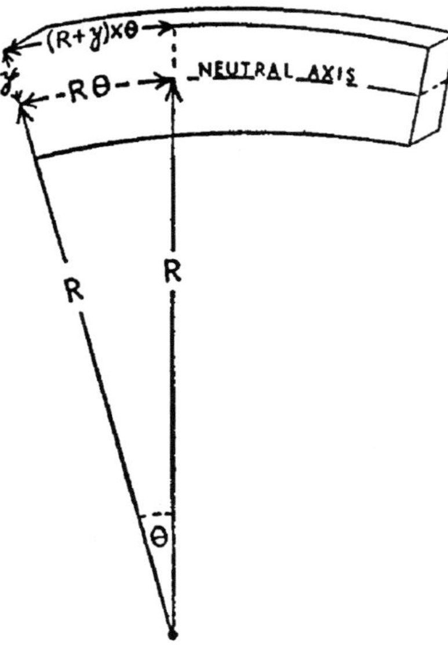

Fig. 184

If y represents the distance from the neutral axis to the outer fibres, the radius to the outer fibres is $R + y$, and the strained length of this piece is:

$$\begin{aligned}\text{New length} &= (R + y) \times \theta \\ \text{Change of length} &= \text{New length} - \text{Original length} \\ &= (R + y) \times \theta - R\theta \\ &= R\,\theta + y\theta - R\,\theta \\ &= y\theta \\ \text{Strain} &= \frac{\text{Change of length}}{\text{Original length}} \\ &= \frac{y\theta}{R\theta} = \frac{y}{R} \\ \text{Stress} \div \text{Strain} &= E\end{aligned}$$

Let σ represent the stress at the outer fibres, then,

$$\sigma \div \frac{y}{R} = E$$

$$\therefore \sigma = \frac{E \times y}{R} \qquad \dots\dots\dots \text{(i)}$$

Thus the stress varies directly as the distance from the neutral axis. At the maximum distance of y is the maximum stress. At any intermediate distance between the neutral axis and the outer fibres there is less stress. At h distance from the neutral axis:

$$\text{Stress at } h = \frac{E \times h}{R}$$

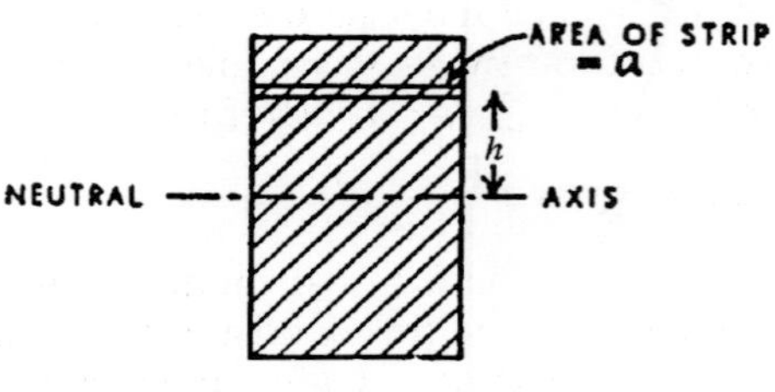

Fig. 185

Now consider a rectangular section (Fig. 185), let a be a strip of area at h from the neural axis.

$$\text{Stress on this strip} = \frac{E \times h}{R}$$

$$\text{Load on this strip} = \text{stress} \times \text{area}$$

$$= \frac{E \times h \times a}{R}$$

This load or force is set up in the material to resist and support the external loading which tends to bend the beam, the moment of resistance of the load carried by this strip of area is, load × its leverage from the neutral axis:

$$\text{Moment of resistance of strip} = \frac{E \times h \times a}{R} \times h$$

$$= \frac{E \times a \times h^2}{R}$$

If the whole area is considered to be made up of such strips, a_1, a_2, a_3 etc., then the total moment of resistance of the whole area will be:

$$\frac{Ea_1h_1^2}{R} + \frac{Ea_2h_2^2}{R} + \frac{Ea_3h_3^2}{R} + \text{etc.}$$

As E and R are common to all strips of area across the one section, this can be written:

Total moment of resistance $= \frac{E}{R}\left\{ a_1h_1^2 + a_2h_2^2 + a_3h_3^2 + \text{etc.} \right\}$

Total moment of resistance $= \frac{E}{R} \sum ah^2$

It was explained in Chapter 7 that the summation of the product of each element of area and the square of its distance from a given point, is the Second Moment of the whole area about that point, and represented by I, thus $\sum ah^2 = I$, therefore:

Total moment of resistance $= \frac{E}{R} \times I$

For equilibrium, the total internal moment resisting bending must be equal in magnitude to the external moment which tends to cause bending, then representing bending moment by M:

$$M = \frac{E}{R} \times I$$

$$\text{or, } \frac{M}{I} = \frac{E}{R} \qquad \dots \quad \dots \quad \dots \text{ (ii)}$$

Expression (i) can be written $\frac{\sigma}{y} = \frac{E}{R}$ then it can be combined with (ii) to produce the **fundamental bending equation**:

$$\frac{M}{I} = \frac{\sigma}{y} = \frac{E}{R}$$

where M = bending moment (= resisting moment),
I = 2nd moment of area about its neutral axis,
σ = stress at outer fibres,
y = distance from neutral axis to outer fibres,
E = modulus of elasticity of the material,
R = radius of curvature.

I values are derived by integration (see Chapter 7). The most common are:

I of a solid rectangular section about its c.g. $= \frac{BD^3}{12}$

I of a symmetrical hollow rectangular section about its c.g. or its equivalent in the form of a channel bar or beam of I section

$$= \frac{BD^3 - bd^3}{12}$$

I of a solid round bar about its diameter $= \dfrac{\pi D^4}{64}$

I of a hollow round bar about its diameter $= \dfrac{\pi}{64}(D^4 - d^4)$

Formulae connecting bending moment and stress can now be derived for some standard cases. The two most common are the solid rectangular and round section beams,

$$\frac{M}{I} = \frac{\sigma}{y}$$

$$\therefore \sigma = \frac{My}{I}$$

For a beam of solid rectangular section where B = breadth and D = depth,

$$y = \frac{D}{2} \text{ and } I = \frac{BD^3}{12}$$

$$\sigma = \frac{My}{I}$$

$$= \frac{M \times D \times 12}{2 \times B \times D^3}$$

$$\sigma = \frac{6M}{BD^2}$$

For a beam of solid round section where D = diameter,

$$y = \frac{D}{2} \text{ and } I = \frac{\pi D^4}{64}$$

$$\sigma = \frac{My}{I}$$

$$= \frac{M \times D \times 64}{2 \times \pi \times D^3}$$

$$\sigma = \frac{32M}{\pi D^3}$$

Example. A cantilever 1·8 m long, of solid rectangular section 75 mm broad by 150 mm deep, carries a uniformly distributed load of 22·5 kN/m over the whole length, a concentrated load of 30 kN at 0·6 m from the wall, and another concentrated load of 20 kN at the free end. Calculate the maximum stress in the beam, and sketch the shearing force and bending moment diagrams.

Maximum stress occurs at the outer fibres of the beam at the wall (that is, at the position of maximum bending moment).

Total distributed load = 22·5 × 1·8 = 40·5 kN

C.G. of this load is at mid-length = 0·9 from wall.

$$\begin{aligned} M \text{ @ wall} &= 40{\cdot}5 \times 0{\cdot}9 + 20 \times 1{\cdot}8 + 30 \times 0{\cdot}6 \\ &= 36{\cdot}45 + 36 + 18 \\ &= 90{\cdot}45 \text{ kN m} \end{aligned}$$

$$\frac{M}{I} = \frac{\sigma}{y} \qquad \sigma = \frac{My}{I}$$

For a regular section:

$$y = \frac{D}{2} \qquad \text{and } I = \frac{BD^3}{12}$$

$$\begin{aligned} \sigma &= \frac{6M}{BD^2} \\ &= \frac{6 \times 90{\cdot}45 \times 10^3}{0{\cdot}075 \times 0{\cdot}15^2} \\ &= 3{\cdot}216 \times 10^8 \text{ N/m}^2 \\ &= 321{\cdot}6 \text{ MN/m}^2 \text{ Ans.} \end{aligned}$$

The bending moment diagram of Fig. 186 is drawn to illustrate the separate u.d.l and point load effects. Strictly the diagram would be an all positive parabola above the zero line (hogging M).

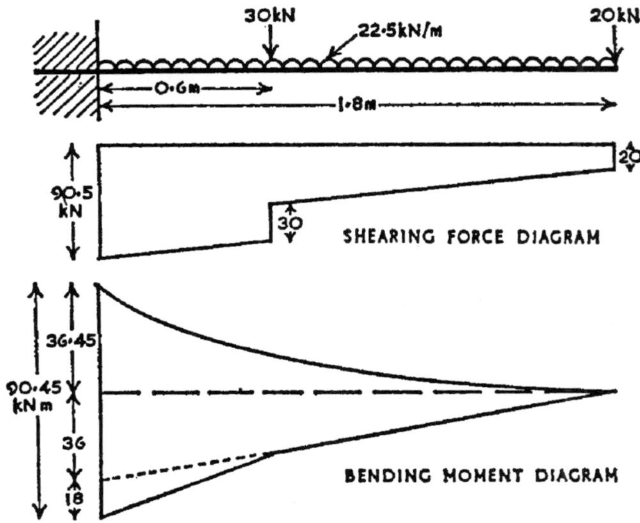

Fig. 186

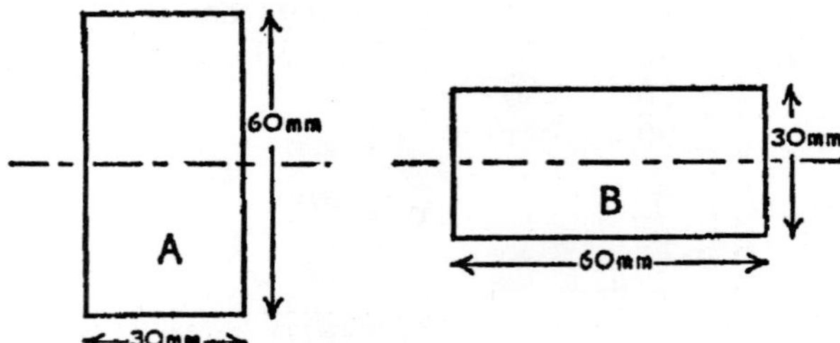

Fig. 187

MODULUS OF SECTION

$$\frac{M}{I} = \frac{\sigma}{y} \qquad\qquad M = \frac{\sigma \times I}{y}$$

As I and y have fixed dimensions for any beam, the stress is therefore proportional to the bending moment. Hence, for a given allowed stress in the material, the bending moment that can be carried by a beam is a measure of the strength of the beam and depends upon the value $I \div y$. This quantity is termed the *modulus of section* (Z).

$$Z = \frac{I}{y}$$

Therefore the strengths of beams are compared by the ratios of their moduli of sections.

Example. Compare the strengths of two solid rectangular beams, one is 30 mm broad and 60 mm deep, the other 60 mm broad and 30 mm deep.

Working in cm units:

$$I \text{ of A} = \frac{BD^3}{12} = \frac{3 \times 6^3}{12} = 54 \text{ cm}^4$$

$$y \text{ of A} = \frac{D}{2} = \frac{6}{2} = 3 \text{ cm}$$

$$\frac{I}{y} \text{ of A} = \frac{54}{3} = 18 \text{ cm}^3 \qquad \ldots \quad \ldots \quad \ldots \text{(i)}$$

$$I \text{ of B} = \frac{BD^3}{12} = \frac{6 \times 3^3}{12} = 13{\cdot}5 \text{ cm}^4$$

$$y \text{ of B} = \frac{D}{2} = \frac{3}{2} = 1{\cdot}5 \text{ cm}$$

$$\frac{I}{y} \text{ of B} = \frac{13{\cdot}5}{1{\cdot}5} = 9 \text{ cm}^3 \qquad \ldots \quad \ldots \quad \ldots \text{(ii)}$$

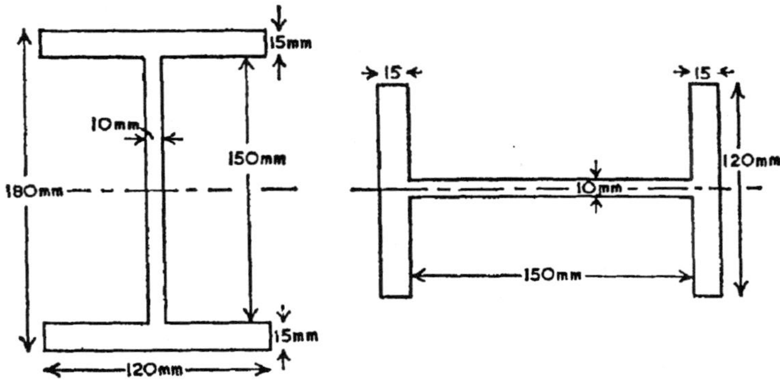

Fig. 188

$$\text{Ratio of strengths} = \text{Ratio of moduli of sections} = 18:9 = 2:1$$

A is twice as strong as B. Ans.

Example. A rolled steel joist has equal flanges 120 mm broad by 15 mm thick, the centre web is 10 mm thick and the overall depth of the joist is 180 mm. Find the ratio of the strength when used (*a*) as an I section with centre web vertical, (*b*) as an H section with centre web horizontal, given that the Second Moment of a rectangle about its centroid is $\frac{BD^3}{12}$

Only the 2nd moment of a rectangle about its centroid is given therefore look for rectangular areas which have their centroids on the same axis as the centroid of the beam section, then add or subtract these values as required to obtain the second moment of the section.

As an I section (Fig. 188). This is composed of a rectangle 120 mm by 180 mm with two rectangular portions each 55 mm by 150 mm removed, all having their centroids on the same axis as the centroid of the beam section.

Working in cm units:

$$\text{As an I section, } I = \frac{12 \times 18^3}{12} - \frac{2 \times 5{\cdot}5 \times 15^3}{12}$$

$$= \frac{12 \times 18^3}{12} - \frac{11 \times 15^3}{12}$$

$$= 2738{\cdot}25 \text{ cm}^4$$

$$y = 18 \div 2 = 9 \text{ cm}$$

$$Z = \frac{I}{y} = \frac{2738{\cdot}25}{9}$$

$$= 304{\cdot}25 \text{ cm}^3 \quad \ldots \quad \ldots \quad \ldots \text{(i)}$$

As an H section (Fig. 188). This is composed of three rectangles which have their centroids on the axis of the centroid of the beam section, two of them are the flanges, now vertical, each 15 mm by 120 mm, the other is the centre web, now horizontal, 150 mm by 10 mm.

$$\text{As an H section, } I = \frac{2 \times 1{\cdot}5 \times 12^3}{12} + \frac{15 \times 1^3}{12}$$

$$= \frac{3 \times 12^3 + 15 \times 1^3}{12}$$

$$= 433{\cdot}25 \text{ cm}^4$$

$$y = 12 \div 2 = 6 \text{ cm}$$

$$Z = \frac{I}{y} = \frac{433{\cdot}25}{6}$$

$$= 72{\cdot}21 \text{ cm}^3 \quad \ldots \quad \ldots \quad \ldots \text{(ii)}$$

$$\text{Ratio of strengths} = \text{Ratio of moduli of section}$$

$$= 304{\cdot}25 : 72{\cdot}21$$

$$= \frac{304{\cdot}25}{72{\cdot}21} : \frac{72{\cdot}21}{72{\cdot}21}$$

$$= 4{\cdot}214 : 1$$

Therefore the beam is 4·214 times as strong when used as an I section compared with it as an H section. Ans.

f DEFLECTION OF BEAMS

Taking the portion of the fundamental bending equation which connects bending moment and radius of curvature:

$$\frac{M}{I} = \frac{E}{R} \qquad R = \frac{EI}{M}$$

it is seen that the radius to which a beam bends is inversely proportional to the bending moment and therefore when the bending moment is different at all parts along the beam (as in the previous examples) then the radius of curvature of the bent beam is different at all parts of its length.

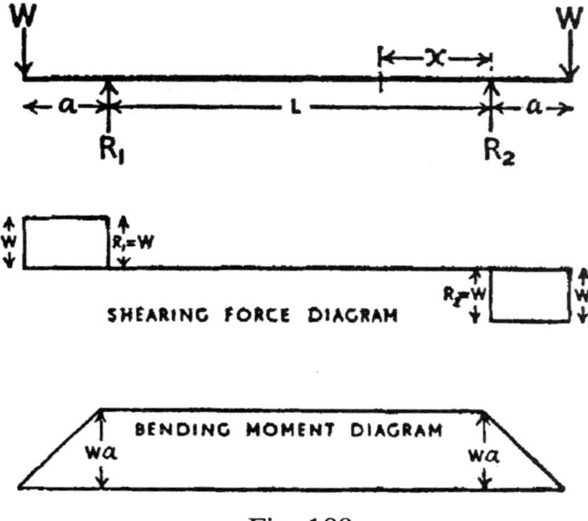

Fig. 189

If, however, the beam is loaded in such a way that the bending moment is constant over part of its length, the radius of curvature must be constant and this part of the beam is bent into a circular arc. As an example of circular bending, Fig. 189 shows a beam simply supported at equal distances from each end and carrying equal concentrated loads at the extreme ends.

Consider a section of the beam between the supports and let x = the distance of this section from R_2,

$$\begin{aligned} M \,@\, x &= W(a+x) - R_2 \times x \\ &= Wa + Wx - Wx \\ &= Wa \end{aligned}$$

Thus the bending moment is independent of x and therefore constant over the length between the supports, the radius of curvature to which the beam is bent is therefore constant and the beam must be bent into an arc of a circle between the supports.

Fig. 190 is an exaggerated diagram of the circular arc. The deflection is the amount that the level of the beam rises or sinks from its horizontal position. The maximum deflection in this case is CO at mid-span and may be calculated thus:

$$\begin{aligned} \text{AB} &= \text{length between supports} = L \\ \text{AO} &= \text{BO} = \tfrac{1}{2}L \end{aligned}$$

By the principle of crossed chords,

$$\begin{aligned} \text{AO} \times \text{BO} &= \text{CO} \times \text{DO} \\ \tfrac{1}{2}L \times \tfrac{1}{2}L &= \text{CO} \times (2R - \text{CO}) \\ \tfrac{1}{4}L^2 &= 2R \times \text{CO} - (\text{CO})^2 \end{aligned}$$

The deflection (CO) is a small quantity, $(\text{CO})^2$ is therefore the second order of smallness which is very small indeed and is a negligible amount to add or subtract from any normal quantity. Therefore neglecting $(\text{CO})^2$:

$$\tfrac{1}{4}L^2 = 2R \times \text{CO}$$

$$\text{CO} = \frac{L^2}{8R}$$

$$\text{maximum deflection} = \frac{L^2}{8R}$$

$$\text{Substituting } R = \frac{EI}{M}$$

$$\text{Maximum deflection} = \frac{ML^2}{8EI}$$

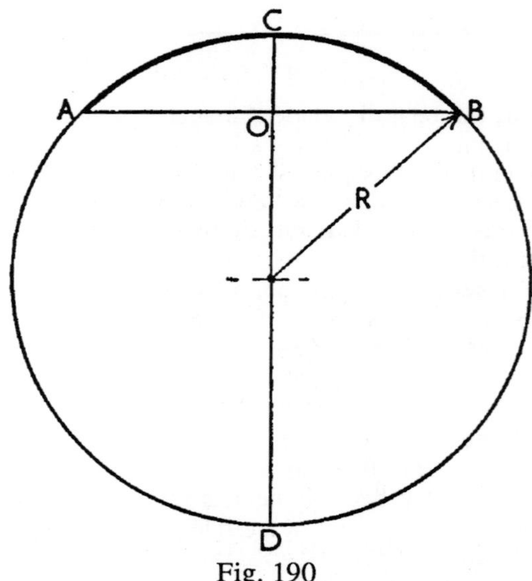

Fig. 190

f Example. A beam of uniform rectangular section is 50 mm broad by 75 mm deep and 4·2 m long. It is simply supported at 0·6 from each end and carries a concentrated load of 2·5 kN at each extreme end. Neglecting the weight of the beam, calculate the radius of curvature between the supports and the deflection at mid-span, taking the modulus of elasticity of the material as 200 GN/m².

Each reaction carries half the total load,

$R_1 = R_2 = 2{\cdot}5$ kN

M @ $R_2 = 2{\cdot}5 \times 0{\cdot}6 = 1{\cdot}5$ kN m

$$\text{Length between supports} = 4{\cdot}2 - (2 \times 0{\cdot}6) = 3 \text{ m}$$

M is constant at $1{\cdot}5 \times 10^3$ Nm between supports.

$$I = \frac{BD^3}{12} = \frac{5 \times 7{\cdot}5^3}{12} \text{ cm}^4 = \frac{5 \times 7{\cdot}5^3}{12} \times 10^{-8} \text{ m}^4$$

$$R = \frac{EI}{M} = \frac{200 \times 10^9 \times 5 \times 7{\cdot}5^3 \times 10^{-8}}{1{\cdot}5 \times 10^3 \times 12}$$

$$= 234{\cdot}4 \text{ m} \quad \text{Ans. (i)}$$

$$\text{Deflection} = \frac{ML^2}{8EI} \text{ or } \frac{L^2}{8R}$$

$$= \frac{3^2}{8 \times 234{\cdot}4} = 4{\cdot}8 \times 10^{-3} \text{ m}$$

$$= 4{\cdot}8 \text{ mm} \quad \text{Ans. (ii)}$$

The maximum deflections of a few other standard cases are given now. Any problem which involves the deflection other than the case of circular bending explained will include the required deflection formula.

(i) Cantilever with concentrated load W at the free end:

$$\text{Maximum deflection} = \frac{WL^3}{3EI}$$

(ii) Cantilever with uniformly distributed load over the entire length, the total load being W:

$$\text{Maximum deflection} = \frac{WL^3}{8EI}$$

(iii) Beam supported simply at each end and carrying a concentrated load of W at mid-span:

$$\text{Maximum deflection} = \frac{WL^3}{48EI}$$

(iv) Beam supported simply at each end and carrying a uniformly distributed load over its entire length, the total of this load being W:

$$\text{Maximum deflection} = \frac{5WL^3}{384EI}$$

Such formulae are derived mathematically from the load expression:

$$EI \frac{d^4x}{dx^4} = w$$

giving F and M expressions (see Page 265). Two more integrations from the curvature expression:

$$EI \frac{d^2y}{dx^2} = M$$

give slope $\frac{dy}{dx}$ and deflection y, *i.e.*

$$EI \frac{dy}{dx} = \int M dx$$

$$EI\, y = \iint M\, dx\, dx$$

COMBINED BENDING AND DIRECT STRESS

It was explained in Chapter 9 that when a direct load is applied on a material it induces a compressive or tensile stress, the intensity of which is expressed as Load ÷ Area of cross-section. If however, a direct load is applied off centre as illustrated in Fig. 191 the material is also subjected to a bending action, the moment causing bending being due to the eccentricity x of the application of the load from the centre of the section. The bending moment on this material is therefore $W \times x$, and this causes compression along the side which is concave, and tension along the side which is convex.

Taking a specimen of rectangular cross-section, let T = thickness (which is equivalent to the depth of the beam), and B = breadth. Let W = load applied, and x = the eccentricity of the line of action of the load from the centre of its thickness on the centreline of the breadth.

$$\text{Stress due to direct load} = \frac{\text{Load}}{\text{Area}}$$

$$= \frac{W}{BT} \quad \ldots \quad \ldots \quad \ldots \text{(i)}$$

$$\text{Stress due to bending} = \frac{6M}{BT^2}$$

$$= \frac{6Wx}{BT^2} \quad \ldots \quad \ldots \quad \ldots \text{(ii)}$$

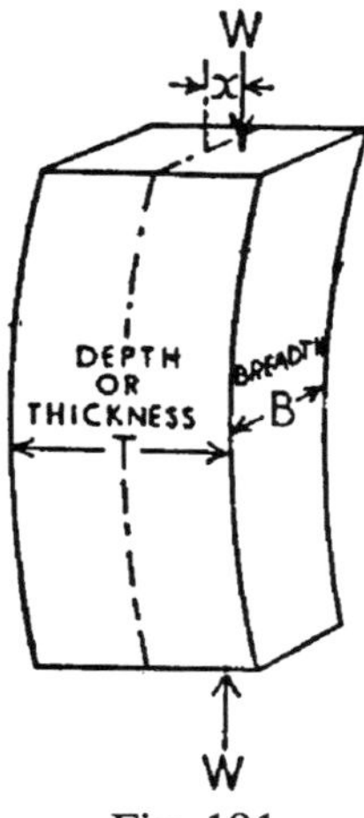

Fig. 191

Assuming W to be compressive, the direct stress is compressive across the whole section. The bending stress will be an additional compressive stress on the concave side therefore the total compressive stress on this side is:

$$\text{Direct stress} + \text{Bending stress} = \frac{W}{BT} + \frac{6Wx}{BT^2}$$

But the bending stress is tensile on the convex side, therefore the compressive stress on this side is relieved somewhat and its value is:

$$\text{Direct stress} - \text{Bending stress} = \frac{W}{BT} - \frac{6Wx}{BT^2}$$

Note that the compressive stress is taken as positive because the applied load is compressive, and tensile stress is expressed as *negative compressive* stress. If the last expression above works out to be a positive value it means that the direct compressive stress is greater than the tensile bending stress, the stress on this side will therefore be compressive and the minimum value across the section. If the result is negative it means that the tensile bending stress is greater than the direct compressive stress and therefore the stress on this side is tensile.

A diagram showing the distribution of stress across the section is given in Fig. 192.

It can be seen that if the eccentricity of the application of W is excessive, the tensile stress due to bending on one side will exceed the direct compressive stress, and this may not be desirable in

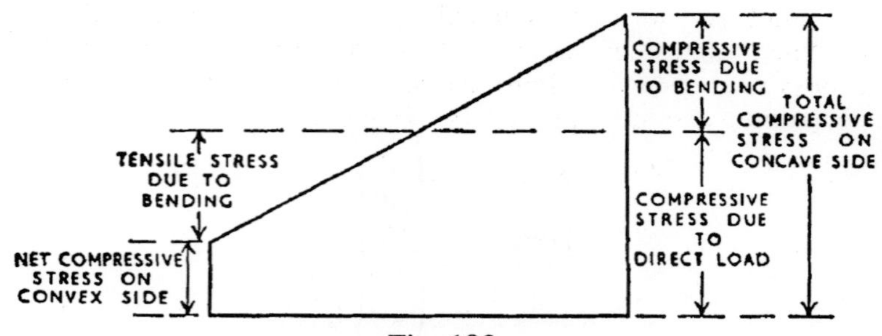

Fig. 192

certain materials which are weak in tension. The limit of eccentricity to avoid any tensile stress will be when the tensile stress due to bending is equal to the direct compressive stress.

Again taking a rectangular section as an example, the limiting value of x will be when:

Tensile stress due to bending = Direct compressive stress.

$$\frac{6Wx}{BT^2} = \frac{W}{BT}$$

$$\therefore\ x = \frac{T}{6}$$

That is, x must not exceed ⅙ of the thickness from the centre of the section.

This is the explanation of the 'Middle Third' rule in masonry which states that the load must be applied within the middle third of the section (see Fig. 193).

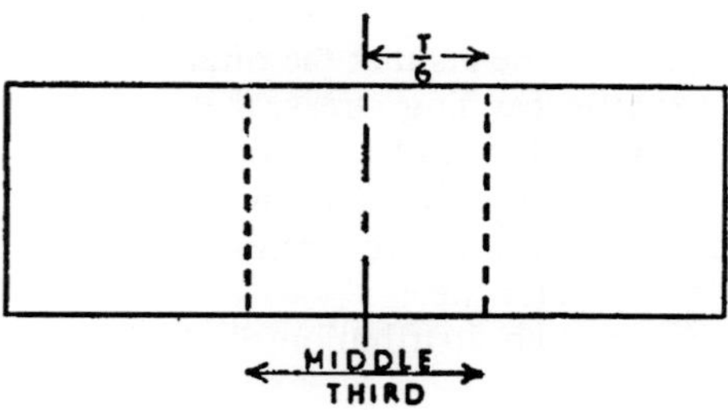

Fig. 193

f Example. A bar of rectangular cross-section is 100 mm thick by 50 mm broad and carries an axial compressive load of 160 kN. If the line of action of the load is 12·5 mm from the centre of the thickness of the section along the centre-line of the breadth, find the maximum and minimum stresses in the bar.

$$\text{Direct stress} = \frac{\text{load}}{\text{area}}$$

$$= \frac{160 \times 10^3}{0{\cdot}1 \times 0{\cdot}05}$$

$$= 3{\cdot}2 \times 10^7 \text{ N/m}^2 = 32 \text{ MN/m}^2$$

$$\text{Bending stress} = \frac{6M}{BT^2}$$

$$= \frac{6 \times 160 \times 10^3 \times 0{\cdot}0125}{0{\cdot}05 \times 0{\cdot}1^2}$$

$$= 2{\cdot}4 \times 10^7 \text{ N/m}^2 = 24 \text{ MN/m}^2$$

Maximum compressive stress

$= 32 + 24 = 56$ MN/m^2 Ans. (i)

Minimum compressive stress

$= 32 - 24 = 8$ MN/m^2 Ans. (ii)

TEST EXAMPLES 10

1. A cantilever 1·5 m long carries a concentrated load of 14 kN at the free end and another of 32 kN at 0·5 m from the wall. Draw the shearing force and bending moment diagrams to scale and measure off the shearing force and bending moment at 0·25 m from the wall.

2. A cantilever 5 m long carries two concentrated loads, one of these is at the free end and the other is at mid-length along the beam. If the bending moments at the wall and at mid-length are 52·5 and 12·5 kN m respectively, find the magnitudes of the two loads.

3. A beam 11 m long is simply supported at each end and carries concentrated loads of 80, 30, 20 and 40 kN at 2·5, 5, 7·5 and 9·5 m respectively from the left support. Draw to scale the shearing force and bending moment diagrams and read off the values of the shearing force and bending moment at the points (i) 3·5 m, and (ii) 8·5 m, from the left support.

4. A beam 15 m long is simply supported at the left end and at 3 m from the right end. It carries concentrated loads of 40 and 20 kN at 6 and 9 m respectively from the left support, and another concentrated load of 50 kN at the extreme right end. Sketch and dimension the shearing force and bending moment diagrams.

5. A beam of uniform section is 10 m long, its mass is 8 Mg, and is simply supported at each end. Calculate the shearing forces and bending moments in the beam at quarter-length and at mid-length due to its own weight. Sketch the shearing force and bending moment diagrams and mark the calculated values on the diagrams.

6. A beam 20 m long is simply supported at 2·5 m from each end. It carries a concentrated load of 20 kN at each extreme end and also a uniformly distributed load over the entire length of 4 kN/m run. Sketch the shearing force and bending moment diagrams. Calculate (i) the bending moment at mid-span, (ii) the bending moment at each support, and (iii) the position along the beam where there is no bending moment *i.e.* point of contraflexure.

7. A beam carries a uniformly distributed load over its entire length which, together with the weight of the beam, amounts to a total of 80 kN. The beam is 10 m long and is simply supported at each end. It also carries concentrated loads of 65 and 55 kN at 2

and 6 m respectively from the left end. Sketch and dimension the shearing force and bending moment diagrams and find the position and magnitude of the maximum bending moment.

8. A cantilever of rectangular section 100 mm broad by 150 mm deep and carries a concentrated load of 15 kN at its free end. Neglecting the weight of the beam, find the distance from the free end where the stress at the outer fibres is 75 MN/m^2.

9. Compare the bending strength of a tube 30 mm outside diameter with a solid round bar 15 mm diameter, if they are of equal cross-sectional area and similar material.

10. A beam of solid rectangular cross-section is 2 m long and 75 mm broad. It is simply supported at each end and is to carry a concentrated load at mid-span. (i) If the maximum bending moment is not to exceed 30 kN m, find the maximum load that can be carried. (ii) If the maximum stress is not to exceed 60 MN/m^2, find the depth of the beam.

*f*11. A cantilever of hollow rectangular box section is constructed of steel of density 7860 kg/m^3. Its dimensions are: outside breadth 78 mm, outside depth 104 mm, length 1·2 m, and it is 12 mm thick throughout. If the maximum stress is not to exceed 35 MN/m^2, calculate the maximum concentrated load that can be carried at the free end of the beam.

*f*12. A beam of symmetrical section about is neutral axis is 3 m long and 380 mm deep, it is simply supported at each end and carries a uniformly distributed load of 12 kN/m run from the left end to 1·2 m from the other end. Neglecting the weight of the beam, calculate the position and magnitude of the maximum stress if the second moment of the section about its neutral axis is $7{\cdot}6 \times 10^3$ cm^4.

*f*13. A solid round steel shaft is 76 mm diameter and rests in short bearings at each end which are 1·27 m apart. Treating it as a simply supported beam and taking the weight of the shaft into consideration, find the maximum concentrated load it can carry at the centre of length to limit the bending stress to 13·8 MN/m^2. Density of steel = 7860 kg/m^3.

*f*14. A tube 40 mm outside diameter, 5 mm thick and 1·5 m long is simply supported at 125 mm from each end and carries a concentrated load of 1 kN at each extreme end. Neglecting the weight of the tube, sketch the shearing force and bending moment diagrams, and calculate the radius of curvature and the deflection at mid-span. Take the modulus of elasticity of the material as 208 GN/m^2.

*f*15. A short solid bar 125 mm diameter supports an axial compressive load of 275 kN, the line of action of this load being 12 mm from the centre of the bar. Calculate the maximum and minimum stresses in the bar.

CHAPTER 11

TORSION OF SHAFTS

The moment of a force applied to a shaft which tends to twist or turn is termed the twisting moment, turning moment, or torque. Its magnitude is the product of the applied force and its effective leverage (like any other moment of force).

TORSION is the state of being twisted, therefore a shaft transmitting torque is said to be in torsion.

When a shaft is twisted, each lamination of cross-section (thin circular discs of the shaft) rotates very slightly relative to the next lamination, thus the shaft suffers shear strain because the particles which constitute the material tend to slide over each other, and shear stress is induced.

THE FUNDAMENTAL TORSION EQUATION follows a similar pattern and expresses similar relationships as the fundamental bending equation explained in the last chapter.

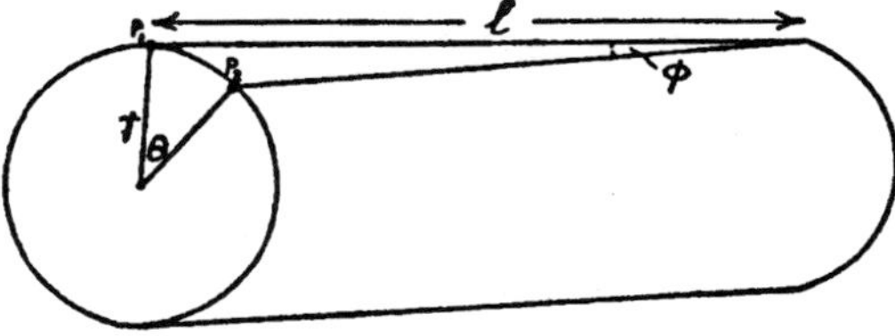

Fig. 194

Consider a shaft of length l twisted as shown in Fig. 194 such that as a result of the strain, a point on the circumference shifts from P_1 to P_2. The amount of distortion is the arc length P_1 to P_2 and this depends upon the length l under twist. As shear strain is expressed by the angle of distortion (see Chapter 9), in this case angle ϕ, then,

$$\text{Shear strain at outer fibres} = \frac{P_1P_2}{l}$$

Let θ = angle of twist at shaft face
and r = radius from centre to outer fibres
then arc P_1P_2 = angle in radians × radius = θr
therefore,

$$\text{Shear strain at outer fibres} = \frac{\theta r}{l}$$

Provided the material is strained within the elastic limit, shear stress and shear strain are proportional to each other, and therefore shear stress divided by shear strain is a constant for any given material. This constant is termed the modulus of rigidity (G).

$$G = \text{shear stress} \div \text{shear strain}$$

$$= \tau \div \frac{\theta r}{l}$$

$$= \frac{\tau l}{\theta r}$$

Thus the shear stress at the outer fibres is expressed by:

$$\tau = \frac{G\theta r}{l} \quad \ldots \ldots \ldots \ldots \ldots \ldots \quad (i)$$

For any given section, θ and l are constant as well as G, the stress is therefore proportional to the radial distance from the centre. At the centre of the shaft the stress is nil, increasing uniformly to a maximum at the outer skin.

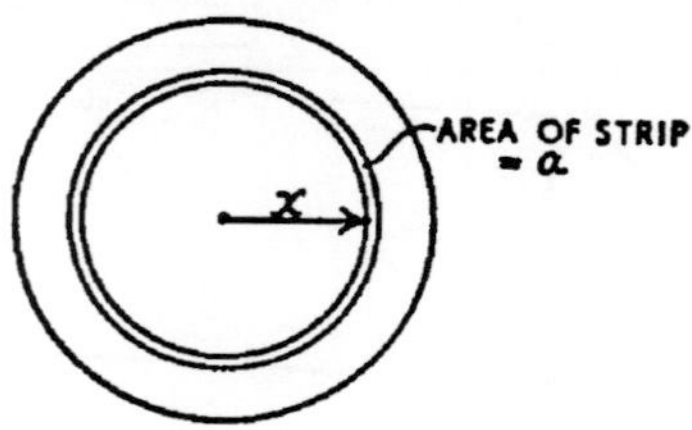

Fig. 195

Consider now a thin ring of the material of area a at the radius x from the centre (Fig. 195).

$$\text{Stress on this ring} = \frac{G\theta r}{l}$$

$$\text{Total load} = \text{stress} \times \text{area}$$

$$\therefore \text{Load on this ring} = \frac{G\theta r}{l} \times a$$

This load or force in the material exists as a resistance to support the externally applied twisting moment. The moment of resistance of the load carried by this ring of area is the product of the load and its leverage from the centre.

$$\text{Moment of resistance of ring} = \frac{G\theta xa}{l} \times x$$

$$= \frac{G\theta ax^2}{l}$$

If the whole area of cross-section is considered to be made up of such thin rings a_1 a_2 a_3 etc., then the total moment of resistance offered by the whole area will be:

$$\frac{G\theta a_1x_1^{\,2}}{l} + \frac{G\theta a_2x_2^{\,2}}{l} + \frac{G\theta a_3x_3^{\,2}}{l} + \text{etc.}$$

As G, θ and l are common to all rings across the section, this can be written:

$$\frac{G\theta}{l}\left\{ a_1x_1^{\,2} + a_2x_2^{\,2} + a_3x_3^{\,2} + \text{etc.} \right\}$$

$$\text{Total moment of resistance} = \frac{G\theta}{l}\,\Sigma ax^2$$

The summation of the products of elements of area and the square of their distances from a given point is the second moment of the whole area about that point. This second moment is taken about the polar centre of the shaft, it is therefore termed the polar second moment and can be represented by Ip or more usually by J.

Also, the total internal moment of resistance in the shaft must be equal to the externally applied twisting moment which is represented by T, therefore:

$$T = \frac{G\theta}{l} \times J$$

$$\text{or,}\ \frac{T}{J} = \frac{G\theta}{l} \quad \dots\ \dots\ \dots\ \dots\ \dots \quad (ii)$$

Equation (i) can be written in a similar manner,

$$\frac{\tau}{r} = \frac{G\theta}{l}$$

and by combining (i) and (ii) gives the **fundamental torsion equation,**

$$\frac{T}{J} = \frac{\tau}{r} = \frac{G\theta}{l}$$

T = twisting moment
J = polar 2nd moment of area
τ = shear stress at outer fibres
r = radius to outer fibres
G = modulus of rigidity
θ = angle of twist
l = length of shaft under twist

J values are derived by integral calculus (see Chapter 7).
For a solid circular section,

$$J = \frac{\pi}{32} D^4 \quad \text{or} \quad \frac{\pi}{2} R^4$$

For a hollow circular section,

$$J = \frac{\pi}{32} \left\{ D^4 - d^4 \right\} \quad \text{or} \quad \frac{\pi}{2} \left\{ R^4 - r^4 \right\}$$

Example. When transmitting a certain power an engine shaft 360 mm diameter is twisted 1° over a length of 4·5 m. If the modulus of rigidity of the shaft material is 103·5 GN/m², find (*i*) the torque transmitted, (*ii*) the stress in the shaft.

$$\frac{T}{J} = \frac{G\theta}{l} \quad T = \frac{JG\theta}{l}$$

$$T = \frac{\pi \times 0{\cdot}36^4 \times 103{\cdot}5 \times 10^9 \times 1}{32 \times 4{\cdot}5 \times 57{\cdot}3}$$

$$= 6{\cdot}619 \times 10^5 \text{ N m} = 661{\cdot}9 \text{ kN m} \qquad \text{Ans. (i)}$$

$$\frac{\tau}{r} = \frac{G\theta}{l} \quad \therefore \tau = \frac{rG\theta}{l}$$

$$\tau = \frac{0{\cdot}18 \times 103{\cdot}5 \times 10^9 \times 1}{4{\cdot}5 \times 57{\cdot}3}$$

$$= 7{\cdot}224 \times 10^7 \text{ N/m}^2$$

$$= 72{\cdot}24 \text{ MN/m}^2 \qquad \text{Ans. (ii)}$$

RELATION BETWEEN TORQUE AND STRESS

$$\frac{T}{J} = \frac{\tau}{r} \qquad T = \frac{J\tau}{r}$$

For a solid round shaft,

$$J = \frac{\pi}{32} D^4 \quad \text{and } r = \tfrac{1}{2} D$$

$$T = \frac{\pi \times D^4 \times 2 \times \tau}{32 \times D}$$

$$T = \frac{\pi D^3 \tau}{16}$$

For a hollow round shaft,

$$J = \frac{\pi}{32}(D^4 - d^4) \qquad \text{and } r = \tfrac{1}{2}D$$

$$T = \frac{\pi\,(D^4 - d^4) \times 2 \times \tau}{32 \times D}$$

$$T = \frac{\pi\,(D^4 - d^4)\,\tau}{16D}$$

The stress in a shaft therefore depends upon the torque transmitted, the stress allowed depends upon the strength of the material, therefore the strength of shafts can be compared by the torque that can be transmitted as expressed in the given formulae.

Example. Compare the strengths and weights of two shafts of similar material, one is solid and the other is hollow of outside diameter equal to the diameter of the solid shaft with a bore of half that diameter.

STRENGTH RATIO:

Strength of solid shaft : Strength of hollow shaft

$$\frac{\pi D^3 \tau}{16} : \frac{\pi\,(D^4 - d^4)\,\tau}{16D}$$

π and 16 are common to both and therefore cancel. As the shafts are of similar material the stress allowed in each case is the same and also cancels, leaving:

$$D^3 : \frac{D^4 - d^4}{D}$$

$$D^4 : D^4 - d^4$$

Substituting $d = \tfrac{1}{2}D$,

$$D^4 : D^4 - (\tfrac{1}{2}D)^4$$

$$D^4 : D^4 - \tfrac{1}{16}D^4$$

$$D^4 : \tfrac{15}{16}D^4$$

Dividing both by D^4,

$$1 : \tfrac{15}{16} \qquad \dots\ \dots\ \dots\ \dots \qquad \text{(i)}$$

To compare their weights,

Let l = length of each shaft (each of same length)
w = specific weight of material (same for each shaft)
Weight = area of section × length × spec. wt.
spec. wt. = 9·81 × density

WEIGHT RATIO:

$$\begin{aligned} \text{Weight of solid shaft} &: \text{Weight of hollow shaft} \\ 0{\cdot}7854 \times D^2 \times l \times w &: 0{\cdot}7854\,(D^2 - d^2) \times l \times w \\ D^2 &: D^2 - d^2 \\ D^2 &: D^2 - (\tfrac{1}{2}D)^2 \\ D^2 &: D^2 - \tfrac{1}{4}D^2 \\ D^2 &: \tfrac{3}{4}D^2 \\ 1 &: \tfrac{3}{4} \quad \ldots\;\ldots\;\ldots\;\ldots \quad \text{(ii)} \end{aligned}$$

From (i) and (ii) it can be seen that by boring a concentric hole through a shaft, the diameter of the hole being half the shaft diameter, only one-sixteenth of the strength is lost but one-quarter of the weight is removed.

EXPERIMENT TO DETERMINE MODULUS OF RIGIDITY

One method of finding the modulus of rigidity of a material is to take a specimen in the form of a straight round bar, grip it at one end, apply various values of twisting moment at the other end and measure the angle of twist produced by each applied twisting moment. These values are plotted on a graph and the plotted points should lie exactly in a straight line because the angle of twist varies directly as the twisting moment. Due to slight errors of observation of measurement during the experiment, some points may not be in line, therefore the graph is drawn as an average between the high and low points in an attempt to obtain the true relationship. All other quantities in that part of the fundamental torsion equation connecting angle of twist and twisting moment being simple measurable dimensions which do not vary (length, diameter, etc.), the modulus of rigidity can be calculated from:

$$\frac{T}{J} = \frac{G\theta}{l} \qquad G = \frac{Tl}{\theta J}$$

Example. The following observations were recorded during a torsion test on a steel bar 12 mm diameter and 1·8 m long. Plot a graph of twisting moment and angle of twist and find the modulus of rigidity of this steel.

Twisting moment (N m)	4	6	8	10	12
Angle of twist (deg.)	2·2	3·3	4·3	5·4	6·5

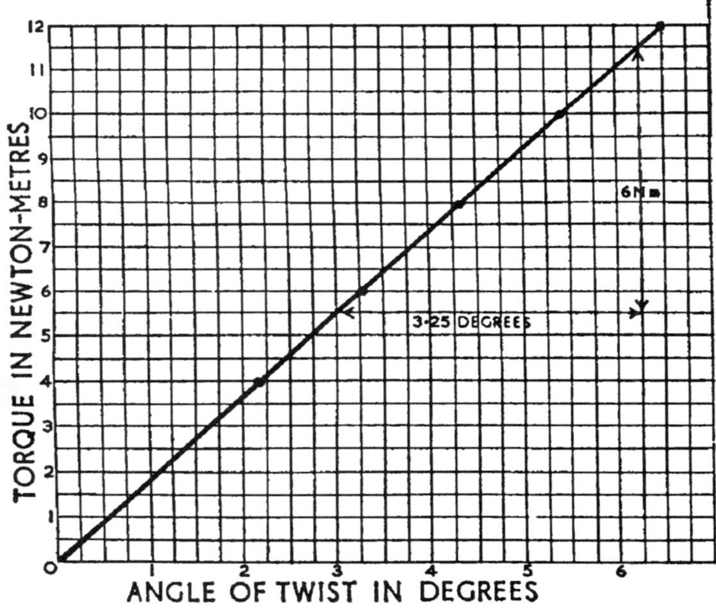

Fig. 196

Choosing two points on the graph such as shown in Fig. 196,

$$T = 6 \text{ N m}$$

$$\text{when } \theta = 3{\cdot}25 \div 57{\cdot}3 \text{ rad}$$

Inserting these values and the given dimensions of the bar into the expression,

$$G = \frac{Tl}{\theta J}$$

$$= \frac{57{\cdot}3 \times 32 \times 6 \times 1{\cdot}8}{3{\cdot}25 \times \pi \times 0{\cdot}012^4}$$

$$= 9{\cdot}354 \times 10^{10} \text{ N/m}^2$$

$$= 93{\cdot}54 \text{ GN/m}^2 \text{ or } 93{\cdot}54 \text{ kN/mm}^2 \text{ Ans.}$$

f TORSIONAL RESILIENCE

Referring to Fig. 197, let the torque be produced by a force F at a leverage of L to twist a shaft through θ rad. The angle of twist is proportional to the twisting moment applied, therefore the force is gradually applied from zero to the maximum of F while it moves through a linear distance of $L \times \theta$.

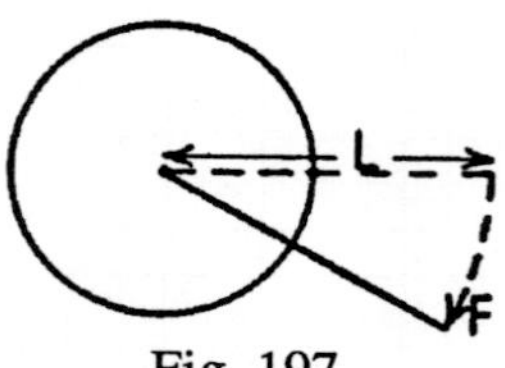

Fig. 197

$$\begin{aligned} \text{Work done} &= \text{Average force} \times \text{distance} \\ &= \frac{O + F}{2} \times L \times \theta \\ &= \frac{FL\theta}{2} \end{aligned}$$

$F \times L$ is the final torque represented by T

$$\text{Work done to twist shaft} = \frac{T\,\theta}{2}$$

Provided the shaft is strained within the elastic limit, the work done to twist it will be stored up as elastic strain energy in the twisted shaft. This is termed *torsional resilience* and, as with direct resilience (see Chapter 9), can be expressed in terms of stress.

$$\begin{aligned} \text{Torsional resilience} &= \text{Work done to twist shaft} \\ &= \frac{T\theta}{2} \end{aligned}$$

From $\dfrac{T}{J} = \dfrac{\tau}{r}$

$$T = \frac{\pi D^3 \tau}{16} \text{ for a solid shaft}$$

From $\dfrac{\tau}{r} = \dfrac{G\theta}{l}$

$$\theta = \frac{\tau l}{rG} = \frac{2\tau l}{DG}$$

Substituting these values for T and θ,

$$\text{Torsional resilience} = \frac{\pi \times D^3 \times \tau \times 2 \times \tau \times l}{16 \times 2 \times D \times G}$$

$$= \frac{\pi \times D^2 \times \tau^2 \times l}{16 \times G}$$

Volume of shaft $= \frac{\pi}{4} D^2 \times l$, therefore these terms can be taken out and replaced by volume.

$$\text{Torsional resilience} = \frac{\tau^2 \times V}{4G}$$

RELATIONSHIP BETWEEN TORQUE AND POWER

Consider a steady force of F applied at a leverage of L turning a shaft,

Work done to turn shaft one revolution

$= \text{force} \times \text{distance}$

$= F \times 2\pi L$ Nm *i.e. J*

If the shaft is running at n rev/s,

Work done per second = power

$= F \times 2\pi L \times n$ J/s *i.e. W*

But, $F \times L$ = Torque

Power = $2\pi Tn$

$2\pi n$ is the angular velocity in rad/s $= \omega$

Power = Torque $\times\ \omega$

$P = Tw$

MAXIMUM AND MEAN TORQUE

It is important to note that the above expression involves the *mean* torque to transmit a given power. In a reciprocating engine the torque varies throughout a revolution, the variation depending upon the cycle of operations within the cylinder and the number and arrangement of the cranks. It is obvious that a shaft must be of sufficient strength to carry the *maximum* torque and the diameter must be calculated from this value, therefore the ratio of maximum to mean torque should be given when when dealing with problems involving reciprocating engines. In turbines the torque is steady and the mean torque is taken to be equal to the maximum.

The expressions previously derived can therefore be written:

$$\text{Maximum torque} = \frac{\pi D^3 \tau}{16} \text{ for a solid shaft}$$

$$= \frac{\pi (D^4 - d^4)\tau}{16D} \text{ for a hollow shaft}$$

$$\text{Mean torque} = \frac{\text{power}}{\omega}$$

$$\text{i.e. } T = \frac{P}{\omega}$$

Example. A solid shaft from a turbine is to transmit 2000 kW at 3000 rev/min. Calculate the diameter of the shaft allowing a stress of 35 MN/m².

$$\frac{3000 \text{ rev/min} \times 2\pi}{60} = 100\pi \text{ rad/s}$$

$$\text{Torque} = \frac{\pi D^3 \tau}{16} = \frac{\text{power}}{\omega}$$

$$\frac{\pi \times D^3 \times 35 \times 10^6}{16} = \frac{2000 \times 10^3}{100 \times \pi}$$

$$D = \sqrt[3]{\frac{2000 \times 103 \times 16}{\pi \times 35 \times 10^6 \times 100 \times \pi}}$$

$$= 0{\cdot}09747 \text{ m} = 97{\cdot}47 \text{ mm} \quad \text{Ans.}$$

Example. Calculate the maximum power that can be transmitted through a solid shaft 380 mm diameter when driven by a reciprocating engine at 110 rev/min if the stress is not to exceed 40 MN/m² and the ratio of maximum to mean torque is 1·4 to 1.

$$\frac{110 \text{ rev/min} \times 2\pi}{60} = 11{\cdot}52 \text{ rad/s}$$

$$\text{Mean torque} = \frac{\text{power}}{\omega}$$

$$\text{Maximum torque} = 1{\cdot}4 \times \text{mean torque}$$

$$= \frac{1{\cdot}4 \times \text{power}}{11{\cdot}52} \quad \ldots \ldots \ldots \quad (i)$$

$$\text{Also, maximum torque} = \frac{\pi D^3 \tau}{16}$$

$$= \frac{\pi \times 0{\cdot}38^3 \times 40 \times 10^6}{16} \quad \ldots \quad (ii)$$

$$\frac{1{\cdot}4 \times \text{power}}{11{\cdot}52} = \frac{\pi \times 0{\cdot}38^3 \times 40 \times 10^6}{16}$$

$$\text{power} = \frac{11{\cdot}52 \times \pi \times 0{\cdot}38^3 \times 40 \times 10^6}{1{\cdot}4 \times 16}$$

$$= 3{\cdot}547 \times 10^6 \text{ W}$$

$$= 3{\cdot}547 \text{ MW or } 3547 \text{ kW} \qquad \text{Ans.}$$

TRANSMISSION OF POWER THROUGH COUPLING BOLTS

The bolts connecting shaft couplings carry the transmitted torque from one shaft to the next, shear stress is induced in them and the magnitude of this stress will depend upon the total sectional area of bolt material carrying the load and the radius of the circle on which the bolts are pitched.

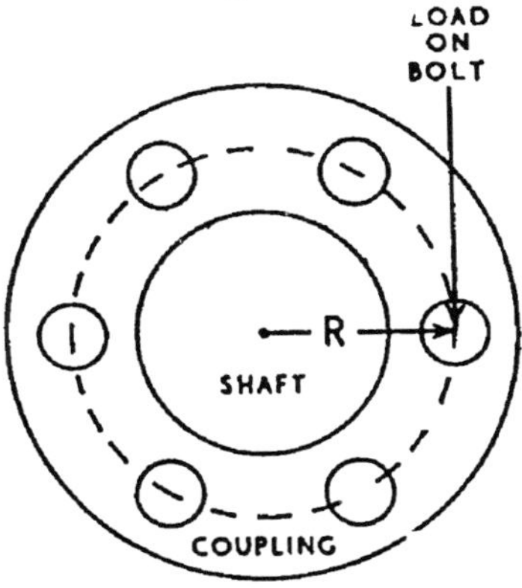

Fig. 198

Let shear stress in bolts $= \tau_B$

Cross-sect. area of each bolt $= \dfrac{\pi}{4}\, d^2$

Radius of pitch circle $= R$

Number of bolts in coupling $= N$

Torque transmitted by each bolt

= Load on one bolt × Radius of pitch circle.

$= \text{Stress in bolt} \times \text{Cross. sect.} \times \text{Radius of pitch circle}$

$= \tau_B \times \frac{\pi}{4} d^2 \times R$

Total torque transmitted by all bolts in coupling

$= \text{Torque transmitted by one bolt} \times \text{number of bolts}$

$= \tau_B \times \frac{\pi}{4} d^2 \times R \times N$

This must be equal to the torque transmitted by the shafting, therefore this can be equated to any one of the formulae for the torque in shafts previously given, depending upon the data given in the problem.

Example. The couplings of a 350 mm diameter solid shaft are 700 mm diameter and there are 8 bolts per coupling on a pitch circle diameter of 525 mm. Under maximum working conditions the stress in the shaft is 42 MN/m^2. Find the diameter of the coupling bolts allowing a stress in them of 35 MN/m^2.

$$\text{Torque transmitted by bolts} = \text{Torque transmitted by shaft}$$

$$\tau_B \times \frac{\pi}{4} d^2 \times R \times N = \frac{\pi}{16} D^3 \times \tau$$

$$35 \times 10^6 \times \frac{\pi d^2}{4} \times 0{\cdot}2625 \times 8 = \frac{\pi}{16} \times 0{\cdot}35^3 \times 42 \times 10^6$$

$$d = \sqrt{\frac{0{\cdot}353 \times 42 \times 4}{35 \times 0{\cdot}2625 \times 8 \times 16}}$$

$$= 0{\cdot}07825 \text{ m} = 78{\cdot}25 \text{ mm Ans.}$$

RECIPROCATING ENGINE MECHANISM

The force on the piston of a reciprocating engine is transmitted through the piston rod to the crosshead and sends a thrust through the connecting rod to the crank pin to exert a turning moment on the shaft.

The method of finding the thrust in the connecting rod for a given piston force and crank angle was explained in Chapter 1.

The connecting rod thrust multiplied by the perpendicular distance from its line of action to the shaft centre, is the turning moment applied to the shaft.

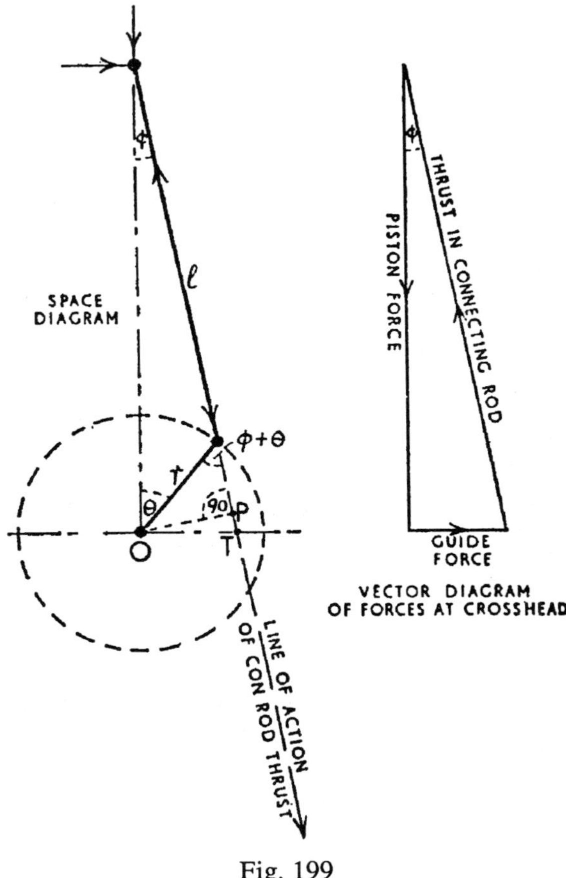

Fig. 199

Fig. 199 shows the crank at θ° past top centre. The perpendicular distance from the line of action of the connecting rod thrust to the shaft centre is marked OP, this is effective leverage, and:

Turning moment = Force × perpendicular distance
= thrust in conn. rod × OP

Example. The effective pressure on the piston of a diesel engine is 17·25 bar (= 17·25 × 10^5 N/m^2) when the crank is 60° past top dead centre. The diameter of the piston is 500 mm, stroke 900 mm, and connecting rod length 1575 mm. Find the turning moment on the crank at this position.

$$\begin{aligned}\text{Piston force} &= \text{pressure} \times \text{area of piston}\\ &= 17{\cdot}25 \times 10^5 \times 0{\cdot}7854 \times 0{\cdot}5^2\\ &= 3{\cdot}387 \times 10^5\ \text{N} = 338{\cdot}7\ \text{kN}\end{aligned}$$

Referring to space diagram of Fig. 199,

$$\begin{aligned}\frac{l}{\sin\theta} &= \frac{r}{\sin\phi}\\ \sin\phi &= \frac{0{\cdot}45 \times \sin 60}{1{\cdot}575} = 0{\cdot}2474\\ \phi &= 14^\circ\ 19'\\ \text{OP} &= r \times \sin(\phi + \theta)\\ &= 0{\cdot}45 \times \sin 74^\circ\ 19' = 0{\cdot}4332\ \text{m}\end{aligned}$$

Referring to vector diagram of Fig. 199,

$$\begin{aligned}\text{Thrust in conn. rod} &= \frac{\text{Piston force}}{\cos\phi}\\ &= \frac{338{\cdot}7}{\cos 14^\circ\ 19'} = 349{\cdot}5\ \text{kN}\\ \text{Turning moment} &= \text{Thrust in conn. rod} \times \text{OP}\\ &= 349{\cdot}5 \times 0{\cdot}4332\\ &= 151{\cdot}5\ \text{kN m} \qquad \text{Ans.}\end{aligned}$$

ALTERNATIVE METHOD

$$\text{Turning moment} = \text{Thrust in conn. rod} \times \text{OP}$$

and,

$$\text{Thrust in conn. rod} = \frac{\text{Piston force}}{\cos\phi}$$

Now, examining the space diagram of Fig. 199, note that angle POT is equal to ϕ, therefore,

$$\text{OP} = \text{OT} \times \cos\phi$$

Substituting,

$$\begin{aligned}\text{Turning moment} &= \text{Thrust in conn. rod} \times \text{OP}\\ &= \frac{\text{Piston force}}{\cos\phi} \times \text{OT} \times \cos\phi\\ &= \text{Piston force} \times \text{OT}\end{aligned}$$

The distance OT is the measurement from the shaft centre to the point where the line of action of the connecting rod thrust cuts the horizontal centre line as shown in Fig. 199 and this measurement can conveniently be found graphically by drawing the crank pin circle and connecting rod to scale.

This is particularly useful because one scale drawing of the crank pin circle can be used for measuring the distances OT for a series of different crank angles, these multiplied by the effective piston forces obtained from the indicator diagram at the points of the stroke relative to the crank angles, produces a series of values of torque which can be plotted as a graph to show the variations of torque throughout a cycle of operations.

CRANK EFFORT

The effects of the thrust of the connecting rod on the crank pin are (*i*) to turn the shaft, (*ii*) to exert either compression or tension in the crank webs. The values of (*i*) and (*ii*) are the rectangular components of the force in the connecting rod as shown in Fig. 200.

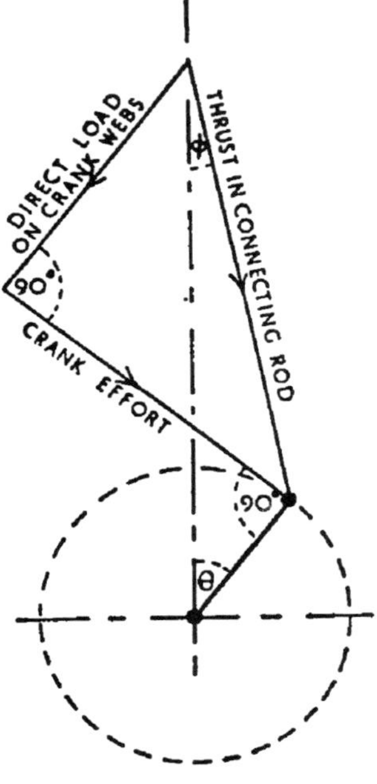

Fig. 200

One component, the force which effectively turns the shaft, is at right angles to the crank. The other component which causes compression or tension in the webs is in direct line with the crank.

The force which effectively turns the shaft is termed the *crank effort* and may be calculated from the force diagram, Fig. 200, alternatively, as the product of Crank Effort and Length of Crank is the Turning Moment, then,

$$\text{Crank Effort} = \frac{\text{turning moment}}{\text{crank length}}$$

HYDRAULIC STEERING GEAR

The arrangement of a hydraulic steering gear is based on the principle of Rapson's Slide. The force applied by the hydraulic ram on the swivel crosshead through which the tiller arm slides is the product of the hydraulic oil pressure and the cross-sectional area of the ram. Let this ram force be represented by F (Fig. 201).

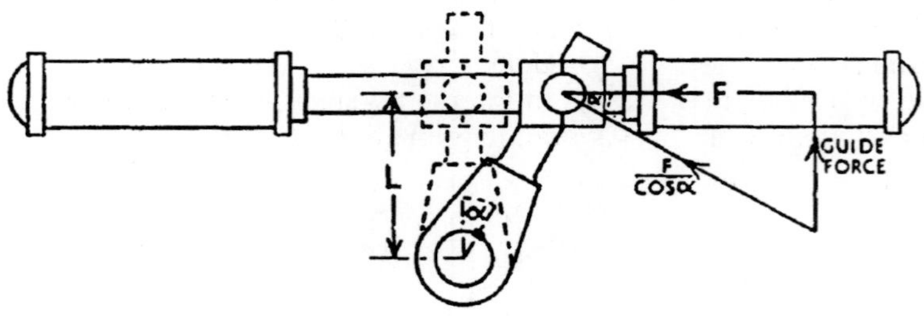

Fig. 201

The ram force F tends to cause the block to slide along the tiller arm towards the rudder stock and this is prevented by the supporting force of the guide acting at right angles to the ram force. The resultant of the ram force and the guide force is the effective force acting perpendicularly on the tiller causing it to rotate and turn the rudder stock.

If the tiller is at $\alpha°$ from mid-position, the resultant force is $F \div \cos\alpha$. If the distance from centre of rudder stock to the centre-line of the rams is L, the distance from rudder stock centre to crosshead is $L \div \cos\alpha$ and this is the perpendicular distance from the application of the effective force to the rudder stock centre.

$$\text{Turning moment} = \frac{F}{\cos\alpha} \times \frac{L}{\cos\alpha} = \frac{FL}{\cos^2\alpha}$$

When α is zero, that is when the rudder is in mid-position, $\cos^2\alpha = 1$ and then the turning moment is simply $F \times L$.

As the rudder moves from mid-position and α increases, the value of $\cos^2\alpha$ decreases, therefore for a given constant working hydraulic pressure the turning moment increases as the rudder moves further from its mid-position. This is one important advantage of the hydraulic steering gear over others because, when the ship is moving ahead the turning moment required to move the rudder becomes greater as the angle of helm is increased.

Fig. 201 is a diagrammatic sketch of the arrangement of a two-ram steering gear in which one ram at a time exerts a turning effect on the rudder. In a four-ram system there is a similar pair on either side of the rudder stock and therefore two diagonally opposite rams act together in exerting the turning moment, this gives twice the turning effect as a two-ram steering gear (a couple).

Example. In a two-ram hydraulic steering gear, the rams are 250 mm diameter, the rudder stock is 400 mm diameter and the distance from centre of rudder stock to centre-line of rams is 800 m. Calculate the stress in the rudder stock when the rudder is in mid-position and the hydraulic pressure is 70 bar.

$$\begin{aligned}\text{Ram force on tiller} &= 70 \times 10^5 \times 0{\cdot}7854 \times 0{\cdot}25^2 \\ &= 3{\cdot}436 \times 10^5 \text{ N} \\ \text{Turning moment} &= 3{\cdot}436 \times 10^5 \times 0{\cdot}8 \text{ N m}\end{aligned}$$

For a solid shaft,

$$T = \frac{\pi}{16} D^3\tau$$

$$\tau = \frac{16 \times T}{\pi \times D^3} = \frac{16 \times 3{\cdot}436 \times 10^5 \times 0{\cdot}8}{\pi \times 0{\cdot}4^3}$$

$$= 2{\cdot}187 \times 10^7 \text{ N/m}^2$$

$$= 21{\cdot}87 \text{ MN/m}^2 \quad \text{Ans.}$$

*f*Example. In a four-ram electric-hydraulic steering gear, the relief by-pass valves are set to lift when the oil pressure is 75 bar. The diameter of the rams is 300 mm, distance from centre of rudder stock to centre-line of each pair of rams is 760 mm and the maximum angle of helm is 35° from mid-position. The tiller arms are parallel round section from the end to 560 mm from the rudder stock centre and the maximum bending stress in them may be taken as occurring at this section. Calculate (*i*) the diameter of the rudder stock to limit the maximum torsional stress to 75 MN/m², and (*ii*) the diameter of the tiller arms to limit the maximum bending stress to 105 MN/m².

$$\text{Max. force on ram} = 75 \times 10^5 \times 0{\cdot}7854 \times 0{\cdot}3^2 \text{ N}$$

Force applied at right angles to tiller

$$= \frac{75 \times 10^5 \times 0{\cdot}7854 \times 0{\cdot}3^2}{\cos 35°}$$

$$= 6{\cdot}471 \times 10^5 \text{ N}$$

Leverage from crosshead to rudder stock centre

$$= \frac{0{\cdot}76}{\cos 35°} = 0{\cdot}9276 \text{ m}$$

Turning moment on rudder stock from two rams

$$= 6{\cdot}471 \times 10^5 \times 0{\cdot}9276 \times 2 \text{ N m}$$

For a solid round section, $T = \frac{\pi}{16} D^3 \tau$

$$\frac{\pi}{16} \times D^3 \times 75 \times 10^6 = 6{\cdot}471 \times 10^5 \times 0{\cdot}9276 \times 2$$

$$D = \sqrt[3]{\frac{16 \times 6{\cdot}471 \times 0{\cdot}9276 \times 2}{\pi \times 75 \times 10}}$$

$$= 0{\cdot}4336 \text{ m} = 433{\cdot}6 \text{ mm} \quad \text{Ans. } (i)$$

Length of tiller arm from crosshead to neck of parallel part which is 0·56 m from rudder stock centre

$$= 0{\cdot}9276 - 0{\cdot}56 = 0{\cdot}3676 \text{ m}$$

Bending moment at this section

$$= 6{\cdot}471 \times 10^5 \times 0{\cdot}3676 \text{ N m}$$

For a round section subject to bending:

$$\sigma = \frac{32M}{\pi D^3}$$

$$D^3 = \frac{32 \times M}{\pi \times \sigma}$$

$$D = \sqrt[3]{\frac{32 \times 6{\cdot}471 \times 10^5 \times 0{\cdot}3676}{\pi \times 105 \times 10^6}}$$

$$= 0{\cdot}2847 \text{ m} = 284{\cdot}7 \text{ mm Ans. } (ii)$$

f DEFLECTION OF A CLOSELY COILED HELICAL SPRING

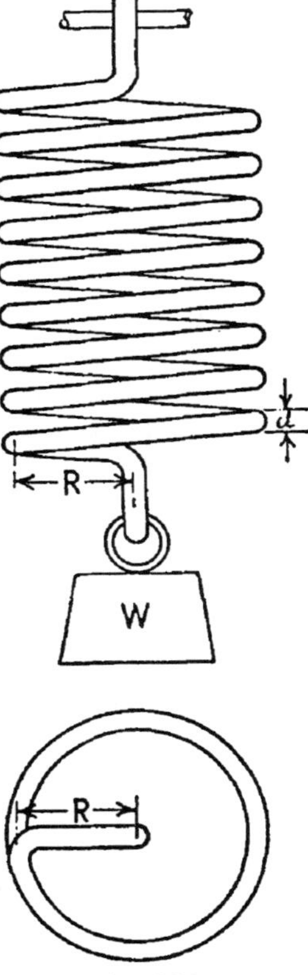

Fig. 202

If the coils of a helical spring are closely pitched, an axial load applied produces pure twist of the spring material, the twisting moment being the product of the load and the mean radius of the coils. It can therefore be considered in the same manner as the twisting of a straight shaft, the length of spring wire under twist being the length of each coil multiplied by the number of coils. The length of one coil can be taken simply as the mean circumference ($2\pi R$) because the spring is closely coiled (Fig. 202).

Let W = axial load,
R = mean radius of coils,
D = mean diameter of coils,
d = diameter of coil wire,

the stress in the wire can be calculated thus:

$$T = \frac{\pi}{16} d^3 \tau$$

$$W \times R = \frac{\pi}{16} d^3 \tau$$

$$\tau = \frac{16WR}{\pi d^3} \text{ or } \frac{8WD}{\pi d^3}$$

The deflection (δ) is the axial stretch of the spring and an expression connecting axial load and axial deflection can be obtained by equating the external work applied to the work required to twist the material.

Let N = number of coils,
δ = axial deflection,
External work applied by W = Internal work to twist wire
Average force $\times$ distance = Average torque $\times$ angle of twist

$$\tfrac{1}{2}(O + W) \times \delta = \tfrac{1}{2}(O + T) \times \theta$$

$$W \times \delta = T \times \theta \quad \ldots \ldots \ldots \ldots \quad (i)$$

Finding suitable equivalents of T and θ for substitution:

$$\frac{T}{J} = \frac{G\theta}{l} \qquad \theta = \frac{Tl}{JG}$$

$$T = W \times R$$

$$l = \text{length of wire} = 2\pi RN$$

$$J = \frac{\pi}{32} d^4$$

$$\theta = \frac{Tl}{JG}$$

$$= \frac{32 \times W \times R \times 2\pi RN}{\pi \times d^4 \times G}$$

Substituting for T and θ into (i)

$$W \times \delta = T \times \theta$$

$$W \times \delta = \frac{W \times R \times 32 \times W \times R \times 2\pi RN}{\pi \times d^4 \times G}$$

$$\delta = \frac{64WR^3N}{Gd^4} \text{ or } \frac{8WD^3N}{Gd^4}$$

f Example. A closely coiled helical spring is made of 6 mm diameter steel wire, the mean diameter of the coils is 60 mm and there are 8 coils. Taking the modulus of rigidity as 100 kN/mm² find the deflection and the stress in the wire when carrying an axial load of 240 N.

$$\delta = \frac{64WR^3N}{Gd^4}$$

Working in mm:

$$\delta = \frac{64 \times 240 \times 30^3 \times 8}{100 \times 10^3 \times 6^4}$$

$$= 25{\cdot}6 \text{ mm} \qquad \text{Ans. (i)}$$

$$T = \frac{\pi}{16} d^3\tau$$

$$W \times R = \frac{\pi}{16} d^3\tau$$

$$\tau = \frac{16 \times W \times R}{\pi \times d^3}$$

$$= \frac{16 \times 240 \times 30}{\pi \times 6^3}$$

$$= 169{\cdot}7 \text{ N/mm}^2$$

$$= 169{\cdot}7 \times 10^6 \text{ N/m}^2$$

$$= 169{\cdot}7 \text{ MN/m}^2 \qquad \text{Ans. (ii)}$$

TEST EXAMPLES 11

1. A hollow shaft is 400 mm diameter outside and 250 mm diameter inside and transmits a torque of 480 kN m. Calculate the shear stress in the material and the angle of twist over a length of 7·5 m. Take $G = 92{\cdot}5$ GN/m^2.

2. A solid shaft is to be replaced by a hollow shaft. The hollow shaft is to be made from a higher quality steel so that the safe working stress allowed can be 20% higher and its outside diameter is to be equal to the diameter of the solid shaft so that the same bearings can be used. Calculate (*i*) the diameter of the bore of the hollow shaft in terms of the outside diameter, (*ii*) the percentage saving in weight assuming that the densities of the steels of both shafts are equal.

3. A solid steel shaft 350 mm diameter has a brass liner shrunk on it over its entire length, the thickness of the liner being 25 mm. Taking G for steel as 85 GN/m^2 and G for brass as 38·5 GN/m^2, calculate the maximum shear stresses in the shaft and the liner when the total torque transmitted between them is 200 kN m.

4. The pinion shaft of a double reduction geared turbine is 140 mm diameter, runs at 3000 rev/min and the stress in it is 48 MN/m^2. The main shaft is 445 mm diameter and runs at 90 rev/min. Assuming that the same power is transmitted by both shafts, find the stress in the main shaft.

5. A torsion-meter on a tunnel shaft 360 mm diameter registers an angle of twist of 0·3° over a length of 2·5 m when running at 115 rev/min. Taking the modulus of rigidity of the shaft material as 93 GN/m^2, find the power transmitted.

6. A propeller shaft 380 mm diameter running at 90 rev/min drives a ship at 19 knots, the total resistance of the ship through the water at this speed is 320 kN. Taking the propeller efficiency as 0·75, find (*i*) the power transmitted by the shaft, (*ii*) the twisting moment in the shaft, (*iii*) the torsional stress in the shaft. 1 knot = 1·852 km/h.

7. An engine develops 4500 kW at 105 rev/min and the ratio of maximum to mean torque is 1·25 to 1.

(*a*) Calculate the diameter of the main shaft coupled direct to the engine to transmit this power allowing a stress of 55 MN/m^2 and assuming the shaft to be (*i*) solid, (*ii*) hollow with a bore equal to half the outside diameter.

(*b*) Calculate the percentage saving in weight by fitting the hollow shaft as compared with the solid shaft.

8. Find the diameter of the coupling bolts to connect shafts which are to transmit 3000 kW when running at 100 rev/min if 8 bolts are to be fitted on a pitch circle diameter of 610 mm and allowing a stress of 38 MN/m^2 in the bolts.

9. The stroke of an internal combustion engine is 630 mm and the connecting rod length is 1200 mm. When the crank is 35° past top centre the load on the piston is 250 kN. Find the turning moment at this point.

10. The rams of a four-ram hydraulic steering gear are 250 mm diameter, the rudder stock is 430 mm diameter and the distance from rudder stock centre to centre-line of rams is 760 mm. Calculate the stress in the rudder stock at the instant the rudder is moving past mid-position and the hydraulic pressure on the rams is 70 bar ($= 70 \times 10^5$ N/m^2).

*f*11. Prove that the elastic strain energy in a hollow shaft subjected to a torsional stress of τ is given by the expression:

$$\frac{\tau^2}{4G} \times \frac{D^2 + d^2}{D^2} \times \text{Volume of shaft.}$$

*f*12. The drive to a centrifuge consists of a solid steel shaft 30 mm in diameter which is a sliding fit in a hollow shaft of outside diameter 38 mm. The two shafts are secured by a shear pin fitted across the diameter of both shafts perpendicular to the shafts axis. The shear stress in the solid shaft is not to exceed 20 MN/m^2. Calculate:

(*a*) the maximum torque that can be transmitted,
(*b*) The maximum shear stress in a hollow shaft when transmitting this maximum torque,
(*c*) a suitable diameter for the pin (which will shear at a stress of 80 MN/m^2) to protect the solid shaft from excessive stress.

*f*13 In a two-ram hydraulic steering gear the diameter of the rams is 280 mm, diameter of rudder stock 350 mm, distance from rudder stock centre to centre-line of rams 860 mm and the maximum angle of helm is 35°. Find the pressure at which the by-pass relief valves should be set to lift in order to limit the maximum shear stress in the rudder stock to 77 MN/m^2.

*f*14. In a close coiled helical spring there are 27 coils of solid round wire 3 mm diameter, the mean diameter of the coils being 30 mm. If the modulus of rigidity of the material is 93 kN/mm^2 find the deflection when carrying an axial load of 50 N. Derive any formula used in the solution.

*f*15. A close coiled helical spring consists of 24 coils of tubular steel, the mean diameter of the coils is 75 mm, the outside diameter of the tube is 12·5 mm and its thickness is 2·5 mm. If the stress in the material is not to exceed 70 N/mm^2 find the maximum axial load the spring can carry and the deflection under this load. Take $G = 90$ $kN.mm^2$.

CHAPTER 12

HYDROSTATICS

The study of fluids at rest.

DENSITY

Density (ρ) is a measure of the mass per unit volume (kg/m^3).

$$t/m^3 = kg/l = g/ml = g/cm^3$$

The relative density of a substance is the ratio of the mass of that substance to the mass of an equal volume of pure water, in other words, it is the ratio of the density of the substance to that of pure water.

Densities of liquids may be measured by a hydrometer. This is an instrument consisting of a graduated stem with a hollow bulb float at the lower end of the stem, and keel beneath the bulb so that the hydrometer floats upright when placed in a liquid. The stem is graduated, usually in g/ml, and this figure is read off at the level of the liquid at the stem.

MIXING OF LIQUIDS OF DIFFERENT DENSITIES

When liquids of different densities are mixed together it is usual to assume that their volumes and masses are not affected due to mixing, that is, the final volume of the whole mixture is equal to the sum of the volumes of each constituent before mixing, and the final mass of the whole mixture is equal to the sum of the masses of each constituent before mixing.

Example. 2 g of oil of relative density 0·8 are mixed with 4 g of oil of relative density 0·9. Find the relative density of the mixture.

Total mass of mixture $= 2 + 4 = 6$ g ... ... (*i*)

As the densities of the oils are 0·8 and 0·9 g/ml respectively, and volume = mass ÷ density, then,

$$\text{Total volume} = \frac{2}{0{\cdot}8} + \frac{4}{0{\cdot}9} \text{ ml}$$

$$= \frac{1{\cdot}8 + 3{\cdot}2}{0{\cdot}72}$$

$$= \frac{5}{0{\cdot}72} \text{ ml} \quad \ldots \quad \ldots \quad \ldots (ii)$$

$$\text{Density of mixture} = \frac{\text{Total mass}}{\text{Total volume}}$$

$$= \frac{6 \times 0{\cdot}72}{5}$$

$$= 0{\cdot}864 \text{ g/ml}$$

$$\text{Relative density} = 0{\cdot}864 \text{ Ans.}$$

Example. Three liquids of relative density 0·75, 0·85 and 0·95 are mixed in the proportion of 2, 3 and 4, by volume. Find the relative density of the mixture.

As the proportion is by volume, let the volumes be 2, 3 and 4 ml respectively.

Densities of the oils are 0·75, 0·85 and 0·95 g/ml respectively.

$$m = V \times \rho$$

$$\text{Total mass} = 2 \times 0{\cdot}75 + 3 \times 0{\cdot}85 + 4 \times 0{\cdot}95$$

$$= 1{\cdot}5 + 2{\cdot}55 + 3{\cdot}8$$

$$= 7{\cdot}85 \text{ g}$$

$$\text{Total volume} = 2 + 3 + 4$$

$$= 9 \text{ ml}$$

$$\text{Density} = \frac{\text{Total mass}}{\text{Total volume}}$$

$$= \frac{7{\cdot}85}{9}$$

$$= 0{\cdot}872 \text{ g/ml}$$

$$\text{Relative density} = 0{\cdot}872 \text{ Ans.}$$

APPARENT LOSS OF WEIGHT OF A SUBMERGED BODY

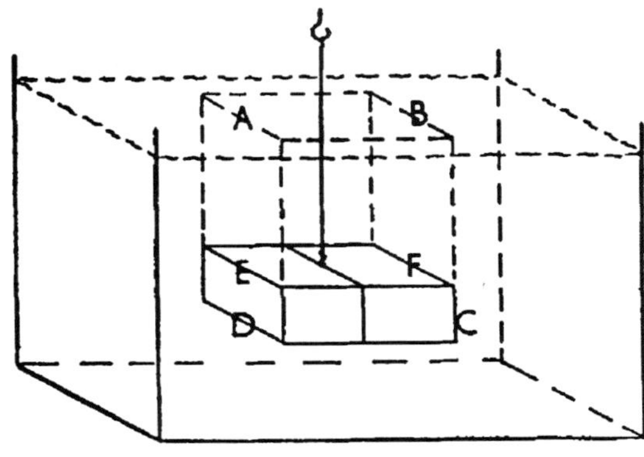

Fig. 203

Consider a vertical column ABCD of the liquid in a tank as shown in Fig. 203 (before a body is submerged). The column of liquid is in equilibrium therefore the upward supporting thrust on its base DC is equal to the downward weight of the column.

When a body is lowered and submerged in the liquid such as EFCD, the upward thrust at the same level DC (that is on the base of the body), must be the same as before but there is also a downward force on the top of the body due to the column of liquid ABFE above it. Therefore the net force on the body is the difference between the upward supporting force on the base and the downward weight of liquid on the top, this is equal to the difference between the columns of liquid ABCD and ABFE which is the weight of the liquid EFCD displaced by the body.

Hence, the upward force of buoyancy on a submerged body is equal to the weight of liquid displaced by the body. This is simply verified experimentally by suspending a regular shaped body (whose volume can be calculated) from a spring balance, noting the weight when the body is in air and again when lowered into water until it is submerged, the difference in weights read from the spring balance is the 'apparent loss of weight' which will be found to be equal to the weight of water displaced (that is the weight of water of equal volume to the body).

This principle also provides a useful method of determining the relative density of a solid heavier than water as shown by the following.

Example. A piece of cast iron registers a weight of 56 N in air and 48·23 N when suspended in fresh water. Find the relative density of this cast iron.

$$\text{Weight of water displaced} = 56 - 48{\cdot}23 = 7{\cdot}77 \text{ N}$$

The volume of water displaced is the same as the volume of the cast iron, therefore,

$$\text{Relative density} = \frac{\text{Wt. of cast iron}}{\text{Wt. of an equal volume of fresh water}} = \frac{56}{7{\cdot}77} = 7{\cdot}207 \quad \text{Ans.}$$

From the above it can be seen that:

$$\text{Relative density} = \frac{\text{Wt. in air}}{\text{Wt. in air} - \text{wt. in fresh water}}$$

Example. A block of aluminium of 11·5 kg mass is suspended from a wire 1·5 mm diameter and lowered until submerged into a tank containing oil of relative density 0·9. Taking the relative density of aluminium as 2·56, find (i) the tension in the wire, and (ii) the stress in the wire.

$$\text{Volume of aluminium (V)} = \frac{\text{mass}}{\text{density}} = \frac{m}{\rho} = \frac{11{\cdot}5 \times 10^3}{2{\cdot}56} \text{ cm}^3$$

$$\text{Mass of an equal vol. of oil} = \text{mass of oil displaced} = V \times \rho = \frac{11{\cdot}5 \times 10^3}{2{\cdot}56} \times 0{\cdot}9 \text{ g} = \frac{11{\cdot}5 \times 0{\cdot}9}{2{\cdot}56} \text{ kg}$$

Wt. of aluminium in oil = wt. in air – wt. of oil displaced

$$= 11{\cdot}5 \times 9{\cdot}81 - \frac{11{\cdot}5 \times 0{\cdot}9 \times 9{\cdot}81}{2{\cdot}56}$$

$$= 11{\cdot}5 \times 9{\cdot}81 \left\{1 - \frac{0{\cdot}9}{2{\cdot}56}\right\}$$

$$= 11{\cdot}5 \times 9{\cdot}81 \left\{\frac{2{\cdot}56 - 0{\cdot}9}{2{\cdot}56}\right\}$$

$$= \frac{11{\cdot}5 \times 9{\cdot}81 \times 1{\cdot}66}{2{\cdot}56}$$

$$= 73{\cdot}16 \text{ N} \quad \text{Ans. (i)}$$

$$\text{Stress} = \frac{\text{load}}{\text{area}}$$

$$= \frac{73{\cdot}16}{0{\cdot}7854 \times 1{\cdot}5^2} = 41{\cdot}4 \text{ N/mm}^2$$

$$= 41{\cdot}4 \times 10^6 \text{ N/m}^2$$

$$= 41{\cdot}4 \text{ MN/m}^2 \quad \text{Ans. (ii)}$$

FLOATING BODIES

A body which floats freely in a liquid is in equilibrium, therefore its downward weight must be equal to the upward force of buoyancy. The upward thrust is equal to the weight of liquid displaced, hence a *floating body displaces an amount of liquid equal to its own weight.*

All cases of floating bodies can be dealt with by the simple law of equilibrium:

$$\text{Total downward force} = \text{Total upward force}$$

Example. A rectangular block of wood 375 mm long by 250 mm broad by 100 mm deep, floats in fresh water. If the density of the wood is 0·75 g/cm³, find the draught at which it floats.

$$\text{Let } d = \text{draught}$$

$$\text{Downward force} = \text{Upward force}$$

$$\text{Weight of wood} = \text{Weight of water displaced}$$

$$37{\cdot}5 \times 25 \times 10 \times 0{\cdot}75 \times 10^{-3} \times 9{\cdot}81$$

$$= 37{\cdot}5 \times 25 \times d \times 1 \times 10^{-3} \times 9{\cdot}81$$

$$10 \times 0{\cdot}75 = d$$

$$d = 7{\cdot}5 \text{ cm} = 75 \text{ mm} \quad \text{Ans.}$$

Example. A solid wood raft 4 m long by 1·6 m broad by 0·6 m deep, floats at a draught of 0·5 m in sea water when carrying a mass of 592 kg on top of the raft. Find the density of the wood, taking the density of sea water as 1025 kg/m³.

$$\text{Downward force} = \text{Upward force}$$

$$\text{Wt. of raft + wt. of mass} = \text{wt. of water displaced}$$

$$4 \times 1{\cdot}6 \times 0{\cdot}6 \times \rho \times 9{\cdot}81 + 592 \times 9{\cdot}81 = 4 \times 1{\cdot}6 \times 0{\cdot}5 \times 1{\cdot}025 \times 10^3 \times 9{\cdot}81$$

$$3{\cdot}84 \times \rho + 592 = 3280$$

$$3{\cdot}84\rho = 2688$$

$$\rho = 700 \text{ kg/m}^3$$

Density of wood = 700 kg/m³ or 0·7 t/m³ or 0·7 g/cm³ Ans.

Example. The mass of a cork lifebuoy is 12·25 kg. Find the maximum mass of cast iron it can support in sea water if the iron is suspended below the buoy. Take the relative density of cork, cast iron, and sea water, as 0·288, 7·21 and 1·025 respectively.

Maximum load will be carried when the buoy is just awash.

$$\text{Volume of buoy} = \frac{\text{mass}}{\text{density}}$$

$$= \frac{12{\cdot}25 \times 10^3}{0{\cdot}288} = 42{\cdot}54 \times 10^3 \text{ cm}^3$$

Mass of sea water displaced by buoy when fully submerged

$$= \text{volume} \times \text{density of sea water}$$

$$= 42{\cdot}54 \times 10^3 \times 1{\cdot}025 \text{ g}$$

$$= 43{\cdot}6 \text{ kg}$$

$$\text{Volume of cast iron} = \frac{\text{mass}}{\text{density}}$$

$$= \frac{m \times 10^3}{7{\cdot}21} \text{ cm}^3$$

Mass of sea water displaced by cast iron

$$= \frac{m \times 10^3}{7{\cdot}21} \times 1{\cdot}025 \text{ g}$$

$$= 0{\cdot}1422\, m \text{ kg}$$

$$\text{Total downward forces} = \text{Total upward forces}$$

$$\text{Weights of buoy \& cast iron} = \text{Weight of water displaced}$$

$$9{\cdot}81\,(12{\cdot}25 + m) = 9{\cdot}81\,(43{\cdot}6 + 0{\cdot}1422m)$$

$$m - 0{\cdot}1422\, m = 43{\cdot}6 - 12{\cdot}25$$

$$0{\cdot}8578\, m = 31{\cdot}35$$

$$m = 36{\cdot}55 \text{ kg} \quad \text{Ans.}$$

PRESSURE HEAD

Consider the equilibrium of a vertical column of liquid of height h m, cross-sectional area A m^2, and of density ρ kg/m^3, as illustrated in Fig. 204.

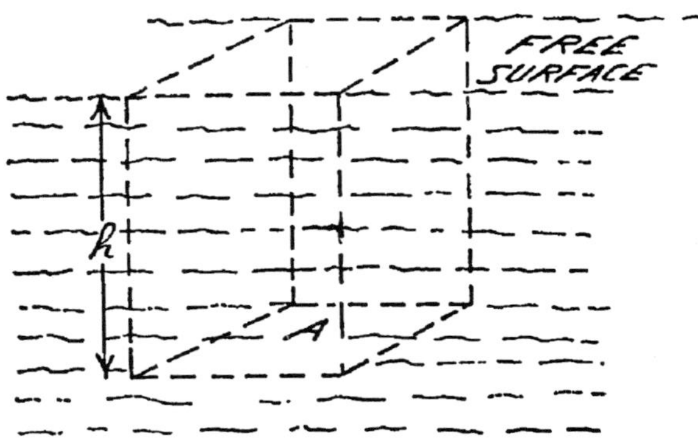

Fig. 204

$$\text{The } \textit{mass} \text{ of the liquid column} = \text{volume} \times \text{density}$$
$$= h\,A \times \rho$$
$$\text{The } \textit{weight} \text{ of the liquid column} = h\,A\,\rho \times g$$

Let p = pressure in the liquid at depth h

For equilibrium:

$$\text{Upward force} = \text{Weight of liquid column}$$
$$p \times A = \rho g h A$$
$$\therefore p = \rho g h$$

Since the density of fresh water is 10^3 kg/m^3 the head of fresh water which will exert a pressure of 1 bar (*i.e.* 10^5 N/m^2) can be calculated:

$$p = \rho g h$$
$$h = \frac{p}{\rho g}$$
$$h = \frac{10^5}{10^3 \times 9{\cdot}81}$$
$$h = 10{\cdot}19 \text{ m}$$

i.e. a head of 10·19 m of fresh water will exert a pressure of 1 bar.

Thus, pumping against a pressure (p) can be considered as lifting the liquid to an equivalent height (h) and the work done or power exerted can be calculated by this method.

Example. An engine developing 3730 kW uses 7·25 kg of steam/kWh. If the boiler pressure is 17 bar (= 17×10^5 N/m^2), calculate the output power of the feed pump.

Equivalent water head of 17 bar $= 17 \times 10{\cdot}19 = 173{\cdot}2$ m

Mass of water pumped into boiler every second

$$= \frac{7{\cdot}25 \times 3730}{3600} = 7{\cdot}511 \text{ kg}$$

$$\text{Force to lift against gravity} = 7{\cdot}511 \times 9{\cdot}81 = 73{\cdot}69 \text{ N}$$

$$\begin{aligned} \text{Power} &= \text{work done per second} \\ &= \text{force} \times \text{height, per second} \\ &= 73{\cdot}69 \times 173{\cdot}2 \\ &= 1{\cdot}277 \times 10^4 \text{ W} = 12{\cdot}77 \text{ kW Ans.} \end{aligned}$$

This is set out as an example of 'equivalent head'. An alternative solution to this type of example, and a more direct approach, is explained in Chapter 4, thus:

$$\begin{aligned} \text{Work done} &= p \times V \\ \text{Work per sec.} &= \text{Pressure} \times \text{Volume flow} \\ \text{Density of water} &= 10^3 \text{ kg/m}^3 \\ \text{Volume flow} &= \frac{7{\cdot}25 \times 3730}{10^3 \times 3600} = 7{\cdot}511 \times 10^{-3} \text{ m}^3\text{/s} \\ \text{Power} &= 17 \times 10^5 \times 7{\cdot}511 \times 10^{-3} \\ &= 1{\cdot}276 \times 10^4 \text{ W} = 12{\cdot}76 \text{ kW} \end{aligned}$$

MANOMETERS

A manometer is a simple pressure measuring device, suitable for measuring small pressures, or small differences in pressure.

The manometer shown in Fig. 205 can be used for measuring air pressure. It consists of a U-tube containing a liquid, one end of the U-tube is connected to an air trunk or duct, the other end is open to the atmosphere.

The difference in the level of liquid in the two limbs of the U-tube registers the difference in pressure between the two ends of the limbs. (*i.e.* it registers the amount by which the pressure in the air trunk is *above* atmospheric pressure.)

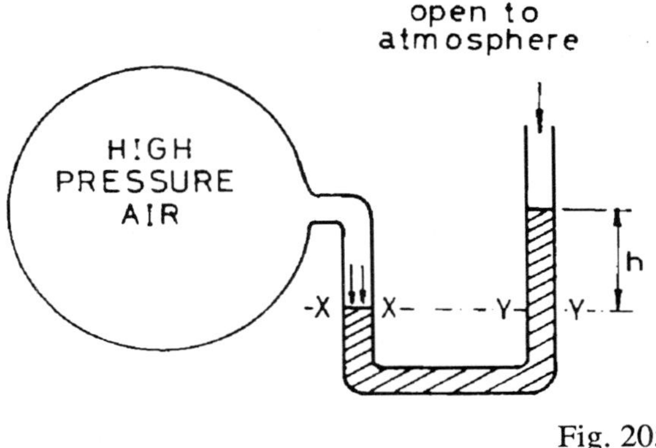

Fig. 205

For equilibrium,

$$\text{Air pressure acting at XX} = \text{pressure due to head } h \text{ of the liquid, at YY.}$$

$$\therefore \text{ Air pressure in the duct} = \rho g h$$

Example. A manometer connected to an air trunk has a head difference of 18 mm of mercury (ρ = 13·6) between the two limbs. Calculate the air pressure in the trunk.

$$\begin{aligned} \text{Air pressure} &= \rho g h \\ &= 13{\cdot}6 \times 10^3 \times 9{\cdot}81 \times 0{\cdot}018 \\ &= 2{\cdot}4 \text{ kN/m}^2 \text{ Ans.} \end{aligned}$$

When a mercury manometer is used for measuring the difference in pressure between two points in a venturi meter, it is sometimes 'submerged' as illustrated in Fig. 206, and each leg of the manometer is full of water above the mercury columns. For this particular case, referring to Fig. 206 and considering the pressure at level LL, the mercury below this level is in equilibrium, therefore the pressure in one leg at level LL is equal to the pressure in the other leg at the same level.

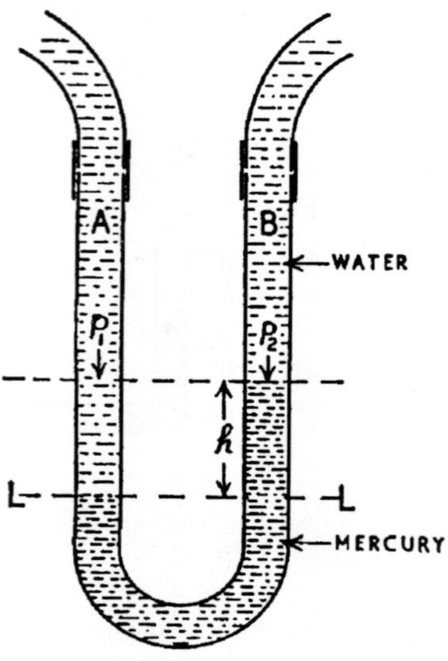

Fig. 206

Let p_1 = pressure of water in leg A

p_2 = pressure of water in leg B

h = difference in mercury level

p_1 + press. due to h metres of water = p_2 + press. due to h metres of mercury.

$p_1 - p_2$ = press. due to h metres of mercury – press. due to h metres of water

$= \rho_m gh - \rho_w gh$

$= 13{\cdot}6 \times 10^3 \times 9{\cdot}81 \times h - 1 \times 10^3 \times 9{\cdot}81 \times h$

$= 9{\cdot}81 \times 10^3 \times h \times (13{\cdot}6 - 1)$

i.e. $p_1 - p_2 = \mathrm{h} \times 9{\cdot}81 \times 12{\cdot}6 \times 10^3$

LOAD ON AN IMMERSED SERVICE

Consider a plane of area A totally immersed in a liquid, lying at some angle θ to the surface of the liquid, as shown in Fig. 207.

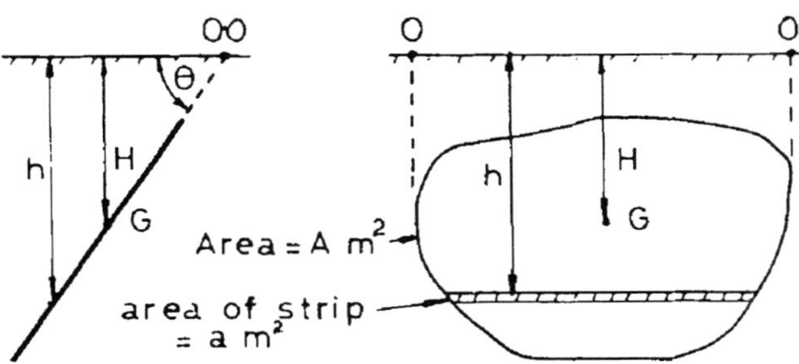

Fig. 207

The area may be divided into a large number of strips, such as the strip of area a at depth h. If the strip is very thin, the variation in pressure with depth can be ignored.

$$\text{Load on one strip} = \text{force} \times \text{area}$$

$$= \rho gh \times a$$

$$\text{Load on plane} = \rho gh_1a_1 + \rho gh_2a_2 + \rho gh_3a_3 + \text{etc.}$$

$$= \rho g \sum ah$$

But, Σah is the first moment of the whole area about the surface at O.

$$i.e.\ \Sigma ah = AH$$

$$\text{Hydrostatic load} = HA\rho g$$

A immersed area, H rectangular depth to its centroid from free surface.

Example. A vertical rectangular bulkhead is 7 m wide and has fresh water to a height of 6 m on one side. Calculate the water load on the bulkhead. (Density of fresh water = 1000 kg/m³).

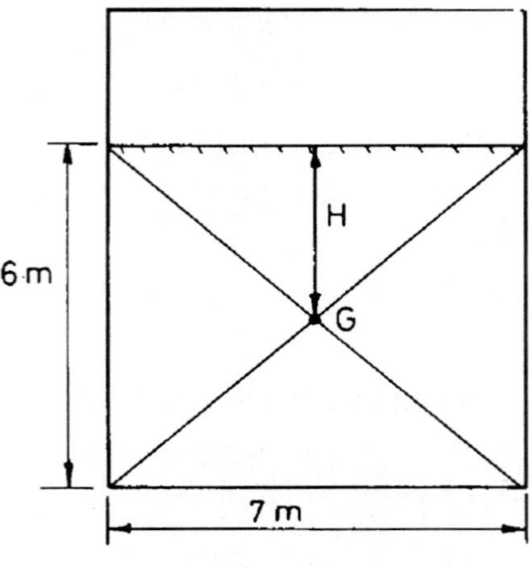

Fig. 208

$$\text{Hydrostatic load} = HA\rho g$$
$$= \frac{6}{2} \times 7 \times 6 \times 10^3 \times 9{\cdot}81$$
$$= 1{\cdot}236 \times 10^6 \text{ N}$$
$$= 1{\cdot}236 \text{ MN} \quad \text{Ans.}$$

Example. A tank 10 m long 4 m wide and 6 m high is filled with oil ($\rho = 0{\cdot}9$) and the oil rises to a height of 5 m up a vent pipe above the top of the tank. Calculate the load on one end plate and on the bottom of the tank.

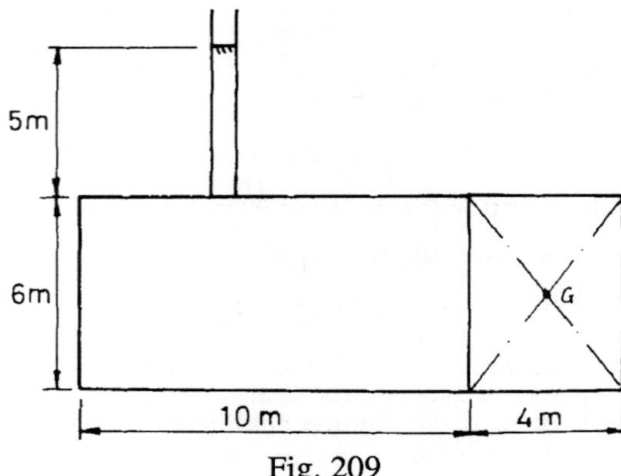

Fig. 209

$$\text{End plate : load} = HA\rho g$$
$$= 8 \times 4 \times 6 \times 900 \times 9{\cdot}81$$
$$= 1{\cdot}695 \text{ MN} \quad \text{Ans.}$$

$$\text{Bottom plate : load} = HA\rho g$$
$$= 11 \times 4 \times 10 \times 900 \times 9{\cdot}81$$
$$= 3{\cdot}885 \text{ MN} \quad \text{Ans.}$$

TRANSMISSION OF POWER

The intensity of fluid pressure at a given depth is the same in all directions and acts perpendicularly (at right angles) to the walls of its container. In closed high pressure systems such as in steering units, hydraulic water-tight door circuits, hydraulic jacks, etc., the difference in pressure due to difference in height is negligible in comparison with the high working pressure and therefore the intensity of pressure can be taken as being uniform throughout the whole system.

The hydraulic jack is a typical example to illustrate transmission of hydraulic pressure and this was dealt with in Chapter 8.

CENTRE OF PRESSURE

The centre of pressure on an immersed area is the point at which the total liquid load may be regarded as acting.

Again considering a plane of area A immersed in a liquid, at an angle θ to the surface, as in Fig. 210:

Let y = distance from the strip to the surface at O.

$$\text{It is seen that } h = y \sin\theta$$
$$\text{Load on strip} = ha\rho g$$
$$= y \sin\theta \, a\rho g$$
$$\text{Load on plane} = \rho g \, \sin\theta \, (a_1y_1 + a_2y_2 + a_3y_3 + \text{etc.})$$
$$= \rho g \, \sin\theta \, \Sigma ay$$

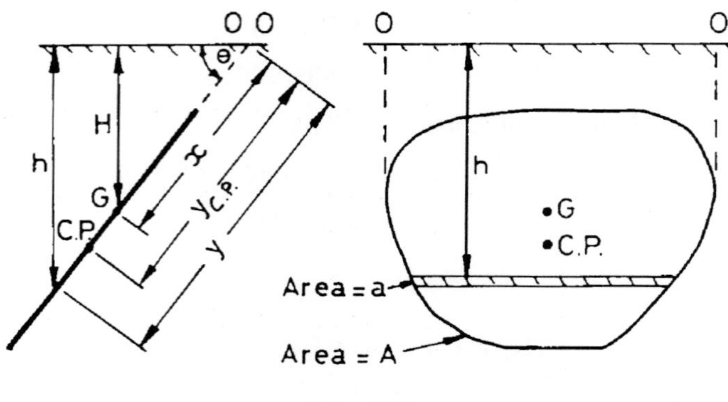

Fig. 210

Taking moments about the surface at O,

Moment of load on the strip $= y \times \rho g a y \sin\theta$

$= \rho g a y^2 \sin\theta$

Moment of load on the plane $= \rho g \sin\theta\,(a_1 y_1 + a_2 y_2 + a_3 y_3 + \text{etc.})$

$= \rho g \sin\theta\,\Sigma a y^2$

$$\text{Distance to C.o.P. from } O = \frac{\text{Moment of load}}{\text{load}}$$

$$= \frac{\rho g \sin\theta\,\Sigma a y^2}{\rho g \sin\theta\,\Sigma a y}$$

$$= \frac{\Sigma a y^2}{\Sigma a y}$$

Now, the term Σay is the *first* moment of the area of the plane about the surface at O and Σay^2 is the *second* moment of the area about O.

$$y_{cp} = \frac{\text{second moment of the area about the surface}}{\text{first moment of the area about the surface}}$$

Where y_{cp} is the projected distance from the centre of pressure to the surface at *O*.

The *second moment* of area about the surface may be calculated using the theorem of parallel axes (see Chapter 7).

$$I_O = I_G + Ax^2$$

where I_O = second moment of the area about the surface at O

I_G = second moment of the area about an axis through its centroid at G

A = immersed area

x = distance between the axis through O and the axis through G

$$\text{hence, } y_{cp} = \frac{I_O}{Ax}$$

$$y_{cp} = \frac{I_G + Ax^2}{Ax}$$

$$y_{cp} = \frac{I_G}{Ax} + x$$

$$x = \frac{H}{\sin\theta}$$

$$y_{cp} = \frac{I_G}{AH/\sin\theta} + \frac{H}{\sin\theta}$$

If the immersed area is *vertical,* then angle = 90°, $\sin\theta = 1$

$$\text{Hence, for a vertical area } y_{cp} = \frac{I_G}{AH} + H$$

Two common shapes for tanks or bulkheads are rectangular and inverted triangular areas.

(i) Rectangular area with its upper edge in the surface of the liquid (Fig. 211):

For a rectangular area as shown,

$$I_G = \frac{BD^3}{12}$$

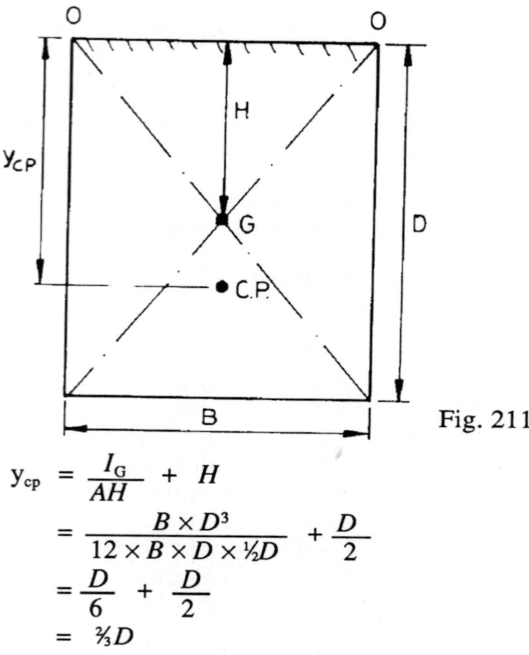

Fig. 211

$$
\begin{aligned}
y_{cp} &= \frac{I_G}{AH} + H \\
&= \frac{B \times D^3}{12 \times B \times D \times \frac{1}{2}D} + \frac{D}{2} \\
&= \frac{D}{6} + \frac{D}{2} \\
&= \tfrac{2}{3}D
\end{aligned}
$$

Centre of pressure is at ⅔ depth below the surface of the liquid.

(ii) Inverted triangular area with its upper edge in the surface of the liquid (Fig. 212).

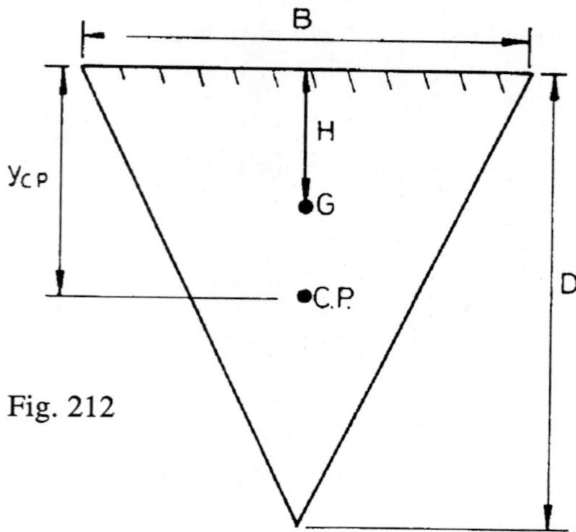

Fig. 212

For a triangular area as shown,

$$I_G = \frac{BD^3}{36}$$

(all such I values are determined by integration).

$$\begin{aligned} y_{cp} &= \frac{I_G}{AH} + H \\ &= \frac{B \times D^3}{36 \times \frac{1}{2} \times B \times D} \\ &= \frac{D}{6} + \frac{D}{3} \\ &= \tfrac{1}{2}D \end{aligned}$$

Centre of pressure is at ½ depth below the surface of the liquid.

Example. A tank with sides 6 m long contains fresh water (ρ = 1000 kg/m³) to a depth of 7 m. Calculate (a) the force on one side of the tank and (b) determine the position of the centre of pressure.

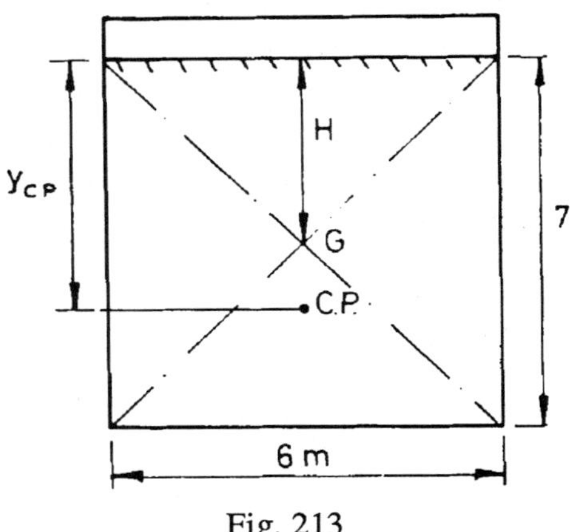

Fig. 213

$$\begin{aligned} \text{(a) Force} &= HA\rho g \\ &= 3{\cdot}5 \times 7 \times 6 \times 10^3 \times 9{\cdot}81 \\ &= 1{\cdot}442 \text{ MN} \qquad \text{Ans.} \end{aligned}$$

$$\begin{aligned} \text{(b)} \quad y_{cp} &= \tfrac{2}{3} \text{ depth below the surface} \\ \text{Position of C.o.P} &= 4\tfrac{2}{3} \text{ m below the liquid surface.} \quad \text{Ans.} \end{aligned}$$

Example. A rectangular plate, 3 m wide and 2 m high is fitted in the vertical side of a tank. The tank contains fresh water to a height of 7 m above the top edge of the plate. Calculate (a) the load on the plate and (b) the position of the centre of pressure.

(Density of F.W. = 1000 kg/m³)

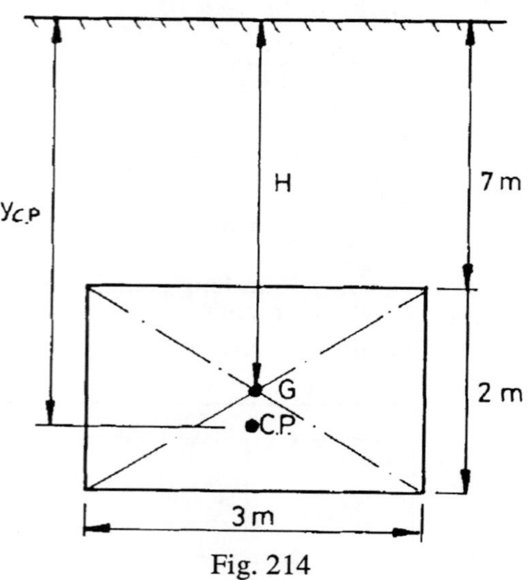

Fig. 214

(a) Load on plate $= HA\rho g$

$= 8 \times 2 \times 3 \times 1000 \times 9{\cdot}81$

$= 470{\cdot}9$ kN Ans.

(b) Note: in the problem, the upper edge of the rectangular area is *not* in the free surface of the liquid.

$$y_{cp} = \frac{I_G}{AH} + H$$

$$= \frac{3 \times 2^3}{12 \times 3 \times 2 \times 8} + 8$$

$$= 8{\cdot}0417$$

Position of C.o.P = 8·0417 m below the surface

f HYDROSTATIC FORCE AND CENTRE OF PRESSURE FOR NON-MIXING LIQUIDS

If a tank contains two non-mixing liquids the resultant force and position of centre of pressure may be determined as shown below.

Example. A tank with sides 2 m long contains 1 m of sea water ($\rho = 1{\cdot}024$) with 0·6 m of oil ($\rho = 0{\cdot}8$) above.

Calculate the force on one side of the tank and the position of the centre of pressure.

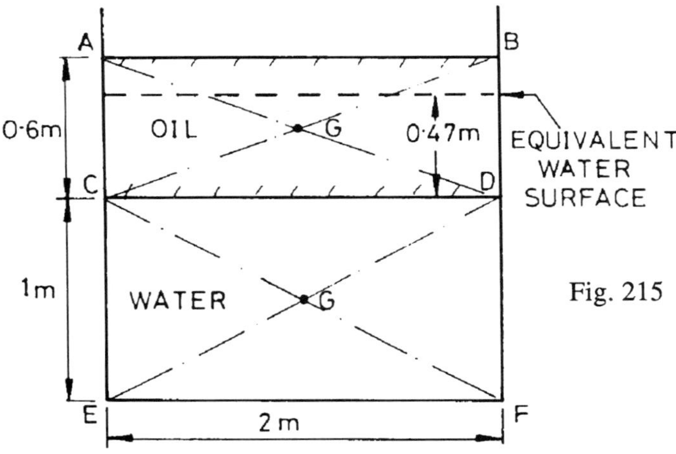

Fig. 215

(a) Area ABCD:

$$\begin{aligned} \text{oil load} &= HA\rho g \\ &= 0{\cdot}3 \times 2 \times 0{\cdot}6 \times 800 \times 9{\cdot}81 \\ &= 2{\cdot}825 \text{ kN} \\ y_{cp} &= \tfrac{2}{3} \text{ depth below the oil surface} \\ &= 0{\cdot}4 \text{ m below the oil surface.} \end{aligned}$$

(b) Area CDEF:

Note: Sea water is acting on the area CDEF and the superimposed liquid (oil) can be converted to an *equivalent depth* of sea water.

$$\text{For the oil, equivalent depth of S.W.} = 0{\cdot}6 \times \frac{800}{1024}$$

$$= 0{\cdot}47 \text{ m}$$

Thus, an equivalent water surface at 0·47 m above CD is used in calculating the water load on CDEF.

$$\text{Water load} = HA\rho g$$
$$= (0{\cdot}5 + 0{\cdot}47) \times 1 \times 2 \times 1024 \times 9{\cdot}81$$
$$= 19{\cdot}49 \text{ kN}$$

$$y_{cp} = \frac{I_G}{AH} + H$$
$$= \frac{2 \times 1^3}{12 \times 1 \times 2 \times 0{\cdot}97} + 0{\cdot}97$$

y_{cp} = 1·056 m below the equivalent water surface.

Position of C.o.P = 1·056 + (0·6 – 0·47)
= 1·186 m below the surface at AB.

(c) Thus, for the total area ABFE,

The total load = 2·825 + 19·49
= 22·315 kN. Ans.

Let y_{cp} = depth to the resultant C.o.P from the surface at AB

Taking moments about the surface:

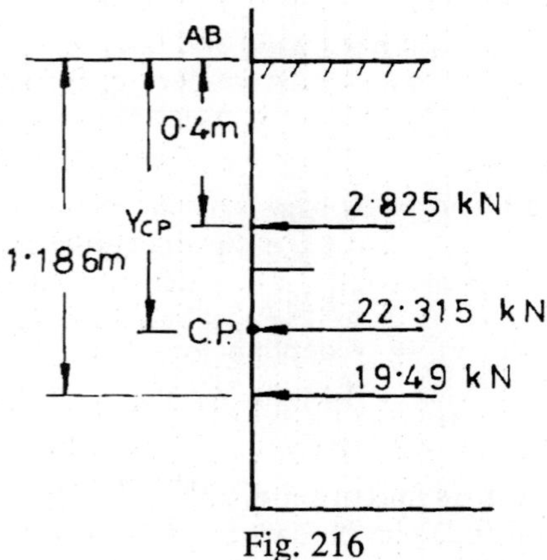

Fig. 216

$$22{\cdot}315 \times y_{cp} = 2{\cdot}825 \times 0{\cdot}4 + 19{\cdot}49 \times 1{\cdot}186$$
$$y_{cp} = 1{\cdot}086 \text{ m}$$

Position of resultant C.o.P = 1·086 m below the surface. Ans.

TEST EXAMPLES 12

1. Three liquids of specific gravities 0·7, 0·8 and 0·9 respectively are mixed in the proportion of 3, 4 and 5 by mass. Find the relative density of the mixture.

2. A piece of brass is suspended by a wire from a spring balance and it registers a weight of 7 N in air. When the brass is lowered into fresh water until submerged the balance shows a weight of 6·17 N. Find the relative density of this brass.

3. A plank of wood of relative density 0·6 is 3 m long by 300 mm broad by 150 mm thick and floats in fresh water at a draught of 125 mm when a casting is carried on top of the plank. Calculate the mass of the casting.

4. A solid wood raft 7·6 m long by 1·22 m wide by 230 mm thick, of relative density 0·75, floats in sea water. Find the draught when a piece of lead of mass 508 kg and relative density 11·4 is suspended beneath the raft. Take density of sea water as 1025 kg/m^3.

5. Find the relative density of an oil if a vertical column of this oil 1·511 m high will balance a vertical column of mercury 10 cm high. Take the relative density of mercury as 13·6.

6. A vertical bulkhead 9·6 m wide and 6 m deep separates two tanks. One tank contains oil of relative density 0·85 to a depth of 5·4 m, and the other tank is empty. Find the total load on the bulkhead in MN and state the position of the centre of pressure.

7. Two tanks are separated by a vertical rectangular bulkhead 7·5 m wide. In one tank there is fresh water to a depth of 7·2 m and in the other there is oil of relative density 0·8 to a depth of 4·2 m. Find the resultant thrust on the bulkhead in MN and the position of the resultant centre of pressure.

8. A rectangular tank is 2 m long, 1·2 m wide and 1m deep, and contains fresh water to a depth of 0·6 m. A block of wood of mass 156 kg is put into the water and floats freely. Find the water load on the bottom, sides and ends of the tank, in KN.

*f*9. Prove that hydrostatic force = $HA\rho g$ on an immersed area A, inclined at any angle below the free surface of a liquid.

*f*10. A thin hollow metal sphere, diameter 2 m, is filled with water (density 1000 kg/m^3). Determine (a) maximum pressure (b) total load (c) resultant force on the interior surface of the sphere.

*f*11. Derive the value of the second moment of area through the centroid about a diameter (I_G) for a circular plate and hence determine the position of the centre of pressure for such a plate with its edge in the surface of a liquid (*xx*).

*f*12. A circular door, 1·5 m diameter, lies with its plane inclined at 35° to the horizontal and its upper edge is 1 m below the surface of the water (density 1000 kg/m³). Calculate the hydrostatic load on the door and the position of the centre of pressure.

*f*13. A bulkhead is in the form of a triangle, 6 m wide at the top and 9 m deep. the tank is filled with sea water (density 1025 kg/m³). Calculate the load on the bulkhead and the position of the centre of pressure relative to the top of the bulkhead if water is: (a) at top of the bulkhead (b) 4 m up a sounding pipe.

*f*14. A plate covers a rectangular hole 30 cm wide by 60 cm high in a vertical side of a tank which contains water (density 1000 kg/m³) which is level with the top edge of the hole. Determine the load on the plate and the position of the centre of pressure relative to the top edge of the hole. If the water level is raised by 1 m what will be the new position of the centre of pressure?

*f*15. A closed cylindrical tank is 1 m diameter and 2 m high. It contains water (density 1000 kg/m³) to a depth of 1·5 m and the space above the water contains air at a pressure of 120 kN/m³.

(a) Calculate the resultant force on the vertical seam of the tank.

(b) Determine the vertical height to the centre of pressure from the bottom of the tank.

CHAPTER 13

HYDRODYNAMICS

The study of fluids in motion

The quantity of liquid flowing through a pipe along a trough, through an orifice, etc., is usually expressed either as the volume flowing per unit time or the mass flowing per unit time. The volume per unit time is termed the volume flow (m^3/s).

The mass flowing per unit time (which is volume flow × density) is termed the *mass flow* (kg/s).

The velocity at which the liquid flows is the 'length' of liquid which passes a given point in unit time. For instance, if the velocity is 2 m/s it can be thought of as a column of liquid 2 m long passing a point every second, therefore:

$$\text{Volume flow } (\dot{V}) = \text{area} \times \text{velocity}$$

Example. Oil of relative density 0·9 flows at full bore through a pipe 75 mm internal diameter at a velocity of 1·2 m/s. Calculate the mass flow in t/h.

$$\begin{aligned}\text{Volume flow} &= 0{\cdot}7854 \times 0{\cdot}075^2 \times 1{\cdot}2\\ &= 5{\cdot}303 \times 10^{-3}\ \text{m}^3/\text{s}\end{aligned}$$

Density of oil of relative density 0·9

$$= 0{\cdot}9 \times 10^3\ \text{kg/m}^3 = 0{\cdot}9\ \text{t/m}^3$$

$$\begin{aligned}\text{Mass flow } (\dot{m}) &= 5{\cdot}303 \times 10^{-3} \times 0{\cdot}9 \times 3600\\ &= 17{\cdot}18\ \text{t/h} \quad \text{Ans.}\end{aligned}$$

FLOW THROUGH VALVES. Referring to Fig. 217, the area of escape is the annular area of the circumferential opening between valve and seat which is, circumference × lift. The lift beyond which it would cause no restriction is when the circumferential area of lift opening is equal to the cross-sectional area of the bore, thus,

$$\text{Circumference} \times \text{Lift} = \text{Cross-sectional area}$$

$$\pi d \times L = \frac{\pi}{4} d^2$$

$$L = \frac{d}{4}$$

Hence a lift equal to one-quarter of the diameter of the valve allows full bore flow.

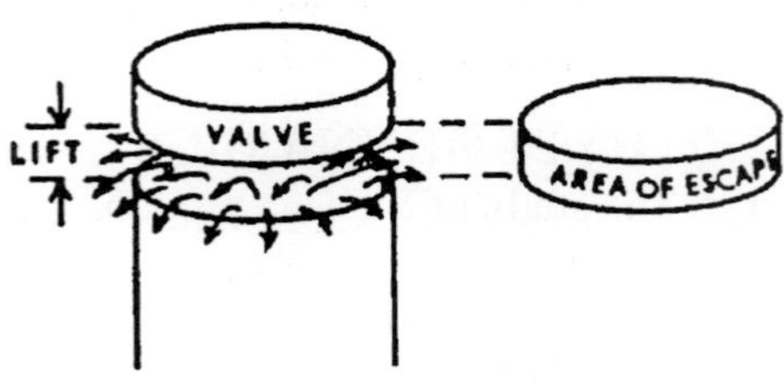

Fig. 217

Example. Calculate the volume flow of water in m³/min through a 150 mm diameter valve when the velocity of the water is 2·5 m/s and the valve lift is (*i*) 30 mm, (*ii*) 45 mm.

Maximum effective lift = $\frac{1}{4} \times 150 = 37{\cdot}5$ mm

when lift is 30 mm:

Volume flow = area × velocity
= circumference × lift × velocity
= $\pi \times 0{\cdot}15 \times 0{\cdot}03 \times 2{\cdot}5 \times 60$
= 2·12 m³/min Ans. (i)

when lift is equal to or more than quarter diameter:

Volume flow = cross-sect. area × velocity
= $0{\cdot}7854 \times 0{\cdot}15^2 \times 2{\cdot}5 \times 60$
= 2·651 m³/min Ans. (ii)

DISCHARGE THROUGH AN ORIFICE

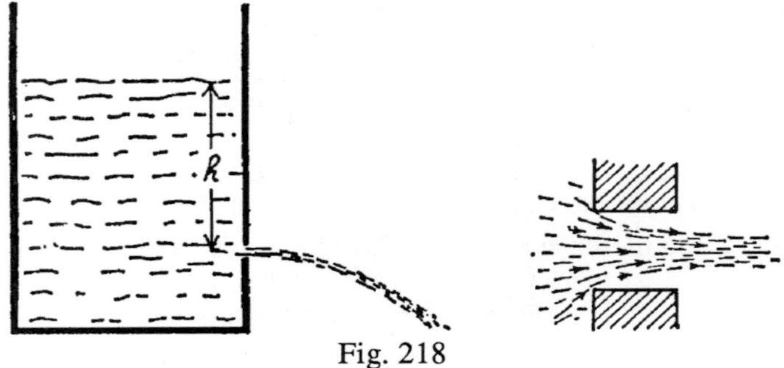

Fig. 218

When water escapes through a hole in the side of a tank (Fig. 218) the potential energy of the water inside the tank due to its head h above the hole is converted into kinetic energy as it flows through the hole, thus,

$$\text{Kinetic energy gained} = \text{Potential energy lost}$$
$$\tfrac{1}{2}mv^2 = mg \times h$$
$$v = \sqrt{2gh}$$

This is the theoretical velocity of the water jet issuing from the hole and this multiplied by the area of the hole will give the theoretical volume flow.

Due to friction, the actual velocity is a little less, the ratio of the actual velocity to the rhetorical velocity is termed the *coefficient of velocity* and represented by C_V hence,

$$\text{Actual velocity} = C_V \sqrt{2gh}$$

Due to eddy currents, the actual area of the water jet is less than the area of the hole, the ratio between the two being termed the *coefficient of reduction of area* and represented by C_A, hence,

$$\text{Area of jet} = C_A \times \text{area of hole.}$$

The actual volume flow is therefore,

$$\text{Volume flow} = C_A \times \text{area of hole} \times C_V \sqrt{2gh}$$

The ratio of the actual quantity discharged to the theoretical quantity is the *coefficient of discharge*, it is represented by C_D and its value is,

$$C_D = C_A \times C_V$$

Example. Water escapes through a hole 20 mm diameter in the side of a tank, the head of water above the hole being 3 m. Taking the coefficient of velocity as 0·97 and the coefficient of reduction of area as 0·64, calculate (*i*) the velocity of the water jet as it leaves the hole, (*ii*) the quantity of water escaping in t/h.

$$\text{Velocity of water jet} = 0{\cdot}97 \times \sqrt{2gh}$$
$$= 0{\cdot}97 \times \sqrt{2 \times 9{\cdot}81 \times 3}$$
$$= 7{\cdot}442 \text{ m/s} \quad \text{Ans. } (i)$$
$$\text{Area of water jet} = 0{\cdot}64 \times \text{area of hole}$$
$$= 0{\cdot}64 \times 0{\cdot}7854 \times 20^2$$
$$= 201 \text{ mm}^2 = 201 \times 10^{-6} \text{ m}^2$$

$$\text{Volume flow} = \text{area} \times \text{velocity}$$
$$= 201 \times 10^{-6} \times 7{\cdot}442 \times 3600$$
$$\dot{V} = 5{\cdot}387 \text{ m}^3\text{/h}$$
$$\text{Mass flow} = \text{vol. flow} \times \text{density}$$
$$= 5{\cdot}387 \times 1$$
$$\dot{m} = 5{\cdot}387 \text{ t/h} \quad \text{Ans. } (ii)$$

Example. In an experiment the water level in a tank was kept constant at 1·25 m above a hole 12 mm diameter in the side of a tank. The jet of water from the hole passed through a ring which was 2·17 m horizontally from the side of the tank and 1 m vertically below the hole. The water discharged into a tank at the rate of 20·84 l/min. From these results calculate the coefficients of velocity, reduction of area and discharge.

$$\text{Theoretical horizontal velocity of the jet} = \sqrt{2gh}$$
$$= \sqrt{2 \times 9{\cdot}81 \times 1{\cdot}25} = 4{\cdot}952 \text{ m/s}$$

Let t = time for water to fall from hole to ring, a vertical distance of 1 m.

$$\text{Initial vertical velocity} = 0$$
$$\text{Vertical velocity after } t = 9{\cdot}81 \times t \text{ m/s}$$
$$\text{Average vertical velocity} = \tfrac{1}{2} \times 9{\cdot}81\, t$$
$$\text{Vertical distance} = \text{average velocity} \times \text{time}$$
$$1 = \tfrac{1}{2} \times 9{\cdot}81 \times t \times t$$
$$t = \sqrt{\frac{2}{9{\cdot}81}} = 0{\cdot}4514 \, s$$
$$\text{Horizontal distance} = \text{horizontal velocity} \times \text{time}$$
$$\text{Horizontal velocity} = \frac{2{\cdot}17}{0{\cdot}4514} = 4{\cdot}807 \text{ m/s}$$
$$\text{Coeff. of velocity} = \frac{\text{Actual velocity}}{\text{Theoretical velocity}}$$
$$= \frac{4{\cdot}807}{4{\cdot}952} = 0{\cdot}9705$$
$$\text{Theo. discharge} = \text{Area of hole} \times \sqrt{2gh}$$
$$= 0{\cdot}7854 \times 0{\cdot}012^2 \times 4{\cdot}952$$
$$= 5{\cdot}602 \times 10^{-4} \text{ m}^3\text{/s}$$

$$(1\text{m}^3 = 10^3 \text{ l})$$

$$\text{Theo. discharge [l/min]} = 5{\cdot}602 \times 10^{-4} \times 10^3 \times 60$$

$$= 33{\cdot}61 \text{ l/min}$$

$$\text{Coeff. of discharge} = \frac{\text{Actual discharge}}{\text{Theoretical discharge}}$$

$$= \frac{20{\cdot}84}{33{\cdot}61} = 0{\cdot}62$$

$$\text{Coeff. of reduction of area} = \frac{\text{Coeff. of discharge}}{\text{Coeff. of velocity}}$$

$$= \frac{0{\cdot}62}{0{\cdot}9705} = 0{\cdot}6389$$

$$\left.\begin{aligned} C_V &= 0{\cdot}9705 \\ C_A &= 0{\cdot}6389 \\ C_D &= 0{\cdot}62 \end{aligned}\right\} \text{Ans.}$$

f BERNOUILLI'S EQUATION

Bernouilli's theorem states that the total energy contained in a given quantity of water is composed of (*i*) potential energy, by virtue of its height, (*ii*) pressure energy, by virtue of its pressure, and (*iii*) kinetic energy, by virtue of its velocity.

If there is no friction loss and no work done between two points 1 and 2 of a water system, then,

Total energy in water at point 1 = Total energy at point 2

$\text{Pot.E.}_1 + \text{Press.E.}_1 + \text{K.E.}_1 = \text{Pot.E.}_2 + \text{Press.E.}_2 + \text{K.E.}_2$

Let m represent the mass of water in kg,

Potential energy is the product of the force of gravity on the mass (*i.e.* its weight) and the height above a given datum level, denoted here by symbol z, thus,

$$\text{Potential energy} = mgz$$

Pressure energy is the product of the pressure and the volume, thus,

$$\text{Potential energy} = pV$$

Kinetic energy is the translational energy of motion, thus,

$$\text{Kinetic energy} = \tfrac{1}{2} mv^2$$

Inserting these values into the equation:

$$mgz_1 + p_1V + \tfrac{1}{2} mv_1^2 = mgz_2 + p_2V + \tfrac{1}{2} mv_2^2$$

Dividing every term by mg, each term will represent its respective energy in J/N weight of the liquid (Nm/N is an equivalent head).

$$z_1 + \frac{p_1 V}{mg} + \frac{v_1^2}{2g} = z_2 + \frac{p_2 V}{mg} + \frac{v_2^2}{2g}$$

mg/V is the specific weight of the liquid (w) N/m³.

$$z_1 + \frac{p_1}{w} + \frac{v_1^2}{2g} = z_2 + \frac{p_2}{w} + \frac{v_2^2}{2g}$$

where z = head of liquid (m)
p = pressure (N/m²)
w = specific weight (N/m³)
v = velocity (m/s)
g = 9·81 (m/s²)

f Example. Part of a vertical fresh water pipe line tapers uniformly from 120 mm diameter at the bottom to 60 mm diameter at the top, the difference in height being 5 m. When the volume flow is 0·0424 m³/s the pressure at the bottom is 160 kN/m², find the pressure at the top.

$$\text{Velocity} = \frac{\text{volume flow}}{\text{area}}$$

$$v_1 = \frac{0{\cdot}0424}{0{\cdot}7854 \times 0{\cdot}12^2} = 3{\cdot}75 \text{ m/s}$$

$$v_2 = \frac{0{\cdot}0424}{0{\cdot}7854 \times 0{\cdot}06^2} = 15 \text{ m/s}$$

Alternatively, as volume flow is constant,

$$\dot{V} = v_1 \times a_1 = v_2 \times a_2$$

$$v_2 = \frac{3{\cdot}75 \times 0{\cdot}7854 \times 0{\cdot}12^2}{0{\cdot}7854 \times 0{\cdot}06^2}$$

$$= 3{\cdot}75 \times 2^2 = 15 \text{ m/s}$$

Taking bottom as datum level, $z_1 = 0$ and $z_2 = 5$
Let p_2 = pressure at top

$$z_1 + \frac{p_1}{w} + \frac{v_1^2}{2g} = z_2 + \frac{p_2}{w} + \frac{v_2^2}{2g}$$

$$0 + \frac{160 \times 10^3}{10^3 \times 9{\cdot}81} + \frac{3{\cdot}75^2}{2 \times 9{\cdot}81} = 5 + \frac{p_2}{10^3 \times 9{\cdot}81} + \frac{15^2}{2 \times 9{\cdot}81}$$

$$160 + \frac{3{\cdot}75^2}{2} = 5 \times 9{\cdot}81 + \frac{p_2}{10^3} + \frac{15^2}{2}$$

$$160 + 7{\cdot}03 = 49{\cdot}05 + \frac{p_2}{10^3} + 112{\cdot}5$$

$$\frac{p_2}{10^3} = 160 + 7{\cdot}03 - 49{\cdot}05 - 112{\cdot}5$$

$$p_2 = 5{\cdot}48 \text{ kN/m}^2 \text{ Ans.}$$

f EFFECT OF FRICTION

In any pipe system, there will be some energy lost by the liquid in overcoming frictional resistance, as the liquid flows through the pipes and valves.

Thus, taking energy loss due to friction into account, between any two points in a system:

Initial energy of liquid	−	Friction losses	=	Final energy of liquid
Initial energy	=	Final energy	+	Friction losses

Hence, referring to Bernoulli's equation,

$$z_1 + \frac{p_2}{w} + \frac{v_1^2}{2g} = z_2 + \frac{p_2}{w} + \frac{v_2^2}{2g} + \text{Friction losses}$$

Note: To conform with the other terms in the equation the energy loss due to friction must be expressed in J/N weight of the liquid. *i.e.* the equivalent m head.

f Example. Fresh water, density 1000 kg/m^3 flows upward through a tapered pipe 6·1 m long 102 mm diameter at the bottom and 204 mm diameter at the top. The pressure at the bottom is 311 kN/m^2, while the velocity at the top is 9·15 m/s. Determine the velocity at the bottom and the pressure at the top, if the energy loss due to friction is equivalent to 32 m head of water.

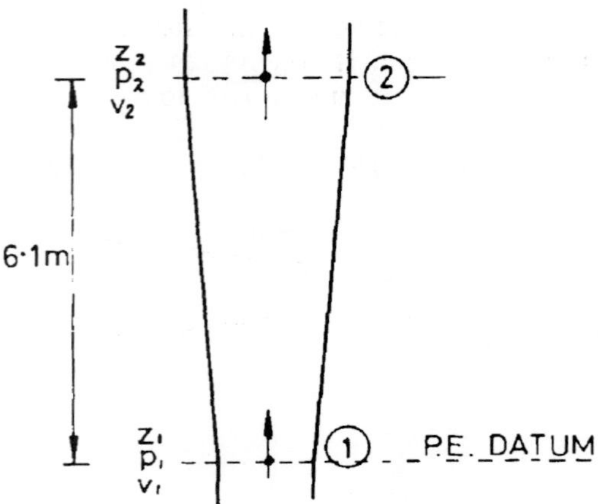

Fig. 219

Denoting conditions at the bottom by suffix 1 and conditions at the top by suffix 2:

Since volumetric flow rate is constant,

$$\text{area}_1 \times \text{velocity}_1 = \text{area}_2 \times \text{velocity}_2$$

$$\frac{\pi}{4} \times d_1^2 \times v_1 = \frac{\pi}{4} \times d_2^2 \times v_2$$

$$0{\cdot}102^2 \times v_1 = 0{\cdot}204^2 \times 9{\cdot}15$$

$$v_1 = 36{\cdot}6 \text{ m/s.} \qquad \text{Ans.}$$

From Bernoulli,

$$z_1 + \frac{p_1}{w} + \frac{v_1^2}{2g} = z_2 + \frac{p_2}{w} + \frac{v_2^2}{2g} + \text{Friction loss}$$

$$\text{O} + \frac{311 \times 10^3}{10^3 \times 9{\cdot}81} + \frac{36{\cdot}6^2}{2 \times 9{\cdot}81} = 6{\cdot}1 + \frac{p_2}{10^3 \times 9{\cdot}81} + \frac{9{\cdot}15^2}{2 \times 9{\cdot}81} + 32$$

$$\therefore\ 31{\cdot}7 + 68{\cdot}27 = 6{\cdot}1 + \frac{p_2}{10^3 \times 9{\cdot}81} + 4{\cdot}27 + 32$$

$$\therefore\ p_2 = 57{\cdot}6 \times 10^3 \times 9{\cdot}81$$

$$\text{Pressure at the top} = 565 \text{ kN/m}^2. \qquad \text{Ans.}$$

f VENTURI METER

A venturi meter is a device for measuring the rate of flow of liquid through a pipeline. It consists of a short length of pipe which tapers from full bore at each end to a smaller bore (termed the throat) at the middle, as illustrated in Fig. 220.

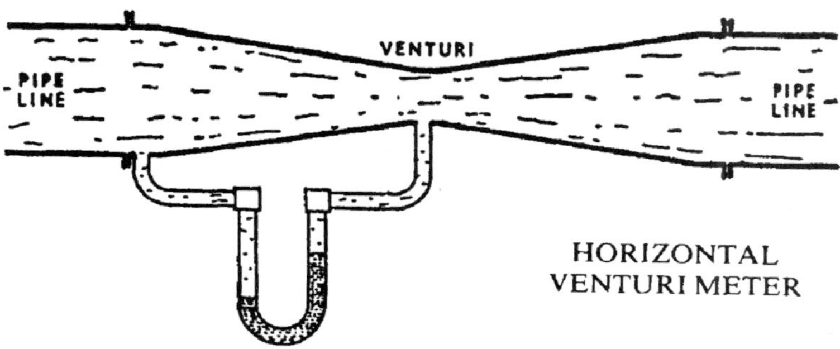

Fig. 220

The areas at entrance and throat are known, and the pressures at these two points, or difference in pressure, is measured, then Bernouilli's equation can be applied to find the velocity of the liquid and hence the quantity flowing.

However, it is usual to connect a venturi meter horizontally in a horizontal part of the pipeline, and this simplifies calculations as there is no difference in head between the two points, that is, $z_1 = z_2$ and cancels. Let the entrance be denoted by point number 1 and the throat by point number 2, for a *horizontal* meter:

$$\text{press. energy}_1 + \text{kinetic energy}_1 = \text{press. energy}_2 + \text{kinetic energy}_2$$
$$\text{press. energy}_1 - \text{press. energy}_2 = \text{kinetic energy}_2 - \text{kinetic energy}_1$$

which is, in effect,

$$\text{loss of pressure energy} = \text{gain in kinetic energy}$$
$$p_1V - p_2V = \tfrac{1}{2}mv_2^2 - \tfrac{1}{2}mv_1^2$$
$$V(p_1 - p_2) = \tfrac{1}{2}m(v_2^2 - v_1^2)$$

but, mass m ÷ volume V = density ρ, therefore, dividing throughout by V,

$$p_1 - p_2 = \tfrac{1}{2}\rho(v_2^2 - v_1^2)$$

From the above equation, the velocity (v_1) of the liquid flowing in the pipe may be calculated.

$$\text{Flow rate} = \text{cross sect. area of pipe} \times \text{velocity in pipe}$$

f METER COEFFICIENT

Taking frictional resistance into account, as the liquid flows through the meter, it is found that the loss of pressure energy is *not* equal to the gain in kinetic energy. Some of the pressure energy loss is due to work done against friction. As a result, the calculated flow rate must be corrected by a factor called a *meter coefficient* or *discharge coefficient*. This meter coefficient is determined by comparing the calculated or *theoretical* flow rate and the *actual* flow rate through the meter.

Actual flow rate = Theoretical flow rate × meter coefficient.

f Example. The diameters at entrance and throat of a horizontal venturi meter are 450 mm and 225 mm respectively, and the difference in pressure between these two points is equivalent to 381 mm head of water. Calculate the mass flow of fresh water through the meter ($\rho = 1000$ kg/m³).

Denoting conditions at entrance by suffix 1, and at throat by 2, since the volume flow rate at any point is constant, then,

$$\text{area}_1 \times \text{velocity}_1 = \text{area}_2 \times \text{velocity}_2$$

$$v_2 = \frac{a_1 \times v_1}{a_2}$$

$$= \frac{0{\cdot}7854 \times 0{\cdot}45^2}{0{\cdot}7854 \times 0{\cdot}225^2} \times v_1$$

$$v_2 = 4v_1$$

$$\text{Difference in pressure} = 381 \text{ mm water}$$

$$p_1 - p_2 = \rho g h$$

$$= 10^3 \times 9{\cdot}81 \times 0{\cdot}381$$

$$= 3738 \text{ N/m}^2$$

$$p_1 - p_2 = \tfrac{1}{2}\rho\,(v_2^2 - v_1^2)$$

$$\text{Substituting } p_1 - p_2 = 3738 \text{ N/m}^2$$

$$v_2 = 4v_1$$

$$3738 = \tfrac{1}{2} \times 10^3 \times \{(4v_1)^2 - v_1^2\}$$

$$3738 = \tfrac{1}{2} \times 10^3 \times 15v_1^2$$

$$v_1 = \sqrt{\frac{2 \times 3738}{10^3 \times 15}}$$

$$= 0{\cdot}706 \text{ m/s}$$

$$\text{Volume flow} = \text{area} \times \text{velocity}$$
$$\dot{V} = 0{\cdot}7854 \times 0{\cdot}45^2 \times 0{\cdot}706$$

$$\text{Mass flow} = \text{volume flow} \times \text{density}$$
$$= 0{\cdot}7854 \times 0{\cdot}45^2 \times 0{\cdot}706 \times 10^3$$
$$\dot{m} = 112{\cdot}3 \text{ kg/s} \quad \text{Ans.}$$

f Example. A pipe 300 mm diameter contains a venturi meter with a throat diameter of 100 mm. The difference in pressure head between inlet and throat is 250 mm of mercury, measured on a U-tube gauge containing water and mercury. The meter coefficient is 0·95. Calculate the discharge rate through the pipe, in m³/s. (Density of water is 1000 kg/m³.)

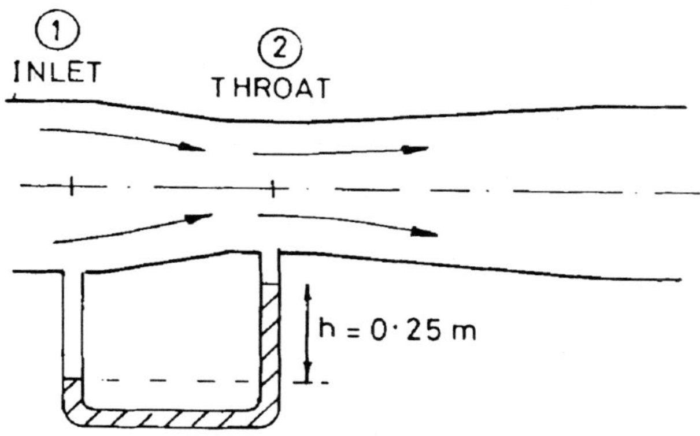

Fig. 221

$$p_1 - p_2 = \tfrac{1}{2}\rho\,(v_2^2 - v_1^2) \ldots \quad \ldots \quad \ldots \text{(i)}$$

For a differential manometer, containing water and mercury,

$$p_1 - p_2 = h \times 9{\cdot}81 \times 12{\cdot}6 \times 10^3$$
$$= 0{\cdot}25 \times 9{\cdot}81 \times 12{\cdot}6 \times 10^3 \text{ N/m}^2$$
$$= 30{\cdot}9 \text{ kN/m}^2$$

Flow rate volume ($\dot{V}$) through the meter is constant,

$$a_1 v_1 = a_2 v_2$$
$$\frac{\pi}{4} \times 0{\cdot}3^2 \times v_1 = \frac{\pi}{4} \times 0{\cdot}1^2 \times v_2$$
$$v_2 = \left(\frac{0{\cdot}3}{0{\cdot}1}\right)^2 \times v_1$$
$$v_2 = 9v_1$$

Substituting in equation (i)

$$30{\cdot}9 \times 10^3 = \tfrac{1}{2} \times 1000\,[(9v_1)^2 - v_1^2]$$

$$30{\cdot}9 \times 10^3 = 500\,[81v_1^2 - v_1^2]$$

$$30{\cdot}9 \times 10^3 = 500 \times 80v_1^2$$

$$v_1 = 0{\cdot}879 \text{ m/s}$$

$$\text{Volume flow rate} = \begin{array}{c}\text{theoretical}\\ \text{flow rate}\end{array} \times \begin{array}{c}\text{meter}\\ \text{coefficient}\end{array}$$

$$\dot{V} = \frac{\pi}{4} \times 0{\cdot}3^2 \times 0{\cdot}879 \times 0{\cdot}95$$

$$= 0{\cdot}059 \text{ m}^3/\text{s}. \quad \text{Ans.}$$

f FRICTION IN PIPES

The laws of friction relating to the flow of liquid through pipes are as follows:-

Fluid friction (*i*) is proportional to the interior wetted surface area of the pipe.

(*ii*) is proportional to the (velocity)^n of flow where n is approximately, and usually taken as 2,

(*iii*) depends upon the roughness of the internal surface of the pipe,

(*iv*) is inversely proportional to the cross-sectional area of the pipe bore,

(*v*) is independent of the intensity of fluid pressure.

Thus, taking water flowing full bore at a velocity of v through a pipe of internal diameter d and length l,

$$\text{Friction } \alpha \ \frac{\text{wetted surface area} \times \text{velocity}^2}{\text{cross-sectional area}}$$

$$\text{Friction } \alpha \ \frac{\pi d l \times v^2}{\frac{\pi}{4} d^2}$$

$$\text{Friction } \alpha \ \frac{4\,l v^2}{d}$$

$$\text{Loss due to friction} = a \text{ constant} \times \frac{4\,l v^2}{d}$$

It is usual to state the effect of friction as a factor of $\frac{v^2}{2g}$, that is expressing it as a loss of head, representing this factor by f,

$$\text{Loss of head due to friction } (h_f) = \frac{4fl}{d} \times \frac{v^2}{2g}$$

f Example. Fresh water flows at the rate of 1·88 m²/h through a pipe 19 mm internal diameter and 6 m long. Taking the friction factor as 0·01, find the loss due to friction expressed in units of (*i*) head, (*ii*) pressure.

$$\dot{V} = av$$

$$\text{Velocity} = \frac{1{\cdot}88}{3600 \times 0{\cdot}7854 \times 0{\cdot}019^2}$$

$$= 1{\cdot}842 \text{ m/s}$$

$$\text{Loss of head due to friction} = \frac{4fl}{d} \times \frac{v^2}{2g}$$

$$h_f = \frac{4 \times 0{\cdot}01 \times 6 \times 1{\cdot}842^2}{0{\cdot}019 \times 2 \times 9{\cdot}81}$$

$$= 2{\cdot}184 \text{ m} \quad \text{Ans. } (i)$$

$$p = \rho g h$$

$$= 10^3 \times 9{\cdot}81 \times 2{\cdot}184$$

$$= 2{\cdot}143 \times 10^4 \text{ N/m}^2$$

$$= 21{\cdot}43 \text{ kN/m}^2 \text{ or } 0{\cdot}2143 \text{ bar}$$

Ans. (*ii*)

f IMPACT OF WATER JET

ON A STATIONARY PERPENDICULAR PLATE

Consider a jet of water of A cross-sectional area moving with a velocity of v striking a stationary flat plate which is perpendicular to the direction of the water jet, as in Fig. 222

If ρ is the density of the water:

Mass of water striking plate every second

$$= \text{area} \times \text{velocity} \times \text{density}$$

$$= Av\rho$$

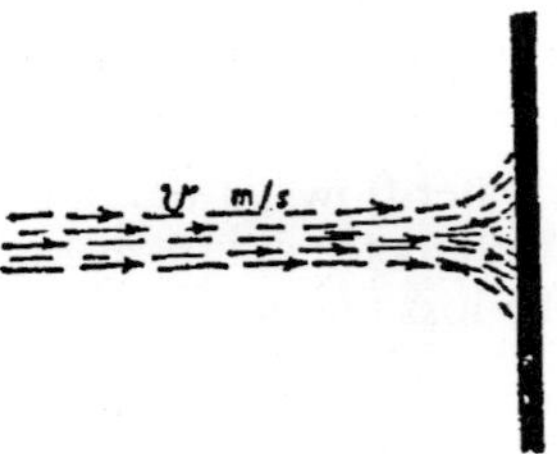

Fig. 222

Assuming that the water has no velocity after striking the plate, its change of velocity is $v - 0 = v$.

$$\begin{aligned} \text{Change of momentum} &= \text{mass} \times \text{change of velocity} \\ \text{Change of momentum / s} &= \text{mass /s} \times \text{change of velocity} \\ &= Av\rho \times v \\ &= A\rho v^2 \end{aligned}$$

Change of momentum per second is equal to the force required to cause that change, hence the force applied by the plate on the jet of water is,

$$\text{Force} = A\rho v^2$$

and in accordance with Newton's third law of motion this is also the force applied by the jet on the plate.

ON A MOVING PERPENDICULAR PLATE. In the case of a jet striking a flat plate which is perpendicular to the jet and moving at u m/s in the same direction of the jet:

$$\text{Relative velocity of jet to plate} = (v - u)$$

Mass of water striking plate every second

$$= (v - u)\,A\rho$$

Assuming that the water does not splash off the plate with any velocity,

$$\begin{aligned} \text{Change of velocity} &= (v - u) \\ \text{Force of jet} &= \text{Change of momentum / s} \\ &= \text{mass / s} \times \text{change of velocity} \\ &= (v - u) \times A \times \rho \times (v - u) \\ &= A\rho\,(v - u)^2 \end{aligned}$$

In this, flat plates perpendicular to the jet were taken as examples and the assumption made that the water trickled down the plate after striking, without any splashing or rebound of the water. If curved plates were used such as turbine blades, the water would leave the blade with a certain amount of velocity in a direction depending upon the exit angle of the blade. The change of velocity of the water jet would then be calculated as for any other body subjected to a change of velocity due to combined change of speed and direction, from which the change of momentum per second and therefore the force of the jet can be found.

f CENTRIFUGAL PUMP

This is a rotary pump which works on the principle of centrifugal force. It consists of a rotating impeller within a stationary casing. The impeller is a hollow disc wheel with internal curved vanes, mounted on a shaft which is driven by an electric motor. Openings in the sides of the impeller near the shaft communicate with the suction branch, liquid (oil, water, etc.) enters the rotating impeller through these ports and, due to the circular motion given to the liquid, it is thrown by centrifugal force to the open periphery of the impeller where it is discharged and enters the space between the outer circumference of the impeller and the casing and directed to the outlet branch, see Fig. 223.

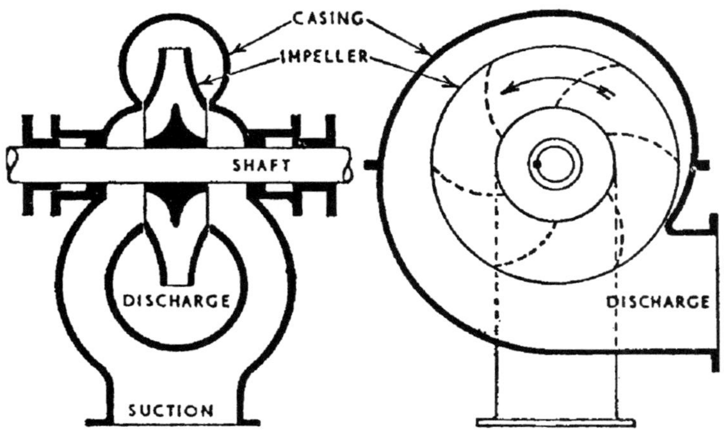

CENTRIFUGAL PUMP

Fig. 223

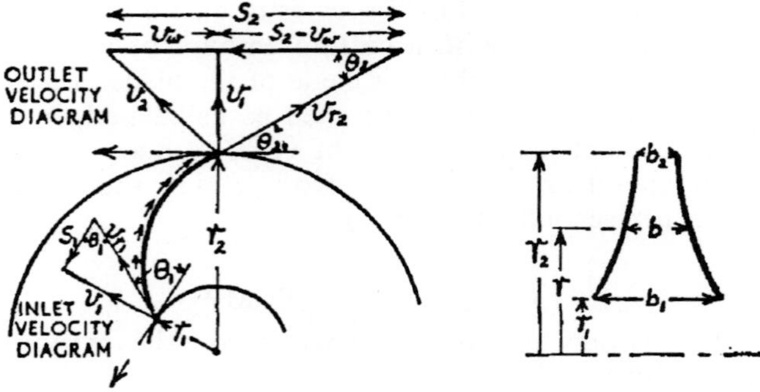

Fig. 224

Referring to the inlet velocity diagram at entrance to the impeller, Fig. 224, it is assumed that the liquid at entrance to the vanes has radial velocity only.

Let s_1 = linear velocity of the vanes at entrance

v_1 = radial velocity of the liquid

v_{r1} = relative velocity of the liquid to the vanes

If the liquid is to enter the vanes without shock, the slope of the vanes at entrance should be the same as the direction of the relative velocity, that is, at θ_1 to the tangent of the inner circle, hence:

$$\tan \theta = \frac{v_1}{s_1} \quad \ldots\ldots\ldots (i)$$

Referring to the outlet velocity diagram at exit from the impeller, after flowing over the surface of the curved vanes, the liquid leaves at a relative velocity of v_{r2},

Let s_2 = linear velocity of the vanes at exit

v_2 = absolute velocity of the liquid leaving the vanes

v_w = velocity of whirl of the liquid

v_1 = radial velocity of the liquid which is assumed to be the same at exit as at entrance, and constant on its way through the impeller.

v_1 and v_w are the rectangular components of v_2. Representing the exit angle of the vanes by θ_2 then:

$$\tan \theta_2 = \frac{v_1}{s_2 - v_w} \quad \dots \dots \dots (ii)$$

The volume flow of the liquid through the impeller is the product of the effective area (circumferential cross-section) of the impeller and the radial velocity of the liquid. Since the volume flow and radial velocity are constant throughout the impeller, then the area of section is constant.

Let A = area of circumferential section of the impeller at radius r where the breadth is b.

then $A = 2\pi rb$

Area being constant, $rb = r_1b_1 = r_2b_2$

If the liquid enters the vanes without shock, the tangential velocity of the liquid at entrance is nil. At exit the tangential velocity of the liquid is v_w therefore, if $\dot{m}$ = mass flow per second, the tangential momentum given to water per second is $\dot{m}\,v_w$.

Since change of momentum per second = force, then the tangential force given to the water is $\dot{m}\,v_w$.

Work /s = force × velocity

Work /s imparted to the liquid = $\dot{m}\,v_w\,s_2$

Also, Work /s = force to lift against gravity × height

$= \dot{m}\,g \times h$

$\dot{m}\,gh = \dot{m}\,v_w\,s_2$

Theoretical head given to water:

$$h = \frac{\dot{m}\,v_w\,s_2}{\dot{m}\,g}$$

$$h = \frac{v_w\,s_2}{g}$$

f MANOMETRIC HEAD

The manometric head is the *effective* or actual energy imparted to the liquid by a pump.

Liquid enters the system at ①, the suction inlet, and is discharged at ②, the discharge pipe outlet. (see Fig. 225)

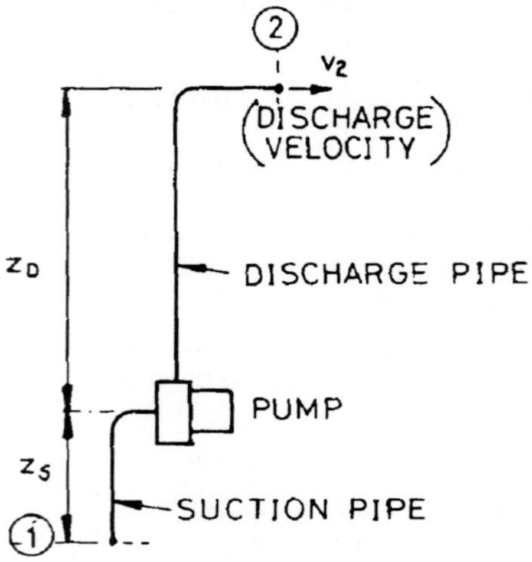

Fig. 225

Manometric head = suction head + discharge head + velocity energy at discharge + friction losses.

$$h_m = z_s + z_d + \frac{v_2^2}{2g} + \text{friction losses}$$

f PUMP POWER

Taking the total energy head into account:

Pump output power = effective work done per second

= weight of liquid flow / s × effective head

i.e. Pump output power $= \dot{m} \times g \times h_m$ W

where $\dot{m}$ = mass flow rate of liquid

g = 9·81

h_m = manometric head

f MANOMETRIC EFFICIENCY

$$\text{Manometric efficiency} = \frac{\text{effective head imparted to liquid}}{\text{theoretical head imparted by pump}}$$

$$= \frac{h_m}{\dfrac{v_w\ s_2}{g}}$$

$$= \frac{g h_m}{v_w\ s_2}$$

f Example. The outer and inner diameters of the impeller of a centrifugal pump are 450 and 150 mm diameter respectively, and the width of the impeller at exit is 40 mm. When running at 500 rev/min it discharges 6·8 m³/min against an effective head of 8 m. Calculate the width of the impeller at inlet and, if the water enters the vanes without shock, calculate the angle of the vanes at inlet. The angle of the vanes at exit is 45°, calculate the theoretical head and the manometric efficiency.

Referring to Fig. 224,

$$r_1 \times b_1 = r_2 \times b_2$$

$$75 \times b_1 = 225 \times 40$$

$$b_1 = 120 \text{ mm} \quad \text{Ans. } (i)$$

$$\text{Radial velocity} = \frac{\text{volume flow}}{\text{area of flow}}$$

$$= \frac{6{\cdot}8}{60 \times \pi \times 0{\cdot}45 \times 0{\cdot}04}$$

$$v_1 = 2{\cdot}004 \text{ m/s}$$

Linear velocity of vanes at entrance,

$$s_1 = \frac{\pi \times 0{\cdot}15 \times 500}{60}$$

$$= 3{\cdot}927 \text{ m/s}$$

$$\tan\theta_1 = \frac{v_1}{s_1} = \frac{2{\cdot}004}{3{\cdot}927} = 0{\cdot}5101$$

$$\text{Angle at entrance } \theta_1 = 27°\ 2' \quad \text{Ans. } (ii)$$

Linear velocity of vanes at exit,

$$s_2 = \frac{\pi \times 0{\cdot}45 \times 500}{60}$$

$$= 3 \times s_1 = 11{\cdot}78 \text{ m/s}$$

$$\frac{v_1}{s_2 - v_w} = \tan \theta_2$$

$$s_2 - v_w = \frac{v_1}{\tan \theta_2}$$

$$= \frac{2{\cdot}004}{\tan 45^\circ} = 2{\cdot}004 \text{ m/s}$$

$$v_w = s_2 - (s_2 - v_w)$$

$$= 11{\cdot}78 - 2{\cdot}004 = 9{\cdot}776 \text{ m/s}$$

$$\text{Theoretical head} = \frac{v_w\, s_2}{g}$$

$$= \frac{9{\cdot}776 \times 11{\cdot}78}{9{\cdot}81}$$

$$= 11{\cdot}74 \text{ m} \quad \text{Ans. } (iii)$$

$$\text{Manometric efficiency} = \frac{\text{effective head}}{\text{theoretical head}}$$

$$= \frac{8}{11{\cdot}74} = 0{\cdot}6816 \quad \text{Ans. } (iv)$$

f ORIFICE CONTROL

Orifice plates with specific characteristics are often inserted into fluid flow lines to control and attemperate the mass flow design performance. Questions usually require the solution and manipulation of given formulae. A worked example illustrates the practice as follows:

f Example. The flow of air to a compressor is given by the equation,

$$m \text{ (kg/s)} = C\sqrt{h}$$

where $C = 0{\cdot}816$

h = the differential pressure head (mm of water) across an orifice in the pipeline.

(*a*) The air flow is regulated within the limits of ± 5% of the normal flow rate of 10 kg/s.

Calculate the values of h for maximum, normal and minimum flow conditions.

(*b*) The regulation of the air flow is performed by a pneumatic control system where the control signals are 100 kN/m² and 30 kN/m² at maximum and minimum flow conditions respectively. The equation for the signal is:

$$S = K_1 + K_2 h$$

where S is the control signal pressure (kN/m²)

(*i*) Determine the value of the constants K_1 and K_2.

(*ii*) Calculate the control signal pressure when the air flow rate is normal.

Maximum air flow rate = 1·05 × 10 = 10·5 kg/s

Minimum air flow rate = 0·95 × 10 = 9·5 kg/s

$$\dot{m} = C\sqrt{h}$$

$$h = \left(\frac{\dot{m}}{C}\right)^2$$

$$\left.\begin{aligned} &\text{At max. flow } h = \left(\frac{10{\cdot}5}{0{\cdot}816}\right)^2 \quad h = 165{\cdot}6 \text{ mm} \\ &\text{At normal flow } h = \left(\frac{10}{0{\cdot}816}\right)^2 \quad h = 150{\cdot}2 \text{ mm} \\ &\text{At min. flow } h = \left(\frac{9{\cdot}5}{0{\cdot}816}\right)^2 \quad h = 135{\cdot}5 \text{ mm} \end{aligned}\right\} \quad \text{Ans.}(a)$$

$$S = K_1 + K_2 h$$

$$\therefore \text{ At max. flow: } 100 = K_1 + (K_2 \times 165{\cdot}6)$$

$$\text{At min. flow: } 30 = K_1 + (K_2 \times 135{\cdot}5)$$

$$70 = 30{\cdot}1\,K_2$$

$$K_2 = 2{\cdot}326$$

$$30 = K_1 + (2{\cdot}326 \times 135{\cdot}5)$$

$$K_1 = -285{\cdot}2 \text{ and } K_2 = 2{\cdot}326 \qquad \text{Ans. } (b)\ (i)$$

$$\text{At normal flow: } S = -285{\cdot}2 + (2{\cdot}326 \times 150{\cdot}2)$$

$$S = 64{\cdot}16 \text{ kN/m}^2 \qquad \text{Ans. } (b)\ (ii)$$

TEST EXAMPLES 13

1. A pump discharges sea water through a 150 mm diameter pipe at a velocity of 1·75 m/s to a height of 9 m. If the efficiency of the pump is 72%, find the input power.Take the density of sea water as 1025 kg/m^3.

2. The diameter of a feed check valve is 60 mm. Find the volume flow in m^3/h when the lift of the valve is 10 mm and the velocity of the water is 1·8 m/s.

3. A fire main contains sea water (density 1025 kg/m^3) which flows with a velocity of 6 m/s. The main supplies three branches in which water flows at 8 m/s. Branches are 60 mm diameter supplying nozzles of 12 mm diameter. Calculate:

(a) the necessary diameter of the fire main;

(b) the mass flow rate;

(c) the velocity of discharge.

4. Derive a formula for the mass flow of liquid through a valve in t/h, using the symbols, d = diameter in mm, l = lift in mm, v = velocity of water in m/s, ρ = density of liquid in g/ml, and assuming one-sixth of the area of opening is blocked by the wings of the valve.

Using the derived formula, find the mass flow (t/h) of water through a 76 mm diameter valve when the lift is 13 mm, velocity of the water 3 m/s, and the density of the water 0·997 g/ml.

5. A rivet falls out of the side of a tank and water escapes through the rivet hole which is 19 mm diameter. If the head of water in the tank above the hole is 2·5 m, calculate (*i*) the velocity of the water as it issues from the hole, (*ii*) the litres escaping every second. Take the coefficient of velocity as 0·97 and coefficient of reduction of area as 0·64.

6. Water flows through a 10 mm diameter orifice in the side of a tank which rests on a horizontal ground. The hole is 0·6 m above ground level and the head of water above the hole is 1·5 m. The jet strikes the ground at a horizontal distance of 1·84 m from the side of the tank and the rate of discharge is 15·85 l/min. Find the coefficients of velocity, reduction of area, and discharge.

*f*7. Fresh water flows through a short smooth horizontal pipe of tapered bore. At the large end the velocity of the water is 1·5 m/s.

At the small end the pressure is 14 kN/m^2 less, calculate the velocity at this end.

*f*8. Water flows through a smooth bore horizontal venturi meter which is 375 mm diameter at entrance and 125 mm diameter at the throat. If the difference in pressure between these two points is equivalent to 457 mm head of water, calculate the mass flow in kg/s.

*f*9. A vertical pipe 200 mm diameter at the top and 100 mm diameter at the bottom is 6 m long. Sea water (density 1025 kg/m^3) enters at the bottom at 9 m/s where the pressure is 300 kN/m^3 and flows upwards. Head loss due to friction is given by:

$$h_f = 0{\cdot}25\ (v_1^2 - v_2^2)\ /\ 9\ \text{m}$$

where v_1 is the velocity at inlet and v_2 is the velocity at exit. Determine the water pressure at the top of the pipe.

*f*10. (*a*) Identify the three forms of energy that may be possessed by a liquid in motion stating energy forms in J/N units.

(*b*) Water flows upward through a vertical venturi meter, 5 cm diameter inlet, 2·5 cm diameter throat. The throat is 15 cm above the inlet and differential pressure between inlet and throat is 250 mm of water. Calculate the mass flow rate of water (density 1000 kg/m^3) through the meter whose coefficient is 0·95.

*f*11. A pipe 20 cm diameter and 100 m long supplies 0·1 m^3/s. Determine the head lost in friction over the pipe length and the slope of the hydraulic gradient. Assume $f = 0{\cdot}005$.

*f*12. A horizontal jet of fresh water 40 mm diameter impinges on a stationary flat plate which is perpendicular to the jet, and the direct force to prevent the plate moving is measured to be 225 N. Assuming that there is no splash back of the water, find the initial velocity of the jet.

*f*13. The outer and inner diameters of the impeller of a centrifugal pump are 600 and 300 mm respectively, and the width at the impeller entrance is 120 mm. Calculate the width of the impeller at exit and at mid-radius so that the radial velocity of the water through the impeller is constant. Calculate also the angle of the vanes at entrance so that the water enters without shock when the pump is running at 420 rev/min and the radial velocity is 1·65 m/s.

*f*14. A pump draws sea water from 3 m below its own level and discharges to a height of 76 m above its level. The flow rate is

108 m^3/h and the diameter of the discharge pipe is 130 mm. If the friction head loss in the pipes is 13·7 m and the pump efficiency is 65%, calculate the power required to drive the pump. (Density of sea = 1025 kg/m^3.)

*f*15. (*a*) The flow rate of air through an orifice is given by:

$$\dot{V} = A \times C_D \times \sqrt{2gh}$$

An orifice plate is used to control the air flow to a ventilation duct. When the orifice plate diameter is 350 mm the air flow rate is 5100 m^3/h. C_D is 0·6. Determine the differential pressure across the orifice plate in mm of water.

(*b*) The flow is to be reduced to one third of its original value. Calculate the diameter of the orifice required assuming the differential pressure remains constant. $C_D = 0{\cdot}58$.

Density: air = 1·22 kg/m^3, water = 1000 kg/m^3.

SOLUTIONS TO TEST EXAMPLES 1

1.

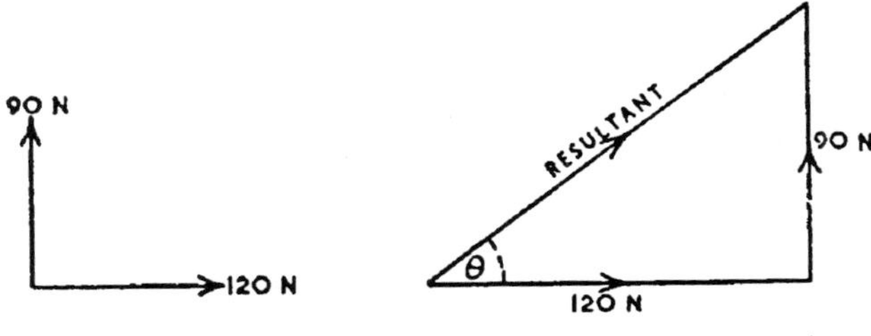

Referring to vector diagram:

$$\text{Resultant} = \sqrt{120^2 + 90^2}$$

$$= 150 \text{ N}$$

$$\tan\theta = \frac{90}{120} = 0{\cdot}75$$

$$\theta = 36^\circ\ 52'$$

Resultant = 150 N at 36° 52′ to the horizontal. Ans.

2.

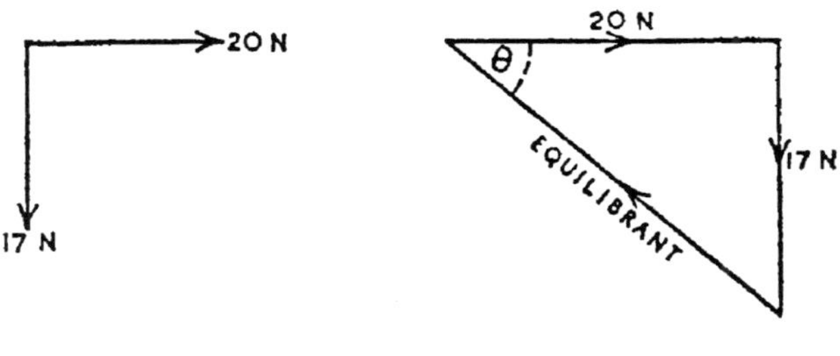

Referring to vector diagram:

$$\text{Equilibrant} = \sqrt{20^2 + 17^2}$$
$$= 26{\cdot}25 \text{ N}$$
$$\tan\theta = \frac{17}{20} = 0{\cdot}85$$
$$\theta = 40^\circ\ 22'$$

Equilibrant = 26·25 N to the left at 40° 22′ above the horizontal. Ans.

3.

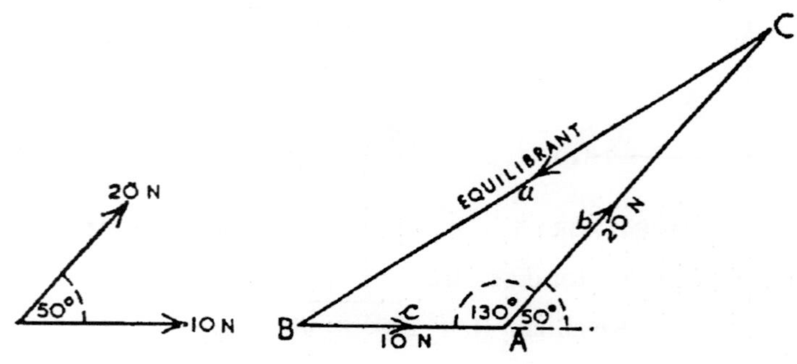

SPACE DIAGRAM VECTOR DIAGRAM

Referring to vector diagram:

$$a^2 = b^2 + c^2 - 2bc\cos A$$
$$= 20^2 + 10^2 - 2 \times 20 \times 10 \times \cos 130^\circ$$
$$= 400 + 100 + 257{\cdot}12$$
$$a = \sqrt{27{\cdot}51}$$
$$= 27{\cdot}51$$

$$\frac{a}{\sin A} = \frac{b}{\sin B}$$
$$\sin B = \frac{20 \times \sin 130^\circ}{27{\cdot}51}$$
$$= 0{\cdot}5569$$
$$\text{angle B} = 33^\circ\ 50'$$

Equilibrant = 27·51 N at 33° 50′ to the horizontal. Ans.

4.

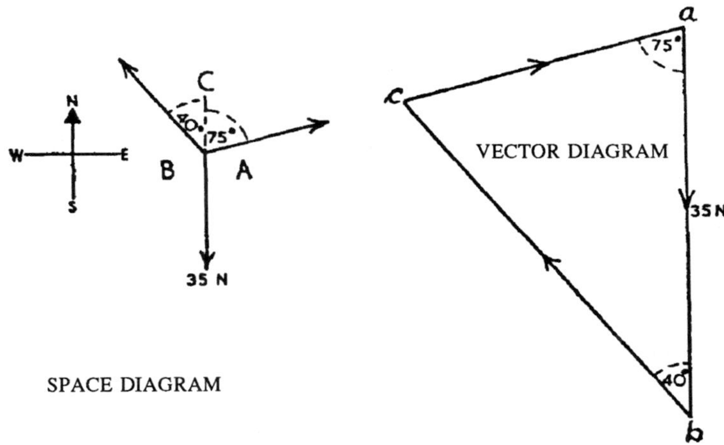

Referring to vector diagram:

$$\text{Angle } c = 180° - (75° + 40°) = 65°$$

$$\frac{ab}{\sin c} = \frac{ac}{\sin b}$$

$$ac = \frac{35 \times \sin 40°}{\sin 65°}$$

$$= 24{\cdot}82 \text{ N}$$

$$\frac{ab}{\sin c} = \frac{bc}{\sin a}$$

$$bc = \frac{35 \times \sin 75°}{\sin 65°}$$

$$= 37{\cdot}3 \text{ N}$$

The magnitudes of the other two forces are 24·82 and 37·3 N. Ans.

5.

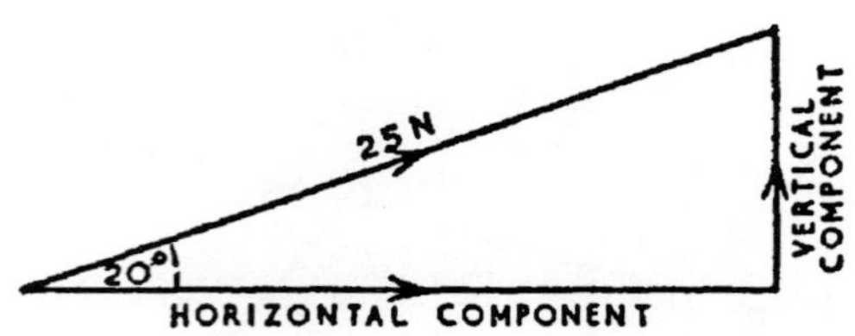

Vertical component $= 25 \times \sin 20° = 8{\cdot}55$N Ans. (*i*)
Horizontal component $= 25 \times \cos 20° = 23{\cdot}49$ N Ans. (*ii*)

6.

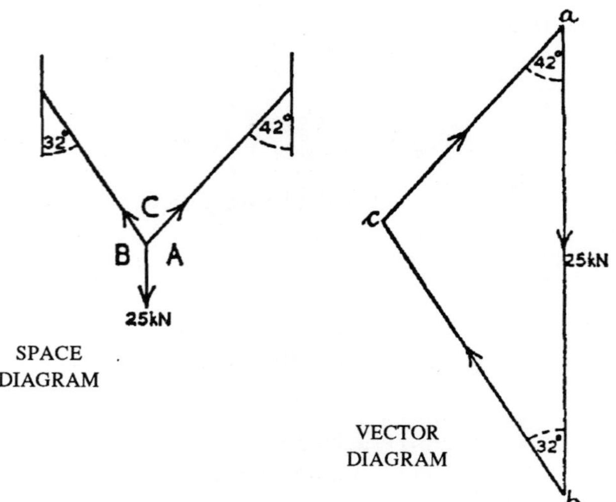

Referring to vector diagram:

Angle $c = 180° - (42° + 32°) = 106°$

$$\frac{ab}{\sin c} = \frac{ac}{\sin b}$$

$$ac = \frac{25 \times \sin 32°}{\sin 106°}$$

$$= 13{\cdot}78 \text{ kN}$$ Ans. (*i*)

$$\frac{ab}{\sin c} = \frac{bc}{\sin a}$$

$$bc = \frac{25 \times \sin 42^\circ}{\sin 106^\circ}$$

$$= 17{\cdot}4 \text{ kN} \qquad \text{Ans } (ii)$$

7.

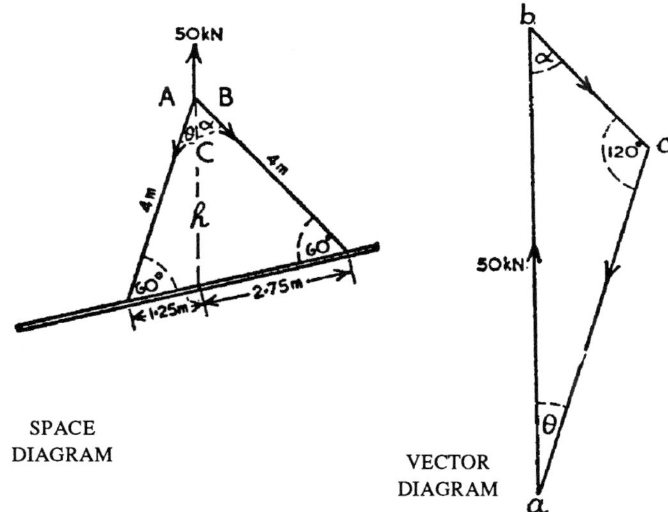

SPACE DIAGRAM

VECTOR DIAGRAM

As the lengths of the chains and distance between connections on the shaft are all equal they form an equilateral triangle.

The centre of gravity of the shaft must lie vertically below the crane hook.

Let h represent height from shaft to crane hook, then:

$$h^2 = 4^2 + 1{\cdot}25^2 - 2 \times 4 \times 1{\cdot}25 \times \cos 60^\circ$$

$$= 16 + 1{\cdot}563 - 5$$

$$h = \sqrt{12{\cdot}563}$$

$$= 3{\cdot}544 \text{ m}$$

$$\frac{3{\cdot}544}{\sin 60°} = \frac{1{\cdot}25}{\sin \theta}$$

$$\sin \theta = \frac{1{\cdot}25 \times 0{\cdot}866}{3{\cdot}544}$$

$$= 0{\cdot}3054$$

$$\therefore \theta = 17° \, 47'$$

$$\alpha = 60° - 17° \, 47'$$

$$= 42° \, 13'$$

$$\text{Mass of shaft} = 5{\cdot}097 \times 10^3 \text{ kg}$$

$$\text{Load on crane hook} = 5{\cdot}097 \times 10^3 \times 9{\cdot}81$$

$$= 50 \text{ kN}$$

Referring to vector diagram:

$$\frac{ab}{\sin 120°} = \frac{bc}{\sin 17° \, 47'}$$

$$bc = \frac{50 \times 0{\cdot}3055}{0{\cdot}866}$$

$$= 17{\cdot}64 \text{ kN} \qquad \text{Ans. } (i)$$

$$\frac{ab}{\sin 120°} = \frac{ac}{\sin 42° \, 13'}$$

$$ac = \frac{50 \times 0{\cdot}6719}{0{\cdot}866}$$

$$= 38{\cdot}8 \text{ kN} \qquad \text{Ans. } (ii)$$

8.

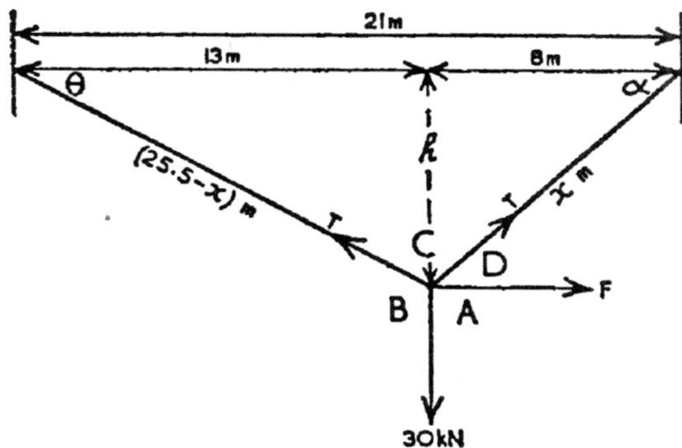

If the space diagram is divided into two right angles as shown, the vertical height (h) is the same for both, therefore:

$$h^2 = (25{\cdot}5 - x)^2 - 13^2$$

$$\text{also, } h^2 = x^2 - 8^2$$

$$\therefore (25{\cdot}5 - x)^2 - 13^2 = x^2 - 8^2$$

$$650{\cdot}25 - 51x + x^2 - 169 = x^2 - 64$$

$$-51x = -545{\cdot}25$$

$$x = 10{\cdot}69 \text{ m}$$

$$\text{and } 25{\cdot}5 - x = 14{\cdot}81 \text{ m}$$

$$\cos\theta = \frac{13}{14{\cdot}81} = 0{\cdot}8776$$

$$\therefore \theta = 28^\circ\, 39'$$

$$\cos\alpha = \frac{8}{10{\cdot}69} = 0{\cdot}7485$$

$$\therefore \alpha = 41^\circ\, 33'$$

As the wire rope is continuous the tension in it must be the same throughout its length, let the tension be represented by T, and let the horizontal force be applied be represented by F.

For vertical equilibrium,

Total upward force = Total downward force

Vertical component of BC + vert. comp. of CD = Downward load

$$T\sin\theta + T\sin\alpha = 30$$

$$0{\cdot}4795T + 0{\cdot}6633T = 30$$

$$1{\cdot}1428T = 30$$

$$T = 26{\cdot}25 \text{ kN} \qquad \text{Ans. } (i)$$

For horizontal equilibrium,

Total forces pulling to right = Total forces pulling to left

F + horiz. component of CD = Horiz. component of BC

$$F + T\cos\alpha = T\cos\theta$$

$$F + 26{\cdot}25 \times 0{\cdot}7485 = 26{\cdot}25 \times 0{\cdot}8776$$

$$F = 26{\cdot}25\,(0{\cdot}8776 - 0{\cdot}7485)$$

$$= 26{\cdot}25 \times 0{\cdot}1291$$

$$= 3{\cdot}388 \text{ kN} \qquad \text{Ans.}(ii)$$

9.

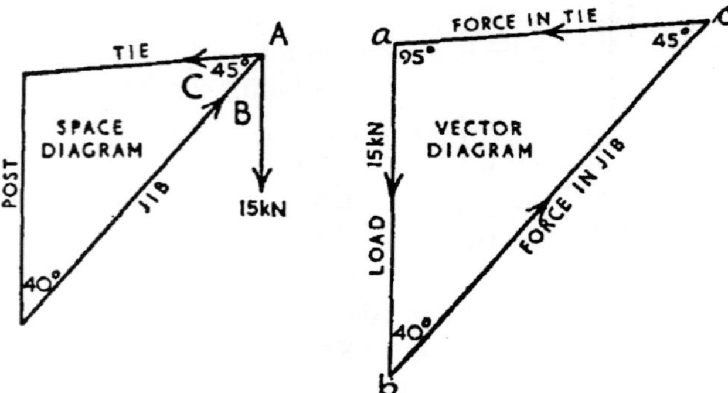

Referring to vector diagram

Angle a = 180° – (40° + 45°) = 95°

$$\frac{\text{Force in jib}}{\sin 95^\circ} = \frac{15}{\sin 45^\circ}$$

$$\text{Force in jib} = \frac{15 \times 0{\cdot}9962}{0{\cdot}7071}$$

$$= 21{\cdot}13 \text{ kN} \qquad \text{Ans. }(i)$$

$$\frac{\text{Force in tie}}{\sin 40^\circ} = \frac{15}{\sin 45^\circ}$$

$$\text{Force in tie} = \frac{15 \times 0{\cdot}6428}{0{\cdot}7071}$$

$$= 13{\cdot}63 \text{ kN} \qquad \text{Ans. }(ii)$$

10.

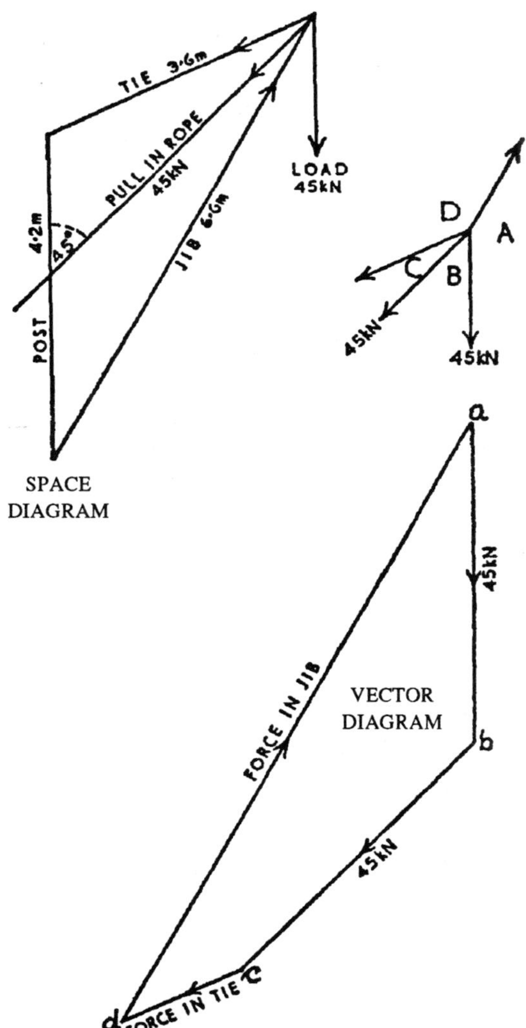

SPACE DIAGRAM

VECTOR DIAGRAM

The space diagram is drawn to scale from the given dimensions, this is rearranged so that the two known forces (*i.e.* 45 kN in each part of the rope) are next to each other. The vector diagram is then constructed as shown by drawing *ab* to scale 45 kN, *bc* at 45° and to scale 45 kN, then *cd* and *ad* parallel to the tie and jib respectively. Measuring *cd* and *da*:

$$\left.\begin{aligned}\text{Force in tie} &= 17{\cdot}7\text{ kN}\\ \text{Force in jib} &= 97\ \ \text{kN}\end{aligned}\right\}\ \text{Ans.}$$

11.

Length of crank = ½ stroke = 300 mm = 0·3 m

Referring to space diagram of Fig 16:

$$\frac{1{\cdot}25}{\sin 60°} = \frac{0{\cdot}3}{\sin \phi}$$

$$\sin \phi = \frac{0{\cdot}866 \times 0{\cdot}3}{1{\cdot}25}$$

$$= 0{\cdot}2078$$

$$\therefore \phi = 12°$$

Referring to vector diagram of Fig 16:

$$\text{Guide load} = \text{Piston effort} \times \tan \phi$$

$$= 180 \times \tan 12°$$

$$= 38{\cdot}26\text{ kN} \quad \text{Ans. } (i)$$

$$\text{Force in conn. rod} = \frac{\text{Piston effort}}{\cos \phi}$$

$$= \frac{180}{\cos 12°}$$

$$= 184{\cdot}1\text{ kN} \quad \text{Ans. } (ii)$$

12.

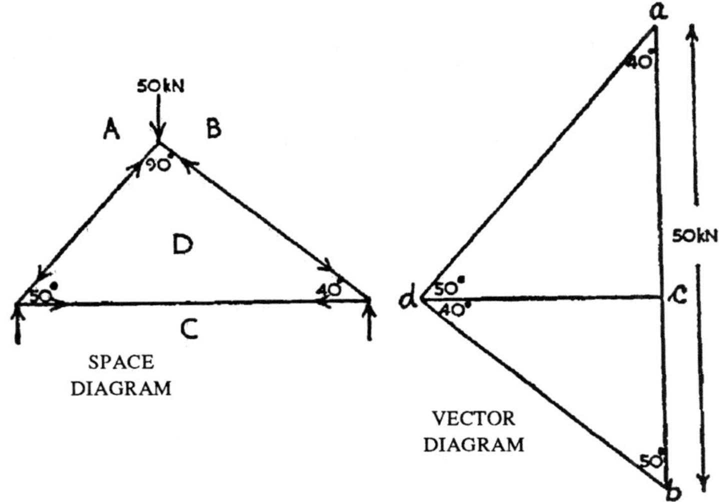

Apex angle of structure = 180° − (50° + 40°) = 90°

Considering whole vector diagram shown, *adb* is a right angled triangle, therefore,

$$ad = ab \times \sin 50°$$
$$= 50 \times 0{\cdot}766 = 38{\cdot}3 \text{ kN}$$
$$bd = ab \times \cos 50°$$
$$= 50 \times 0{\cdot}6428 = 32{\cdot}14 \text{ kN}$$

Considering the top triangle of the vector diagram, *i.e.* the right angled triangle *acd*,

$$cd = ad \times \sin 40°$$
$$= 38{\cdot}3 \times 0{\cdot}6428 = 24{\cdot}61 \text{ kN}$$
$$ac = ad \times \cos 40°$$
$$= 38{\cdot}3 \times 0{\cdot}766 = 29{\cdot}34 \text{ kN}$$
$$bc = ab - ac$$
$$= 50 - 29{\cdot}34 = 20{\cdot}66 \text{ kN}$$

Tabulating results:

MEMBER	MAGNITUDE OF FORCE	NATURE OF FORCE
AD	38·3 kN	Compression
BD	32·14 ,,	Compression
CD	24·61 ,,	Tension
AC	29·34 ,,	Upward reaction (left)
BC	20·66 ,,	Upward reaction (right)

13.

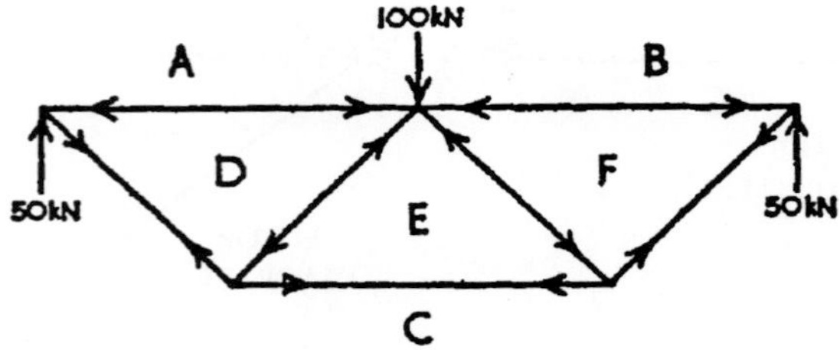

SPACE DIAGRAM

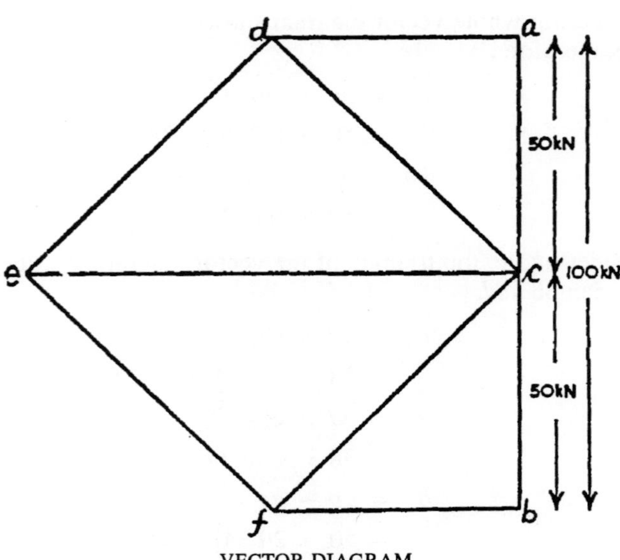

VECTOR DIAGRAM

As the loading is symmetrical each reaction carries half the load. By measurement of the vector diagram:

MEMBER	MAGNITUDE OF FORCE	NATURE OF FORCE
AD	50 kN	Compression
BF	50 ,,	Compression
DC	70·7 ,,	Tension
FC	70·7 ,,	Tension
DE	70·7 ,,	Compression
FE	70·7 ,,	Compression
CE	100 ,,	Tension

14.

Refer to Figs. 23, 24 and 25 which illustrate the picture, space and vector diagrams in a similar way as applies in this question:

Length of imaginary leg $= \sqrt{10^2 - 5^2} = 8{\cdot}661$ m

Angles of the vector diagram (Fig. 25) are here:

$a = 30° \, 14'$; $b = 114° \, 38'$; $c = 35° \, 8'$.

$$\text{Force in back stay} = \frac{10 \times 9{\cdot}81 \times \sin 30° \, 14'}{\sin 35° \, 8'}$$

$$= 85{\cdot}53 \text{ kN}$$

$$\text{Force in imaginary leg} = \frac{9{\cdot}81 \times \sin 114° \, 38'}{\sin 35° \, 8'}$$

$$= 154{\cdot}6 \text{ kN}$$

Force in each real front leg (see Fig. 25):

$$= \frac{154{\cdot}6}{2} \times \frac{10}{8{\cdot}661}$$

$$= 89{\cdot}25 \text{ kN}$$

$$\left.\begin{aligned} \text{Force in back stay} &= 85{\cdot}53 \text{ kN} \\ \text{Force in each front leg} &= 89{\cdot}25 \text{ kN} \end{aligned}\right\} \text{Ans.}$$

15.

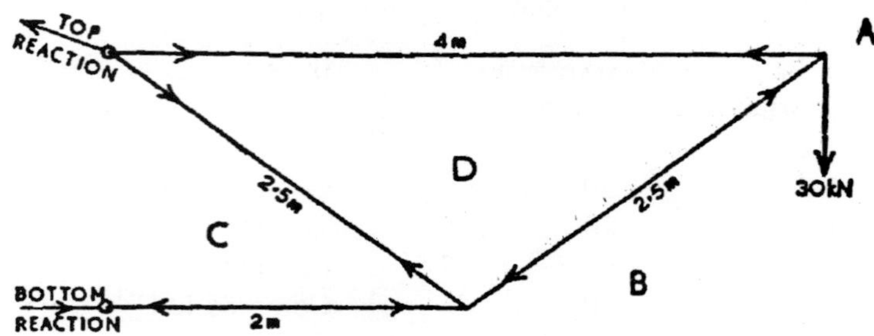

SPACE DIAGRAM

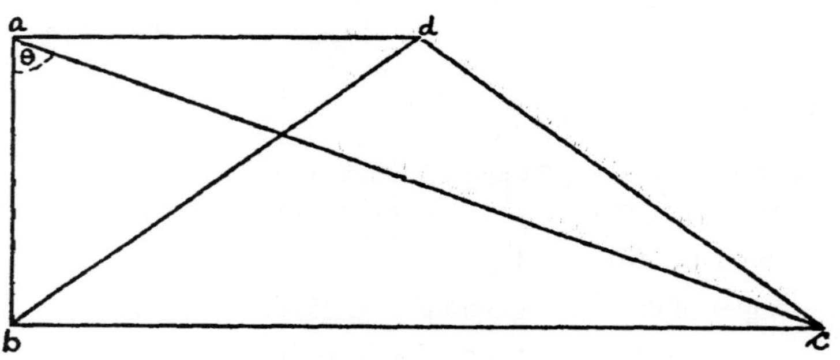

VECTOR DIAGRAM

Referring to space diagram,

Vertical height between the two wall connections

$$= \sqrt{2{\cdot}5^2 - 2^2} = 1{\cdot}5 \text{ m}$$

By moments about top wall connection,

Clockwise moments = Anticlockwise moments

$$30 \times 4 = \text{Force BC} \times 1{\cdot}5$$

$$\text{BC} = 80 \text{ kN}$$

Referring to vector diagram,

$$ac = \sqrt{(ab)^2 + (bc)^2}$$

$$= \sqrt{30^2 + 80^2}$$

$$= 85{\cdot}44 \text{ kN}$$

$$\cos\theta = \frac{ab}{ac}$$

$$= \frac{30}{85{\cdot}44} = 0{\cdot}3511$$

$$\therefore \theta = 69^\circ\ 27'$$

Bottom reaction = 80 kN horizontally
Top reaction = 85·44 kN at 69° 27′ to vertical } Ans.

SOLUTIONS TO TEST EXAMPLES 2

1.

SPACE DIAGRAM

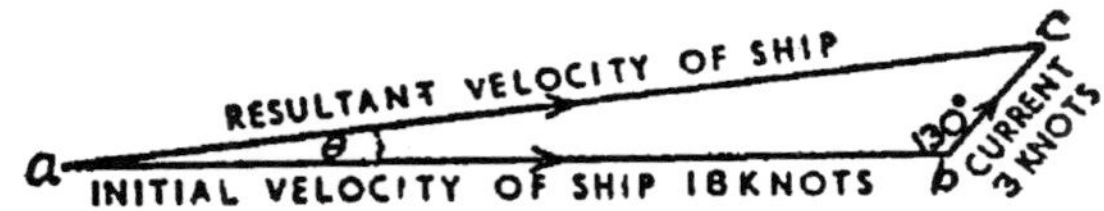

VECTOR DIAGRAM

Referring to vector diagram:

$$\begin{aligned} (ac)^2 &= (ab)^2 + (bc)^2 - 2 \times (ab) \times (bc) \times \cos b \\ &= 18^2 + 3^2 - 2 \times 18 \times 3 \times \cos 130° \\ &= 324 + 9 + 69{\cdot}42 \\ ac &= \sqrt{402{\cdot}42} \\ &= 20{\cdot}06 \text{ knots} \end{aligned}$$

$$\frac{20{\cdot}06}{\sin 130°} = \frac{3}{\sin \theta} \quad \text{so } \sin \theta = \frac{3 \times 0{\cdot}766}{20{\cdot}06}$$

$$\left.\begin{aligned} \text{Resultant speed} &= 20{\cdot}06 \text{ knots} \\ \text{Resultant direction} &= 6° \, 35' \text{ N of E} \end{aligned}\right\} \quad \text{Ans.}$$

2.

Let v = maximum speed

During acceleration period from rest,

$$\begin{aligned} \text{Distance} &= \text{average speed} \times \text{time} \\ 0{\cdot}5 &= \frac{0 + v}{2} \times \frac{1}{60} \\ v &= 60 \text{ km/h} \qquad \text{Ans } (i) \end{aligned}$$

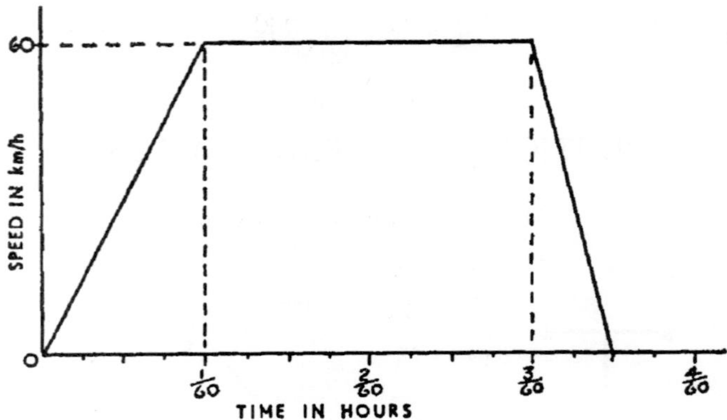

Distance travelled (= area of speed-time graph),

$$= 0{\cdot}5 + 60 \times \frac{2}{60} + \left\{ \frac{60 + 0}{2} \right\} \times \frac{1}{2 \times 60}$$

$$= 0{\cdot}5 + 2 + 0{\cdot}25$$

$$= 2{\cdot}75 \text{ km} \quad \text{Ans. } (ii)$$

3.

$$u = \frac{20 \times 1000}{3600} \text{ m/s}, \quad v = \frac{74 \times 1000}{3600} \quad t = 12 \text{ s}$$

$$u = 5{\cdot}55 \text{ m/s} \quad v = 20{\cdot}55 \text{ m/s.}$$

$$v = u + at$$

$$20{\cdot}55 = 5{\cdot}55 + 12a$$

$$a = \frac{15}{12}$$

$$\text{Acceleration} = 1{\cdot}25 \text{ m/s}^2 \qquad \text{Ans.}$$

$$s = \tfrac{1}{2}(u + v)\,t$$

$$s = \tfrac{1}{2}(5{\cdot}55 + 20{\cdot}55) \times 12$$

$$= 156{\cdot}6 \text{ m} \qquad \text{Ans.}$$

4.

$$u = \frac{400 \times 1000}{3600} \qquad v = \frac{500 \times 1000}{3600} \qquad s = 1500 \text{ m}$$

$$u = 111{\cdot}1 \text{ m/s} \qquad v = 138{\cdot}9 \text{ m/s}$$

$$s = \tfrac{1}{2}(u + v)t$$

$$1500 = \tfrac{1}{2}(111{\cdot}1 + 138{\cdot}9) \times t$$

$$t = \frac{1500}{125}$$

$$\text{Time} = 12 \text{ s} \qquad \text{Ans.}$$

$$v = u + at$$

$$138{\cdot}9 = 111{\cdot}1 + 12a$$

$$a = \frac{27{\cdot}8}{12}$$

$$\text{Acceleration} = 2{\cdot}317 \text{ m/s}^2 \qquad \text{Ans.}$$

5.

$$v^2 = u^2 - 2ah$$

$$24^2 = 36^2 - 2 \times 9{\cdot}81 \times h$$

$$19{\cdot}62s = 1296 - 576$$

$$h = \frac{720}{19{\cdot}62}$$

$$= 36{\cdot}69 \text{ m} \qquad \text{Ans. } (i)$$

$$h = \left\{ \frac{u + v}{2} \right\} \times t$$

$$36{\cdot}69 = \left\{ \frac{36 + 24}{2} \right\} \times t$$

$$t = \frac{36{\cdot}69}{30}$$

$$\text{Time} = 1{\cdot}223 \text{ s} \qquad \text{Ans. } (ii)$$

6.

Using the formula, $h = ut \pm \frac{1}{2}gt^2$

Time to get to passing point is the same for both bodies,

Distance travelled by rising projectile $= 35t - 4{\cdot}905t^2$ (*i*)

Distance travelled by falling object $= 0 + 4{\cdot}905t^2$ (*ii*)

The sum of these distances is 70 m,

$$35t - 4{\cdot}905t^2 + 4{\cdot}905t^2 = 70$$
$$35t = 70$$
$$t = 2\text{ s}$$

Substituting value of t into equation (*i*),

$$h = 35 \times 2 - 4{\cdot}905 \times 2^2$$
$$= 70 - 19{\cdot}62$$
$$= 50{\cdot}38\text{ m}$$

The bodies will pass each other 2 s after release at a height of 50·38 m Ans.

7.

$$\text{Displacement} = \text{Average velocity} \times \text{Time}$$
$$1800 = \text{Average velocity} \times 1{\cdot}5$$
$$\text{Average velocity} = 1200\text{ rev/min}$$
$$\text{Also, average velocity} = \frac{\text{Initial velocity} + \text{Final velocity}}{2}$$
$$1200 = \frac{\text{Initial velocity} + 0}{2}$$
$$\text{Initial velocity} = 2400\text{ rev/min} \qquad \text{Ans.}$$

$$\omega_1 = \frac{2\pi \times 2400}{60} \qquad \omega_2 = 0 \qquad \theta = 2\pi \times 1800$$
$$\omega_1 = 251{\cdot}3\text{ rad/s} \qquad \theta = 11310\text{ rad}$$
$$\omega_2^2 = \omega_1^2 + 2\alpha\theta$$
$$0 = 251{\cdot}3^2 + 2 \times \alpha \times 11310$$
$$22620\,\alpha = -63152$$
$$\therefore \alpha = -2{\cdot}792$$
$$\text{Retardation} = 2{\cdot}792\text{ rad/s}^2 \qquad \text{Ans.}$$

8.

For the load, $u = 0 \qquad s = 3 \text{ m} \qquad t = 8 \text{ s}$

$$s = ut + \tfrac{1}{2}at^2$$

$$3 = 0 + \tfrac{1}{2} \times a \times 8^2$$

$$a = \frac{3}{32}$$

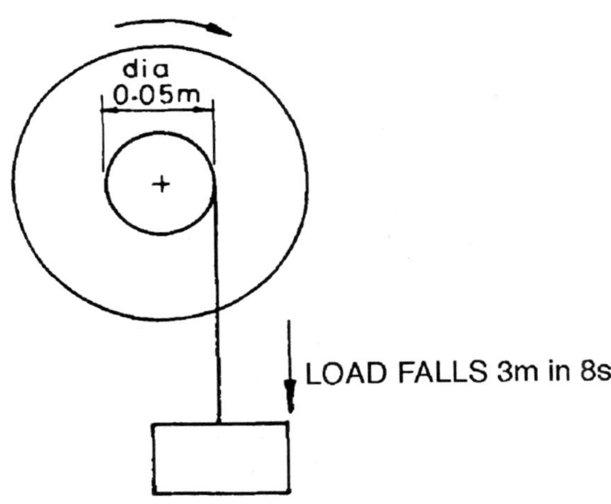

Linear acceleration of load $= 0{\cdot}09375 \text{ m/s}^2$ Ans.

$$a = \alpha r$$

$$\alpha = \frac{0{\cdot}09375}{0{\cdot}025}$$

Angular acceleration of wheel and axle $= 3{\cdot}75 \text{ rad/s}^2$ Ans

9.

$$s = 0{\cdot}2t^2 + 10{\cdot}4$$

$$v = \frac{ds}{dt} = 0{\cdot}4t$$

Velocity $= 2 \text{ m/s at } 5 \text{ s}$ Ans.

$$a = \frac{dv}{dt} = 0{\cdot}4 \text{ i.e. constant (uniform)}$$

$$\text{Acceleration} = 0{\cdot}4 \text{ m/s}^2 \text{ at } 5 \text{ s} \qquad \text{Ans.}$$

$$t = 0, \quad v = 0, \quad s = 10{\cdot}4$$

$$t = 10, \quad v = 4, \quad s = 30{\cdot}4$$

$$\text{Average velocity} = \frac{\text{Total displacement}}{\text{Total time}}$$

$$= \frac{20}{10}$$

$$= 2 \text{ m/s} \qquad \text{Ans.}$$

10.

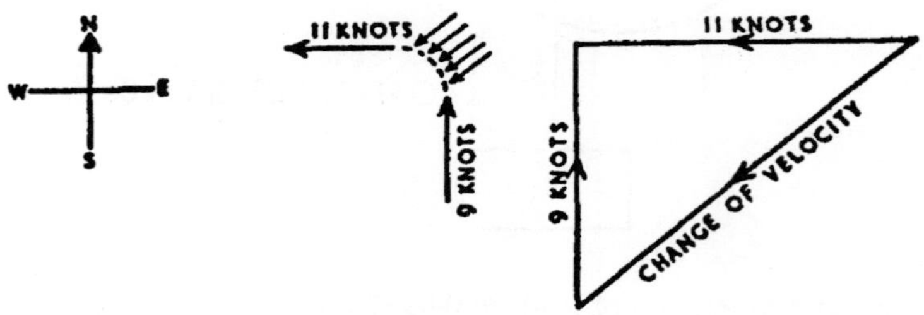

$$\text{Change of velocity} = \sqrt{9^2 + 11^2} = 14{\cdot}21 \text{ knots}$$

$$= \frac{14{\cdot}21 \times 1{\cdot}852 \times 10^3}{3600} \text{ m/s}$$

$$\text{Acceleration} = \frac{\text{Change of velocity}}{\text{Time to change}}$$

$$= \frac{14{\cdot}21 \times 1{\cdot}852 \times 10^3}{3600 \times 30}$$

$$= 0{\cdot}2436 \text{ m/s}^2 \qquad \text{Ans.}$$

11.

Speed of one train relative to the other

$$= 50 + 100 = 150 \text{ km/h}$$

Distance to travel to completely pass each other

$$= 20 + 40 = 60 \text{ m} = 0{\cdot}06 \text{ km}$$

$$\text{Time} = \frac{\text{Distance}}{\text{Speed}} = \frac{0{\cdot}06}{150} \text{ h}$$

$$= \frac{0{\cdot}06 \times 3600}{150} \text{ s}$$

$$= 1{\cdot}44 \text{ s}$$ Ans

12.

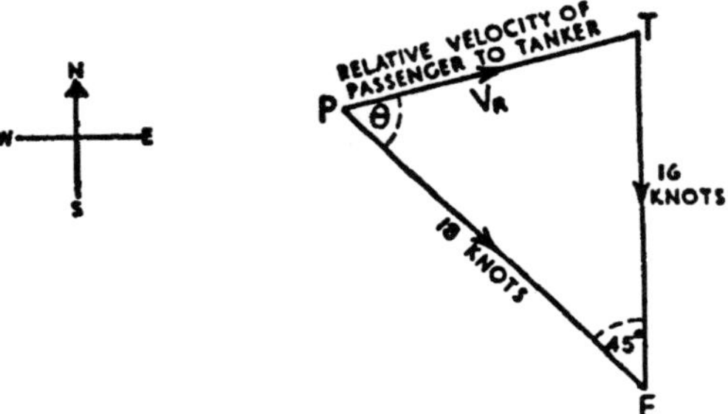

$$V_R^2 = 18^2 + 16^2 - 2 \times 18 \times 16 \times \cos 45°$$

$$= 324 + 256 - 407{\cdot}2$$

$$V_R = \sqrt{172{\cdot}8}$$

$$= 13{\cdot}14 \text{ knots}$$

$$\frac{16}{\sin\theta} = \frac{13{\cdot}14}{\sin 45°}$$

$$\sin\theta = \frac{16 \times 0{\cdot}7071}{13{\cdot}14}$$

$$= 0{\cdot}8610$$

$$\theta = 59° \, 26'$$

Direction of relative velocity = 59° 26′ to course of passenger ship
59° 26′ − 45° = 14° 26′ North of East

Now imagine being on the tanker and that it is stationary, seeing the passenger ship 7·5 nautical miles due West and apparently moving at a speed of 13·14 knots in the direction 14° 26′ North of East. Sketch is a space diagram representing these apparent conditions, nearest approach is indicated by distance *pt*.

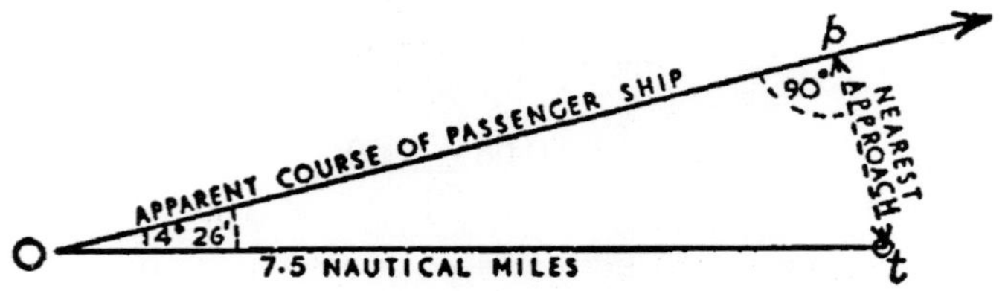

$op = 7{\cdot}5 \times \cos 14^\circ 26' = 7{\cdot}264$ naut. miles

Time to travel 7·264 naut. miles at 13·14 knots

$$= \frac{7{\cdot}264}{13{\cdot}14} = 0{\cdot}5527 \text{ h}$$

$$= 33{\cdot}16 \text{ min}$$

$pt = 7{\cdot}5 \times \sin 14^\circ 26' = 1{\cdot}87$ naut. miles.

Therefore ships will be nearest together at 12-33·16 p.m. and their distance apart will then be 1·87 nautical miles. Ans.

13.

See Fig. 48. Use scale measurement or calculation.

Linear velocity of crank pin, $v_Q = \omega r$

$$= \frac{1200 \times 2\pi \times 0{\cdot}5}{60}$$

$$= 20\pi \text{ m/s} \qquad \text{Ans.}$$

Angular velocity of connecting rod, $\Omega = v_a / IQ$

$$= \frac{20\pi}{1{\cdot}725}$$

$$= 36{\cdot}5 \text{ rad/s} \qquad \text{Ans.}$$

Linear velocity of piston, $v_P = \Omega\, IP$

$$= 36{\cdot}5 \times 1{\cdot}125$$

$$= 32{\cdot}4 \text{ m/s} \qquad \text{Ans.}$$

14.

Initial vertical velocity = nil

Vertical acceleration = $9{\cdot}81\ \text{m/s}^2$

Vertical distance fallen = 60 m

$$h = ut + \tfrac{1}{2}gt^2$$

$$60 = 0 + 4{\cdot}905t^2$$

$$t = \sqrt{\frac{60}{4{\cdot}905}}$$

$= 3{\cdot}497$ s Ans. (*i*)

Horizontal distance = horizontal velocity × time

$= 12 \times 3{\cdot}497$

$= 41{\cdot}96$ m Ans. (*ii*)

15.

Vertical component of initial velocity

$= 600 \times \sin 30° = 300$ m/s

At maximum height, vertical velocity is nil,

time to reach maximum height $= \dfrac{300}{9{\cdot}81}$ s

Total time in flight $= \dfrac{2 \times 300}{9{\cdot}81}$ s

Horizontal component of initial velocity

$= 600 \times \cos 30° = 519{\cdot}6$ m/s

Horizontal distance = horizontal velocity × time

$$= \frac{519{\cdot}6 \times 2 \times 300}{9{\cdot}81}$$

$= 31\,780$ m

$= 31{\cdot}78$ km Ans.

SOLUTIONS TO TEST EXAMPLES 3

1.

$$\text{Acceleration} = 5 \text{ knots in } 10 \text{ min}$$
$$= \frac{5 \times 1{\cdot}852 \times 10^3}{3600 \times 10 \times 60} \text{ m/s}^2$$
$$10\,000 \text{ t} = 10\,000 \times 10^3 \text{ kg} = 10^7 \text{ kg}$$
$$F = ma$$
$$= \frac{10^7 \times 5 \times 1{\cdot}852 \times 10^3}{3600 \times 10 \times 60}$$
$$= 42{\cdot}86 \times 10^3 \text{ N}$$
$$= 42{\cdot}86 \text{ kN} \qquad \text{Ans.}$$

2.

$$\text{Force to overcome resistances} = 155 \text{ N for each t}$$
$$= 310 \text{ N for two t}$$
$$\text{Accelerating force} = 1000 - 310$$
$$= 690 \text{ N}$$
$$\text{Acceleration} = \frac{\text{accelerating force}}{\text{mass}}$$
$$= \frac{690}{2 \times 10^3}$$
$$= 0{\cdot}345 \text{ m/s}^2 \qquad \text{Ans. } (i)$$
$$\text{Velocity after 60 s} = 0{\cdot}345 \times 60$$
$$= 20{\cdot}7 \text{ m/s}$$
$$\frac{20{\cdot}7 \times 3600}{10^3} = 74{\cdot}52 \text{ km/h} \qquad \text{Ans. } (ii)$$

3.

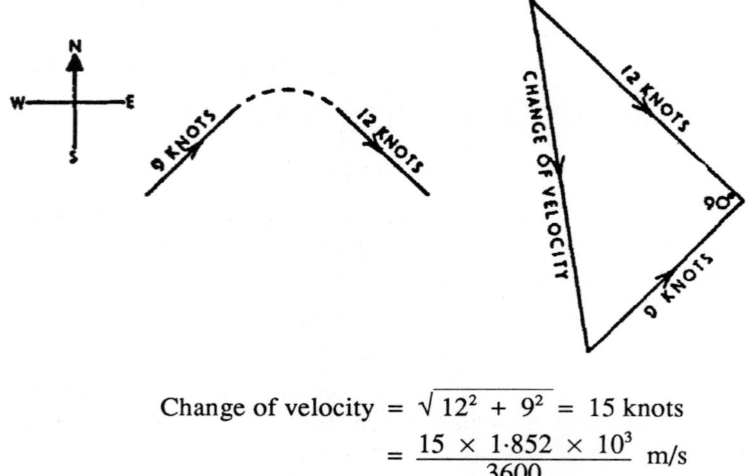

$$\text{Change of velocity} = \sqrt{12^2 + 9^2} = 15 \text{ knots}$$

$$= \frac{15 \times 1{\cdot}852 \times 10^3}{3600} \text{ m/s}$$

$$\text{Acceleration} = \frac{\text{change of velocity}}{\text{time to change}}$$

$$= \frac{15 \times 1{\cdot}852 \times 10^3}{3600 \times 2 \times 60} \text{ m/s}^2$$

$$\text{Accelerating force} = \text{mass} \times \text{acceleration}$$

$$= \frac{5 \times 10^3 \times 15 \times 1{\cdot}852 \times 10^3}{3600 \times 2 \times 60}$$

$$= 321{\cdot}5 \text{ N} \qquad \text{Ans.}$$

4.

$$\text{Accelerating force} = \text{mass} \times \text{acceleration}$$

When accelerating at 2 m/s^2:

$$\text{Accelerating force} = 1134 \times 2 = 2268 \text{ N}$$

When retarding at 1 m/s^2:

$$\text{Retarding force} = 1134 \times 1 = 1134 \text{ N}$$

$$\text{Weight of lift and contents} = 1134 \times 9{\cdot}81$$

$$= 11\ 125 \text{ N}$$

(*i*) When moving down and increasing speed, the force to support the mass acts *upwards*, but acceleration and therefore accelerating force acts *downwards*,

$$\therefore \text{Tension in wire} = \text{supporting force} - \text{accelerating force}$$
$$= 11\,125 - 2268$$
$$= 8857 \text{ N}$$
$$= 8{\cdot}857 \text{ kN} \qquad \text{Ans. } (i)$$

(*ii*) When moving down at steady speed there is no acceleration therefore no accelerating force and the tension in the wire is only the force to support the mass,

$$\text{Tension in wire} = \text{supporting force}$$
$$= 11\,125 \text{ N} = 11{\cdot}125 \text{ kN} \quad \text{Ans.}(ii)$$

(*iii*) When moving down and reducing speed, the retarding force pulls upwards to slow down the speed, this force is in the same direction as the upward supporting force,

$$\text{Tension in wire} = \text{supporting force} + \text{retarding force}$$
$$= 11\,125 + 1134$$
$$= 12\,259 \text{ N}$$
$$= 12{\cdot}259 \text{ kN} \qquad \text{Ans. } (iii)$$

5.

$$\text{Total mass accelerated} = 2{\cdot}5 + 2{\cdot}5 + 0{\cdot}25 = 5{\cdot}25 \text{ kg}$$
$$\text{Accelerating force} = 0{\cdot}25 \times 9{\cdot}81 \text{ N}$$
$$\text{Acceleration} = \frac{\text{Accelerating force}}{\text{mass}}$$
$$= \frac{0{\cdot}25 \times 9{\cdot}81}{5{\cdot}25}$$
$$= 0{\cdot}4671 \text{ m/s}^2 \qquad \text{Ans. } (i)$$
$$\text{Increase in velocity} = \text{acceleration} \times \text{time}$$
$$\therefore \text{Velocity after 3 s} = 0{\cdot}4671 \times 3$$
$$= 1{\cdot}4013 \text{ m/s} \qquad \text{Ans. } (ii)$$

6.

$$\text{Total mass accelerated} = 5{\cdot}44 + 0{\cdot}22 = 5{\cdot}66 \text{ kg}$$

$$\text{Accelerating force} = 0{\cdot}22 \times 9{\cdot}81 \text{ N}$$

$$\text{Accelerating force} = \text{mass} \times \text{acceleration}$$

$$\text{Acceleration} = \frac{\text{acceleration force}}{\text{mass}}$$

$$= \frac{0{\cdot}22 \times 9{\cdot}81}{5{\cdot}66}$$

$$= 0{\cdot}3814 \text{ m/s}^2 \qquad \text{Ans. } (i)$$

$$\text{Tension in cord} = \text{accelerating force on the } 5{\cdot}44 \text{ kg mass of the carriage}$$

$$= \text{mass} \times \text{acceleration}$$

$$= 5{\cdot}44 \times 0{\cdot}3814$$

$$= 2{\cdot}074 \text{ N} \qquad \text{Ans. } (ii)$$

7.

$$\text{Force} = \frac{\text{change of momentum}}{\text{time}}$$

$$= \frac{\text{mass} \times \text{change of velocity}}{\text{time}}$$

$$= \frac{3 \times 7}{1/100}$$

$$= 2100 \text{ N} = 2{\cdot}1 \text{ kN} \qquad \text{Ans.}$$

8.

$$\text{Linear momentum before impact} = \text{Linear momentum after impact}$$

$$(2 \times 10) + (10 \times 16) = (2 + 10) \times v$$

$$20 + 160 = 12 \times v$$

$$v = \frac{180}{12}$$

$$= 15 \text{ km/h} \qquad \text{Ans.}$$

9.

$$\begin{aligned}
\text{Angular acceleration} &= 230 \text{ rev/min in } 10 \text{ s} \\
&= \frac{230 \times 2\pi}{60 \times 10} \text{ rad/s}^2 \\
\text{Accelerating torque} &= I\alpha \\
T &= mk^2\alpha \\
&= \frac{544 \times 0{\cdot}5^2 \times 230 \times 2\pi}{60 \times 10} \\
&= 327{\cdot}5 \text{ Nm} \qquad \text{Ans.}
\end{aligned}$$

10.

$$\begin{aligned}
\text{Retarding torque} &= I\alpha = mk^2\alpha \\
\alpha &= \frac{T}{mk^2} \\
&= \frac{24}{68 \times 0{\cdot}229^2} \\
&= 6{\cdot}732 \text{ rad/s}^2 \qquad \text{Ans. } (i) \\
\text{Change of angular velocity} &= \frac{2800 \times 2\pi}{60} \text{ rad/s} \\
\text{Time} &= \frac{\text{change of velocity}}{\text{retardation}} \\
&= \frac{2800 \times 2\pi}{60 \times 6{\cdot}732} \\
&= 43{\cdot}55 \text{ s} \qquad \text{Ans. } (ii)
\end{aligned}$$

11.

$$\begin{aligned}
I &= mk^2 \\
&= 1000 \times 1{\cdot}5^2 \\
&= 2250 \; kgm^2 \\
\theta &= 2{\cdot}1 - 3{\cdot}2\text{t} + 4{\cdot}8\text{t}^2 \\
\omega = \frac{d\theta}{dt} &= -3{\cdot}2 + 9{\cdot}6t \\
\alpha = \frac{d\omega}{dt} = \frac{d^2\theta}{dt^2} &= 9{\cdot}6 \;\; i.e. \text{ constant} \\
T &= I\alpha \\
&= 2250 \times 9{\cdot}6 \\
\text{Accelerating torque} &= 21\,600 \text{ Nm at } 1{\cdot}5 \text{ s} \qquad \text{Ans}
\end{aligned}$$

12.

Total piston force = effective pressure × area of piston
$= 6 \times 10^5 \times 0{\cdot}7854 \times 0{\cdot}25^2$
$= 29\,450$ N

Weight of reciprocating parts $= 317{\cdot}5 \times 9{\cdot}81$
$= 3115$ N

Total downward force $= 29\,450 + 3115$
$= 32\,565$ N

Force to accelerate = mass × acceleration
$= 317{\cdot}5 \times 21$
$= 6668$ N

The accelerating force is used to accelerate the parts, the remainder of the total force is the thrust transmitted to the cross-head.

Thrust on crosshead $= 32\,565 - 6668$
$= 25\,897$ N
$= 25{\cdot}897$ kN Ans.

13.

Total mass accelerated $= 1{\cdot}9 + 1{\cdot}8 = 3{\cdot}7$ kg

Accelerating force $= (1{\cdot}9 - 1{\cdot}8) \times 9{\cdot}81$
$= 0{\cdot}981$ N

$$\text{Acceleration} = \frac{\text{accelerating force}}{\text{mass}}$$
$$= \frac{0{\cdot}981}{3{\cdot}7}$$
$= 0{\cdot}2652$ m/s^2 Ans. (*i*)

Velocity after 4 s $= 0{\cdot}2652 \times 4$
$= 1{\cdot}0608$ m/s Ans. (*ii*)

Distance = average velocity × time
$= \frac{1}{2}(0 + 1{\cdot}0608) \times 4$
$= 2{\cdot}1216$ m Ans. (*iii*)

Considering side carrying the 1·8 kg

Tension in cord = force supporting 1·8 kg + force accelerating 1·8 kg

$= 1{\cdot}8 \times 9{\cdot}81 + 1{\cdot}8 \times 0{\cdot}2652$

$= 17{\cdot}66 + 0{\cdot}4773$

$= 18{\cdot}1373$ N Ans. (*iv*)

The total load on the bearings, neglecting weight of the pulley, is the pull on one side of the cord plus the pull on the other side. These of course are of equal values because the tension in the cord is the same throughout its length, hence,

Load on bearings = $2 \times$ tension in cord

$= 2 \times 18{\cdot}1373$

$= 36{\cdot}2746$ N Ans. (*v*)

14.

Total mass accelerated $= 9 + 0{\cdot}9 + 0{\cdot}45 = 10{\cdot}35$ kg

Accelerating force $= 0{\cdot}45 \times 9{\cdot}81$ N

$$\text{Acceleration} = \frac{\text{accelerating force}}{\text{mass}}$$

$$= \frac{0{\cdot}45 \times 9{\cdot}81}{10{\cdot}35}$$

$= 0{\cdot}4266\ \text{m/s}^2$ Ans. (*i*)

Velocity after 2·5 s = acceleration $\times$ time

$= 0{\cdot}4266 \times 2{\cdot}5$

$= 1{\cdot}066$ m/s Ans. (*ii*)

Distance = average velocity $\times$ time

$= \frac{1}{2}(0 + 1{\cdot}066) \times 2{\cdot}5$

$= 1{\cdot}332$ m Ans. (*iii*)

Considering horizontal part of cord pulling the 9 kg block along:

Pull in cord = friction force + force to accelerate

$= 0{\cdot}9 \times 9{\cdot}81 + 9 \times 0{\cdot}4266$

$= 8{\cdot}829 + 3{\cdot}839$

$= 12{\cdot}668$ N Ans. (*iv*)

15.

The acceleration of the rope is $0{\cdot}6\ \text{m/s}^2$ and this is the linear acceleration of the winding surface of the drum.

$$\text{Radius of drum} = 0{\cdot}75\ \text{m}$$

$$\text{Angular acceleration of drum} = \frac{\text{linear acceleration}}{\text{radius}}$$

$$= \frac{0{\cdot}6}{0{\cdot}75} = 0{\cdot}8\ \text{rad/s}^2$$

The total torque required is the sum of:

(i) torque to overcome friction,
(ii) " " support the load,
(iii) " " accelerate the load,
(iv) " " accelerate the rotating parts.

$$\text{Torque to overcome friction} = 190\ \text{N m} \quad \ldots\ \ldots\ \ldots \quad (i)$$

$$\text{Torque to support load} = \text{load} \times \text{radius of drum}$$

$$= 450 \times 9{\cdot}81 \times 0{\cdot}75$$

$$= 3311\ \text{N m} \quad \ldots\ \ldots\ \ldots \quad (ii)$$

$$\text{Torque to accelerate load} = \text{accelerating force} \times \text{radius of drum}$$

$$= \text{mass} \times \text{acceleration} \times \text{radius}$$

$$= 450 \times 0{\cdot}6 \times 0{\cdot}75$$

$$= 202{\cdot}5\ \text{N m} \quad \ldots\ \ldots\ \ldots \quad (iii)$$

Torque to accelerate rotating parts

$$= mk^2\alpha$$

$$= 1225 \times 0{\cdot}53^2 \times 0{\cdot}8$$

$$= 275{\cdot}4\ \text{Nm} \quad \ldots\ \ldots\ \ldots \quad (iv)$$

$$\text{Total torque} = 190 + 331 + 202{\cdot}5 + 275{\cdot}4$$

$$= 3978{\cdot}9\ \text{N m} \qquad \text{Ans.}$$

SOLUTIONS TO TEST EXAMPLES 4

1.

When compressed 50 mm,

Force on spring = 88 × 50 = 4400 N

When compressed 80 mm,

Force on spring = 88 × 80 = 7040 N

Average force to compress spring from 50 to 80 mm,

= ½ (4400 + 7040) = 5720 N

Work done = average force × distance

= 5720 × 0·03

= 171·6 J Ans.

2.

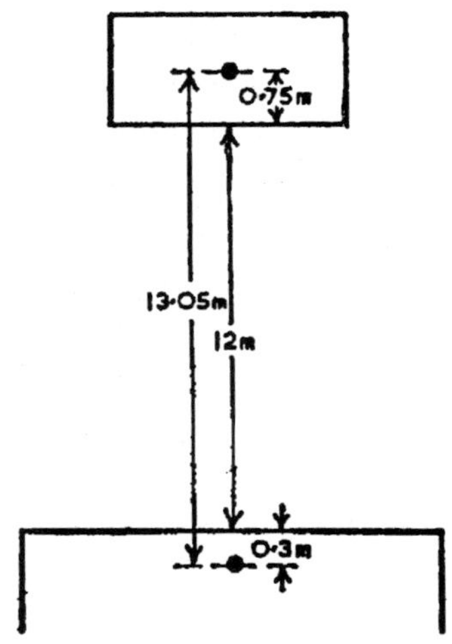

Volume of deck tank = 3·6 × 3 × 1·5 = 16·2m^3

∴ Mass of water moved = 16·2 t Ans. (*i*)

Depth of water taken out of double bottom-tank

$$= \frac{\text{volume}}{\text{length} \times \text{breadth}}$$

$$= \frac{16{\cdot}2}{6 \times 4{\cdot}5} = 0{\cdot}6 \text{ m} \qquad \text{Ans. } (ii)$$

Height through which centre of gravity of this water is raised:

$$= \tfrac{1}{2} \times 0{\cdot}6 + 12 + \tfrac{1}{2} \times 1{\cdot}5 = 13{\cdot}05 \text{ m}$$

$$\text{Work done} = \text{weight} \times \text{height}$$

$$= 16{\cdot}2 \times 10^3 \times 9{\cdot}81 \times 13{\cdot}05$$

$$= 2073 \times 10^3 \text{ J}$$

$$= 2073 \text{ kJ} \qquad \text{Ans. } (iii)$$

3.

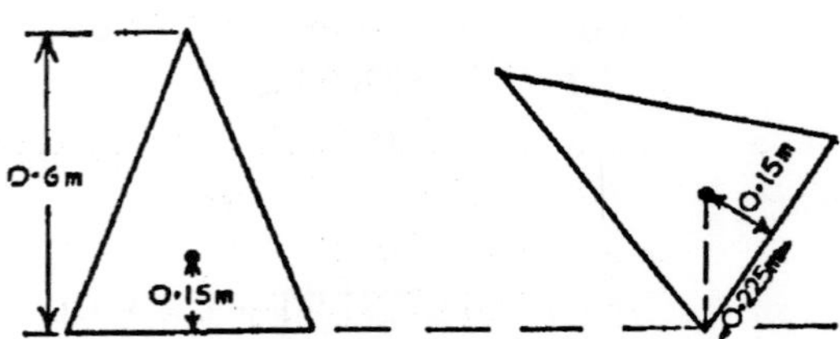

$$\text{Volume of cone} = \tfrac{1}{3} (\text{area of base} \times \text{perp. height})$$

$$\text{Density} = 7{\cdot}86 \times 10^3 \text{ kg/m}^3$$

$$\text{Weight of cone} = \tfrac{1}{3} \times \pi \times 0{\cdot}225^2 \times 0{\cdot}6 \times 7{\cdot}86 \times 10^3 \times 9{\cdot}81$$

$$= 2453 \text{ N}$$

$$\text{Height of C.G. at first} = \tfrac{1}{4} \times 0{\cdot}6 = 0{\cdot}15 \text{ m}$$

Height of C.G. when tilted until the centre of gravity is vertically above point of contact of the ground

$$= \sqrt{0{\cdot}225^2 + 0{\cdot}15^2} = 0{\cdot}2705 \text{ m}$$

Height C.G. is raised

$$= 0{\cdot}2705 - 0{\cdot}15 = 0{\cdot}1205 \text{ m}$$

$$\text{Work done} = \text{weight} \times \text{vertical height C.G. is raised}$$

$$= 2453 \times 0{\cdot}1205$$

$$= 295{\cdot}6 \text{ J} \qquad \text{Ans.}$$

4.

$$\text{Diameter of drum} = 0{\cdot}38 \text{ m}$$

$$\text{Angular velocity} = 40 \div 60 = \tfrac{2}{3} \text{ rev/s}$$

$$\text{Linear velocity of rope} = \tfrac{2}{3} \times \pi \times 0{\cdot}38 \text{ m/s}$$

$$\text{Power} = \text{weight} \times \text{height lifted per second}$$

$$= 544 \times 9{\cdot}81 \times \tfrac{2}{3} \times \pi \times 0{\cdot}38$$

$$= 4247 \text{ W} = 4{\cdot}247 \text{ kW}$$

$$\text{Input power} = \frac{4{\cdot}247}{0{\cdot}7}$$

$$= 6{\cdot}067 \text{ kW} \qquad \text{Ans.}$$

5.

$$\dot{m} = \dot{V}\rho$$

$$= 0{\cdot}7854 \times 0{\cdot}05^2 \times 1{\cdot}5 \times 1024$$

$$= 3{\cdot}016 \text{ kg/s.}$$

Taking into account the translational K.E. imparted to the water due to its velocity at discharge,

$$\text{Power} = mgh + \tfrac{1}{2} mv^2$$

$$= 3{\cdot}016 \times 9{\cdot}81 \times 9 + \tfrac{1}{2} \times 3{\cdot}016 \times 1{\cdot}5^2$$

$$= 266{\cdot}3 + 3{\cdot}4$$

$$= 269{\cdot}7 \text{ W}$$

$$\therefore \text{Input power} = \frac{269{\cdot}7}{0{\cdot}6}$$

$$= 449{\cdot}5 \text{ W} \qquad \text{Ans.}$$

6.

$$\text{Supply pressure} = 80 \times 10^5 \text{ N/m}^2 = 80 \times 10^2 \text{ kN/m}^2$$

$$\text{Volume flow} = \frac{9000}{3600 \times 10^3} \text{ m}^3\text{/s}$$

$$\text{Power} = \text{pressure} \times \text{vol. flow}$$

$$= \frac{80 \times 10^2 \times 9000}{3600 \times 10^3}$$

$$= 20 \text{ kW} \qquad \text{Ans. } (i)$$

$$\text{Output power} = 0{\cdot}75 \times 20$$

$$= 15 \text{ kW} \qquad \text{Ans. } (ii)$$

7.

Effective diameter of pulley is from mid-thickness of belt on one side to mid-thickness of belt on other side.

$$= 500 + 8 = 508 \text{ mm} = 0{\cdot}508 \text{ m}$$

$$\text{Speed of belt} = \text{circumference} \times \text{rev/s}$$

$$= \pi \times 0{\cdot}508 \times \frac{450}{60}$$

$$\text{Power} = (F_1 - F_2) \times \text{speed of belt}$$

$$F_1 - F_2 = \frac{4{\cdot}5 \times 10^3 \times 60}{\pi \times 0{\cdot}508 \times 450}$$

$$= 375{\cdot}9 \text{ N}$$

$$\text{but, } F_1 = 2{\cdot}25 \times F_2$$

$$2{\cdot}25\, F_2 - F_2 = 375{\cdot}9$$

$$1{\cdot}25\, F_2 = 375{\cdot}9$$

$$F_2 = \frac{375{\cdot}9}{1{\cdot}25}$$

$$F_1 = \frac{2{\cdot}25 \times 375{\cdot}9}{1{\cdot}25}$$

$$= 676{\cdot}6 \text{ N}$$

Allowing 7 N per mm width of belt,

$$\text{width} = \frac{676{\cdot}6}{7}$$

$$= 96{\cdot}66 \text{ mm} \qquad \text{Ans.}$$

8.

$$36 \text{ km/h} = \frac{36 \times 10^3}{3600} = 10 \text{ m/s}$$

$$\text{Translational kinetic energy} = \tfrac{1}{2}mv^2$$

$$= \tfrac{1}{2} \times 240 \times 10^2$$

$$= 12 \times 10^3 \text{ J or } 12 \text{ kJ} \qquad \text{Ans. } (i)$$

Kinetic energy is proportional to $(\text{velocity})^2$

$$\therefore \text{Kinetic energy at 18 km/h} = 12 \times (^{18}/_{36})^2$$

$$= 12 \times \tfrac{1}{4} = 3 \text{ kJ}$$

$$\text{Change in kinetic energy} = 12 - 3$$

$$= 9\text{kJ} \qquad \text{Ans. } (ii)$$

9.

At 450 m/s,

Translational kinetic energy $= \frac{1}{2}mv^2$

$= \frac{1}{2} \times 0{\cdot}028 \times 450^2$

$= 2835$ J

At 250 m/s,

Translational kinetic energy $= \frac{1}{2} \times 0{\cdot}028 \times 250^2$

$= 875$ J

Change in kinetic energy $= 2835 - 875$

$= 1960$ J

$$\text{Average force} = \frac{\text{change in kinetic energy}}{\text{distance}}$$

$$= \frac{1960}{0{\cdot}1}$$

$= 19\,600$ N or $19{\cdot}6$ kN Ans. (*i*)

To bring the bullet to rest, change in kinetic energy is to be 2835 J

$$\text{Distance} = \frac{\text{change in kinetic energy}}{\text{average force}}$$

$$= \frac{2835}{19\,600}$$

$= 0{\cdot}1446$m $= 144{\cdot}6$ mm Ans. (*ii*)

10.

$$\frac{100 \text{ rev/min} \times 2\pi}{60} = 10{\cdot}47 \text{ rad/s}$$

Kinetic energy at 100 rev/min $= \frac{1}{2}mk^2\omega^2$

$= \frac{1}{2} \times 109 \times 0{\cdot}38^2 \times 10{\cdot}47^2$

Rotational K.E. $= 863$ J Ans. (*i*)

At 300 rev/min,

Rotational K.E. $= 863 \times (^{300}/_{100})^2$

$= 863 \times 9 = 7767$ J

or $7{\cdot}767$ kJ Ans. (*ii*)

11.

$$\text{Radius of gyration, } k = \frac{r}{\sqrt{2}}$$

$$k^2 = \frac{r^2}{2} = \frac{0{\cdot}7^2}{2} = 0{\cdot}245 \text{ m}^2$$

$$\begin{aligned}
\text{Change in K.E.} &= \tfrac{1}{2} mk^2\omega_1^{\,2} - \tfrac{1}{2} mk^2\omega_2^{\,2} \\
&= \tfrac{1}{2} mk^2 (\omega_1^{\,2} - \omega_2^{\,2}) \\
&= \tfrac{1}{2} mk^2 (\omega_1 + \omega_2)(\omega_1 - \omega_2) \\
&= \tfrac{1}{2} \times 1{\cdot}25 \times 10^3 \times 0{\cdot}245 \times 18\pi \times 2\pi \\
&= 54\,430 \text{ J} \\
&= 54{\cdot}43 \text{ kJ} \qquad \text{Ans.}
\end{aligned}$$

12.

$$\frac{200 \text{ rev/min} \times 2\pi}{60} = 20{\cdot}94 \text{ rad/s}$$

$$\text{Angular acceleration} = \frac{20{\cdot}94}{60} = 0{\cdot}349 \text{ rad/s}^2$$

$$\frac{180 \text{ rev/min} \times 2\pi}{60} = 18{\cdot}85 \text{ rad/s}$$

$$\begin{aligned}
\text{Accelerating torque} &= I\alpha \text{ or } mk^2\alpha \\
&= 1220 \times 0{\cdot}58^2 \times 0{\cdot}349 \\
&= 143{\cdot}2 \text{ N m} \qquad \text{Ans. } (i)
\end{aligned}$$

Change in kinetic energy from 200 to 180 rev/min

$$\begin{aligned}
&= \tfrac{1}{2} mk^2 (\omega_1^{\,2} - \omega_2^{\,2}) \\
&= \tfrac{1}{2} mk^2 (\omega_1 + \omega_2)(\omega_1 - \omega_2) \\
&= \tfrac{1}{2} \times 1220 \times 0{\cdot}58^2 \times 39{\cdot}79 \times 2{\cdot}09 \\
&= 17\,060 \text{ J}
\end{aligned}$$

$$\begin{aligned}
\text{Average force} &= \frac{\text{change in kinetic energy}}{\text{distance}} \\
&= \frac{17\,060}{0{\cdot}15} \\
&= 113\,700 \text{ N} = 113{\cdot}7 \text{ kN} \qquad \text{Ans. } (ii)
\end{aligned}$$

13.

$$\frac{150 \text{ rev/min} \times 2\pi}{60} = 5\pi \text{ rad/s}$$

Rotational kinetic energy $= \tfrac{1}{2}mk^2\omega^2$

$= \tfrac{1}{2} \times 2{\cdot}54 \times 10^3 \times 0{\cdot}686^2 \times (5\pi)^2$

$= 147{\cdot}4 \times 10^3$ J

$= 147{\cdot}4$ kJ Ans. (*i*)

Work = torque × angle turned = $T\theta$

Work lost to friction in one revolution (i.e. 2π rad)

$= 27 \times 2\pi = 169{\cdot}6$ J Ans. (*ii*)

No. of revs to come to rest $= \dfrac{\text{total change in energy}}{\text{change in energy per rev}}$

$= \dfrac{147{\cdot}4 \times 10^3}{169{\cdot}6}$

$= 869$ Ans. (*iii*)

Average velocity while coming to rest

$= \tfrac{1}{2}(150 + 0) = 75$ rev/min

Time $= \dfrac{\text{distance}}{\text{average velocity}}$

$= \dfrac{\text{total revs. turned}}{\text{average rev/min}}$

$= \dfrac{869}{75} = 11{\cdot}59$ min Ans. (*iv*)

14.

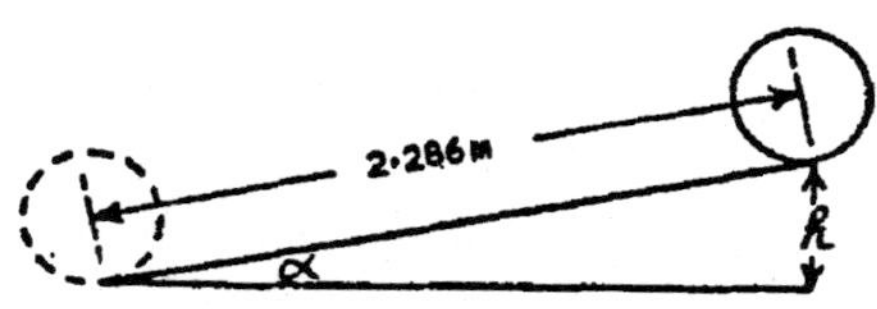

Average linear velocity down the incline

$= \dfrac{\text{distance}}{\text{time}} = \dfrac{2{\cdot}286}{3} = 0{\cdot}762$ m/s

Final linear velocity $= 2 \times 0{\cdot}762 = 1{\cdot}524$ m/s

When the roller loses height by rolling down the incline, it loses potential energy and gains an equal amount of kinetic energy. Note however, that the roller acquires angular velocity as well as linear velocity and therefore the total kinetic energy gained consists of kinetic energy of translation plus kinetic energy of rotation.

$$\text{Potential energy lost} = \text{Kinetic energy gained}$$

$$mgh = \tfrac{1}{2}mv^2 + \tfrac{1}{2}mk^2\omega^2$$

m cancels, $\omega^2 = \dfrac{v^2}{r^2}$ $\qquad k^2 = \dfrac{r^2}{2}$

$$9{\cdot}81 \times h = \tfrac{1}{2}\left\{ v^2 + \frac{r^2 \times v^2}{2 \times r^2} \right\}$$

$$9{\cdot}81 \times h = \tfrac{1}{2}v^2(1 + \tfrac{1}{2})$$

$$9{\cdot}81 \times h = 0{\cdot}75 \times 1{\cdot}524^2$$

$$h = 0{\cdot}1775 \text{ m}$$

$$\sin\alpha = \frac{0{\cdot}1775}{2{\cdot}286} = 0{\cdot}07764$$

$$\text{Angle of incline} = 4^\circ\ 27' \qquad \text{Ans.}$$

15.

$$\text{Mean speed} = \frac{2\pi \times 1000}{60} = 104{\cdot}7 \text{ rad/s}$$

$$\text{Max speed} = 104{\cdot}7 \times 1{\cdot}015 = 106{\cdot}3 \text{ rad/s}$$

$$\text{Min speed} = 104{\cdot}7 \times 0{\cdot}985 = 103{\cdot}1 \text{ rad/s}$$

$$\text{The fluctuations of energy} = 0{\cdot}9 \times \text{the work done per revolution}$$

$$= 0{\cdot}9 \times \frac{\text{work done/min}}{\text{rev/min}}$$

$$= 0{\cdot}9 \times \frac{10 \times 10^3 \times 60}{1000}$$

$$= 540\,J$$

$$\text{Also, fluctuation of energy} = I\omega\,(\omega_1 - \omega_2)$$

$$\therefore 540 = I \times 104{\cdot}7\,(106{\cdot}3 - 103{\cdot}1)$$

$$\therefore I = 1{\cdot}612 \text{ kg m}^2. \quad \text{Ans.}$$

SOLUTIONS TO TEST EXAMPLES 5

1. $$\frac{75 \text{ rev/min} \times 2\pi}{60} = 7{\cdot}854 \text{ rad/s}$$

$$\begin{aligned}
\text{Centrifugal force} &= m\omega^2 r \\
&= 1{\cdot}2 \times 7{\cdot}854^2 \times 0{\cdot}6 \\
&= 44{\cdot}42 \text{ N}
\end{aligned}$$

Maximum tension occurs when the mass is passing bottom dead centre,

$$\begin{aligned}
\text{Weight of mass} &= 1{\cdot}2 \times 9{\cdot}81 \\
&= 11{\cdot}77 \text{ N} \\
\text{Maximum tension} &= 44{\cdot}42 + 11{\cdot}77 \\
&= 56{\cdot}19 \text{ N} \quad \text{Ans. } (i)
\end{aligned}$$

Minimum tension occurs when the mass is passing top dead centre,

$$\begin{aligned}
\text{Centrifugal force} - \text{weight} &= 0 \\
\text{centrifugal force} &= \text{weight} = 11{\cdot}77 \text{ N} \\
\text{C.F.} &= m\omega^2 r \\
11{\cdot}77 &= 1{\cdot}2 \times \omega^2 \times 0{\cdot}6 \\
\omega &= \sqrt{\frac{11{\cdot}77}{1{\cdot}2 \times 0{\cdot}6}} = 4{\cdot}042 \text{ rad/s} \\
\frac{4{\cdot}042 \times 60}{2\pi} &= 38{\cdot}6 \text{ rev/min} \quad \text{Ans. } (ii)
\end{aligned}$$

2. The disc would balance about its centre before the hole was bored. Therefore after the hole is cut, the out-of-balance force is equal to the centrifugal force that would be caused if the cut-out piece of material was rotated at the given speed of 240 rev/min at the radius of 0·038 m from its centre of gravity to the centre of rotation.

$$\frac{240 \text{ rev/min} \times 2\pi}{60} = 8\pi \text{ rad/s}$$

$$\text{Centrifugal force} = m\omega^2 r$$
$$= 0{\cdot}8 \times (8\pi)^2 \times 0{\cdot}038$$
$$= 19{\cdot}21 \text{ N} \quad \text{Ans.}$$

3. $$\frac{48 \text{ km/h} \times 10^3}{3600} = \frac{40}{3} \text{ m/s}$$

For conditions of skidding,

$$\text{Friction force} = \text{Centrifugal force}$$
$$\mu \times \text{force between surfaces} = \frac{mv^2}{r}$$
$$\mu = \frac{m \times 40^2}{m \times 9{\cdot}81 \times 3^2 \times 30}$$
$$= 0{\cdot}604 \quad \text{Ans.}$$

4. Referring to Fig. 67, moments about 0,

$$\text{Centrifugal force} \times h = \text{weight} \times \tfrac{1}{2} \text{ track}$$
$$\frac{mv^2}{r} \times h = mg \times \tfrac{1}{2}\text{track}$$
$$v = \sqrt{\frac{9{\cdot}81 \times 0{\cdot}84 \times 23}{0{\cdot}98}}$$
$$= 13{\cdot}91 \text{ m/s}$$
$$\frac{13{\cdot}91 \times 3600}{10^3} = 50{\cdot}06 \text{ km/h} \quad \text{Ans.}$$

5. $$\frac{75 \text{ km/h} \times 10^3}{3600} = 20{\cdot}83 \text{ m/s}$$

$$\text{tan of angle of banking} = \frac{\text{centrifugal force}}{\text{weight}}$$
$$= \frac{m \times v^2}{r \times mg}$$
$$= \frac{20{\cdot}83^2}{250 \times 9{\cdot}81} = 0{\cdot}1769$$
$$\text{Angle} = 10^\circ\, 2'$$
$$\text{Super elevation} = \text{track} \times \sin 10^\circ\, 2'$$
$$= 1{\cdot}435 \times 0{\cdot}1742$$
$$= 0{\cdot}25 \text{ m} \quad \text{Ans.}$$

6.

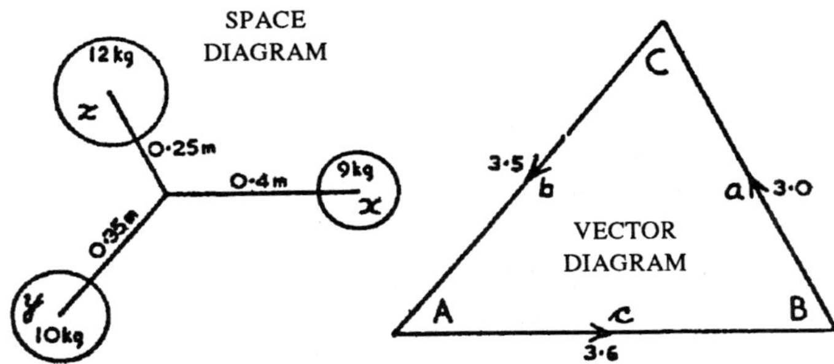

Centrifugal forces are represented by $m \times r$,

$$x = 9 \times 0{\cdot}4 = 3{\cdot}6$$
$$y = 10 \times 0{\cdot}35 = 3{\cdot}5$$
$$z = 12 \times 0{\cdot}25 = 3{\cdot}0$$

Referring to vector diagram:

$$\cos A = \frac{b^2 + c^2 - a^2}{2bc}$$
$$= \frac{3{\cdot}5^2 + 3{\cdot}6^2 - 3^2}{2 \times 3{\cdot}5 \times 3{\cdot}6}$$
$$= \frac{16{\cdot}21}{25{\cdot}2} = 0{\cdot}6433$$
$$\text{Angle } A = 49°58'$$
$$\frac{a}{\sin \text{A}} = \frac{b}{\sin \text{B}}$$
$$\sin \text{B} = \frac{3{\cdot}5 \times \sin 49°58'}{3} = 0{\cdot}8933$$
$$\text{Angle B} = 63°17'$$
$$\text{Angle C} = 180° - (49°58' + 63°17')$$
$$= 66°45'$$

$$\left.\begin{aligned} \text{Angle between } x \text{ and } y &= 180° - 49°58' = 130°2' \\ \text{Angle between } y \text{ and } z &= 180° - 66°45' = 113°15' \\ \text{Angle between } z \text{ and } x &= 180° - 63°17' = 116°43' \end{aligned}\right\} \text{Ans}$$

7.
$$\text{Stress} = \rho\omega^2 r^2 \text{ N/m}^2$$
$$15 \times 10^6 = 7{\cdot}21 \times 10^3 \times \omega^2 \times 0{\cdot}6^2$$
$$\omega = \sqrt{\frac{15 \times 10^6}{7{\cdot}21 \times 10^3 \times 0{\cdot}6^2}} = 76{\cdot}02 \text{ rad/s}$$
$$\frac{76{\cdot}02 \times 60}{2\pi} = 725{\cdot}7 \text{ rev/min} \quad \text{Ans.}$$

8.
$$\frac{60 \text{ rev/min} \times 2\pi}{60} = 2\pi \text{ rad/s}$$
$$\frac{80 \text{ rev/min} \times 2\pi}{60} = \frac{8\pi}{3} \text{ rad/s}$$
$$\text{Height} = \frac{g}{\omega^2}$$
$$h_1 = \frac{9{\cdot}81}{(2\pi)^2} = 0{\cdot}2485 \text{ m}$$
$$h_2 = \frac{9{\cdot}81 \times 3^2}{8^2 \times \pi^2} = 0{\cdot}1397 \text{ m}$$
$$\text{Change in height} = 0{\cdot}2485 - 0{\cdot}1397$$
$$= 0{\cdot}1088 \text{ m} = 108{\cdot}8\text{mm} \quad \text{Ans.}$$

9.
$$\frac{240 \text{ rev/min} \times 2\pi}{60} = 25{\cdot}13 \text{ rad/s} = \omega_1$$
$$\frac{270 \text{ rev/min} \times 2\pi}{60} = 28{\cdot}27 \text{ rad/s} = \omega_2$$
$$h = \frac{g}{\omega^2}\left\{\frac{m + M}{m}\right\} = \frac{g}{\omega^2}\left\{1 + \frac{M}{m}\right\}$$
$$h_1 - h_2 = g\left\{1 + \frac{M}{m}\right\}\left\{\frac{1}{\omega_1^2} - \frac{1}{\omega_2^2}\right\}$$

Note, $\frac{1}{\omega_1^2} - \frac{1}{\omega_2^2} = \frac{\omega_2^2 - \omega_1^2}{\omega_1^2 \times \omega_2^2} = \frac{(\omega_2 + \omega_1)(\omega_2 - \omega_1)}{\omega_1^2 \times \omega_2^2}$

$$\therefore 0{\cdot}025 = 9{\cdot}81\left\{1 + \frac{M}{2{\cdot}25}\right\} \times \frac{53{\cdot}4 \times 3{\cdot}14}{25{\cdot}13^2 \times 28{\cdot}27^2}$$
$$1 + \frac{M}{2{\cdot}25} = \frac{0{\cdot}025 \times 25{\cdot}13^2 \times 28{\cdot}27^2}{9{\cdot}81 \times 53{\cdot}4 \times 3{\cdot}14}$$

$$\frac{M}{2{\cdot}25} = 7{\cdot}67 - 1$$
$$M = 6{\cdot}67 \times 2{\cdot}25$$
$$= 15{\cdot}01 \text{ kg Ans.}$$

10. $$\frac{90 \text{ rev/min} \times 2\pi}{60} = 3\pi \text{ rad/s}$$

Amplitude = ½ travel = 0·025 m

Displacement from mid-travel = 0·025 – 0·006
= 0·019 m

$$\cos\theta = \frac{0{\cdot}019}{0{\cdot}025} = 0{\cdot}76$$
$$\theta = 40°32'$$

Velocity = $\omega r \sin\theta$
= $3\pi \times 0{\cdot}025 \times 0{\cdot}6498$
= 0·1531 m/s Ans. (*i*)

Acceleration = $\omega^2 r \cos\theta$
= $\omega^2 \times$ displacement
= $(3\pi)^2 \times 0{\cdot}019$
= 1·688 m/s² Ans. (*ii*)

11. $$\frac{120 \text{ rev/min} \times 2\pi}{60} = 4\pi \text{ rad/s}$$

r = ½ stroke = 0·55 m

Acceleration at beginning of stroke
= $\omega^2 r = (4\pi)^2 \times 0{\cdot}55$ m/s²

Accelerating force = mass × acceleration
= $1524 \times 4^2 \times \pi^2 \times 0{\cdot}55$
= $1{\cdot}323 \times 10^5$ N = 132·3 kN

Total downward force = press. on piston + weight of parts
= $800 + 1524 \times 9{\cdot}81 \times 10^{-3}$ kN
= 814·95 kN

Of this, 132·3 kN is required to accelerate the parts, the remainder being the effective thrust,

Effective thrust = 814·95 – 132·3
= 682·65 kN Ans.

12. (a) Time for one revolution of the cam $= \frac{1}{3}$ s

Time taken for the follower to lift $= \frac{1}{3} \times \frac{1}{3}$

$= \frac{1}{9}$ s

Periodic time for the follower $= 2 \times \frac{1}{9}$

i.e. $t = \frac{2}{9}$ s

$$\omega = \frac{2\pi}{\frac{2}{9}}.$$

$$\omega = 28{\cdot}3 \text{ rad/s}$$

Maximum acceleration occurs at each end of the travel,

where $x = 30$ mm

Max. acceleration $= \omega^2 x$

$= 28{\cdot}3^2 \times 0{\cdot}03$

$= 24{\cdot}03$ m/s^2. Ans. (a)

(b) The maximum force between cam and follower occurs at the bottom end of the followers travel,

where force on follower = force to overcome weight + force to accelerate

Maximum force $= mg + ma$

$= (2 \times 9{\cdot}81) + (2 \times 24{\cdot}03)$

$= 67{\cdot}68$ N. Ans. (b)

13. 120 osc/min = 2 osc/s

$\therefore$ periodic time $= \frac{1}{2}$ s

$$t = 2\pi\sqrt{\frac{l}{g}}$$

$$(\tfrac{1}{2})^2 = 2^2 \times \pi^2 \times \frac{1}{9{\cdot}81}$$

$$1 = \frac{9{\cdot}81}{2^2 \times 2^2 \times \pi^2}$$

$= 0{\cdot}06213$ m $= 62{\cdot}13$ mm Ans.

14.

$$\text{Spring stiffness} = 1500 \text{ N/m}$$

$$t = 2\pi \sqrt{\frac{m}{S}}$$

$$= 2\pi \sqrt{\frac{6{\cdot}5}{1500}}$$

$$= 0{\cdot}4136 \text{ s}$$

$$\text{Frequency} = \frac{1}{0{\cdot}4136}$$

$$= 2{\cdot}418 \text{ osc/s}$$

$$= 145 \text{ vib/min} \quad \text{Ans.}$$

15.

$$t = 2\pi \sqrt{\frac{m}{S}}$$

$$= 2\pi \sqrt{\frac{100}{25 \times 10^3}}$$

$$\text{Periodic time} = 0{\cdot}3974 \text{ s} \qquad \text{Ans. } (i)$$

$$\text{Since } t = \frac{2\pi}{\omega}$$

$$\omega = \frac{2\pi}{0{\cdot}3974}$$

$$\omega = 15{\cdot}81 \text{ rad/s}$$

$$\text{Amplitude } r = 0{\cdot}05 \text{ m}$$

$$\text{When displacement } x = 0{\cdot}03 \text{ m}$$

$$\text{Instantaneous velocity} = \omega \sqrt{r^2 - x^2}$$

$$= 15{\cdot}81 \sqrt{0{\cdot}05^2 - 0{\cdot}03^2}$$

$$= 0{\cdot}632 \text{ m/s}$$

$$\text{Instantaneous acceleration} = \omega^2 x$$

$$= 15{\cdot}81^2 \times 0{\cdot}03$$

$$= 7{\cdot}5 \text{ m/s}^2 \qquad \text{Ans. } (ii)$$

SOLUTIONS TO TEST EXAMPLES 6

1. Force to overcome friction $= \mu \times$ force between surfaces
$= 0{\cdot}32 \times 750$
$= 240$ N Ans. (*i*)

Work done $=$ force $\times$ distance
$= 240 \times 15$
$= 3600$ J or $3{\cdot}6$ kJ Ans. (*ii*)

2. Friction force $= \mu \times$ force between surfaces
$= 0{\cdot}04 \times 25 \times 10^3$
$= 1000$ N or 1 kN Ans. (*i*)

Work/rev $=$ force at skin of shaft $\times$ circumference
$= 1000 \times \pi \times 0{\cdot}1$
$= 314{\cdot}2$ J Ans. (*ii*)

Work/s $=$ work/rev $\times$ rev/s
$= 314{\cdot}2 \times 50$
$= 15\,710$ J/s or $15{\cdot}71$ kJ/s Ans. (*iii*)

Power $=$ work/s
$= 15{\cdot}71$ kW Ans. (*iv*)

3.

Force to overcome tractive resistance $= 53 \times 100 = 5300$ N

Force available for acceleration $= 25\,000 - 5300 = 19\,700$ N

$$\text{Acceleration} = \frac{\text{accelerating force}}{\text{mass}}$$

$$= \frac{19\,700}{100 \times 10^3} = 0{\cdot}197 \text{ m/s}^2$$

Velocity after 120 s $= 0{\cdot}197 \times 120$
$= 23{\cdot}64$ m/s

$$\frac{23{\cdot}64 \times 3600}{10^3} = 85{\cdot}1 \text{ km/h} \quad \text{Ans.}$$

4. $\tan \phi = \mu = 0{\cdot}25$

$\therefore$ friction angle $\phi = 14°2'$

Referring to Fig. 83,

Angle opposite $W = 90° - 14°2' + 30° = 105°58'$

$$\frac{F}{\sin 14°2'} = \frac{200}{\sin 105°58'}$$

$$F = \frac{200 \times 0{\cdot}2425}{0{\cdot}9615}$$

$$= 50{\cdot}44 \text{ N} \quad \text{Ans.}$$

5. $\tan \phi = \mu = 0{\cdot}15$

friction angle $\phi = 8°32'$

Referring to Fig. 84,

Least force $= W \sin \phi$

$= 400 \times \sin 8°32'$

$= 59{\cdot}36$ N Ans. (*i*)

Direction of least force $= 8°32'$ to the horizontal. Ans. (*ii*)

6. Force_{up} = force of gravity + friction force

$= W \sin \alpha + \mu W \cos \alpha$

$= 100 \times \sin 30° + 0{\cdot}2 \times 100 \times \cos 30°$

$= 50 + 17{\cdot}32$

$= 67{\cdot}32$ N Ans. (*i*)

Work done = force × distance

$= 67{\cdot}32 \times 5$

$= 336{\cdot}6$ J Ans. (*ii*)

Note that the same amount of work would be done in hauling the body along the horizontal and then hoisting it vertically upwards to get to the same position, as follows:

Horizontal distance $= 5 \times \cos 30° = 4{\cdot}33$ m

Horizontal force exerted $= \mu W = 0{\cdot}2 \times 100 = 20$ N

Work done on the level $= 20 \times 4{\cdot}33 = 86{\cdot}6$ J

Vertical distance lifted $= 5 \times \sin 30° = 2{\cdot}5$ m

Vertical force exerted $= 100$ N

Work done to lift body $= 100 \times 2{\cdot}5 = 250$ J

Total work done $= 86{\cdot}6 + 250$

$= 336{\cdot}6$ J (as before)

7. Referring to Fig. 86,

$$\tan\phi = \mu = 0{\cdot}2$$
$$\therefore \phi = 11°19'$$
$$90° - \phi = 78°41'$$
$$\frac{F}{\sin(\phi + \alpha)} = \frac{W}{\sin(90 - \phi)}$$
$$\frac{336{\cdot}6}{\sin(\phi + \alpha)} = \frac{500}{\sin 78°41'}$$
$$\sin(\phi + \alpha) = \frac{336{\cdot}6 \times 0{\cdot}9806}{500}$$
$$= 0{\cdot}6601$$
$$\text{Angle } (\phi + \alpha) = 41°19'$$
$$\alpha = 41°19' - 11°19'$$
$$= 30° \text{ Ans.}$$

8. Sine of angle of incline $= \frac{1}{4} = 0{\cdot}25$

$$\text{Angle} = 14°29'$$
$$\cos\alpha = 0{\cdot}9682$$
$$\text{Force}_{\text{down}} = \text{friction force} - \text{force of gravity}$$
$$= \mu W \cos\alpha - W \sin\alpha$$
$$= 0{\cdot}35 \times 700 \times 0{\cdot}9682 - 700 \times 0{\cdot}25$$
$$= 237{\cdot}2 - 175$$
$$= 62{\cdot}2 \text{ N Ans.}$$

9. Weight of mass $= 500 \times 9{\cdot}81 = 4905$ N

$$\text{Force of gravity} = W \sin\alpha$$
$$= 4905 \times \sin 18°12'$$
$$= 1532 \text{ N}$$
$$\text{Friction force} = \mu W \cos\alpha$$
$$= 0{\cdot}27 \times 4905 \times \cos 18°12'$$
$$= 1258 \text{ N}$$

Force of gravity acting down the plane exceeds the friction force opposing motion by:

$$1532 - 1258 = 274 \text{ N}$$

and this force produces acceleration,

$$\text{Acceleration} = \frac{\text{accelerating force}}{\text{mass}}$$

$$= \frac{274}{500}$$

$$= 0{\cdot}548 \text{ m/s}^2 \qquad \text{Ans. } (i)$$

$$\text{Change of velocity} = \text{acceleration} \times \text{time}$$

$$\text{Velocity after 6 s from rest} = 0{\cdot}548 \times 6$$

$$= 3{\cdot}288 \text{ m/s} \qquad \text{Ans. } (ii)$$

$$\text{Average velocity during the 6 s} = \tfrac{1}{2}(0 + 3{\cdot}288)$$

$$= 1{\cdot}644 \text{ m/s}$$

$$\text{Distance} = \text{average velocity} \times \text{time}$$

$$= 1{\cdot}644 \times 6$$

$$= 9{\cdot}864 \text{ m} \qquad \text{Ans. } (iii)$$

10. The question infers that if the two masses were not connected, the mass with the smaller coefficient of friction would run down the plane and the other with its higher coefficient of friction would remain at rest. The effect of connecting the two by a cord is therefore to prevent the former from sliding and to be just on the verge of pulling the latter down.

Regarding the 25 N body:

$$\text{Force}_{\text{hold}} = \text{force of gravity} - \text{friction force}$$

$$= W \sin \alpha - \mu W \cos \alpha$$

$$= 25 \sin \alpha - 0{\cdot}15 \times 25 \times \cos \alpha$$

$$= 25 \sin \alpha - 3{\cdot}75 \cos \alpha \;\ldots\ldots\; (i)$$

Regarding the 50 N body:

$$\text{Force}_{\text{down}} = \text{friction force} - \text{force of gravity}$$

$$= \mu W \cos \alpha - W \sin \alpha$$

$$= 0{\cdot}3 \times 50 \times \cos \alpha - 50 \sin \alpha$$

$$= 15 \cos \alpha - 50 \sin \alpha \;\ldots\ldots\; (ii)$$

When (*i*) is equal to (*ii*) is the critical angle of the plane when the system is just about to slide:

$$25 \sin \alpha - 3{\cdot}75 \cos \alpha = 15 \cos \alpha - 50 \sin \alpha$$

$$25 \sin \alpha + 50 \sin \alpha = 15 \cos \alpha + 3{\cdot}75 \cos \alpha$$

$$75 \sin \alpha = 18{\cdot}75 \cos \alpha$$

$$\frac{\sin \alpha}{\cos \alpha} = \frac{18{\cdot}75}{75}$$

$$\tan \alpha = 0{\cdot}25$$

$$\alpha = 14° \, 2' \quad \text{Ans. } (a)$$

The tension in the cord is the force to pull the 50 N body down the plane (which is equal to the force to hold the 25 N body to prevent it sliding down), substituting the value of the angle of the incline in (*ii*):

$$\text{Tension in cord} = 15 \cos 14° \, 2' - 50 \sin 14° \, 2'$$

$$= 14{\cdot}55 - 12{\cdot}13$$

$$= 2{\cdot}42 \text{ N} \quad \text{Ans. } (b)$$

11.

$$\text{Force}_{up} = \mu W \cos \alpha + W \sin \alpha$$

$$\text{Force}_{down} = \mu W \cos \alpha - W \sin \alpha$$

$$36 = 0{\cdot}4W \cos \alpha + W \sin \alpha \; \ldots \ldots \ldots (i)$$

$$2 = 0{\cdot}4W \cos \alpha - W \sin \alpha \; \ldots \ldots \ldots (ii)$$

$$38 = 0{\cdot}8W \cos \alpha \; \ldots \ldots \ldots (iii)$$

$$34 = 2W \sin \alpha \; \ldots \ldots \ldots (iv)$$

$$\frac{34}{38} = \frac{2W \sin \alpha}{0{\cdot}8W \cos \alpha}$$

$$\tan \alpha = \frac{34 \times 0{\cdot}8}{38 \times 2}$$

$$= 0{\cdot}3580$$

$$\alpha = 19°42' \quad \text{Ans. } (a)$$

Substituting value of angle into (*iv*),

$$2 \times W \times \sin 19°42' = 34$$

$$W = \frac{34}{2 \times 0{\cdot}3371}$$

$$= 50{\cdot}43 \text{ N} \quad \text{Ans. } (b)$$

12.

$$\tan\phi = \mu = 0{\cdot}24$$

$$\text{friction angle } \phi = 13°30'$$

$$\phi + \alpha = 13°30' + 15° = 28°30'$$

$$\text{Least force} = W \sin(\phi + \alpha)$$

$$= 224 \times \sin 28°30'$$

$$\left.\begin{aligned} &= 106{\cdot}9 \text{ N} \\ \text{Direction} &= 13°30' \text{ to plane} \end{aligned}\right\} \text{Ans. } (i)$$

$$\text{Horizontal force} = W \tan(\phi + \alpha)$$

$$= 224 \times \tan 28°30'$$

$$= 121{\cdot}6 \quad \text{Ans. } (ii)$$

13. The taper is equally divided between the two sides, therefore on each side the taper is ½ over a length of 10.

$$\text{Tan of angle of incline } \frac{\frac{1}{2}}{10} = 0{\cdot}05$$

$$\alpha = 2°52'$$

$$\tan\phi = \mu = 0{\cdot}18,$$

$$\phi = 10°12'$$

$$\phi + \alpha = 10°12' + 2°52' = 13°4'$$

$$\phi - \alpha = 10°12' - 2°52' = 7°20'$$

$$\text{Force to drive cotter in} = 2W \tan(\phi + \alpha)$$

$$500 = 2 \times W \times \tan 13°4'$$

$$W = \frac{500}{2 \times 0{\cdot}2321}$$

$$= 1077 \text{ N} \quad \text{Ans. } (i)$$

$$\text{Force to drive cotter out} = 2W \tan(\phi - \alpha)$$

$$= 2 \times 1077 \times \tan 7°20'$$

$$= 277{\cdot}3 \text{ N} \quad \text{Ans. } (ii)$$

14.

$$\text{Tan of angle of incline} = \frac{1}{40} = 0{\cdot}025$$

$$\alpha = 1°26'$$

$$\tan\phi = \mu = 0{\cdot}2,$$

$$\phi = 11°19'$$

$$\text{Force to drive wedge in} = W\tan(\phi+\alpha)+\mu W$$
$$= W\tan 12°45' + 0{\cdot}2W$$
$$= W\ (0{\cdot}2263 + 0{\cdot}2)$$
$$1{\cdot}5 = 0{\cdot}4263W$$
$$W = \frac{1{\cdot}5}{0{\cdot}4263}$$
$$= 3{\cdot}519\ \text{kN} \quad \text{Ans.}$$

15.

$$\text{Tan}\ \phi = \mu = 0{\cdot}15$$
$$\text{Angle}\ \phi = 8{\cdot}53°$$

$$\text{Mean circumference} = \pi d$$
$$= \pi \times 50$$
$$= 157{\cdot}1\ \text{mm}$$

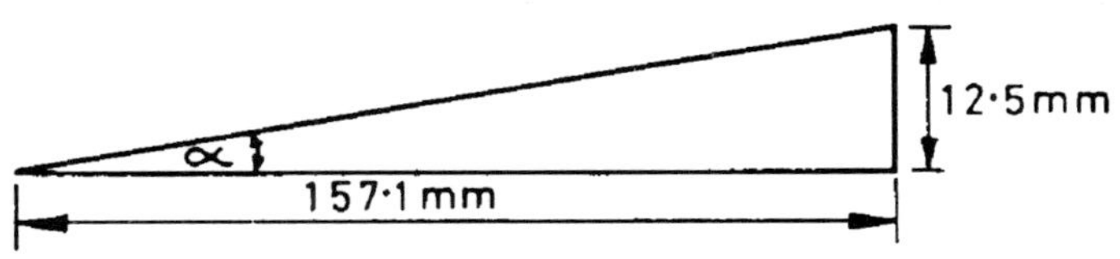

$$\text{Tan}\ \alpha = \frac{12{\cdot}5}{157{\cdot}1}$$
$$\text{Angle}\ \alpha = 4{\cdot}55°$$
$$\text{Efficiency} = \frac{\tan\alpha}{\tan(\phi+\alpha)} \times 100$$
$$= \frac{\tan 4{\cdot}55°}{\tan 13{\cdot}08°} \times 100$$
$$= 34{\cdot}2\%. \qquad \text{Ans.}$$

$$\text{Force to raise load} = W \times \tan(\phi+\alpha)$$
$$= 4{\cdot}5 \times \tan 13{\cdot}08°$$
$$= 1{\cdot}045\ \text{kN}$$
$$\text{Torque} = \text{force} \times \text{mean radius}$$
$$= 1{\cdot}045 \times 10^3 \times 0{\cdot}025$$
$$= 26{\cdot}14\ \text{Nm}. \qquad \text{Ans.}$$

SOLUTIONS TO TEST EXAMPLES 7

1.

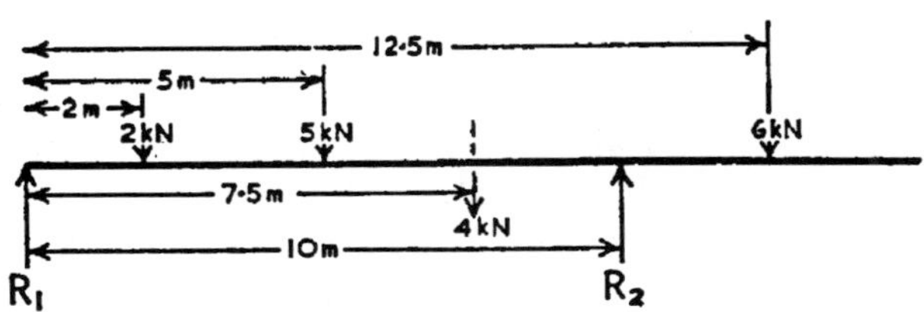

Being uniform in section, the centre of gravity of the beam is at its mid-length, which is 7·5 m from the end.

Moments about R_1,

$$\text{Clockwise moments} = \text{Anticlockwise moments}$$

$$(2 \times 2) + (5 \times 5) + (4 \times 7{\cdot}5) + (6 \times 12{\cdot}5) = R_2 \times 10$$

$$4 + 25 + 30 + 75 = R_2 \times 10$$

$$134 = R_2 \times 10$$

$$R_2 = 13{\cdot}4 \text{ kN}$$

$$\text{Upward forces} = \text{Downward forces}$$

$$R_1 + 13{\cdot}4 = 2 + 5 + 4 + 6$$

$$R_1 = 17 - 13{\cdot}4$$

$$= 3{\cdot}6 \text{ kN}$$

Reactions are 3·6 and 13·4 kN Ans.

2.

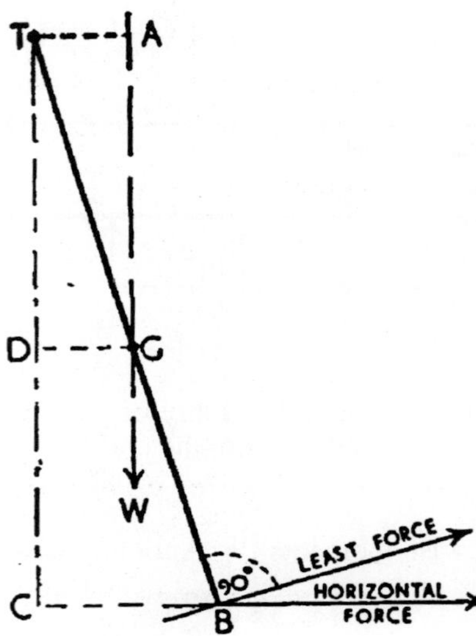

Taking moments about T. The least force is at the maximum perpendicular distance, that is, when the line of action of the force is at right angles to the rod and the perpendicular distance is the length of the rod.

Perpendicular distance from line of action of weight W to the top T is TA, and TA = DG.

By similar triangles,

$$\frac{DG}{TG} = \frac{CB}{TB}$$

$$DG = \frac{1{\cdot}17 \times 0{\cdot}7}{2{\cdot}1} = 0{\cdot}39 \text{ m}$$

$$\text{Clockwise moments} = \text{Anticlockwise moments}$$

$$9{\cdot}24 \times 0{\cdot}39 = \text{Least force} \times 2{\cdot}1$$

$$\text{Least force} = \frac{9{\cdot}24 \times 0{\cdot}39}{2{\cdot}1}$$

$$= 171{\cdot}6 \text{ kN} \quad \text{Ans. } (i)$$

When the force is applied horizontally, the perpendicular distance from the line of action of the force to the top end is TC.

$$\text{TC} = \sqrt{2{\cdot}1^2 - 0{\cdot}7^2} = 1{\cdot}98 \text{ m}$$

Clockwise moments = Anticlockwise moments

$$9{\cdot}24 \times 0{\cdot}39 = \text{Horizontal force} \times 1{\cdot}98$$

$$\text{Horizontal force} = \frac{9{\cdot}24 \times 0{\cdot}39}{1{\cdot}98}$$

$$= 1{\cdot}82 \text{ kN} \quad \text{Ans. } (ii)$$

3.

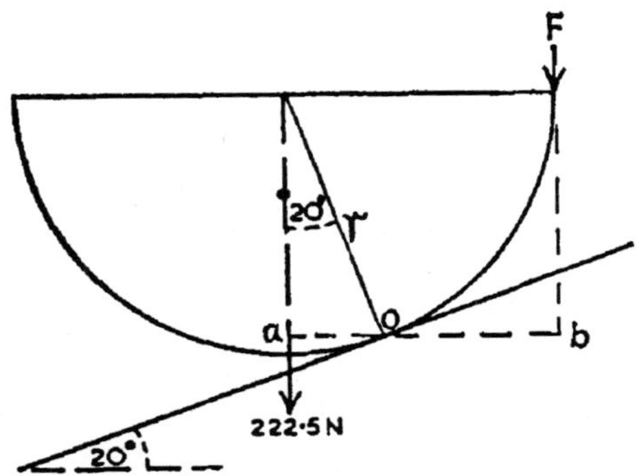

Perpendicular distance of weight of hemisphere from point of contact $o = ao = r \sin 20° = 0{\cdot}342r$.

Perpendicular distance of applied force F from point of contact $= bo = r - r \sin 20° = 0{\cdot}658r$.

Weight of hemisphere = 22·68 × 9·81 = 222·5 N

Moments about o,

Clockwise moments = Anticlockwise moments

$$F \times 0{\cdot}658r = 222{\cdot}5 \times 0{\cdot}342r$$

$$F = \frac{222{\cdot}5 \times 0{\cdot}342}{0{\cdot}658}$$

$$= 115{\cdot}6 \text{ N} \quad \text{Ans.}$$

4. Working in cm, and taking moments about the centre of sheave arranged vertically:

$$\bar{y} = \frac{\Sigma \text{ moments of areas}}{\Sigma \text{ areas}}$$

$$= \frac{0{\cdot}7854 \times 50^2 \times 0 - 0{\cdot}7854 \times 12{\cdot}5^2 \times 10}{0{\cdot}7854 \times 50^2 - 0{\cdot}7854 \times 12{\cdot}5^2}$$

$$\bar{y} = \frac{0 - 10}{4^2 - 1} = \frac{-10}{15} = -0{\cdot}667\text{cm}$$

The minus sign means that the centre of gravity is in the opposite direction to the hole, from the centre of the sheave,

$\therefore$ CG = 6·67 mm from centre. Ans.

Alternatively, if moments are taken about the thin edge of the sheave,

$$\bar{y} \text{ from edge} = \frac{0{\cdot}7854 \times 50^2 \times 25 - 0{\cdot}7854 \times 12{\cdot}5^2 \times 15}{0{\cdot}7854 \times 50^2 - 0{\cdot}7854 \times 12{\cdot}5^2}$$

$$= \frac{4^2 \times 25 - 15}{4^2 - 1} = \frac{385}{15}$$

= 25·667 cm from edge

= 6·67 mm from centre

5.

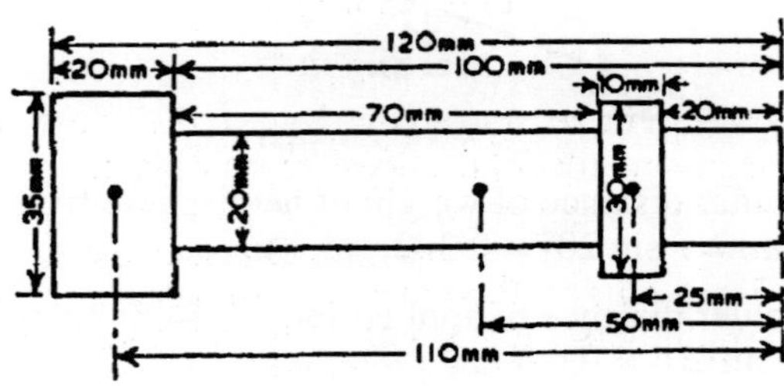

Length of shank = 120 – 20 = 100 mm

C.G. of shank from end = 50 mm

C.G. of head from end = 100 + 10 = 110 mm

C.G. of ring from end = 20 + 5 = 25 mm

$$\bar{x} = \frac{\Sigma \text{ moments of weights}}{\Sigma \text{ weights}}$$

The conversion factor from grammes mass to newtons weight, and 0.7854, are common to each term and cancel out.

Working in cm and taking moments about end of shank:

$$\bar{x} = \frac{3{\cdot}5^2 \times 2 \times 7{\cdot}86 \times 11 + 2^2 \times 10 \times 7{\cdot}86 \times 5 + (3^2 - 2^2) \times 1 \times 8{\cdot}4 \times 2{\cdot}5}{3{\cdot}5^2 \times 2 \times 7{\cdot}86 + 2^2 \times 10 \times 7{\cdot}86 + (3^2 - 2^2) \times 1 \times 8{\cdot}4}$$

$$= \frac{2118 + 1572 + 105}{192{\cdot}6 + 314{\cdot}4 + 42} = \frac{3795}{549}$$

$= 6{\cdot}911$ cm $= 69{\cdot}11$ mm from end. Ans.

6.

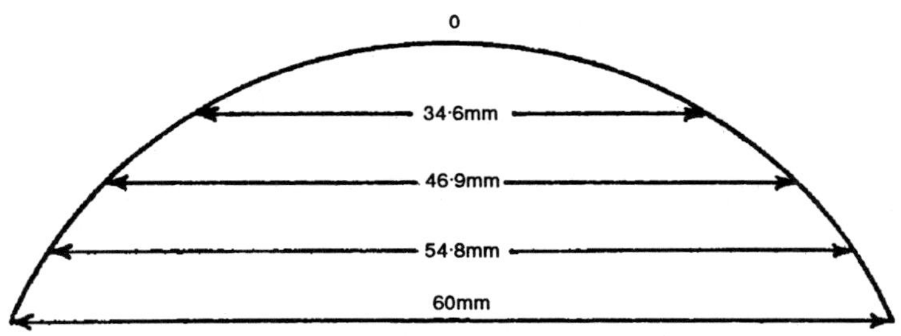

By crossed chords, AO × BO=CO × DO

$$20 \times BO = 30 \times 30$$

$$BO = 45 \text{ mm}$$

$$\therefore \text{ Diameter} = 45 + 20 = 65 \text{ mm}$$

Drawing the segment to scale (the larger the more accurate), dividing the segment into four slices, measuring the ordinates as shown and applying Simpson's rule:

Ordinates	Simpson's Multipliers	Products for Areas	Distance of Ordinates from Chord	Product for Moment
60	1	60	0	0
54·8	4	219·2	5	1096
46·9	2	93·8	10	938
34·6	4	138·4	15	2076
0	1	0	20	0
		Sum = 511·4		Sum = 4110

Area $= 511{\cdot}4 \times \frac{1}{3} \times 5 = 852{\cdot}3$ mm². Ans. (*i*)

C.G. $= \dfrac{4110}{511{\cdot}4} = 8{\cdot}037$ mm from chord. Ans. (*ii*)

7. Working in cm:

Area of rectangle ABED	$= 9 \times 4$	$= 36$ cm^2
C.G. " "	from AB	$= 2$ cm
C.G. " "	BC	$= 4{\cdot}5$ cm
Area of triangle DEC	$= \frac{1}{2} \times 9 \times 3$	$= 13{\cdot}5$ cm^2
C.G. " "	from AB	$= 4 + 1 = 5$ cm
C.G. " "	BC	$= 3$ cm

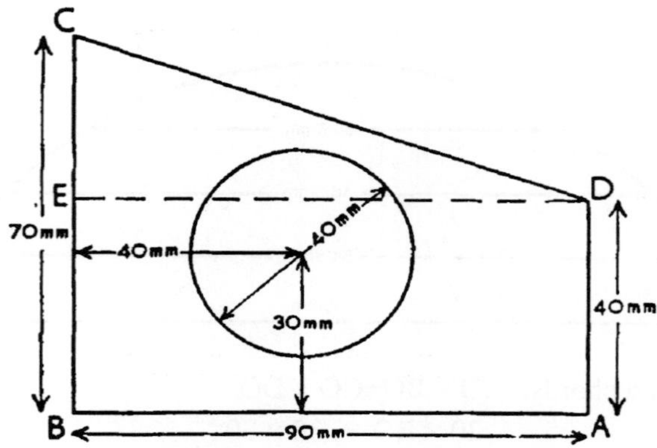

Area of hole $= 0{\cdot}7854 \times 4^2 = 12{\cdot}57$ cm^2

Net area of plate $= 36 + 13{\cdot}5 - 12{\cdot}57 = 36{\cdot}93$ cm^2

Moments about AB,

$$\bar{y} = \frac{\Sigma \text{ moments of areas}}{\Sigma \text{ areas}}$$

$$= \frac{36 \times 2 + 13{\cdot}5 \times 5 - 12{\cdot}57 \times 3}{36 + 13{\cdot}5 - 12{\cdot}57}$$

$$= \frac{101{\cdot}79}{36{\cdot}93} = 2{\cdot}756 \text{ cm}$$

$= 27{\cdot}56$ mm from AB. Ans. (*ia*)

Moments about BC,

$$\bar{x} = \frac{36 \times 4{\cdot}5 + 13{\cdot}5 \times 3 - 12{\cdot}57 \times 4}{36{\cdot}93}$$

$$= \frac{152{\cdot}22}{36{\cdot}93} = 4{\cdot}121 \text{ cm}$$

$= 41{\cdot}21$ mm from BC. Ans. (*ib*)

Moments of forces about AB,

$$\text{Upward moments} = \text{Downward moments}$$
$$\text{Force at C} \times 7 = 46{\cdot}6 \times 2{\cdot}756$$
$$\text{Force at C} = 18{\cdot}34 \text{ N. Ans. } (ii)$$

8.

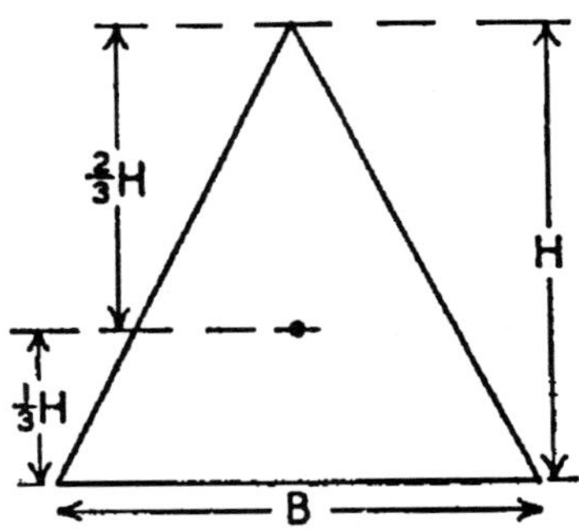

$$\text{From the rule } I_{xx} = I_{CG} + Ah^2$$

$$I_{CG} = I_{base} - Ah^2$$
$$= \frac{BH^3}{12} - \frac{BH}{2} \times \left\{\frac{H}{3}\right\}^2$$
$$= \frac{BH^3}{12} - \frac{BH^3}{18}$$
$$= \frac{BH^3}{36} \text{ Ans. } (i)$$

$$I_{apex} = I_{CG} - Ah^2$$
$$= \frac{BH^3}{36} + \frac{BH}{2} \times \left\{\frac{2H}{3}\right\}$$
$$= \frac{BH^3}{36} - \frac{2BH^3}{9}$$
$$= \frac{BH^3}{4} \text{ Ans. } (ii)$$

9.

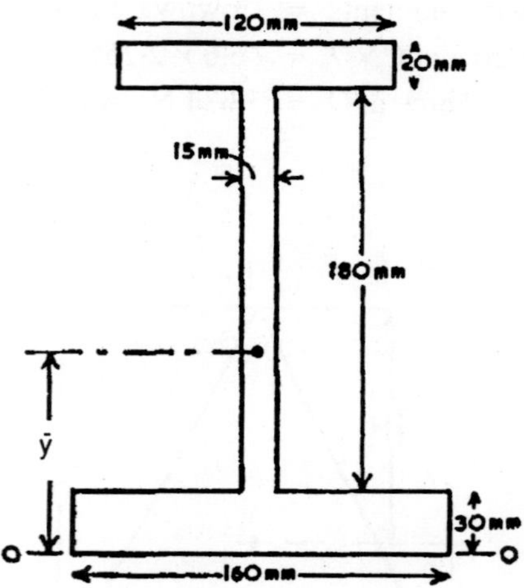

Working in cm:

I of top flange about its own C.G.,

$$= \frac{BD^3}{12} = \frac{12 \times 2^3}{12} = 8 \text{ cm}^4$$

$$\begin{aligned} I_{OO} \text{ of top flange} &= I_{CG} + Ah^2 \\ &= 8 + 12 \times 2 \times 22^2 \\ &= 11\,624 \text{ cm}^4 \quad \ldots \ldots \ldots \ldots \ldots (i) \end{aligned}$$

I of centre web about its own C.G.,

$$= \frac{BD^3}{12} = \frac{1{\cdot}5 \times 18^3}{12} = 729 \text{ cm}^4$$

$$\begin{aligned} I_{OO} \text{ of centre web} &= I_{CG} + Ah^2 \\ &= 729 + 1{\cdot}5 \times 18 \times 12^2 \\ &= 4617 \text{ cm}^4 \quad \ldots \ldots \ldots \ldots \ldots (ii) \end{aligned}$$

I of bottom flange about its own C.G.,

$$= \frac{BD^3}{12} = \frac{16 \times 3^3}{12} = 36 \text{ cm}^4$$

$$\begin{aligned} I_{OO} \text{ of bottom of flange} &= I_{CG} + Ah^2 \\ &= 36 + 16 \times 3 \times 1{\cdot}5^2 \\ &= 144 \text{ cm}^4 \quad \ldots \ldots \ldots \ldots \ldots (iii) \end{aligned}$$

I of whole section about base oo

$$= 11\,624 + 4617 + 144$$

$$= 16\,385 \text{ cm}^4. \quad \text{Ans. } (a)$$

$$\bar{y} = \frac{\Sigma \text{ moments of areas}}{\Sigma \text{ areas}}$$

$$= \frac{12 \times 2 \times 22 + 1{\cdot}5 \times 18 \times 12 + 16 \times 3 \times 1{\cdot}5}{12 \times 2 + 1{\cdot}5 \times 18 + 16 \times 3}$$

$$= \frac{924}{99} = 9\tfrac{1}{3} \text{ cm from base}$$

I of whole section about its own C.G.,

$$I_{CG} = I_{OO} - Ah^2$$

$$= 16\,385 - 99 \times (9\tfrac{1}{3})^2$$

$$= 16\,385 - 8624$$

$$= 7761 \text{ cm}^4. \quad \text{Ans. } (b)$$

10. Moments about axis through mid-depth

$$I \text{ of rectangle} = \frac{BD^3}{12}$$

$$I \text{ of square} = \frac{S \times S^3}{12} = \frac{S^4}{12}$$

Moments about axis through mid-breadth

$$I \text{ of rectangle} = \frac{DB^3}{12}$$

$$I \text{ of square} = \frac{S \times S^3}{12} = \frac{S^4}{12}$$

$$J = I_{xx} + I_{yy}$$

$$= \frac{S^4}{12} + \frac{S^4}{12}$$

$$= \frac{S^4}{6} \quad \text{Ans.}$$

SOLUTIONS TO TEST EXAMPLES 8

1.
$$\text{V.R.} = \frac{2R}{r_1 - r_2} = \frac{2 \times 240}{55 - 40} = 32 \quad \text{Ans. } (i)$$

$$\text{M.A.} = \frac{\text{Load}}{\text{Effort}} = \frac{1120}{80} = 14 \quad \text{Ans. } (ii)$$

$$\mathfrak{I} = \frac{\text{M.A.}}{\text{V.R.}} = \frac{14}{32} = 0{\cdot}4375 = 43{\cdot}75\% \quad \text{Ans. } (iii)$$

2.
$$\text{V.R.} = \frac{2 \times \text{no. of teeth in large pulley}}{\text{difference in no. of teeth in the two pulleys}}$$

$$= \frac{2 \times 27}{27 - 24} = 18$$

$$\text{M.A.} = \text{V.R.} \times \mathfrak{I}$$

$$= 18 \times 0{\cdot}35 = 6{\cdot}3$$

$$\text{Effort} = \frac{\text{Load}}{\text{M.A.}}$$

$$= \frac{1890}{6{\cdot}3} = 300 \text{ N} \quad \text{Ans.}$$

3.
$$\text{M.A.} = \frac{\text{Load}}{\text{Effort}} = \frac{560}{50} = 11{\cdot}2$$

$$\text{V.R.} = \frac{\text{M.A.}}{\mathfrak{I}} = \frac{11{\cdot}2}{0{\cdot}4} = 28$$

$$\text{V.R.} = \frac{2D}{D - d}$$

$$28 = \frac{2D}{D - 130}$$

$$28(D - 130) = 2D$$

$$28D - 3640 = 2D$$

$$26D = 3640$$

$$D = 140 \text{ mm} \quad \text{Ans.}$$

4. $$\text{V.R.} = \frac{2DN}{d} = \frac{2 \times 200 \times 40}{125} = 128$$

$$\text{M.A.} = \frac{W}{P} = \frac{6720}{150} = 44{\cdot}8$$

$$\mathfrak{I} = \frac{\text{M.A.}}{\text{V.R.}} = \frac{44{\cdot}8}{128} = 0{\cdot}35 \text{ or } 35\% \qquad \text{Ans. } (i)$$

$$\text{Ideal effort} = \frac{\text{Load}}{\text{V.R.}} = \frac{6720}{128} = 52{\cdot}5 \text{ N} \qquad \text{Ans. } (ii)$$

$$\text{Friction effort} = \text{Actual effort} - \text{Ideal effort}$$
$$= 150 - 52{\cdot}5$$
$$= 97{\cdot}5 \text{ N} \qquad \text{Ans } (iii)$$

5. If only the one toggle bar of 500 mm length was used:

$$\text{V.R.} = \frac{2\pi L}{\text{Pitch}} = \frac{2\pi \times 500}{12}$$

$$\text{M.A.} = \text{V.R.} \times \mathfrak{I} = \frac{2\pi \times 500 \times 0{\cdot}35}{12}$$

$$\text{Load lifted} = 3 \times 10^3 \times 9{\cdot}81 \text{ N}$$

$$\text{Effort} = \frac{\text{Load}}{\text{M.A.}}$$
$$= \frac{3 \times 10^3 \times 9{\cdot}81 \times 12}{2\pi \times 500 \times 0{\cdot}35}$$
$$= 321{\cdot}1 \text{ N}$$

The force actually applied on the end of the 500 mm toggle is 220 N which is 101·1 N less than that required to lift the load.

Therefore extra turning moment required

$$= 101{\cdot}1 \times 500 \text{ N mm}$$

If F = force required on the 450 mm toggle,

$$F \times 450 = 101{\cdot}1 \times 500$$
$$F = 112{\cdot}3 \text{ N} \qquad \text{Ans.}$$

6.
$$\text{V.R.} = \frac{\pi DN}{\text{Pitch}}$$
$$550 = \frac{\pi \times 100 \times \text{N}}{16}$$
$$N = \frac{550 \times 16}{\pi \times 100}$$
$$= 28 \text{ teeth} \qquad \text{Ans. }(i)$$
$$\text{M.A.} = \text{V.R.} \times \Im$$
$$= 550 \times 0{\cdot}3$$
$$\text{Effort} = \frac{\text{Load}}{\text{M.A.}}$$
$$= \frac{50 \times 10^3}{550 \times 0{\cdot}3}$$
$$= 303 \text{ N} \qquad \text{Ans. }(ii)$$

7.
$$\text{V.R.} = \frac{2\pi L}{\text{pitch}_R + \text{pitch}_L}$$
$$= \frac{2\pi \times 250}{4 + 4} = \frac{2\pi \times 250}{8}$$
$$\text{M.A.} = \text{V.R.} \times \Im$$
$$= \frac{2\pi \times 250 \times 0{\cdot}25}{8}$$
$$\text{Load} = \text{M.A.} \times \text{Effort}$$
$$= \frac{2\pi \times 250 \times 0{\cdot}25 \times 90}{8}$$
$$= 4{\cdot}418 \times 10^3 \text{ N} = 4{\cdot}418 \text{ kN} \qquad \text{Ans.}$$

8. Effective radius of lifting drum = radius of drum + radius of rope
$$= 100 + 12{\cdot}5 = 112{\cdot}5 \text{ mm}$$
$$\text{V.R.} = \frac{L}{r} \times \frac{\text{Product of teeth in followers}}{\text{Product of teeth in drivers}}$$
$$= \frac{300 \times 90 \times 125}{112{\cdot}5 \times 25 \times 30}$$
$$= 40 \qquad \text{Ans. }(i)$$
$$\text{M.A.} = \frac{\text{Load}}{\text{Effort}} = \frac{1000 \times 9{\cdot}81}{350}$$
$$\Im = \frac{\text{M.A.}}{\text{V.R.}} = \frac{1000 \times 9{\cdot}81}{350 \times 40}$$
$$= 0{\cdot}7007 \text{ or } 70{\cdot}07\% \qquad \text{Ans. }(ii)$$

9.

$$\text{V.R.} = \frac{\text{Area of load ram}}{\text{Area of effort plunger}} \times \text{leverage}$$
$$= \frac{0{\cdot}7854 \times 50^2}{0{\cdot}7854 \times 20^2} \times 18$$
$$= 112{\cdot}5$$
$$\text{M.A.} = \text{V.R.} \times \mathfrak{I}$$
$$= 112{\cdot}5 \times 0{\cdot}8 = 90$$
$$\text{Load} = \text{Effort} \times \text{M.A.}$$
$$= 120 \times 90$$
$$= 10\,800 \text{ N} = 10{\cdot}8 \text{ kN} \quad \text{Ans.}$$

10.

Calculating $\text{M.A.} = \dfrac{\text{Load}}{\text{Effort}}$ and $\% \mathfrak{I} = \dfrac{\text{M.A.}}{\text{V.R.}} \times 100$

and tabulating results:

W	7	14	21	28	35	42
P	3·5	4·6	5·7	6·7	7·7	8·8
M.A.	2	3·04	3·68	4·18	4·55	4·77
% $\mathfrak{I}$	16·7	25·3	30·7	34·8	37·8	39·88

The graph is plotted as shown:

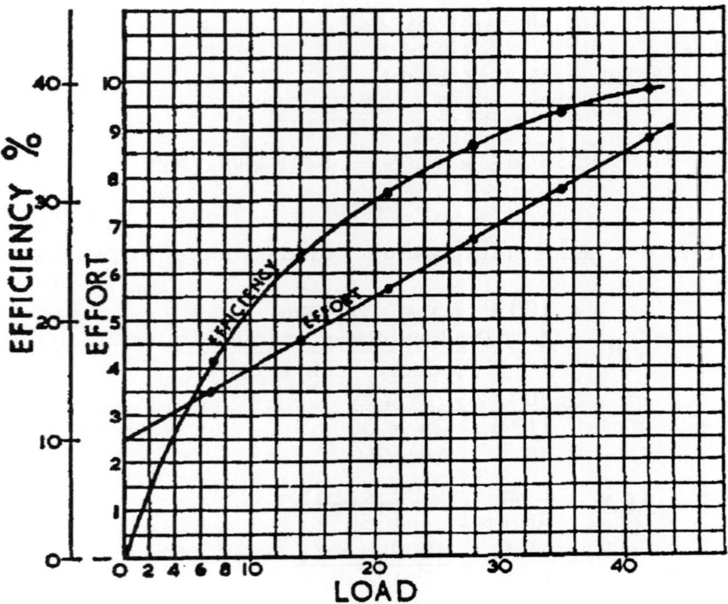

Choosing two points on the Effort line, say,

$$\text{when } W = 40, \; P = 8{\cdot}5$$

$$\text{and when } W = 10, \; P = 4$$

$$\text{Slope of line } b = \frac{8{\cdot}5 - 4}{40 - 10} = \frac{4{\cdot}5}{30} = 0{\cdot}15$$

Where effort line cuts zero load, $a = 2{\cdot}5$

Linear law is $P = 2{\cdot}5 + 0{\cdot}15\,W$ Ans. (*i*)

When effort is 7 N the load lifted is 30 N, reading from the load base line to the efficiency graph, efficiency reads 35·7% when the load lifted is 30 N.

Therefore,

When effort is 7 N, efficiency = 35·7% Ans. (*ii*)

SOLUTIONS TO TEST EXAMPLES 9

1.

$$\text{Working stress} = \frac{\text{tensile strength}}{\text{factor of safety}}$$

$$= \frac{462}{12} = 38{\cdot}5 \text{ MN/m}^2$$

Let d = diameter in mm:

$$\text{Area} = \frac{\text{load}}{\text{stress}}$$

$$0{\cdot}7854 \times d^2 = \frac{11{\cdot}12 \times 10^3}{38{\cdot}5}$$

$$d = \sqrt{\frac{11{\cdot}12 \times 10^3}{0{\cdot}7854 \times 38{\cdot}5}}$$

$$= 19{\cdot}17 \text{ mm} \quad \text{Ans.}$$

2.

$$\text{Safe load per stud} = \text{safe stress} \times \text{area}$$

$$= 35 \times 580{\cdot}2$$

$$= 2{\cdot}031 \times 10^4 = 20{\cdot}31 \text{ kN} \quad \text{Ans. } (i)$$

$$\text{Total load on cover} = \text{pressure} \times \text{area}$$

$$= 42 \times 10^5 \times 0{\cdot}7854 \times 0{\cdot}38^2 \text{ N}$$

$$\text{Number of studs} = \frac{\text{total load}}{\text{load per stud}}$$

$$= \frac{42 \times 10^5 \times 0{\cdot}7854 \times 0{\cdot}38^2}{20{\cdot}31 \times 10^3}$$

$$= 23{\cdot}45, \text{ say 24 studs} \quad \text{Ans. } (ii)$$

3.

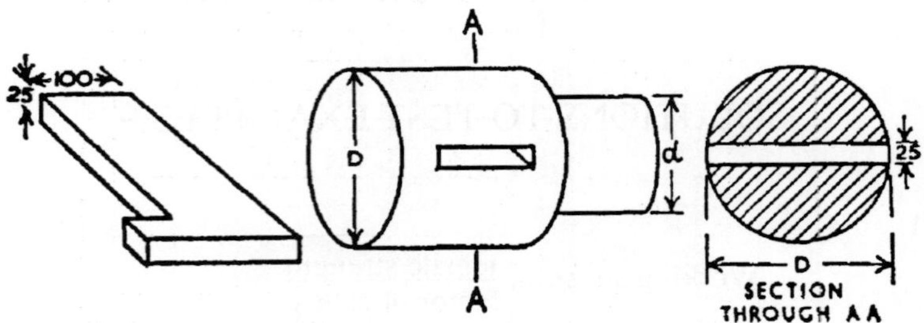

Let d = diameter of stay in mm

$$\text{Breaking load of stay} = \text{Breaking load of cotter}$$

$$440 \times 10^6 \times 0{\cdot}7854 \times d^2 \times 10^{-6} = 385 \times 10^6 \times 1{\cdot}8 \times 100 \times 25 \times 10^{-6}$$

$$d = \sqrt{\frac{385 \times 1{\cdot}8 \times 100 \times 25}{440 \times 0{\cdot}7854}}$$

$$= 70{\cdot}8 \text{ mm} \qquad \text{Ans. } (i)$$

Let D = diameter of swelled end in mm, for equal strength:

$$\text{Area of swelled end} = \text{Area of stay}$$

$$0{\cdot}7854D^2 - (D \times 25) = 0{\cdot}7854 \times 70{\cdot}8^2$$

$$D^2 - \frac{D \times 25}{0{\cdot}7854} = 70{\cdot}8^2$$

$$D^2 - 31{\cdot}83D = 5103$$

Solving this quadratic,

$$D = 88{\cdot}5 \text{ mm} \qquad \text{Ans. } (ii)$$

4.

Choosing two points on the straight line graph say P_1 and P_2, the increase of load is 300 N for a corresponding increase of extension 1·8 mm.

$$E = \frac{\text{stress}}{\text{strain}} \text{ or } \frac{\text{increase of stress}}{\text{increase of strain}}$$

$$E = \frac{\text{increase of load} \times \text{length}}{\text{area} \times \text{increase in extension}}$$

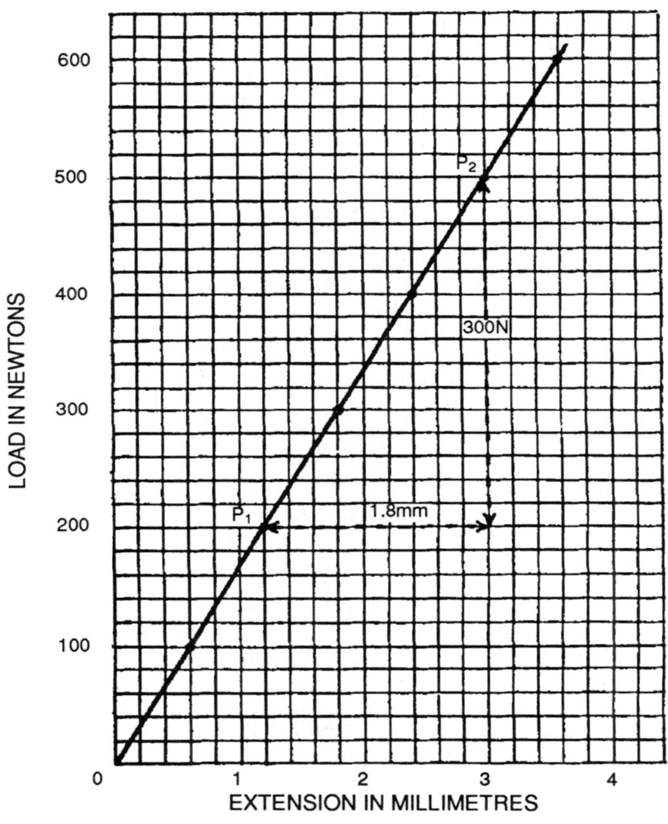

$$E = \frac{300 \times 4 \times 10^3}{0{\cdot}7854 \times 2^2 \times 1{\cdot}8} = 2{\cdot}121 \times 10^5 \text{ N/mm}^2$$

$= 212{\cdot}1 \text{ kN/mm}^2$ or $212{\cdot}1 \text{ GN/m}^2$ Ans.

5. Original cross-sectional area

$= 0{\cdot}7854 \times 20^2 = 314{\cdot}2 \text{ mm}^2 = 314{\cdot}2 \times 10^{-6}\text{m}^2$

$$\text{Yield stress} = \frac{\text{yield load}}{\text{area}}$$

$$= \frac{85{\cdot}4 \times 10^3}{314{\cdot}2 \times 10^{-6}}$$

$$= 2{\cdot}718 \times 10^3 \text{ N/m}^2$$

$= 271{\cdot}8 \text{ MN/m}^2$ Ans. (*i*)

$$\text{U.T.S.} = \frac{\text{breaking load}}{\text{original area}}$$

$$= \frac{143{\cdot}6 \times 10^3}{314{\cdot}2 \times 10^{-6}}$$

$$= 4{\cdot}571 \times 10^8 \text{ N/m}^2$$

$$= 457{\cdot}1 \text{ MN/m}^2 \qquad \text{Ans. } (ii)$$

$$\% \text{ Elongation} = \frac{\text{total extension}}{\text{original length}} \times 100$$

$$= \frac{127 - 100}{100} \times 100$$

$$= 27 \qquad \text{Ans. } (iii)$$

$$\% \text{ Contraction of area} = \frac{\text{reduction of area}}{\text{original area}} \times 100$$

$$= \frac{20^2 - 14{\cdot}8^2}{20^2} \times 100$$

$$= 45{\cdot}25 \qquad \text{Ans. } (iv)$$

6. (a) It is only necessary to consider the conditions for the longitudinal seams, where the stress is greatest.

Riveted joint:

$$\text{Working pressure} = \frac{2t \times \text{working stress}}{d} \times F_{\text{LONG}}$$

$$10 \times 10^5 = \frac{2t \times 80 \times 10^6}{3} \times 0{\cdot}7$$

$$\text{Plate thickness} = 0{\cdot}02678 \text{ m}$$

$$= 26{\cdot}78 \text{ mm} \qquad \text{Ans.}$$

(b) Welded joint:

joint strength is the same as the plate strength.

$$\text{Working stress} = \frac{pd}{2t}$$

$$80 \times 10^6 = \frac{10 \times 10^5 \times 3}{2 \times t}$$

$$\text{Plate thickness} = 18{\cdot}75 \text{ mm} \qquad \text{Ans.}$$

7.

$$\text{Load} = 203{\cdot}9 \times 9{\cdot}81 = 2000 \text{ N}$$

Cross-sectional area of steel wire

$$= 0{\cdot}7854 \times 4^2 = 12{\cdot}57 \text{ mm}^2$$
$$= 12{\cdot}57 \times 10^{-6}\text{m}^2$$

Cross-sectional area of brass wire

$$= 0{\cdot}7854 \times 5^2 = 19{\cdot}63 \text{ mm}^2$$
$$= 19{\cdot}63 \times 10^{-6}\text{m}^2$$

Strains of all wires are equal, therefore,

$$\text{Strain of steel} = \text{Strain of brass}$$

$$\frac{\text{Stress}_S}{E_S} = \frac{\text{Stress}_B}{E_B}$$

$$\frac{\text{Stress}_S}{200 \times 10^9} = \frac{\text{Stress}_B}{100 \times 10^9}$$

$$\text{Stress}_S = 2 \times \text{Stress}_B$$

Load carried by steel + Load carried by brass = total load

$$\text{Stress}_S \times \text{Area}_S + \text{Stress}_B \times \text{Area}_B = 2000$$

Substituting Stress_S from (i) and inserting values of areas,

$$2 \times \text{Stress}_B \times (2 \times 12{\cdot}57 \times 10^{-6}) + \text{Stress}_B \times 19{\cdot}63 \times 10^{-6} = 2000$$

$$50{\cdot}28 \text{ Stress}_B + 19{\cdot}63 \text{ Stress}_B = 2000 \times 10^6$$

$$69{\cdot}91 \text{ Stress}_B = 200 \times 10^6$$

$$\text{Stress}_B = 2{\cdot}861 \times 10^7 \text{ N/m}^2$$

Stress in brass wire $= 28{\cdot}61 \text{ MN/m}^2$ Ans. (*ia*)

Stress in steel wires $= 57{\cdot}22 \text{ MN/m}^2$ Ans. (*ib*)

Load carried by brass wire $= \text{stress} \times \text{area}$

$$= 2{\cdot}861 \times 10^7 \times 19{\cdot}63 \times 10^{-6}$$

$= 561{\cdot}4$ N Ans. (*iia*)

Load carried by each steel wire $= \frac{1}{2}(2000 - 561{\cdot}4)$

$= 719{\cdot}3$ N Ans. (*iib*)

$$\text{Elongation} = \text{strain} \times \text{length}$$

$$= \frac{\text{stress}}{E} \times \text{length}$$

$$= \frac{28{\cdot}61 \times 10^6 \times 4{\cdot}5}{100 \times 10^9}$$

$$= 0{\cdot}001287 \text{ m}$$

$= 1{\cdot}287$ mm Ans. (*c*)

8. If expansion is totally restricted, compression = 2 mm

$$\varepsilon = \frac{x}{l} \qquad = \frac{2 \times 10^{-3}}{1{\cdot}5}$$

$$E = \frac{\sigma}{\varepsilon}$$

$$\text{Compressive stress} = \text{strain} \times E$$

$$= \frac{2 \times 10^{-3}}{1{\cdot}5} \times 210 \times 10^9$$

$$= 280 \times 10^6 \text{ N/m}^2 = 280 \text{ MN/m}^2$$

and this is to be relieved by giving an initial tensile stress so that the compressive stress is limited to 177 MN/m^2.

$$\text{Initial tensile stress} = 280 - 177$$

$$= 103 \text{ MN/m}^2 \qquad \text{Ans.}$$

9.

$$x = \varepsilon \times l$$

$$= \frac{\sigma}{E} \times l$$

$$= \frac{75 \times 10^6}{85 \times 10^9} \times 1{\cdot}25$$

$$= 0{\cdot}001104 \text{ m or } 1{\cdot}104 \text{ mm} \qquad \text{Ans.}(i)$$

$$W = \sigma \times A$$

$$= 75 \times 10^6 \times 0{\cdot}7854 \times 50^2 \times 10^{-6}$$

$$= 1{\cdot}472 \times 10^5 \text{ N}$$

$$\text{Work done} = \text{average load} \times \text{extension}$$

$$= \tfrac{1}{2} Wx$$

$$= \tfrac{1}{2} \times 1{\cdot}472 \times 10^5 \times 0{\cdot}001104$$

$$= 81{\cdot}25 \text{ J} \qquad \text{Ans. } (ii)$$

Alternatively

$$\text{Work done} = \frac{\sigma^2 V}{2E}$$

$$= \frac{(75 \times 10^6)^2 \times 0{\cdot}7854 \times 50^2 \times 10^{-6} \times 1{\cdot}25}{2 \times 85 \times 10^9}$$

$$= 81{\cdot}23 \text{ J}$$

10.

$$\text{Outward pull of copper} = \text{Inward pull of steel}$$

$$\text{Stress}_C \times \text{Area}_C = \text{Stress}_S \times \text{Area}_S$$

$$\text{Stress}_C \times 1 = \text{Stress}_S \times 2$$

$$\text{Stress}_C = 2 \times \text{Stress}_S \quad \ldots \ldots \ldots \ldots \text{(i)}$$

$$\text{Strain}_C + \text{Strain}_S = \text{difference in free expansion per unit length}$$

$$\frac{\text{Strain}_C}{E_C} + \frac{\text{Strain}_C}{E_S} = \alpha_C\theta - \alpha_S\theta$$

$$\frac{2 \times \text{Stress}_s}{103 \times 10^9} + \frac{\text{Stress}_s}{206 \times 10^9} = 100 \times (17 - 12) \times 10^{-6}$$

$$2 \times 2 \times \text{Stress}_s + \text{Stress}_s = 100 \times 5 \times 10^{-6} \times 206 \times 10^9$$

$$5 \times \text{Stress}_s = 100 \times 5 \times 206 \times 10^3$$

$$\text{Stress}_s = 2{\cdot}06 \times 10^7 \text{ N/m}^2$$

$$= 20{\cdot}6 \text{ MN/m}^2 \qquad \text{Ans } (i)$$

$$\text{Stress}_c = 2 \times \text{Stress}_s$$

$$= 41{\cdot}2 \text{ MN/m}^2 \qquad \text{Ans } (ii)$$

11.

$$\text{Side of hexagon} = \tfrac{1}{2} \times 50 = 25 \text{ mm}$$

$$\text{Area of hexagon} = 6 \times 0{\cdot}433 \times \text{side}^2$$

$$= 2{\cdot}598 \times 25^2$$

$$= 1624 \text{ mm}^2$$

$$\text{Area of hole} = 0{\cdot}7854 \times 30^2$$

$$= 706{\cdot}8 \text{ mm}^2$$

Cross-sectional area of drilled hexagon

$$= 1624 - 706{\cdot}8$$

$$= 917{\cdot}2 \text{ mm}^2$$

$$\text{Resilience} = \frac{\sigma^2 V}{2E}$$

Ratio, Resilience of solid bar : Resilience of hollow bar

$$\frac{W^2 \times 1624 \times l}{1624^2 \times 2 \times E} : \frac{W^2 \times 917{\cdot}2 \times l}{917{\cdot}2^2 \times 2 \times E}$$

$$\frac{1}{1624} : \frac{1}{917{\cdot}2}$$

$$\frac{1624}{1624} : \frac{1624}{917{\cdot}2}$$

$$1 : 1{\cdot}77$$

$$100 : 177$$

Increase in resilience = 77% Ans.

12.

$$\begin{matrix}\text{Potential Energy lost} \\ \text{by falling mass}\end{matrix} = \begin{matrix}\text{Elastic Strain Energy} \\ \text{taken up by bar}\end{matrix}$$

$$W(h + x) = \frac{\sigma^2 V}{2E}$$

$$x = \text{compression} = \varepsilon \times l$$

$$= \frac{\sigma}{E} \times l$$

$$W\left(h + \frac{\sigma \times l}{E}\right) = \frac{\sigma^2 \times V}{2E}$$

$$102 \times 9{\cdot}81 \left(0{\cdot}012 + \frac{\sigma \times 10^6 \times 1{\cdot}5}{207 \times 109}\right)$$

$$= \frac{(\sigma \times 10^6)^2 \times 0{\cdot}7854 \times 0{\cdot}025^2 \times 1{\cdot}5}{2 \times 207 \times 10^9}$$

$$12 + 0{\cdot}007251\,\sigma = 0{\cdot}01778\,\sigma^2$$

Simplify and rearrange:

$$\sigma^2 - 0{\cdot}4078\,\sigma - 67{\cdot}5 = 0$$

Solving this quadratic, $\sigma = 84{\cdot}2$ MN/m². Ans.

An approximate value of the stress can be obtained by neglecting the small compression x as explained in the text, thus, working in basic units:

$$W \times h = \frac{\sigma^2 V}{2E}$$

$$\sigma = \sqrt{\frac{2EWh}{V}}$$

$$= \sqrt{\frac{2 \times 207 \times 10^9 \times 102 \times 9{\cdot}81 \times 0{\cdot}012}{0{\cdot}7854 \times 0{\cdot}025^2 \times 1{\cdot}5}}$$

$$= 8{\cdot}217 \times 10^7 \text{ N/m}^2$$

$$= 82{\cdot}17 \text{ MN/m}^2$$

13.

Static load: $\sigma = \dfrac{W}{A}$

$$= \frac{2000 \times 9{\cdot}81}{0{\cdot}7854 \times 0{\cdot}015^2}$$

Stress $= 111 \text{ MN/m}^2$ Ans.

Sudden stop:

Strain energy gained by wire = KE loss + PE loss

$$\frac{\sigma^2 V}{2E} = \tfrac{1}{2}mv^2 + mgx$$

$$\sigma = \frac{xE}{l}$$

$$\frac{x^2E^2Al}{2El^2} = \tfrac{1}{2}mv^2 + mgx$$

$$\frac{x^2 \times 210 \times 10^9 \times 0{\cdot}7854 \times 0{\cdot}015^2}{2 \times 10} =$$

$$\tfrac{1}{2}(2000 \times 0{\cdot}2^2) + (2000 \times 9{\cdot}81 \times x)$$

$$1{\cdot}855 \times 10^6 \times x^2 = 40 + 19620x$$

$$46375x^2 - 490{\cdot}5 - 1 = 0$$

$$x = 0{\cdot}1233$$

Extension $= 12{\cdot}33$ mm Ans. (*a*)

$$\sigma = \frac{xE}{l}$$

$$= \frac{0{\cdot}01233 \times 210 \times 10^9}{10}$$

Add. stress $= 259 \text{ MN/m}^2$ Ans. (*b*)

14.

Direct stress $= \dfrac{\text{load}}{\text{cross-sect. area}}$

$$= \frac{50 \times 10^3}{0{\cdot}7854 \times 20^2 \times 10^{-6}}$$

$$= 1{\cdot}592 \times 10^8 \text{ N/m}^2 = 159{\cdot}2 \text{ MN/m}^2$$

$$\sigma_n = \sigma_1 \cos^2\theta$$

$$= 159{\cdot}2 \times 0{\cdot}7071^2$$

$$= 79{\cdot}6 \text{ MN/m}^2 \qquad \text{Ans. } (i)$$

$$\tau = \tfrac{1}{2}\sigma_1 \sin 2\theta$$

$$= \tfrac{1}{2} \times 159{\cdot}2 \times \sin 90^\circ$$

$$= 79{\cdot}6 \text{ MN/m}^2 \qquad \text{Ans. } (ii)$$

15.

A shear stress is always accompanied by another shear stress of equal magnitude acting on a perpendicular plane, known as complementary shear stress. See Fig. 158 and proof.

Referring to Fig. 158:

Normal to plane:

$$\sigma_n = 45 \sin 60^\circ \cos 60^\circ + 45 \cos 60^\circ \sin 60^\circ$$

$$\sigma_n = 45 \sin 120^\circ$$

$$= 39 \text{ MN/m}^2 \qquad \text{Ans.}$$

Tangential to plane:

$$\tau_\theta = 45 \sin^2 60^\circ - 45 \cos^2 60^\circ$$

$$= 22{\cdot}5 \text{ MN/m}^2 \qquad \text{Ans.}$$

SOLUTIONS TO TEST EXAMPLES 10

1.

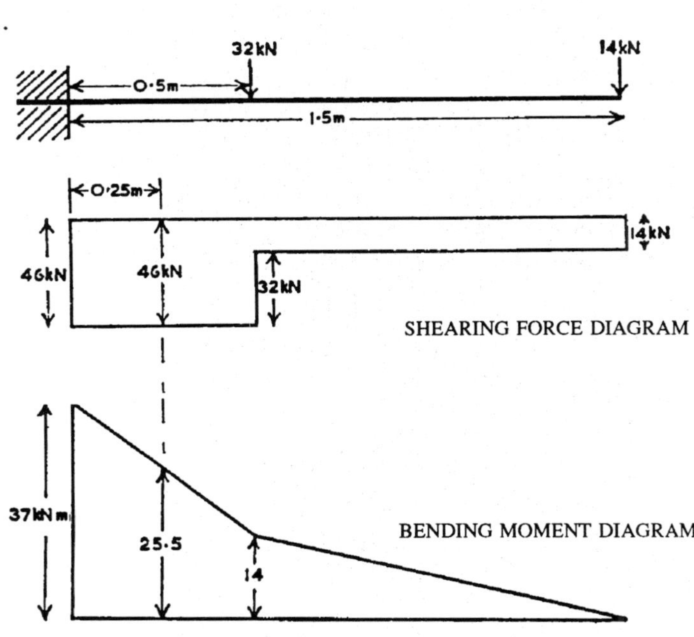

$$M \text{ @ } 0{\cdot}5 \text{ m from wall} = 14 \times 1 = 14 \text{ kN m}$$

$$M \text{ @ wall} = 32 \times 0{\cdot}5 + 14 \times 1{\cdot}5 = 37 \text{ kN m}$$

The shearing force and bending moment diagrams are drawn as shown. Reading these diagrams as 0·25 m from the wall, the values measure:

$$\left.\begin{aligned} \text{Shearing force} &= 46 \text{ kN} \\ \text{Bending moment} &= 25{\cdot}5 \text{ kN m} \end{aligned}\right\} \text{Ans.}$$

2.

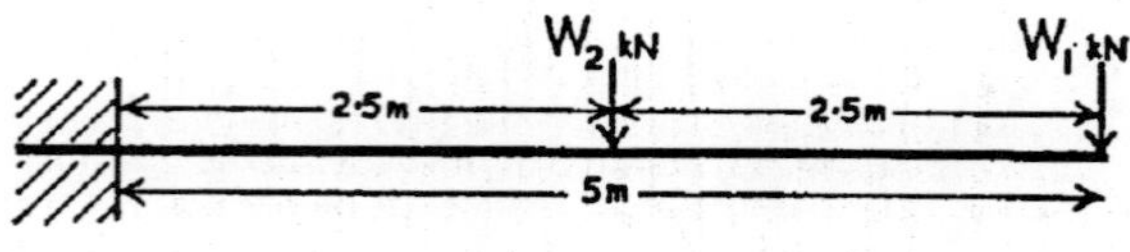

$$
\begin{aligned}
M \text{ @ mid-length} &= W_1 \times 2{\cdot}5 \\
12{\cdot}5 &= W_1 \times 2{\cdot}5 \\
W_1 &= 5 \text{ kN} \\
M \text{ @ wall} &= W_1 \times 5 + W_2 \times 2{\cdot}5 \\
52{\cdot}5 &= 5 \times 5 + W_2 \times 2{\cdot}5 \\
52{\cdot}5 - 25 &= W_2 \times 2{\cdot}5 \\
27{\cdot}5 &= W_2 \times 2{\cdot}5 \\
W_2 &= 11 \text{ kN}
\end{aligned}
$$

Concentrated loads are:

At the free end = 5 kN
At mid-length = 11 kN } Ans.

3.

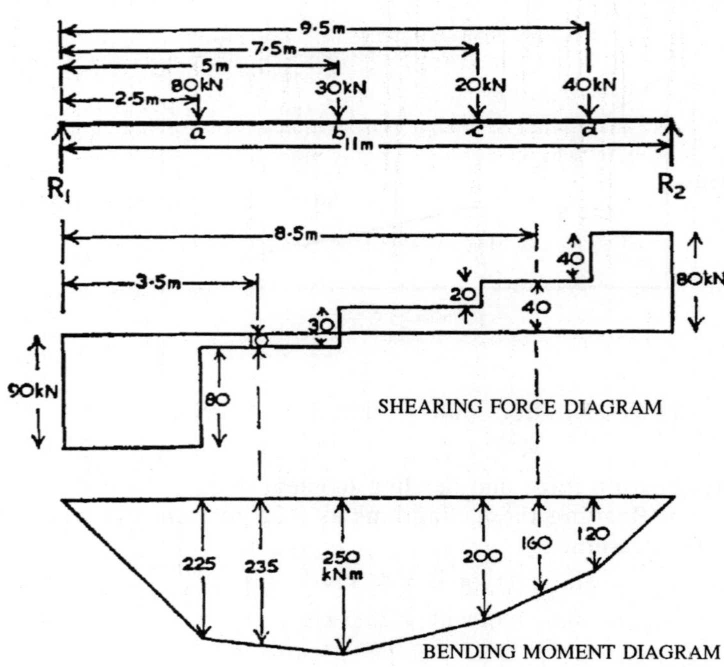

SHEARING FORCE DIAGRAM

BENDING MOMENT DIAGRAM

Moments about R_1

$$\text{Clockwise moments} = \text{Anticlockwise moments}$$
$$(80 \times 2{\cdot}5) + (30 \times 5) + (20 \times 7{\cdot}5) + (40 \times 9{\cdot}5) = R_2 \times 11$$
$$200 + 150 + 150 + 380 = R_2 \times 11$$
$$880 = R_2 \times 11$$
$$R_2 = 80 \text{ kN}$$

$$\text{Upward forces} = \text{Downward forces}$$
$$R_1 + R_2 = 80 + 30 + 20 + 40$$
$$R_1 = 170 - 80$$
$$R_1 = 90 \text{ kN}$$

M @ a (to the left) $= -90 \times 2{\cdot}5 = -225$ kN m (– is sagging)

M @ $b = -90 \times 5 + 80 \times 2{\cdot}5 = -250$ kN m

M @ c (to the right) $= -80 \times 3{\cdot}5 + 40 \times 2 = -200$ kN m

M @ $d = -80 \times 1{\cdot}5 = -120$ kN m

The shearing force and bending moment diagrams are drawn to represent the above quantities to some suitable scale. Reading the values at the points stated in the question:

F @ 3·5 m from left end = –10 kN
M " " " = –235 kN m
F @ 8·5 m from left end = 40 kN
M " " " = –160 kN m

} Ans.

4. Moments about R_1

$$\text{Clockwise moments} = \text{Anticlockwise moments}$$
$$40 \times 6 + 20 \times 9 + 50 \times 15 = R_2 \times 12$$
$$240 + 180 + 750 = R_2 \times 12$$
$$1170 = R_2 \times 12$$
$$\therefore R_2 = 97{\cdot}5 \text{ kN}$$

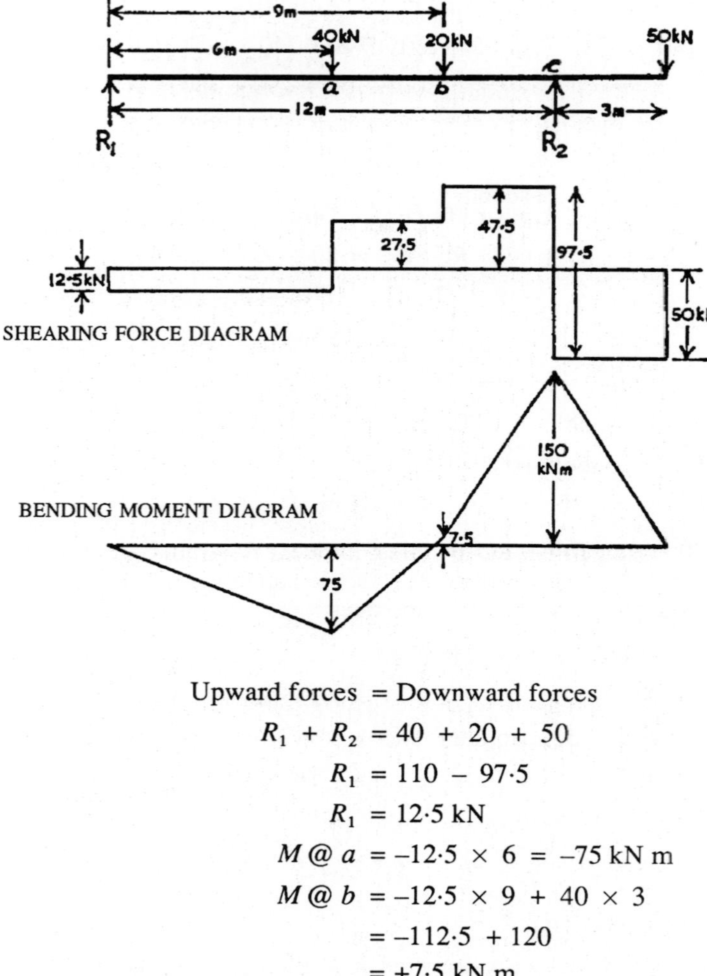

$$\text{Upward forces} = \text{Downward forces}$$

$$R_1 + R_2 = 40 + 20 + 50$$

$$R_1 = 110 - 97{\cdot}5$$

$$R_1 = 12{\cdot}5 \text{ kN}$$

$$M @ a = -12{\cdot}5 \times 6 = -75 \text{ kN m}$$

$$M @ b = -12{\cdot}5 \times 9 + 40 \times 3$$

$$= -112{\cdot}5 + 120$$

$$= +7{\cdot}5 \text{ kN m}$$

$$M @ c = 50 \times 3$$

$$= 150 \text{ kN m}$$

The shearing force and bending moment diagrams are shown with the above values inserted. Note that negative values of bending moment indicate sagging, and positive values indicate hogging bending moment.

5.

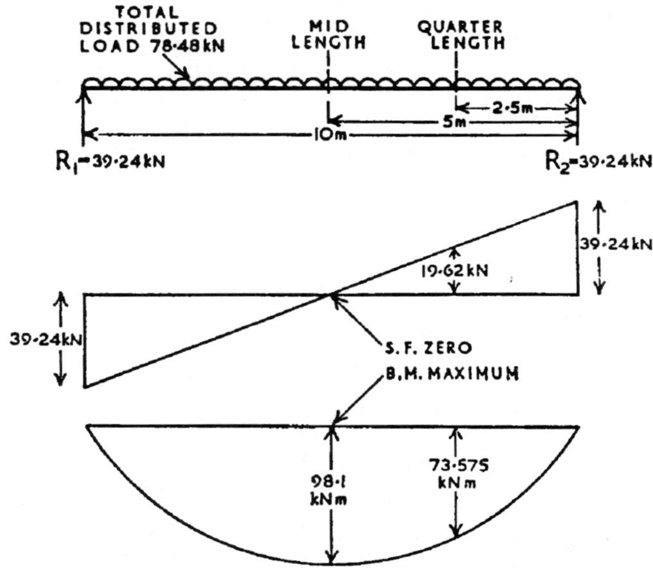

$$\text{Mass of beam} = 8 \text{ Mg} = 8 \times 10^3 \text{ kg}$$
$$\text{Weight of beam} = 8 \times 10^3 \times 9{\cdot}81 \text{ N}$$
$$= 78{\cdot}48 \text{ kN}$$

Weight of beam is a uniformly distributed load of 78·48 kN which is $78{\cdot}48 \div 10 = 7{\cdot}848$ kN per metre of length.

Each reaction carries half the total load,

$$R_1 = R_2 = \tfrac{1}{2} \times 78{\cdot}48 = 39{\cdot}24 \text{ kN}$$

At quarter-length:

$$\left.\begin{aligned} F &= 39{\cdot}24 - 7{\cdot}848 \times 2{\cdot}5 = 19{\cdot}62 \text{ kN} \\ M &= -39{\cdot}24 \times 2{\cdot}5 + 7{\cdot}848 \times 2{\cdot}5 \times 1{\cdot}25 = -73{\cdot}575 \text{ kN m} \end{aligned}\right\} \text{Ans. } (i)$$

At mid-length:

$$\left.\begin{aligned} F &= 39{\cdot}24 - 7{\cdot}848 \times 5 = \text{Nil} \\ M &= -39{\cdot}24 \times 5 + 7{\cdot}848 \times 5 \times 2{\cdot}5 = -98{\cdot}1 \text{ kN m} \end{aligned}\right\} \text{Ans. } (ii)$$

The dimensioned diagrams are shown above.

6.

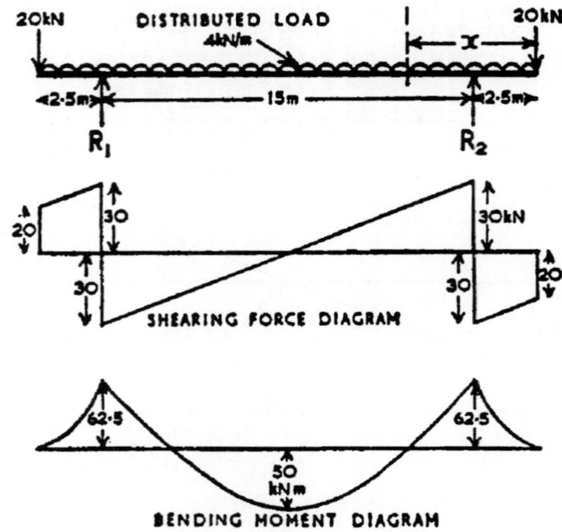

$$\text{Total distributed load} = 4 \times 20 = 80 \text{ kN}$$

$$\text{Total load on beam} = 80 + 20 + 20 = 120 \text{ kN}$$

$$\text{Each reaction carries } \tfrac{1}{2} \text{ of } 120 = 60 \text{ kN}$$

$$\begin{aligned} \text{M @ centre} &= -60 \times 7{\cdot}5 + 4 \times 10 \times 5 + 20 \times 10 \\ &= -450 + 200 + 200 \\ &= -50 \text{ kN m} \quad \text{Ans. } (i) \end{aligned}$$

$$\begin{aligned} \text{M @ supports} &= 4 \times 2{\cdot}5 \times 1{\cdot}25 + 20 \times 2{\cdot}5 \\ &= 12{\cdot}5 + 50 \\ &= 62{\cdot}5 \\ &= 62{\cdot}5 \text{ kN m} \quad \text{Ans. } (ii) \end{aligned}$$

Note: The negative value of the bending moment at the centre indicates that the curvature of the beam there is of a sagging nature, and the positive value at the supports indicates that the bending there is of a hogging nature.

The bending moment diagram may be drawn as a combined diagram composed of two parts, one part due to the distributed load only and the other due to the concentrated loads only (as explained in the text), or as a single diagram representing the values for the combined loading as shown here.

Let x = distance from end of beam where the value of the bending moment is zero.

$$M \text{ @ } x = -60 \times (x - 2{\cdot}5) + 4x \times 0{\cdot}5x + 20x$$
$$= -60x + 150 + 2x^2 + 20x$$
$$= -40x + 150 + 2x^2$$

and this is to be equal to zero:

$$2x^2 - 40x + 150 = 0$$
$$\text{or} \quad x^2 - 20x + 75 = 0$$
$$\therefore x = 5 \text{ or } 15$$

Therefore bending moment is zero (points of contraflexure) at 5 m from each end. Ans. (*iii*)

7.

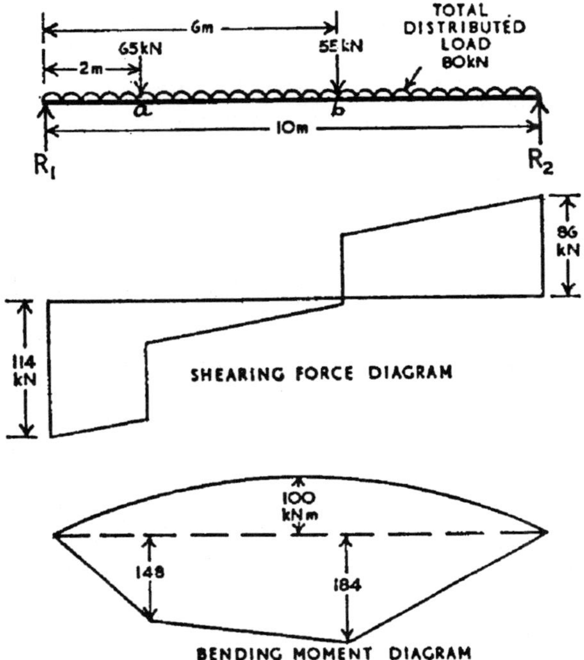

Moments about R_1

$$\text{Clockwise moments} = \text{Anticlockwise moments}$$
$$65 \times 2 + 55 \times 6 + 80 \times 5 = R_2 \times 10$$
$$130 + 330 + 400 = R_2 \times 10$$
$$860 = R_2 \times 10$$
$$R_2 = 86 \text{ kN}$$
$$\text{Upward forces} = \text{Downward forces}$$
$$R_1 + R_2 = 65 + 55 + 80$$
$$R_1 = 200 - 86$$
$$R_1 = 114 \text{ kN}$$

Here separate diagrams for the u.d.l and concentrated loads are shown in the one diagram (two parts). Strictly the combined loading would give one diagram, a parabola under the zero line (sagging M).

For sketching and dimensioning bending moment diagram:

$$\text{Distribution load} = 80 \text{ kN over } 10 \text{ m}$$
$$= 8 \text{ kN per m}$$

Considering distributed load only,

$$\text{Each reaction} = 40 \text{ kN}$$
$$M \text{ @ centre} = -40 \times 5 + 40 \times 2{\cdot}5$$
$$= -100 \text{ kN m} \quad \textit{i.e.} \text{ sagging}$$

Considering concentrated loads only,

$$R_1 = 114 - 40 = 74 \text{ kN}$$
$$R_2 = 86 - 40 = 46 \text{ kN}$$
$$M \text{ @ } a = -74 \times 2 = -148 \text{ kN m} \quad \textit{i.e.} \text{ sagging}$$
$$M \text{ @ } b = -46 \times 4 = -184 \text{ kN m} \quad \textit{i.e.} \text{ sagging}$$

Maximum bending moment occurs in the beam at the position where the shearing force diagram crosses its base line, this is at 4 m from R_2 (or 6 m from R_1).

$$M_{max} = -86 \times 4 + 8 \times 4 \times 2$$
$$= -344 + 64$$
$$= -280 \text{ kN m} \quad \textit{i.e.} \text{ sagging}$$

Hence, maximum bending moment is 280 kN m under the 55 kN load. Ans.

8. Let L = distance in m from free end where stress is 75×10^6 N/m^2

$$M \text{ @ this section} = 15 \times 10^3 \times L \quad \text{N m}$$

$$\frac{M}{I} = \frac{\sigma}{y} \qquad M = \frac{\sigma \times I}{y}$$

For a rectangular section,

$$I = \frac{BD^3}{12} \text{ and } y = \frac{D}{2}$$

$$M = \frac{\sigma \times B \times D^2}{6}$$

$$15 \times 10^3 \times L = \frac{75 \times 10^6 \times 0.1 \times 0.15^2}{6}$$

$$L = \frac{75 \times 10^6 \times 0.1 \times 0.15^2}{15 \times 10^3 \times 6}$$

$$L = 1{\cdot}875 \text{ m} \quad \text{Ans.}$$

9. Bar and tube have equal cross-sectional areas, therefore, letting d = inside diameter of tube, and working in cm:

$$0{\cdot}7854 \times 1{\cdot}5^2 = 0{\cdot}7854\,(3^2 - d^2)$$

$$d^2 = 3^2 - 1{\cdot}5^2$$

$$d^2 = 6{\cdot}75$$

$$\text{Modulus of section of tube} = \frac{I}{y} = z$$

$$= \frac{\pi\,(D^4 - d^4) \times 2}{64 \times D}$$

$$= \frac{\pi\,(D^4 - d^4)}{32 \times D}$$

$$= \frac{\pi\,(D^2 + d^2)\ (D^2 - d^2)}{32 \times D}$$

$$= \frac{\pi\,(9 + 6{\cdot}75)\ (9 - 6{\cdot}75)}{32 \times 3}$$

$$= \frac{\pi \times 15{\cdot}75 \times 2{\cdot}25}{32 \times 3} \qquad \ldots\ \ldots\ \ldots(i)$$

$$\text{Modulus of section of bar} = \frac{I}{y} = z$$

$$= \frac{\pi \times D^4 \times 2}{64 \times D}$$

$$= \frac{\pi \times D^3}{32}$$

$$= \frac{\pi \times 1{\cdot}5^3}{32} \quad \ldots \ldots \ldots (ii)$$

Ratio of strengths = Ratio of moduli of sections

Strength of Tube : Strength of Bar

$$\frac{\pi \times 15{\cdot}75 \times 2{\cdot}25}{32 \times 3} : \frac{\pi \times 1{\cdot}5^3}{32}$$

Dividing both sides by strength of bar,

$$\frac{\pi \times 15{\cdot}75 \times 2{\cdot}25 \times 32}{32 \times 3 \times 1{\cdot}5^3 \times \pi} : 1$$

$$3{\cdot}5 : 1$$

tube is 3½ times strength of solid bar. Ans.

10. Let W = load in kN

$$R_1 = R_2 = \tfrac{1}{2}W$$

Maximum bending moment is at mid-span,

$$M \text{ @ centre} = \frac{W}{2} \times \frac{L}{2} = \frac{WL}{4} \text{ (sagging)}$$

$$\frac{W \times 2}{4} = 30$$

$$W = 60 \text{ kN} \quad \text{Ans. } (i)$$

$$\frac{M}{I} = \frac{\sigma}{y} \quad \therefore \sigma = \frac{My}{I}$$

For a rectangular section,

$$I = \frac{BD^3}{12} \quad \text{and } y = \frac{D}{2}$$

$$\sigma = \frac{6M}{BD^2}$$

$$60 \times 10^6 = \frac{6 \times 30 \times 10^3}{0{\cdot}075 \times D^2}$$

$$D = \sqrt{\frac{6 \times 30 \times 10^3}{60 \times 10^6 \times 0{\cdot}075}}$$

$$= 0{\cdot}2 \text{ m or } 200 \text{ mm} \quad \text{Ans. } (ii)$$

11. Working initially in cm to obtain mass and I, changing later into m for calculations involving bending moment and stress:

$$\text{Inside breadth} = 7{\cdot}8 - (2 \times 1{\cdot}2) = 5{\cdot}4 \text{ cm}$$

$$\text{Inside depth} = 10{\cdot}4 - (2 \times 1{\cdot}2) = 8 \text{ cm}$$

$$\text{Mass of beam} = (7{\cdot}8 \times 10{\cdot}4 - 5{\cdot}4 \times 8) \times 120 \times 7{\cdot}86 \times 10^{-3}$$

$$= 35{\cdot}76 \text{ kg}$$

$$\text{Weight of beam} = 35{\cdot}76 \times 9{\cdot}81 = 350{\cdot}9 \text{ N}$$

Centre of gravity of beam is at mid-length = 0·6m from wall

Let W = concentrated load at free end,

Maximum bending moment which induces maximum stress is at the wall,

$$M \text{ @ wall} = 350{\cdot}9 \times 0{\cdot}6 + W \times 1{\cdot}2$$

$$= 210{\cdot}5 + 1{\cdot}2W \text{ N m} \quad \ldots \; \ldots \; \ldots (i)$$

$$I \text{ of section} = \frac{BD^3 - bd^3}{12}$$

$$= \frac{7{\cdot}8 \times 10{\cdot}4^3 - 5{\cdot}4 \times 8^3}{12} \text{ cm}^4$$

$$= 500{\cdot}6 \text{ cm}^4 = 500{\cdot}6 \times 10^{-8} \text{ m}^4$$

$$y = \tfrac{1}{2} \text{ outside depth} = 0{\cdot}052 \text{ m}$$

$$\frac{M}{I} = \frac{\sigma}{y} \qquad M = \frac{\sigma I}{y}$$

$$M_{max} = \frac{35 \times 10^6 \times 500{\cdot}6 \times 10^{-8}}{0{\cdot}052}$$

$$= 3370 \text{ N m} \quad \ldots \; \ldots \; \ldots \; \ldots \; \ldots (ii)$$

From (*i*) and (*ii*):

$$210{\cdot}5 + 1{\cdot}2W = 3370$$

$$1{\cdot}2W = 3159{\cdot}5$$

$$W = 2633 \text{ N or } 2{\cdot}633 \text{ kN} \quad \text{Ans.}$$

12.

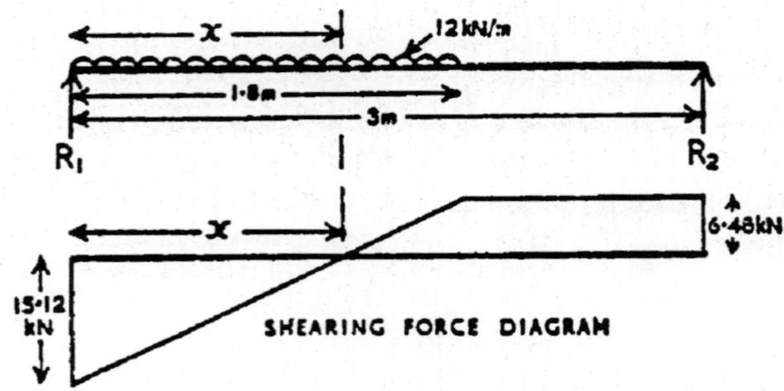

$$\text{Total load on beam} = 12 \times 1{\cdot}8 = 21{\cdot}6 \text{ kN}$$

Centre of gravity of this load is at its own mid-length of 0·9 m from the left end.

Moments about R_1

$$\text{Clockwise moments} = \text{Anticlockwise moments}$$

$$21{\cdot}6 \times 0{\cdot}9 = R_2 \times 3$$

$$R_2 = 6{\cdot}48 \text{ kN}$$

$$\text{Upward forces} = \text{Downward forces}$$

$$R_1 + R_2 = 21{\cdot}6$$

$$R_1 = 21{\cdot}6 - 6{\cdot}48$$

$$R_1 = 15{\cdot}12 \text{ kN}$$

Maximum bending moment (and maximum stress) occurs at the section of the beam where the shearing force is zero. A sketch of the shearing force diagram shows that this is somewhere along the first 1·8 m. Let x = this distance in m from the left end.

$$SF \text{ @ } x \text{ from } R_1 = -15{\cdot}12 + 12x$$

and this is to be zero,

$$12x = 15{\cdot}12$$

$$x = 1{\cdot}26 \text{ m}$$

$$M \text{ @ } 1{\cdot}26 \text{ m from } R_1 = -15{\cdot}12 \times 1{\cdot}26 + 12 \times 1{\cdot}26 \times 0{\cdot}63$$

$$= -9{\cdot}526 \text{ kN m } (\textit{sagging})$$

The section being symmetrical, distance from neutral axis to outer fibres is at half-depth, *i.e.*, $y = 0{\cdot}19$ m.

$$I = 7{\cdot}6 \times 10^3 \text{ cm}^4 = 7{\cdot}6 \times 10^3 \times 10^{-8} \text{ m}^4 = 7{\cdot}6 \times 10^{-5} \text{ m}^4$$

$$\frac{M}{I} = \frac{\sigma}{y} \qquad \sigma = \frac{My}{I}$$

$$= \frac{9{\cdot}526 \times 10^3 \times 0{\cdot}19}{7{\cdot}6 \times 10^{-5}}$$

$$= 2{\cdot}381 \times 10^7 \text{ N/m}^2$$

$$= 23{\cdot}81 \text{ MN/m}^2$$

Position of max. stress from left end = 1·26 m
Magnitude of max. stress = 23·81 MN/m² } Ans.

13. Mass of shaft $= 0{\cdot}7854 \times 7{\cdot}6^2 \times 127 \times 7{\cdot}86 \times 10^{-3}$

$$= 45{\cdot}28 \text{ kg}$$

$$\text{Weight} = 45{\cdot}28 \times 9{\cdot}81 = 444{\cdot}2 \text{ N}$$

Let W = central concentrated load, each reaction carries half of total load:

$$R_1 = R_2 = \tfrac{1}{2}(W + 444{\cdot}2)$$

Maximum bending moment (and maximum stress) is at mid-length,

$$M \text{ @ centre} = -0{\cdot}5\,(W + 444{\cdot}2) \times 0{\cdot}635 + 222{\cdot}1 \times 0{\cdot}3175$$

$$= -0{\cdot}3175\,(W + 444{\cdot}2) + 222{\cdot}1 \times 0{\cdot}3175$$

$$= -0{\cdot}3175\,(W + 222{\cdot}1) \text{ N m } (\textit{sagging})$$

$$\frac{M}{I} = \frac{\sigma}{y} \qquad M = \frac{\sigma \times I}{y}$$

For a solid circular section,

$$I = \frac{\pi}{64} D^4 \qquad \text{and } y = \frac{D}{2}$$

$$M = \frac{\sigma \times \pi D^3}{32}$$

$$0{\cdot}3175\ (W + 222{\cdot}1) = \frac{13{\cdot}8 \times 10^6 \times \pi \times 0{\cdot}076^3}{32}$$

$$W + 222{\cdot}1 = \frac{13{\cdot}8 \times 10^6 \times \pi \times 0{\cdot}076^3}{32 \times 0{\cdot}3175}$$

$$W + 222{\cdot}1 = 1874$$

$$W = 1651{\cdot}1 \text{ N} = 1{\cdot}6519 \text{ kN} \quad \text{Ans.}$$

14.

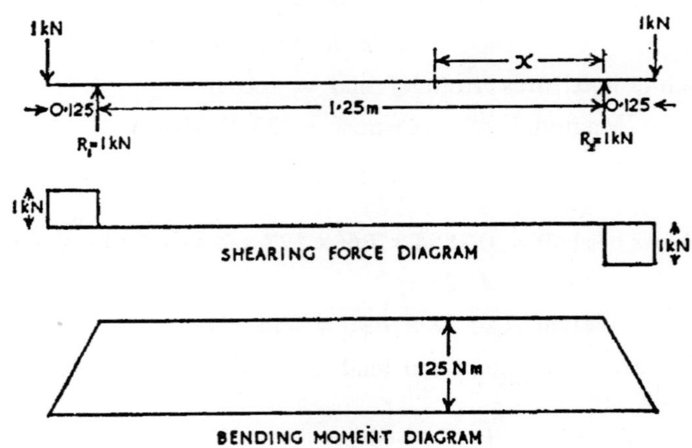

$$R_1 + R_2 = 1 \text{ kN}$$

M @ x m from support, towards centre,

$$= -R_2 \times x + W(x + 0{\cdot}125)$$

$$= -1 \times x + 1 \times (x + 0{\cdot}125)$$

$$= -x + x + 0{\cdot}125$$

$$= 0{\cdot}125 \text{ kN m}$$

i.e., 125 N m of hogging moment, and this is constant over the whole span between the supports.

$$\frac{M}{I} = \frac{E}{R} \qquad \therefore R = \frac{EI}{M}$$

For a hollow circular section,

$$I = \frac{\pi}{64}(D^4 - d^4)$$

$$I = \frac{\pi}{64}(4^4 - 3^4)\ \text{cm}^4$$

$$= \frac{\pi}{64} \times 175 \times 10^{-8}\ \text{m}^4$$

$$R = \frac{EI}{M}$$

$$= \frac{208 \times 10^9 \times \pi \times 175 \times 10^{-8}}{64 \times 125}$$

$$= 143\ \text{m} \quad \text{Ans. } (i)$$

$$\text{Deflection} = \frac{L^2}{8R} \quad \text{or} \quad \frac{ML^2}{8EI}$$

$$= \frac{1{\cdot}25^2}{8 \times 143}$$

$$= 1{\cdot}366 \times 10^{-3}\ \text{m} = 1{\cdot}366\ \text{mm Ans. } (ii)$$

15. Direct compressive stress $= \dfrac{\text{load}}{\text{area}}$

$$= \frac{275 \times 10^3}{0{\cdot}7854 \times 0{\cdot}125^2}\ \text{N/m}^2$$

$$= 2{\cdot}241 \times 10^7\ \text{N/m}^2 = 22{\cdot}41\ \text{MN/m}^2 \quad (i)$$

$$\text{Bending moment} = 275 \times 10^3 \times 0{\cdot}12$$

$$= 3{\cdot}3 \times 10^3\ \text{N m}$$

$$\frac{M}{I} = \frac{\sigma}{y} \qquad \sigma = \frac{My}{I}$$

For a solid circular section,

$$I = \frac{\pi}{64} D^4 \qquad \text{and } y = \frac{D}{2}$$

$$\sigma = \frac{32M}{\pi D^3}$$

$$\text{Bending stress} = \frac{32 \times 3{\cdot}3 \times 10^3}{\pi \times 0{\cdot}125^3}$$

$$= 1{\cdot}721 \times 10^7 \text{ N/m}^2 = 17{\cdot}21 \text{ MN/m}^2 \quad (ii)$$

Max. compressive stress = 22·41 + 17·21 = 39·62 MN/m^2 Ans.

Min. compressive stress = 22·41 – 17·21 = 5·2 MN/m^2 Ans.

SOLUTIONS TO TEST EXAMPLES 11

1.

$$J = \frac{\pi}{32}(D^4 - d^4)$$

$$= \frac{\pi}{32}(0{\cdot}4^4 + 0{\cdot}25^4)(0{\cdot}4^4 - 0{\cdot}25^4)$$

$$= \frac{\pi}{32} \times 0{\cdot}2225 \times 0{\cdot}0975 \text{ m}^4$$

$$r = \tfrac{1}{2}D = 0{\cdot}2 \text{ m}$$

$$\frac{T}{J} = \frac{\tau}{r} \qquad \tau = \frac{Tr}{J}$$

$$\tau = \frac{32 \times 480 \times 10^3 \times 0{\cdot}2}{\pi \times 0{\cdot}2225 \times 0{\cdot}0975}$$

$$= 4{\cdot}506 \times 10^7 \text{ N/m}^2$$

$$= 45{\cdot}06 \text{ MN/m}^2 \quad \text{Ans. } (i)$$

$$\frac{\tau}{r} = \frac{G\theta}{l} \qquad \theta = \frac{\tau l}{rG} \text{ rad}$$

$$\theta = \frac{45{\cdot}06 \times 10^6 \times 7{\cdot}5 \times 57{\cdot}3}{0{\cdot}2 \times 92{\cdot}5 \times 10^9} \text{ deg.}$$

$$= 1{\cdot}047° \quad \text{Ans. } (ii)$$

2.

T transmitted by solid shaft = T transmitted by hollow shaft

$$\frac{\pi D^3 \tau_1}{16} = \frac{\pi (D^4 - d^4)\tau_2}{16D}$$

Substituting $\tau_2 = 1{\cdot}2\,\tau_1$ cancelling π and 16, and multiplying both sides by D,

$$D^4 \times \tau_1 = (D^4 - d^4) \times 1{\cdot}2\,\tau_1$$

$$D^4 = (D^4 - d^4) \times 1{\cdot}2$$

$$D^4 = 1{\cdot}2\,D^4 - 1{\cdot}2\,d^4$$

$$1{\cdot}2\,d^4 = 1{\cdot}2\,D^4 - D^4$$

$$d^4 = \frac{0{\cdot}2\,D^4}{1{\cdot}2}$$

$$d = \frac{D}{\sqrt[4]{6}}$$

$$d = 0{\cdot}639\,D \quad \text{Ans. } (i)$$

RATIO OF WEIGHTS

$$\text{Weight of solid shaft} : \text{Weight of hollow shaft}$$

$$0{\cdot}7854\,D^2 \times l \times w : 0{\cdot}7854\,(D^2 - d^2) \times l \times w$$

$$D^2 : D^2 - (0{\cdot}639\,D)^2$$

$$D^2 : D^2 - 0{\cdot}4083\,D^2$$

$$D^2 : 0{\cdot}5917\,D^2$$

$$1 : 0{\cdot}5917$$

$$\text{Saving in weight} = \frac{1 - 0{\cdot}5917}{1} \times 100$$

$$= 40{\cdot}83\% \quad \text{Ans. } (ii)$$

3.

$$J \text{ for liner} = \frac{\pi}{32}\,(0{\cdot}4^4 - 0{\cdot}35^4)$$

$$= \frac{\pi}{32}\,(0{\cdot}4^2 + 0{\cdot}35^2)\,(0{\cdot}4^2 - 0{\cdot}35^2)$$

$$= 1{\cdot}04 \times 10^{-3}\ \text{m}^4$$

$$J \text{ for shaft} = \frac{\pi}{32} \times 0{\cdot}35^4$$

$$= 1{\cdot}474 \times 10^{-3} \text{ m}^4$$

$$\frac{T}{J} = \frac{\tau}{r}$$

$$T \text{ for liner} = \frac{J\tau}{r} = \frac{1{\cdot}04 \times 10^{-3} \times \tau_1}{0{\cdot}2}$$

$$= 5{\cdot}2 \times 10^{-3} \times \tau_1 \quad \text{N m}$$

$$T \text{ for shaft} = \frac{J\tau}{r} = \frac{1{\cdot}474 \times 10^{-3} \times \tau_S}{0{\cdot}175}$$

$$= 8{\cdot}424 \times 10^{-3} \times \tau_S \quad \text{N m}$$

$$\text{Total torque transmitted} = 200 \times 10^3 \text{ N m}$$

$$5{\cdot}2 \times 10^{-3}\, \tau_1 + 8{\cdot}424 \times 10^{-3}\, \tau_S = 200 \times 10^3$$

$$5{\cdot}2\, \tau_1 + 8{\cdot}424\, \tau_S = 200 \times 10^6 \quad \ldots\ \ldots(i)$$

$$\frac{\tau}{r} = \frac{G\theta}{l} \qquad \theta = \frac{\tau l}{Gr}$$

Liner and shaft have same angle of twist and also the same length,

$$\theta \text{ for liner} = \theta \text{ for shaft}$$

$$\frac{\tau_1 \times l}{38{\cdot}5 \times 10^9 \times 0{\cdot}2} = \frac{\tau_S \times l}{85 \times 10^9 \times 0{\cdot}175}$$

$$\tau_S = \frac{85 \times 0{\cdot}175 \times \tau_1}{38{\cdot}5 \times 0{\cdot}2}$$

$$\tau_S = 1{\cdot}932\, \tau_1 \quad \ldots\ \ldots(ii)$$

Substituting value of τ_S from equation (*ii*) into (*i*),

$$5{\cdot}2\, \tau_1 + 8{\cdot}424\, \tau_S = 200 \times 10^6$$

$$5{\cdot}2\, \tau_1 + 8{\cdot}424 \times 1{\cdot}932\, \tau_1 = 200 \times 10^6$$

$$21{\cdot}48\, \tau_1 = 200 \times 10^6$$

$$\tau_1 = 9{\cdot}311 \times 10^6 \text{ N/m}^2$$

$$= 9{\cdot}311 \text{ MN/m}^2$$

Substituting value of τ_1 into equation (*ii*),

$$\tau_S = 1{\cdot}932 \times 9{\cdot}311$$
$$= 17{\cdot}99 \text{ MN/m}^2$$

Maximum stresses in shaft and liner are,

$$17{\cdot}99 \text{ and } 9{\cdot}311 \text{ MN/m}^2 \quad \text{Ans.}$$

4.

$$P = T\omega = \frac{\pi}{16} D^3 \tau \times \omega$$

for constant power, $D^3 \times \tau \times \text{speed} = \text{constant}$

$$D_1{}^3 \times \tau_1 \times \text{speed}_1 = D_2{}^3 \times \tau_2 \times \text{speed}_2$$

$$\tau_2 = \frac{0{\cdot}14^3 \times 48 \times 3000}{0{\cdot}445^3 \times 90}$$
$$= 49{\cdot}79 \text{ MN/m}^2 \quad \text{Ans.}$$

5.

$$P = T\omega$$

$$\frac{T}{J} = \frac{G\theta}{l} \qquad T = \frac{JG\theta}{l}$$

$$\text{Power} = \frac{J \times G \times \theta \times \omega}{l}$$

$$= \frac{\pi \times 0{\cdot}36^4 \times 93 \times 10^9 \times 0{\cdot}3 \times 115 \times 2\pi}{32 \times 2{\cdot}5 \times 57{\cdot}3 \times 60}$$

$$= 3{\cdot}868 \times 10^6 \text{ W}$$

$$= 3{\cdot}868 \text{ MW or } 3868 \text{ kW} \quad \text{Ans.}$$

6.

$$\text{Speed of ship} = \frac{19 \times 1{\cdot}852 \times 10^3}{3600} = 9{\cdot}774 \text{ m/s}$$

$$\text{Work done} = \text{force} \times \text{distance}$$

$$\text{Work done/s} = \text{force} \times \text{speed}$$

$$\text{Propeller power} = 320 \times 9{\cdot}774$$
$$= 3127 \text{ kW}$$

and this is 0·75 of the shaft power

$$\text{Shaft power} = \frac{3127}{0{\cdot}75} = 4170 \text{ kW} \quad \text{Ans. } (i)$$

$$T = \frac{P}{\omega}$$
$$= \frac{4170 \times 60}{90 \times 2\pi}$$
$$= 442{\cdot}4 \text{ kN m} \quad \text{Ans. } (ii)$$

$$T = \frac{\pi D^3 \tau}{16} \qquad \tau = \frac{16T}{\pi D^3}$$

$$T = \frac{16 \times 442{\cdot}4 \times 10^3}{\pi \times 0{\cdot}38^3}$$
$$= 4{\cdot}105 \times 10^7 \text{ N/m}^2$$
$$= 41{\cdot}05 \text{ MN/m}^2 \quad \text{Ans. } (iii)$$

7.

$$\text{Mean torque} = \frac{\text{power}}{\omega}$$
$$\text{Maximum torque} = \text{mean torque} \times 1{\cdot}25$$
$$= \frac{4500 \times 10^3 \times 60}{105 \times 2\pi} \times 1{\cdot}25$$
$$= 5{\cdot}114 \times 10^5 \text{ N m}$$

For a solid shaft,

$$\text{Max. torque} = \frac{\pi D^3 \tau}{16} \qquad D = \sqrt[3]{\frac{16T}{\pi \tau}}$$

$$D = \sqrt[3]{\frac{16 \times 5{\cdot}114 \times 10^5}{\pi \times 55 \times 10^6}}$$

$$= 0{\cdot}3617 \text{ m} = 361{\cdot}7 \text{ mm Ans. } (ai)$$

For a hollow shaft,

$$\text{Max. torque} = \frac{\pi (D^4 - d^4)\,\tau}{16D}$$

$$\text{When } d = \tfrac{1}{2}D, \quad \frac{D^4 - d^4}{D} = \frac{15D^3}{16}$$

$$\text{Max. torque} = \frac{\pi \times 15 \times D^3 \times \tau}{16 \times 16}$$

$$D = \sqrt[3]{\frac{16 \times 16 \times 5{\cdot}114 \times 10^5}{\pi \times 15 \times 55 \times 10^6}}$$

$$= 0{\cdot}3696 \text{ m} = 369{\cdot}6 \text{ mm Ans. } (aii)$$

RATIO OF WEIGHTS:

$$\text{Weight of hollow shaft} : \text{Weight of solid shaft}$$

$$0{\cdot}7854\,(0{\cdot}3696^2 - 0{\cdot}1848^2) \times l \times w : 0{\cdot}7854 \times 0{\cdot}3617^2 \times l \times w$$

$$0{\cdot}5544 \times 0{\cdot}1848 : 0{\cdot}3617^2$$

$$0{\cdot}1024 : 0{\cdot}1308$$

$$\text{Saving in weight} = \frac{0{\cdot}1308 - 0{\cdot}1024}{0{\cdot}1308} \times 100$$

$$= 21{\cdot}72\% \quad \text{Ans. } (b)$$

8. Twisting moment transmitted by bolts

$= \text{stress} \times \text{area of each bolt} \times \text{P.C. radius} \times \text{no. of bolts}$

$= 38 \times 10^6 \times 0{\cdot}7854\, d^2 \times 0{\cdot}305 \times 8 \text{ N m}$

$$T = \frac{P}{\omega}$$

$$38 \times 10^6 \times 0{\cdot}7854\, d^2 \times 0{\cdot}305 \times 8 = \frac{3000 \times 10^3 \times 60}{100 \times 2\pi}$$

$$d = \sqrt{\frac{3000 \times 10^3 \times 60}{38 \times 10^6 \times 0{\cdot}7854 \times 0{\cdot}305 \times 8 \times 100 \times 2\pi}}$$

$$= 0{\cdot}06272 \text{ m} = 62{\cdot}72 \text{ mm} \quad \text{Ans.}$$

9. Referring to space diagram of Fig. 199

$$\frac{r}{\sin \phi} = \frac{l}{\sin \theta}$$

$$\sin \phi = \frac{315 \times \sin 35°}{1200} = 0{\cdot}1506$$

$$\phi = 8°40'$$

Perpendicular distance of line of action of connecting rod force from centre of shaft =

$$\begin{aligned} OP &= r \times \sin(\phi + \theta) \\ &= 0{\cdot}315 \times \sin(8°40' + 35°) \\ &= 0{\cdot}315 \times \sin 43°40' \\ &= 0{\cdot}2174 \text{ m} \end{aligned}$$

Referring to vector diagram of Fig. 199,

$$\begin{aligned} \text{Thrust in conn. rod} &= \frac{\text{Piston force}}{\cos \phi} \\ &= \frac{250}{\cos 8°40'} \\ &= 252{\cdot}7 \text{ kN} \end{aligned}$$

$$\begin{aligned} \text{Twisting moment} &= \text{Thrust in conn. rod} \times \text{perp. distance} \\ &= 252{\cdot}7 \times 0{\cdot}2174 \\ &= 54{\cdot}94 \text{ kN m} \quad \text{Ans.} \end{aligned}$$

10.

Total force on each ram $= 70 \times 10^5 \times 0{\cdot}7854 \times 0{\cdot}25^2$ N

Turning moment exerted by two rams

$$= 70 \times 10^5 \times 0{\cdot}7854 \times 0{\cdot}25^2 \times 0{\cdot}76 \times 2$$

$$T = \frac{\pi}{16} D^3 \tau \qquad \tau = \frac{16\,T}{\pi\,D^3}$$

$$\tau = \frac{16 \times 70 \times 10^5 \times 0{\cdot}7854 \times 0{\cdot}25^2 \times 0{\cdot}76 \times 2}{\pi \times 0{\cdot}43^3}$$

$$= 3{\cdot}345 \times 10^7 \text{ N/m}^2$$

$$= 33{\cdot}45 \text{ MN/m}^2 \quad \text{Ans.}$$

11. Torsional resilience $=$ Work done to twist shaft

$$= \tfrac{1}{2} \text{ (Torque} \times \text{Angle of twist in radians)}$$

$$= \frac{T\theta}{2}$$

For a hollow shaft,

$$T = \frac{\pi (D^4 - d^4)\, \tau}{16D}$$

$$= \frac{\pi (D^2 + d^2)(D^2 - d^2)\, \tau}{16D}$$

and from

$$\frac{\tau}{r} = \frac{G\theta}{l}$$

$$\theta = \frac{\tau l}{Gr} = \frac{\tau l \times 2}{G \times D}$$

Substituting values of T and θ,

$$\text{Torsion resilience} = \frac{T\theta}{2}$$

$$= \frac{\pi (D^2 + d^2)(D^2 - d^2) \times \tau \times \tau \times l \times 2}{16 \times D \times 2 \times G \times D}$$

$$= \frac{\pi (D^2 + d^2)(D^2 - d^2) \times \tau^2 \times l}{16 \times G \times D^2}$$

$$\text{Volume of shaft} = \frac{\pi}{4} (D^2 - d^2) \times l,$$

taking these quantities out and replacing by volume,

$$\text{Torsional resilience} = \frac{(D^2 - d^2) \times \tau^2 \times \text{volume}}{4 \times G \times D^2}$$

$$= \frac{\tau^2}{4G} \times \frac{D^2 + d^2}{D^2} \times \text{ volume.} \quad \text{Ans.}$$

12. Solid shaft:

$$J = \frac{\pi \times 0{\cdot}03^4}{32}$$

$$= 7{\cdot}952 \times 10^{-8} \text{ m}^4$$

$$T = \frac{\tau \times J}{r}$$

$$= \frac{20 \times 10^6 \times 7{\cdot}952 \times 10^{-8}}{0{\cdot}015}$$

$$\text{Max. torque} = 106 \text{ Nm} \quad \text{Ans. } (a)$$

Hollow shaft:

$$J = \frac{\pi \times (0{\cdot}038^4 - 0{\cdot}03^4)}{32}$$

$$= 1{\cdot}252 \times 10^{-7} \text{ m}^4$$

$$\tau = \frac{T\,r}{J}$$

$$= \frac{106 \times 0{\cdot}019}{1{\cdot}252 \times 10^{-7}}$$

$$\text{Max. shear stress} = 16{\cdot}09 \text{ MN/m}^2 \quad \text{Ans. } (b)$$

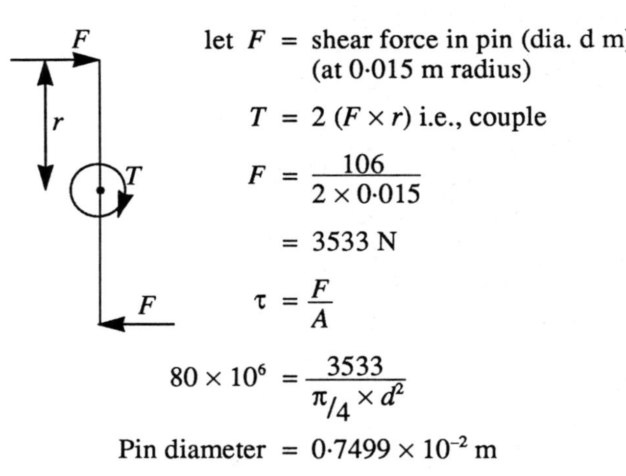

let F = shear force in pin (dia. d m) (at 0·015 m radius)

$$T = 2\,(F \times r) \text{ i.e., couple}$$

$$F = \frac{106}{2 \times 0{\cdot}015}$$

$$= 3533 \text{ N}$$

$$\tau = \frac{F}{A}$$

$$80 \times 10^6 = \frac{3533}{\pi/_4 \times d^2}$$

$$\text{Pin diameter} = 0{\cdot}7499 \times 10^{-2} \text{ m}$$

$$= 7{\cdot}5 \text{ mm} \quad \text{Ans. } (c)$$

13. Let p = maximum pressure in system when relief valves lift, in N/m^2,

$$\text{Max. ram force} = p \times 0{\cdot}7854 \times 0{\cdot}28^2 \text{ N}$$

$$\text{Max. torque on rudder stock} = \frac{p \times 0{\cdot}7854 \times 0{\cdot}28^2 \times 0{\cdot}86}{\cos^2 35°} \text{ Nm}$$

$$T = \frac{\pi D^2 \tau}{16}$$

$$\frac{p \times 0{\cdot}7854 \times 0{\cdot}28^2 \times 0{\cdot}86}{\cos^2 35°} = \frac{\pi \times 0{\cdot}35^2 \times 77 \times 10^6}{16}$$

$$p = \frac{\pi \times 0{\cdot}35^2 \times 77 \times 10^6 \times \cos^2 35°}{16 \times 0{\cdot}7854 \times 0{\cdot}28^2 \times 0{\cdot}86}$$

$$= 8{\cdot}215 \times 10^6 \text{ N/m}^2$$

$$= 82{\cdot}15 \text{ bar} \quad \text{Ans.}$$

14.
$$\delta = \frac{64\, WR^3 N}{G\, d^4} \quad \text{(see Fig. 202 and proof)}$$

Working in mm:

$$= \frac{64 \times 50 \times 15^3 \times 27}{93 \times 10^3 \times 3^4}$$

$$= 38{\cdot}71 \text{ mm} \quad \text{Ans.}$$

15.
$$T = \frac{\pi\,(D^4 - d^4)\,\tau}{16D} = W \times R$$

Working in mm:

$$W \times 37{\cdot}5 = \frac{\pi\,(12{\cdot}5^2 + 7{\cdot}5^2)(12{\cdot}5^2 - 7{\cdot}5^2) \times 70}{16 \times 12{\cdot}5}$$

$$W = \frac{\pi \times 212{\cdot}5 \times 100 \times 70}{16 \times 12{\cdot}5 \times 37{\cdot}5}$$

$$= 623 \text{ N} \quad \text{Ans. } (i)$$

$$\delta = \frac{64\,WR^3\,N}{G\,(D^4 - d^4)}$$

$$= \frac{64 \times 623 \times 37{\cdot}5^3 \times 24}{90 \times 10^3 \times 212{\cdot}5 \times 100}$$

$$= 26{\cdot}38 \text{ mm} \quad \text{Ans. } (ii)$$

SOLUTIONS TO TEST EXAMPLES 12

1. As the proportion is by mass, let the masses of the three liquids be 3, 4 and 5 grammes respectively.

$$\text{Total mass} = 3 + 4 + 5 = 12 \text{ g}$$

As density in g/cm^3 is numerically the same as relative density, their densities are 0·7, 0·8 and 0·9 g/cm^3 respectively.

$$\text{Volume} = \frac{\text{mass}}{\text{density}}$$

$$\text{Total volume} = \frac{3}{0{\cdot}7} + \frac{4}{0{\cdot}8} + \frac{5}{0{\cdot}9}$$

$$= \frac{2{\cdot}16 + 2{\cdot}52 + 2{\cdot}8}{0{\cdot}7 \times 0{\cdot}8 \times 0{\cdot}9}$$

$$= \frac{7{\cdot}48}{0{\cdot}504} \text{ cm}^3$$

$$\text{Density of mixture} = \frac{\text{total mass}}{\text{total volume}}$$

$$= \frac{12 \times 0{\cdot}504}{7{\cdot}48}$$

$$= 0{\cdot}8085 \text{ g/cm}^3$$

$$\text{Relative density} = 0{\cdot}8085 \quad \text{Ans.}$$

2. $$\text{Relative density} = \frac{\text{wt. in air}}{\text{wt. in air} - \text{wt. in water}}$$

$$= \frac{7}{7 - 6{\cdot}17}$$

$$= \frac{7}{0{\cdot}83}$$

$$= 8{\cdot}433 \quad \text{Ans.}$$

3. Let m = mass of casting

Downward force = Upward force

Wt. of plank + wt. of casting

= wt. of water displaced

$$3 \times 0{\cdot}3 \times 0{\cdot}15 \times 0{\cdot}6 \times 10^3 \times 9{\cdot}81 + m \times 9{\cdot}81 = 3 \times 0{\cdot}3 \times 0{\cdot}125 \times 10^3 \times 9{\cdot}81$$

$$3 \times 0{\cdot}3 \times 0{\cdot}15 \times 0{\cdot}6 \times 10^3 + m = 3 \times 0{\cdot}3 \times 0{\cdot}125 \times 10^3$$

$$81 + m = 112{\cdot}5$$

$$m = 31{\cdot}5 \text{ kg} \quad \text{Ans.}$$

4.

$$\text{Volume of lead} = \frac{\text{mass}}{\text{density}}$$

$$= \frac{508}{11{\cdot}4 \times 10^3} \text{ m}^3$$

Mass of sea water displaced by lead

$$= \text{volume} \times \text{density}$$

$$= \frac{508}{11{\cdot}4 \times 10^3} \times 1{\cdot}025 \times 10^3$$

$$= 45{\cdot}67 \text{ kg}$$

$$\text{Mass of raft} = 7{\cdot}6 \times 1{\cdot}22 \times 0{\cdot}23 \times 0{\cdot}75 \times 10^3$$

$$= 1600 \text{ kg}$$

Let d = draught of raft in m

Mass of sea water displaced by raft

$$= 7{\cdot}6 \times 1{\cdot}22 \times d \times 1{\cdot}025 \times 10^3$$

$$= 9504d \text{ kg}$$

$$\text{Downward forces} = \text{Upward forces}$$

$$\text{Wt. of raft + wt. of lead} = \text{wt. of water displaced}$$

$$1600 \times 9{\cdot}81 \times 508 \times 9{\cdot}81 = 9504d \times 9{\cdot}81 + 45{\cdot}67 \times 9{\cdot}81$$

$$1600 + 508 = 9504d + 45{\cdot}67$$

$$9504d = 2062{\cdot}23$$

$$d = 0{\cdot}217 \text{ m} = 217 \text{ mm} \quad \text{Ans.}$$

5. The pressures due to the head of oil and the head of mercury must be equal.

$$\text{Pressure} = \rho g h$$

$$\rho_o g h_o = \rho_m g h_m$$

Density of oil × head of oil = Density of mercury × head of mercury

$$\text{Density of oil} \times 1{\cdot}511 = 13{\cdot}6 \times 10^3 \times 0{\cdot}1$$

$$\text{Density of oil} = 900 \text{ kg/m}^3$$

$$\text{Relative density of oil} = 0{\cdot}9 \quad \text{Ans.}$$

6. Centre of gravity of wetted plate below free surface

$$= \tfrac{1}{2} \times 5{\cdot}4 = 2{\cdot}7 \text{ m}$$

$$\text{Area of wetted plate} = 9{\cdot}6 \times 5{\cdot}4 \text{ m}^2$$

$$\text{Density of oil} = 0{\cdot}85 \times 10^3 \text{ kg/m}^3$$

$$\text{Load} = HA\rho g$$

$$= 2{\cdot}7 \times 9{\cdot}6 \times 5{\cdot}4 \times 0{\cdot}85 \times 10^3 \times 9{\cdot}81$$

$$= 1{\cdot}168 \times 10^6 \text{ N}$$

$$= 1{\cdot}168 \text{ MN} \quad \text{Ans. } (i)$$

Centre of pressure on a rectangular area is at one-third of the height from the base.

$$\text{Centre of pressure} = \tfrac{1}{3} \times 5{\cdot}4$$

$$= 1{\cdot}8 \text{ m from bottom.} \quad \text{Ans. } (ii)$$

7. $$\text{Water thrust on one side} = HA\rho g$$

$$= 3{\cdot}6 \times 7{\cdot}5 \times 7{\cdot}2 \times 10^3 \times 9{\cdot}81$$

$$= 1{\cdot}907 \times 10^6 \text{ N} = 1{\cdot}907 \text{ MN}$$

$$\text{Centre of water pressure} = \tfrac{1}{3} \times 7{\cdot}2$$

$$= 2{\cdot}4 \text{ m from bottom}$$

$$\text{Oil thrust on other side} = HA\rho g$$

$$= 2{\cdot}1 \times 7{\cdot}5 \times 4{\cdot}2 \times 0{\cdot}8 \times 10^3 \times 9{\cdot}81$$

$$= 5{\cdot}192 \times 10^5 \text{ N} = 0{\cdot}5192 \text{ MN}$$

$$\text{Centre of oil pressure} = \tfrac{1}{3} \times 4{\cdot}2$$

$$= 1{\cdot}4 \text{ m from bottom}$$

$$\text{Resultant thrust} = \text{thrust on one side} - \text{thrust on other side}$$

$$= 1{\cdot}907 - 0{\cdot}5192$$

$$= 1{\cdot}3878 \text{ MN} \quad \text{Ans. } (i)$$

Let y = distance of resultant centre of pressure from bottom.
Taking moments about bottom,

$$1{\cdot}3878 \times y = 1{\cdot}907 \times 2{\cdot}4 - 0{\cdot}5192 \times 1{\cdot}4$$

$$1{\cdot}3878 \times y = 4{\cdot}577 - 0{\cdot}7296$$

$$1{\cdot}3878 \times y = 3{\cdot}8501$$

$$y = 2{\cdot}774 \text{ m from bottom.} \quad \text{Ans. } (ii)$$

8. The floating wood displaces an amount of water equal to its own mass, therefore,

$$\text{Mass of water displaced} = 156 \text{ kg}$$

$$\text{Volume of water displaced} = \frac{\text{mass}}{\text{density}} = \frac{156}{10^3} = 0{\cdot}156 \text{ m}^3$$

$$\text{Rise of water level in tank} = \frac{\text{volume}}{\text{area}}$$

$$= \frac{0{\cdot}156}{2 \times 1{\cdot}2} = 0{\cdot}065 \text{ m}$$

$$\text{New depth of water} = 0{\cdot}6 + 0{\cdot}065 = 0{\cdot}665 \text{ m}$$

$$\text{Load on bottom} = HA\rho g$$

$$= 0{\cdot}665 \times 2 \times 1{\cdot}2 \times 10^3 \times 9{\cdot}81$$

$$= 1{\cdot}566 \times 10^4 \text{ N} = 15{\cdot}66 \text{ kN} \quad \text{Ans. } (i)$$

$$\text{Load on each side} = 0{\cdot}3325 \times 2 \times 0{\cdot}665 \times 10^3 \times 9{\cdot}81$$

$$= 4{\cdot}338 \times 10^3 \text{ N} = 4{\cdot}338 \text{ kN} \quad \text{Ans. } (ii)$$

$$\text{Load on each side} = 0{\cdot}3325 \times 1{\cdot}2 \times 0{\cdot}665 \times 10^3 \times 9{\cdot}81$$

$$= 2{\cdot}602 \times 10^3 \text{ N} = 2{\cdot}602 \text{ kN} \quad \text{Ans. } (iii)$$

9. Proof – see Text of Chapter 12 (including Fig. 207). Ans.

10.

$$p = \rho g h_{max}$$

$$= 1000 \times 9{\cdot}81 \times 2$$

$$\text{Max. pressure} = 19620 \text{ N/m}^2 \quad \text{Ans. } (a)$$

Hydrostatic load $= HA\rho g$

$$A = 4\pi r^2 \text{ and } H = \frac{d}{2}$$

Hydrostatic level $= 1 \times 4\pi \times 1^2 \times 1000 \times 9{\cdot}81$

$= 123{,}214$ N Ans. (*b*)

Resultant force on spherical surface

$=$ wt. of fluid $=$ specific wt. $\times$ vol.

$$= \frac{4}{3} \times \rho g \times \pi r^3$$

$$= 1000 \times 9{\cdot}81 \times 1{\cdot}333 \times \pi \times 1^3$$

Resultant force $= 41{,}000$ N Ans. (c)

11. Proof – see Text of Chapter 7 (including Figs. 120 and 122). Ans.

$$J = \frac{\pi D^4}{32}$$

$$I_{xx} = \frac{\pi D^4}{64} = I_G$$

$$I_{xx} = I_G + AH^2$$

$$y_{cp} = \frac{I_G}{AH} + H$$

$$= \frac{\frac{\pi}{64} D^4}{\frac{\pi}{4} D^2 \times \frac{1}{2} D} + \frac{D}{2}$$

$$y_{cp} = \frac{5}{8} D \quad \text{Ans.}$$

12.

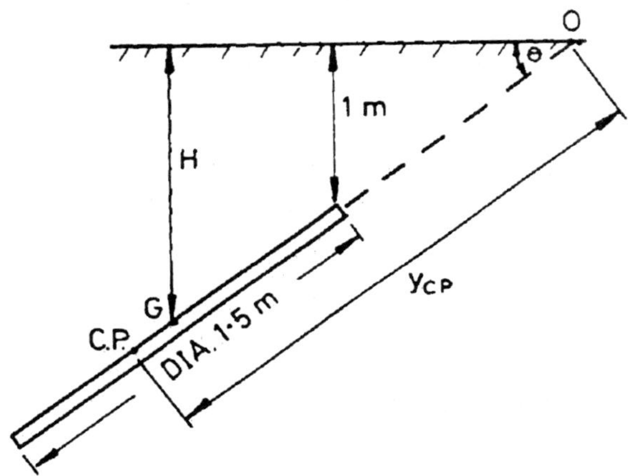

Referring to the above figure:

$$H = 1 + 0{\cdot}75 \sin 35°$$

$$= 1{\cdot}4302 \text{ m}$$

$$y_{cp} = \frac{I_G}{A\,H / \sin\theta} + \frac{H}{\sin\theta}$$

$$= \frac{\pi \times 1{\cdot}5^4 \times \sin 35°}{64 \times 0{\cdot}7854 \times 1{\cdot}5^2 \times 1{\cdot}4302} + \frac{1{\cdot}4302}{\sin 35°}$$

$$= 2{\cdot}55$$

i.e. Centre of pressure is 2·55 m from the surface at 0. Ans.

Load on door $= HA\rho g$

$$= 1{\cdot}4302 \times 0{\cdot}7854 \times 1{\cdot}5^2 \times 10^3 \times 9{\cdot}81$$

$$= 24{\cdot}8 \text{ kN}$$ Ans.

13. (*a*) Hydrostatic load $= HA\rho g$

$$= \frac{9}{3} \times \frac{6 \times 9}{2} \times 1{\cdot}025 \times 9{\cdot}81$$

$$= 814{\cdot}5 \text{ kN} \quad \text{Ans. } (a)$$

C.o.P. from top $= \frac{D}{2}$

$$= \frac{9}{2}$$

$$= 4{\cdot}5 \text{ m} \quad \text{Ans. } (b)$$

(*b*) Hydrostatic load $= \left(\frac{9}{3} + 4\right) \times \frac{6 \times 9}{2} \times 1{\cdot}025 \times 9{\cdot}81$

$$= 1901 \text{ kN} \quad \text{Ans. } (b)$$

C.o.P. from surface $= \frac{I_G}{AH} + H$

$$= \frac{\frac{1}{36} \times 6 \times 9^3}{\frac{1}{2} \times 6 \times 9 \times 7} + 7$$

$$= 7{\cdot}642 \text{ m}$$

C.o.P. from top $= 7{\cdot}642 - 4$

$$= 3{\cdot}642 \text{ m} \quad \text{Ans. } (b)$$

14. Hydrostatic load $= HA\rho g$

$$= 0{\cdot}3 \times 0{\cdot}6 \times 0{\cdot}3 \times 1000 \times 9{\cdot}81$$

$$= 529{\cdot}7 \text{ N} \quad \text{Ans.}$$

$$y_{cp} = \frac{I_G}{A\,H} + H$$

$$= \frac{0{\cdot}3 \times 0{\cdot}6^3}{12 \times 0{\cdot}3 \times 0{\cdot}6 \times 0{\cdot}3} + 0{\cdot}3$$

$$\text{C.o.P.} = 0{\cdot}4 \text{ m} \quad \text{Ans.}$$

$$y_{cp} = \frac{0{\cdot}3 \times 0{\cdot}6^3}{12 \times 0{\cdot}3 \times 0{\cdot}6 \times 1{\cdot}3} + 1{\cdot}3$$

$$= 1{\cdot}323 \text{ m}$$

$$\text{New C.o.P. from top} = 0{\cdot}323 \text{ m} \quad \text{Ans.}$$

15.

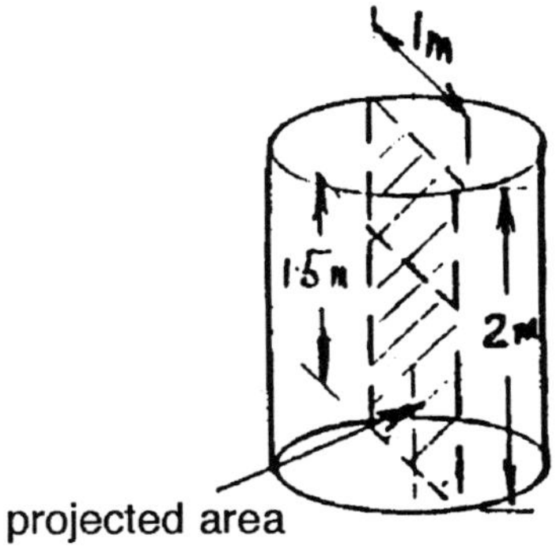

Due to internal pressure:

$$\text{force} = \text{pressure} \times \text{projected area}$$

$$= 120 \times 10^3 \times 2 \times 1$$

$$= 240 \text{ kN (C.o.P. at 1 m above base)}$$

Due to hydrostatic load:

$$\text{force} = HA\rho g$$

$$= 10^3 \times g \times 0{\cdot}75 \times 1{\cdot}5 \times 1$$

$$= 11{\cdot}036 \text{ kN (C.o.P. at 0.5 m above base)}$$

$$\text{Resultant force} = 240 + 11{\cdot}036 = 251{\cdot}036 \text{ kN}$$

$$\text{Force on a seam} = 251{\cdot}036 \div 2 = 125{\cdot}5 \text{ kN} \quad \text{Ans. (a)}$$

Take moments about bottom of tank:

(let y_{cp} = vertical distance to resultant C.o.P.)

$$251{\cdot}036 \times y_{cp} = (240 \times 1) + (11{\cdot}036 \times 0{\cdot}5)$$

$$y_{cp} = 0{\cdot}978$$

i.e., Centre of pressure at 0·978 m from bottom. Ans. (*b*)

SOLUTIONS TO TEST EXAMPLES 13

1.

$$\text{Volume flow} = \text{area} \times \text{velocity}$$
$$= 0{\cdot}7854 \times 0{\cdot}15^2 \times 1{\cdot}75$$
$$\dot{V} = 0{\cdot}03092 \text{ m}^3/\text{s}$$

$$\text{Mass flow rate} = \text{volume flow rate} \times \text{density}$$
$$= 0{\cdot}03092 \times 1{\cdot}025 \times 10^3$$
$$\dot{m} = 31{\cdot}75 \text{ kg/s}$$

Force to lift 31·7 kg against gravity

$$= 31{\cdot}7 \times 9{\cdot}81 = 311 \text{ N}$$

$$\text{Power} = \text{work done/s}$$
$$= \text{force} \times \text{height/s}$$
$$= 311 \times 9$$
$$P = 2799 \text{ W or } 2{\cdot}799 \text{ kW}$$

$$\text{Input power} = \frac{\text{output power}}{\text{efficiency}}$$
$$= \frac{2{\cdot}799}{0{\cdot}72} = 3{\cdot}888 \text{ kW} \quad \text{Ans.}$$

Alternatively,

$$\text{Work done/s} = \text{pressure} \times \text{volume flow}$$
$$= 9 \times 1{\cdot}025 \times 10^3 \times 9{\cdot}81 \times 0{\cdot}03092$$
$$= 2{\cdot}799 \times 10^3 \text{ W}$$

2. $$\text{Area of escape} = \text{circumference} \times \text{lift}$$
$$= \pi \times 60 \times 10 \text{ mm}^2$$
$$= \pi \times 60 \times 10 \times 10^{-6} \text{ m}^2$$
$$\dot{V} = av$$
$$= \pi \times 60 \times 10 \times 10^{-6} \times 1{\cdot}8 \times 3600$$
$$= 12{\cdot}22 \text{ m}^3/\text{h} \quad \text{Ans.}$$

3. $$Q_M = Q_B$$
$$A_1 v_1 = A_2 v_2$$
$$0{\cdot}7854 \times D^2 \times 6 = 0{\cdot}7854 \times 60^2 \times 3 \times 8$$
$$\text{Fire main dia.} = 120 \text{ mm} \quad \text{Ans. } (a)$$
$$Q = Av$$
$$= 0{\cdot}7854 \times 0{\cdot}12^2 \times 6$$
$$\dot{m} = 0{\cdot}7854 \times 0{\cdot}12^2 \times 6 \times 1025$$
$$\text{Mass flow rate} = 69{\cdot}55 \text{ kg/s} \quad \text{Ans. } (b)$$
$$Q_B = Q_N$$
$$0{\cdot}7854 \times 60^2 \times 8 = 0{\cdot}7854 \times 12^2 \times v_N$$
$$\text{Discharge velocity} = 200 \text{ m/s} \quad \text{Ans. } (c)$$

4. $$\text{Area of escape} = {}^5\!/_6 \times \text{circumference} \times \text{lift}$$
$$= {}^5\!/_6 \times \pi \times d \times l \times 10^{-6} \text{ m}^2$$
$$\text{Volume flow rate} = \text{area} \times \text{velocity}$$
$$\dot{V} = {}^5\!/_6 \times \pi \times d \times l \times 10^{-6} \times v \times 3600$$
$$\text{Density} = \rho \text{ g/ml} = \rho \text{ t/m}^3$$

$$\text{Mass flow rate} = {}^5/_6 \times \pi \times d \times l \times 10^{-6} \times v \times 3600 \times \rho$$

$$\dot{m} = 3\pi \times 10^{-3}\, d\, l\, v\, \rho$$

$$\text{or } 9{\cdot}425 \times 10^{-3}\, d\, l\, v\, \rho \quad \text{Ans. } (i)$$

Inserting given values:

$$\text{Mass flow rate} = 9{\cdot}425 \times 10^{-3} \times 76 \times 13 \times 3 \times 0{\cdot}997$$

$$\dot{m} = 27{\cdot}86\ t/\text{h} \quad \text{Ans. } (ii)$$

5. $$\text{Velocity of jet} = C_v \sqrt{2\,gh}$$

$$= 0{\cdot}97 \times \sqrt{2 \times 9{\cdot}81 \times 2{\cdot}5}$$

$$= 6{\cdot}794 \text{ m/s} \quad \text{Ans. } (i)$$

$$\dot{V} = a\, v$$

$$= 0{\cdot}64 \times 0{\cdot}7854 \times 0{\cdot}019^2 \times 6{\cdot}794$$

$$= 1{\cdot}233 \times 10^{-3} \text{ m}^3\text{/s}$$

$$= 1{\cdot}233\ l\text{/s} \quad \text{Ans. } (ii)$$

6. $$\text{Theoretical velocity} = \sqrt{2gh}$$

$$= \sqrt{2 \times 9{\cdot}81 \times 1{\cdot}5}$$

$$= 5{\cdot}425 \text{ m/s}$$

Let t secs = time to fall 0·6 m, initial vertical velocity = 0

$$h = {}^1/_2\, at^2$$

$$t = \sqrt{\frac{2h}{a}} = \sqrt{\frac{2 \times 0{\cdot}6}{9{\cdot}81}} = 0{\cdot}3497 \text{ s}$$

$$\text{Horizontal distance} = \text{horizontal velocity} \times \text{time}$$

$$1{\cdot}84 = \text{horizontal velocity} \times 0{\cdot}3497$$

$$\text{Velocity of jet} = \frac{1{\cdot}84}{0{\cdot}3497} = 5{\cdot}261 \text{ m/s}$$

$$\text{Coeff. of velocity} = \frac{\text{actual velocity}}{\text{theoretical velocity}}$$

$$= \frac{5{\cdot}261}{5{\cdot}425} = 0{\cdot}97$$

$$\text{Theoretical discharge rate} = \text{area of hole} \times \text{theoretical velocity}$$

$$= 0{\cdot}7854 \times 0{\cdot}01^2 \times 5{\cdot}425$$

$$= 4{\cdot}261 \times 10^{-4} \text{ m}^3\text{/s}$$

$$l\text{/min} = 4{\cdot}261 \times 10^{-4} \times 10^3 \times 60$$

$$= 25{\cdot}57$$

$$\text{Coeff. of discharge} = \frac{\text{actual velocity}}{\text{theoretical velocity}}$$

$$= \frac{15{\cdot}85}{25{\cdot}57} = 0{\cdot}62$$

$$\text{Coeff. of reduction of area} = \frac{\text{Coeff. of discharge}}{\text{Coeff. of velocity}}$$

$$= \frac{0{\cdot}62}{0{\cdot}97} = 0{\cdot}6392$$

$$\left.\begin{aligned}\text{Coeff. of velocity} &= 0{\cdot}97\\ \text{Coeff. of area} &= 0{\cdot}6392\\ \text{Coeff. of discharge} &= 0{\cdot}62\end{aligned}\right\} \text{Ans.}$$

7. $$\text{Loss of Pressure} = \text{Gain in kinetic energy}$$

$$p_1V - p_2V = {}^1\!/\!_2\, mv_2^{\,2} - {}^1\!/\!_2\, mv_1^{\,2}$$

$$V(p_1 - p_2) = {}^1\!/\!_2\, m\,(v_2^{\,2} - v_1^{\,2})$$

$$\frac{m}{V} = \text{density} = \rho$$

$$p_1 - p_2 = \tfrac{1}{2}\,\rho\,(v_2^2 - v_1^2)$$

$$14 \times 10^3 = \tfrac{1}{2} \times 10^3\,(v_2^2 - 1{\cdot}5^2)$$

$$v_2^2 = 14 \times 2 + 2{\cdot}25$$

$$v_2 = \sqrt{30{\cdot}25}$$

$$= 5{\cdot}5 \text{ m/s} \quad \text{Ans.}$$

8. Pressure difference $= 457$ mm water

$$= 457 \times 9{\cdot}81 \text{ N/m}^2$$

$$p_1 - p_2 = \tfrac{1}{2}\,\rho\,(v_2^2 - v_1^2)$$

$$457 \times 9{\cdot}81 = \tfrac{1}{2} \times 10^3\,(v_2^2 - v_1^2) \quad \text{Ans. } (i)$$

Volume flow rate at any point is constant

$$\text{area}_1 \times \text{velocity}_1 = \text{area}_2 \times \text{velocity}_2$$

$$v_2 = \frac{0{\cdot}7854 \times 0{\cdot}375^2}{0{\cdot}7854 \times 0{\cdot}125^2} \times v_1$$

$$v_2 = 3^2 \times v_1 = 9\,v_1$$

$$v_2^2 = 81\,v_1^2$$

Substituting for v_2^2 into (i),

$$457 \times 9{\cdot}81 = \tfrac{1}{2} \times 10^3\,(81v_1^2 - v_1^2)$$

$$457 \times 9{\cdot}81 = \tfrac{1}{2} \times 10^3 \times 80v_1^2$$

$$v_1 = \sqrt{\frac{457 \times 9{\cdot}81 \times 2}{10^3 \times 80}}$$

$$= 0{\cdot}3347 \text{ m/s}$$

$$\text{Volume flow rate} = 0{\cdot}7854 \times 0{\cdot}375^2 \times 0{\cdot}3347$$

$$= 0{\cdot}03697 \text{ m}^3\text{/s}$$

$$\text{Mass flow rate} = \text{volume flow} \times \text{density}$$

$$= 0{\cdot}03697 \times 10^3$$

$$= 36{\cdot}97 \text{ kg/s} \quad \text{Ans.}$$

9.

$$v_1\, d_1^{\,2} = v_2\, d_2^{\,2}$$

$$v_2 = 9 \times \left(\frac{0{\cdot}1}{0{\cdot}2}\right)^2$$

$$= 2{\cdot}25 \text{ m/s}$$

$$\text{Head loss} = \frac{0{\cdot}25\,(9^2 - 2{\cdot}25^2)}{9{\cdot}81}$$

$$= 1{\cdot}935 \text{ m}$$

$$\frac{p_1}{\rho g} + \frac{v_1^{\,2}}{2g} + z_1 = \frac{p_2}{\rho g} + \frac{v_2^{\,2}}{2g} + z_2 + hf$$

$$\frac{300 \times 10^3}{1025 \times g} + \frac{9^2}{2g} + 0 = \frac{p_2}{1025 \times g} + \frac{2{\cdot}25^2}{2g} + 6 + 1{\cdot}935$$

$$p_2 = 259{\cdot}1 \times 10^3$$

$$\text{Pressure at top} = 259{\cdot}1 \text{ kN/m}^2 \quad \text{Ans.}$$

10. (*a*) The three forms of energy are:

pressure energy, velocity energy and potential energy

$$\text{Total energy (J/N } i.e., m) = \frac{p}{\rho g} + \frac{v_2^{\,2}}{2g} + z$$

where p = pressure (N/m^2)
ρ = density (kg/m^3)
v = velocity (m/s)
z = head above P.E. datum (m) Ans. (*a*)

(*b*) For flow continuity:

$$v_1\, d_1^{\,2} = v_2\, d_2^{\,2}$$

$$\therefore \quad v_2 = v_1 \times \left(\frac{0{\cdot}05}{0{\cdot}025}\right)^2$$

$$v_2 = 4v_1$$

$$\frac{p_1}{\rho g} + \frac{v_1^{\,2}}{2g} = \frac{p_2}{\rho g} + \frac{v_2^{\,2}}{2g} + z_2 \text{ (} z_1 \text{ is zero)}$$

$$\frac{p_1}{\rho g} - \frac{p_2}{\rho g} = \frac{v_2^{\,2} - v_1^{\,2}}{2g} + z_2$$

$$h_1 - h_2 = \frac{(4\,v_1)^2 - v_1^{\,2}}{2\,g} + z_2$$

$$0{\cdot}25 = \frac{15\,v_1^{\,2}}{2\,g} + 0{\cdot}15$$

$$v_1 = 0{\cdot}3617 \text{ m/s}$$

$$\dot{m} = 0{\cdot}3617 \times \frac{\pi}{4} \times 0{\cdot}05^2 \times 0{\cdot}95 \times 10^3$$

Mass flow rate of water $= 0{\cdot}675$ kg/s Ans. (*b*)

11.

$$v = \frac{\dot{V}}{A}$$

$$= \frac{0{\cdot}1}{0{\cdot}7854 \times 0{\cdot}2^2}$$

$$= 3{\cdot}18 \text{ m/s}$$

$$h_f = \frac{4fl}{d} \times \frac{v^2}{2g}$$

$$= \frac{4 \times 0{\cdot}005 \times 100 \times 3{\cdot}18^2}{0{\cdot}2 \times 2 \times 9{\cdot}81}$$

$$= 5{\cdot}14 \text{ m}$$

Loss of level due to friction is 5·14 m Ans.

$$\text{Slope} = \frac{5 \cdot 14}{100}$$

$$= 0 \cdot 0514$$

$$\text{Slope of hydraulic gradient} = 0 \cdot 0514 \quad \text{Ans.}$$

12. Let v = velocity of jet

Mass of water striking plate per second

$$= \text{area} \times \text{velocity} \times \text{density}$$

$$= 0 \cdot 7854 \times 0 \cdot 04^2 \times v \times 10^3$$

$$= 1 \cdot 257\, v \text{ kg/s}$$

$$\text{Change of velocity} = v - 0 = v \text{ m/s}$$

$$\text{Force} = \text{change of momentum/s}$$

$$= \text{mass/s} \times \text{change of velocity}$$

$$225 = 1 \cdot 257\, v \times v$$

$$v = \sqrt{\frac{225}{1 \cdot 257}}$$

$$= 13 \cdot 39 \text{ m/s} \quad \text{Ans.}$$

13. Let b_2 = width at exit:

$$r_1 \times b_1 = r_2 \times b_2$$

$$150 \times 120 = 300 \times b_2$$

$$b_2 = 60 \text{ mm} \quad \text{Ans. } (i)$$

Let b_3 = width at mid-radius, which is ½ (150 + 300) = 225 mm

$$r_1 \times b_1 = r_3 \times b_3$$

$$150 \times 120 = 300 \times b_3$$

$$b_3 = 80 \text{ mm} \quad \text{Ans. } (ii)$$

Linear velocity of vanes at entrance,

$$s_1 = \pi \times 0{\cdot}3 \times \frac{420}{60} = 6{\cdot}6 \text{ m/s}$$

$$\tan\theta_1 = \frac{v_1}{s_1} = \frac{1{\cdot}65}{6{\cdot}6} = 0{\cdot}25$$

$$\text{Angle of vanes at entrance} = 14^\circ\, 2' \quad \text{Ans. } (iii)$$

14.

$$\text{Volumetric flow rate} = \frac{108}{3600}$$

$$= 0{\cdot}03 \text{ m}^3\text{/s}$$

$$\text{Velocity of water in discharge pipe} = \frac{\text{flow rate}}{\text{cross sect. area of pipe}}$$

$$= \frac{0{\cdot}03}{0{\cdot}7854 \times 0{\cdot}13^2}$$

$$\text{Velocity at discharge} = 2{\cdot}26 \text{ m/s}$$

Manometric head h_m = suctn. head + disch. head + velocity energy + friction losses

$$= z_S + z_D + \frac{v_2^2}{2g} + \text{friction losses}$$

$$= 3 + 76 + \frac{2{\cdot}26^2}{2g} + 13{\cdot}7$$

$$= 93 \text{ m}$$

$$\text{Pump output power} = mgh_m$$

$$= 0{\cdot}03 \times 1025 \times 9{\cdot}81 \times 93$$

$$= 28{\cdot}05 \text{ kW}$$

$$\text{Pump input power} = \frac{28{\cdot}05}{0{\cdot}65}$$

$$= 43{\cdot}15 \text{ kW} \quad \text{Ans.}$$

15. (*a*)

$$V = A\, C_D \sqrt{2gh}$$

$$\frac{5100}{3600} = \frac{\pi}{4} \times 0{\cdot}35^2 \times 0{\cdot}6 \times \sqrt{2gh}$$

$$2gh = 24{\cdot}54^2$$

$$\text{Diff. press. } h = 30{\cdot}696 \text{ m of air}$$

$$\text{Equivalent water head} = 30{\cdot}696 \times \frac{1{\cdot}22}{1000}$$

$$= 0{\cdot}03745 \text{ m}$$

$$= 37{\cdot}45 \text{ mm of water} \quad \text{Ans. } (a)$$

(b)

$$\text{New volume flow rate} = \frac{5100}{3 \times 3600}$$

$$= 0{\cdot}4722 \text{ m}^2\text{/s}$$

$$0{\cdot}4722 = \frac{\pi}{4}\, d^2 \times 0{\cdot}58 \times \sqrt{2 \times g \times 30{\cdot}696}$$

$$d = 0{\cdot}2055 \text{ m}$$

$$\text{Orifice diameter} = 205{\cdot}5 \text{ mm} \quad \text{Ans. } (b)$$

SELECTION OF EXAMINATION QUESTIONS

CLASS TWO

1. A worm-driven lifting block consists of a single-start worm driving a worm wheel of 80 teeth. The effort chain wheel is 360 mm diameter, the load chain drum is 120 mm diameter, and the load is suspended from a snatch block on the load chain. Find (*i*) the velocity ratio, (*ii*) the ideal effort to lift a mass of 2·04 Mg, (*iii*) the actual effort to lift this mass if the efficiency is 45%.

2. A pump has a suction lift of 4 m and it delivers water to a nozzle located 30 m above the pump. The velocity of discharge from the nozzle is 20 m/s. Friction loss in the pipes and nozzle amounts to 20% of the power absorbed by the pump.

Calculate (*a*) the mass flow rate of water, and (*b*) the time taken to discharge 50 t of water, given that the power supplied to the pump is 5 kW.

3. A vertical sluice gate is 600 mm wide by 900 mm high and its mass is 367 kg. It covers an inlet from the sea, the bottom of the gate being 5 m below water level. Taking the coefficient of friction between the sliding surfaces of the gate and its guides as 0·28, and the density of sea water as 1025 kg/m^3, find the maximum force required to lift the gate.

4. A mild steel tensile test bar originally 100 mm long between gauge points and 20 mm diameter, yielded when the axial tensile force was 87 kN and broke when the force was 145 kN. After fracture the minimum diameter at the waist measured 15·2 mm and length between gauge points 126 mm. Calculate (*i*) stress at yield point, (*ii*) ultimate tensile stress, (*iii*) percentage elongation, and (*iv*) percentage reduction of area. Sketch the stress-strain graph.

5. A specimen 22 mm diameter and 200 mm between gauge points was tested in a torsion machine and found to twist 2° when subjected to a torque of 343 N m and at this loading was just within

the elastic limit. Calculate the modulus of rigidity of the material. Calculate also the maximum power that can be transmitted by a shaft of identical material 175 mm diameter running at 300 rev/min allowing a maximum shear stress of 50% of the stress at the elastic limit.

6. A solid steel disc of uniform thickness is 400 mm diameter and its mass is 80 kg. If a hole 125 mm diameter is cut out, the centre of this hole being 100 mm from the centre of the disc, find the position of the centre of gravity of the holed disc measured from the centre of the hole. It is now rotated at 120 rev/min about the centre of the hole, calculate the centrifugal force set up.

7. A beam 10 m long is simply supported at each end. It carries a uniformly distributed load of 5 kN/m run over its entire length, and a concentrated load of 75 kN at 3 m from the left end. Find what upward force should be exerted on the beam at a point 4 m from the left end so that the bending moment at mid-span is zero.

8. An on-off cock controls the inflow of liquid to the top of a tank, and the liquid flows out through an orifice 300 mm diameter in the bottom, the coefficient of discharge being 0·62. By opening and closing the cock, the depth of the liquid is maintained between 3·5 m and 4 m and at this part of the tank the free surface area of the liquid is 25 m^2. Take the mean flow rate through the orifice as being based on the mean depth and calculate the maximum time the tank can operate under these conditions with the inflow cock shut off.

9. A metal door, of mass 509·7 kg, is supported by two hinges on one of its vertical sides. The distance between the hinge centres is 1·5 m and the centre of gravity of the door is 450 mm from the centre-line of the hinges. The bottom hinge carries the whole weight of the door. Calculate the reactions of the hinges.

10. A cubical tank of 900 mm sides contains fresh water to a depth of 600 mm. A block of steel of volume 0·03 m^3 and density 7860 kg/m^3 is suspended on a rope and completely immersed in the water. Find (*i*) the total force on the tank bottom due to the water before the steel was immersed, (*ii*) the increase of force on the tank bottom after immersing the steel, (*iii*) the tension in the rope.

11. A solid hemisphere of mass 25 kg rests with its curved surface on a rough inclined plane. A downward force of 65 N is exerted on the periphery of the flat surface of the hemisphere to maintain it in equilibrium with its flat surface horizontal. Find the angle of the incline.

12. A gear train consists of a primary wheel of 72 teeth on the input shaft of 100 mm diameter driving a secondary wheel of 24 teeth on the output shaft of 75 mm diameter with an idler of 20 teeth interposed between the primary and secondary wheels. The rotational speed of the input shaft is 100 rev/min and the maximum shear stress in it is 21 MN/m^2. If the efficiency of the gear train is 0·85, calculate the maximum shear stress in the output shaft and the power transmitted by it.

13. A valve is 60 mm diameter, its lift is 4 mm, and water flows through it at a velocity of 2 m/s. Allowing one-ninth of the circumference to be taken up by the wings of the valve, find the mass flow of water through the valve per hour. Take the density of the water as 1000 kg/m^3.

14. A body of mass 150 t is pulled up a gradient inclined at 3° to the horizontal by means of a cable running parallel to the gradient and over a winding drum 1·2 m diameter. The drum shaft is driven through a system of gears of 50 to 1 reduction and 0·75 efficient from an electric motor. The tractive resistance of the body to motion is 8×10^{-3} N/kg. When the body is moving at a velocity of 1·8 m/s, calculate (*i*) the rotational speed of the motor, (*ii*) the torque on the drum shaft, and (*iii*) the power of the electric motor.

15. A bullet, having an initial velocity of 480 m/s is fired into a stationary block of wood and emerges from the opposite side with its velocity reduced by 150 m/s. How many times thicker should the wood have been to just bring the bullet to rest.

16. In a single threaded worm driven jack the screw is prevented from turning as the worm rotates. A mass of 670 kg was lifted by an effort of 22·24 N, the efficiency being 32%. If the diameter of the effort wheel is 72 mm and the pitch of the screw 12 mm, find (*i*) the mechanical advantage, (*ii*) the velocity ratio, (*iii*) the number of teeth in the worm wheel.

17. A straight rigid beam *ABCDE* is 7·6 m long and 400 mm deep and is simply supported at points *B* and *D*. The beam carries uniformly distributed loads of $3{\cdot}3 \times 10^4$ N/m length over the two sections *AB* and *DE* and a concentrated load of 6×10^4 N at *C*. The distances of points *B*, *C* and *D* from *A* are 1·6, 3·2 and 6·4 m respectively. Sketch the shearing force and bending moment diagrams and mark thereon the values of the shearing force and bending moments at points *B*, *C* and *D*. Also calculate (*i*) the positions along the beam from *A* where the bending moment is zero, and (*ii*) the second moment of area of the beam so that the maximum bending stress will not exceed 60 MN/m^2.

18. A vertical sluice gate 1·8 m square covers an outlet to the sea. The gate is hinged along its top edge and opens into the sea. When the level of sea water is 0·9 m above the hinge and dry on the other side of the gate, calculate (*i*) the total water thrust on the gate, (*ii*) the horizontal force to apply at the bottom edge to open the gate. Take the density of the sea water as 1024 kg/m^3.

19. A pulley 500 mm diameter is driven at 375 rev/min by a belt 200 mm wide by 12 mm thick. The tension in the tight side of the belt is three times the tension in the slack side. Calculate the maximum power that can be transmitted if the stress in the belt is not to exceed 400 kN/m^2, neglecting the stress set up in the belt due to centrifugal force.

20. A hollow shaft rotating at 180 rev/min is found to twist 1·5° over a length of 800 mm when the torque is 1·52 kN m. The outside diameter of the shaft is 50 mm and inside diameter 30 mm. Calculate (*i*) the shear stress induced in the shaft, (*ii*) the modulus of rigidity, and (*iii*) the power transmitted.

21. The following four forces pull on a point: 15 N due East, 20 N North 55° East, 10 N North 75° West, and 20 N South 65° West. Calculate the resultant force and its direction.

22. A Warwick screw is used to tighten a guy rope. It has a right-hand thread of 10 mm pitch at the top and a left-hand thread of 5 mm pitch at the bottom, and effective radius of the toggle bar is 336 mm. Find (*i*) the velocity ratio, and (*ii*) the effort required at the end of the toggle to give a tension of 10 kN in the guy rope, taking the efficiency as 0·25.

23. A steel cylindrical pressure vessel, 2 m internal diameter has plates 20 mm thick. The longitudinal joint is welded and the weld is flush with the plates. The U.T.S. of the plate material is 200 MN/m^2 and the U.T.S. of the welded joint is 170 MN/m^2. The working pressure of the vessel is 0·6 MN/m^2.

Calculate (*a*) the percentage efficiency of the welded joint, (*b*) the stress at the longitudinal seam, and (*c*) the factor of safety for the vessel.

24. A four-ram electric-hydraulic steering gear has rams 300 mm diameter, and the rudder stock is 450 mm diameter. The centre lines of the cylinders are 750 mm on either side of the rudder stock. The rudder was struck by a heavy sea when in mid-position and the relief valves lifted at a pressure of 75 bar (= 75×10^5 N/m^2). Calculate the stress set up in the rudder stock.

25. A cubical tank of 200 mm sides contains oil of density 840 kg/m^3 to a depth of 100 mm. A cubical block of wood of 100 mm sides and density 630 kg/m^3 is now put into the oil and allowed to float freely. Find the load on the bottom and sides of the tank.

26. Calculate (*i*) the kinetic energy stored in a mass of 48 kg when moving with a velocity of 25 m/s. If the mass is brought to rest over a distance of 21·25 m, find (*ii*) the time to bring it to rest, (*iii*) the average retardation, and (*iv*) the retarding force.

27. A ballast pump discharges sea water through a pipe 125 mm diameter at a velocity of 1·2 m/s to a height of 5·8 m, the efficiency of the pump being 62%. Taking the density of sea water as 1025 kg/m^3, find the output and input power of the pump.

28. (See following diagram) A light stay rod, AB, is 2 m long and carries three forces at A, all acting in a vertical plane of 100 N, 150 N and 250 N.

Determine (a) the magnitude and direction of the resultant force at A, and (b) the moment of this force about point B.

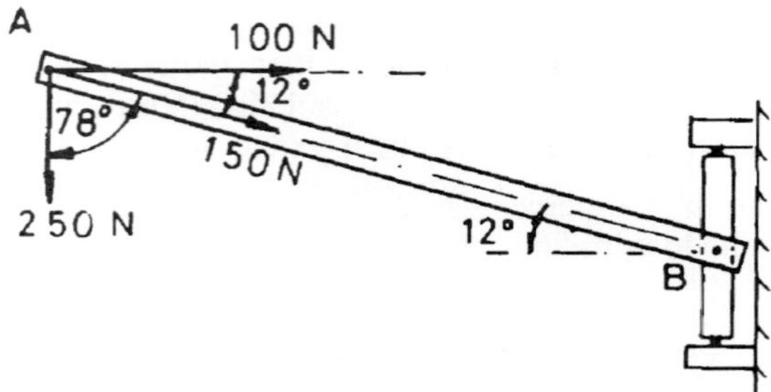

29. In a double geared hand winch the driving pinions have 14 and 20 teeth respectively and the driven wheels have 56 and 125 teeth. The lifting barrel is 300 mm diameter and the rope is 25 mm diameter. If it takes an effort of 740 N on the end of a handle of 600 mm leverage to lift a mass of 5000 kg, find the velocity ratio, mechanical advantage and efficiency.

30. A loaded truck has a tractive resistance on the level of 0·902 N/kg of its mass. The force required to pull the truck up a certain incline is 4·98 kN, and the force to hold it stationary and prevent it from running down is 3·34 kN. Find the rise of the incline and the mass of the loaded truck.

31. The centre of gravity of a shaft *AB* of irregular cross-section is at 1·2 m from *B*. When the shaft is freely suspended from a single lifting hook attached to a point 1·4 m from *A*, a vertical force of 1·8 kN exerted downwards at *A* just maintains the shaft horizontal. When the lifting attachment to the shaft is at 1·5 m from *A* the downward force at *A* to keep the shaft horizontal is 1·26 kN. Calculate the overall length of the shaft and its mass.

32. A motor boat of 4 t displacement changes its velocity from 11 knots due North to 18 knots due East in 30 s. Find the magnitude in m/s^2 and direction of the average acceleration, and also the average accelerating force on the boat. 1 knot = 1·852 km/h.

33. In an oil engine of 0·57 m bore and 1·2 m stroke, the length of the connecting rod is 2·4 m. At the point when the crank is 30° past top dead centre the effective pressure on the piston is $2{\cdot}3 \times 10^6$ N/m^2. Calculate (*i*) the force on the crankpin at this instant, and (*ii*) the torque exerted on the crank shaft.

34. The effective length and diameter of an engine bottom end bolt is 560 mm and 60 mm respectively and the pitch of the screw thread is 4·2 mm. When the bolt is given a further tightening an additional stress of 51·6 MN/m^2 is induced in it, calculate the extra load on the bolt. If the modulus of elasticity of the material is 210 GN/m^2 and assuming 50% of the total deformation takes place in the bolt, calculate the angular movement of the nut, in degrees, during the further tightening process.

35. A motor transmits 2500 kW to a solid shaft through flanges which have 6 coupling bolts of 50 mm diameter on a pitch circle diameter of 350 mm. The diameter of the shaft is 0·25 m, length 3 m, rotational speed 120 rev/min, and the modulus of rigidity is 83 GN/m^2. Calculate (*i*) the maximum shear stress in the shaft, (*ii*) the angle of twist in the shaft, (*iii*) the shear stress in the coupling bolts.

36. Fresh water issues from a hole 16 mm diameter in the side of a tank, the head of water above the hole being 2·14 m. Taking the coefficient of area as 0·64 and the coefficient of velocity as 0·97, calculate the volumetric discharge rate from the hole.

37. Two ships, A and B, leave the same port at the same time. A travels North West at 15 knots, and B travels 30° South of West at 17 knots. Find the velocity of A relative to B, and the time taken for the ships to be 100 nautical miles apart.

38. Two spur wheels in gear transmit a power of 6·7 kW. The driver has 40 teeth of 12 mm pitch and runs at 7 revs/s; find the force on the teeth. If the driven wheel has 240 teeth find the ratio of the torque in the two shafts.

39. A rod of uniform cross-section is loaded at one end so that it will float upright. When placed in a liquid of relative density 0·8 the length of rod exposed above the surface is 150 mm, and in a liquid of relative density 0·9 the exposed length is 200 mm. Calculate the length exposed above the surface when placed in a liquid of relative density 0·86.

40. A beam of uniform cross section is 11·1 m long and is simply supported at each end. It carries a uniformly distributed load over its entire length which, together with the weight of the beam, amounts

to a total of 570 kN, also a concentrated load of 80 kN at 4·3 m from the left-hand support. Find the position to place a concentrated load of 20 kN so that each reaction will carry an equal load.

41. An enclosed tank is 1·2 m wide and 2·5 m deep. It contains fresh water to a depth of 1·5 m and the free volume of the tank is pressurised with air to a pressure of 40 kN/m^2.

Determine the resultant force on the tank side. Density of water is 1000 kg/m^3.

42. A compartment containing water of density 1000 kg/m^3 to a depth of 3 m has an adjoining compartment which is to be filled with sea water of density 1025 kg/m^3.

If the resultant hydrostatic force per metre width of bulkhead is not to exceed 36·3 kN, calculate (*a*) the maximum possible depth of sea water, and (*b*) the minimum possible depth of sea water. (The bottoms of both compartments are at the same level.)

43. A casting of 15 kg mass is fixed to the faceplate of a lathe, the centre of gravity of the casting being 60 mm radially from the lathe centre. Calculate the radius at which a balancing mass of 3 kg should be fixed to counter-balance the casting. If the centre of gravity of the casting is 120 mm axially from the faceplate and that of the counter balance mass 10 mm from the faceplate, calculate the resultant unbalanced couple when the lathe is running at 150 rev/min.

44. A cubical vessel of 120 mm sides is initially half full of fresh water. A solid metal bar of diameter 100 mm, length 160 mm and relative density 2·56 is suspended with its longitudinal axis vertical, from a spring balance graduated in N, and lowered into the water until the bottom end of the bar is 15 mm above the bottom of the vessel. Calculate (*i*) the distance of the new water-level from the top of the vessel, (*ii*) the increase in the thrust on the side of the vessel, (*iii*) the reading of the spring balance.

45. A body is uniformly accelerated from rest for 80 s and travels 1·32 km during this time. It then travels at uniform velocity, and is finally retarded for 50 s to bring it to rest. The total time from start to finish is 5 min. Sketch a velocity-time graph and find the total distance travelled.

46. The couplings of a shafting have 36 slots cut out at regular intervals around the circumference of each coupling, instead of the more usual bolt holes, and plates 10 mm thick and 40 mm deep fitted in the slots. The plates are parallel to the axis of the shaft and the mean radius of all slots is 300 mm from the shaft centre. Calculate the shear stress in the plates when transmitting 1·5 MW of power at a rotational speed of 3 rev/s.

47. The following data were taken during an experiment on a simple lifting machine which has a velocity ratio of 15:

Load (N)	30	40	50	60	70	80	90	100
Effort (N)	7·0	8·5	10·1	11·6	13·1	14·6	16·0	17·5

On a base of load plot the curves of actual effort, ideal effort and efficiency. From the curves find the effort expended in friction when the efficiency is 35%.

48. Two inclined planes, of 60° and 20° respectively, are placed back to back with a frictionless pulley at the common vertex. A body is placed on each plane and connected by a cord running parallel to the planes, and passing over the pulley. The mass of the body on the 60° plane is 10 kg and the coefficient of friction is 0·4. The mass of the body on the 20° plane is 7·5 kg. If the system is just on the point of moving, find the coefficient of friction between the 7·5 kg mass and its plane.

49. A ballast tank 1·25 m deep can be completely run up from the sea in 50 min when the inlet valve is submerged 4m. The tank can be completely pumped out in 2 h 5 min. Commencing with 0·15 m depth of water in the tank and the inlet valve submerged 9 m, the inlet valve is opened and the pump started simultaneously. Find the time to fill the tank.

50. A beam AB is lifted by two slings of unequal lengths, attached at their lower ends to the extreme ends of the beam and at their top ends to a crane hook. The beam is 2 m long, its mass is 7·5 Mg, and its centre of gravity is at 0·5 m from end B. When the end A is just about to clear the ground, the beam makes and angle of 30° to the ground and the vertical distance from ground to crane hook is 2·25 m. Calculate the tension in each sling.

51. A body is projected vertically upwards with an initial velocity of 36 m/s, at the same instant another body is allowed to fall from

rest from a height of 170 m. After what time and at what height above the ground will the bodies pass each other?

52. Two hollow shafts of equal dimensions are connected by a flanged coupling. The outside diameter of the shafts is 350 mm and there are 8 coupling bolts of 67 mm diameter on a P.C.D. of 560 mm. Allowing the same maximum shear stress in the bolts as in the shafts, calculate the inside diameter of the shafts.

53. A vertical bulkhead divides a tank 9 m wide. On one side of the bulkhead there is fresh water (density 1000 kg/m^3) to a height of 6·3 m and on the other side there is oil (density 850 kg/m^3) to a height of 4·5 m. Find the resultant load on the bulkhead and the resultant centre of pressure.

54. The stroke of the piston of an internal combustion engine is 1040 mm and the length of the connecting rod is 1960 mm. At the instant the crank is 45° past the top centre, the total effective force on the piston is 196 kN. Find the force on the guide in this position and the turning moment on the crank.

55. A beam carries a uniformly distributed load $8{\cdot}8 \times 10^4$ N/m over the left half of its length and a uniformly distributed load of $4{\cdot}4 \times 10^4$ N/m over the remaining half. The length of the beam is 6 m and it is simply supported at two points, one at the extreme left end and the other within the right half length at such a position to make the two reactions equal. Calculate the position of the right hand reaction and sketch the shearing force and bending moment diagrams inserting the principal values.

56. A jib is pin-jointed at its bottom end to a vertical bulkhead and the top of the jib is supported by two horizontal wire ties from the bulkhead which, as a plan view, form an isosceles triangle with the bulkhead. The length of the jib is 8 m and it hangs out at an inclination of 30° to the bulkhead. The length of each tie is 5 m. Calculate the forces in the jib and ties when a load of 1570 kg hangs from the top of the jib.

57. In a bell crank lever AOB, the fulcrum shaft is at O. Arm AO is horizontal and to the left, its length is 500 mm. Arm OB is vertically downwards and its length is 200 mm. A force of 60 N pulls on B in the direction 60° to the right of the vertical inclined downwards, and

an upward vertical force is applied at A to maintain static equilibrium. Calculate (*i*) the force at A, and (*ii*) the magnitude and direction of the resultant force at the fulcrum O.

58. A horizontal beam is supported by a hinge at one end and two vertical props of equal length. One prop has a cross-sectional area of 11·6 cm^2, its modulus of elasticity is 180 GN/m^2 and it is located at 1.8 m from the hinge. The other prop has a cross sectional area of 14.8 cm^2, modulus of elasticity 150 GN/m^2 and its location is 3·3 m from the hinge. Calculate the load on each prop and the hinge when a load of 200 kN is placed on the beam at 2·4 m from the hinge, assuming the beam remains rigid and horizontal.

59. A cylindrical tank 1·5 m diameter contains liquid and stands with its longitudinal axis vertical. The liquid flows into the top of the tank at the volume flow rate of 1·2 m^3/min and flows out of the bottom of the tank through an orifice 65 mm diameter. Taking the coefficient of discharge as 0·7, calculate (*i*) the depth in the tank for steady liquid-level conditions, and (*ii*) the time to empty the tank from that level after the supply at the top is shut off, taking the mean volume flow rate through the orifice to be one-third of the maximum.

60. The rim of a cast iron flywheel has a mean radius of 1·8 m and a rectangular cross-section 100 mm radial thickness and 150 mm wide. Taking the density of cast iron as $7{\cdot}2 \times 10^3$ kg/m^3, calculate (*i*) the mean stress in the rim when the rotational speed is 250 rev/min, (*ii*) the energy stored up at this speed, and (*iii*) the reduction in speed if it loses 81 kJ of energy.

61. A steel tube is 5 mm thick, 30 mm bore and 12 m long. A solid steel bar 20 mm diameter, screwed at both ends with a thread of one mm pitch, passes centrally through the tube. The ends of the tube are covered by washers and tightened by nuts screwed on the ends of the bar. If the modulus of elasticity of both tube and bar is 210 GN/m^2, calculate the stress in the tube and bar when the nuts are screwed up until the compressive force in the tube is 12·5 kN. Calculate also the increase in stress in the tube and the bar when one nut is held and the other further tightened by one complete turn.

62. A uniform ladder 5 m long and 14 kg mass is placed against a vertical wall at an angle of 50° to the horizontal ground. The

coefficient of friction between ladder and wall is 0·2, and between ladder and ground it is 0·5. Calculate how far up the ladder a man of mass 63 kg can climb before the ladder slips.

Note: There are 9 examination questions in each examination paper, each question indicates the mark allocation. The remaining 18 questions here follow this practice.

63. A screw down valve of 27 000 mm^2 effective area has a valve spindle with a pitch of 4 mm. It is operated by a handwheel 200 mm diameter working through a reduction gearing of 4 to 1. The overall efficiency of the mechanism is 43%.

Calculate the force required at the handwheel to close the valve against a pressure of 1 MN/m^2. (16)

64. A horizontal steel pipe of 400 m internal diameter, 500 mm external diameter and 9 m length, is simply supported at both ends. Determine for the pipe:
(*a*) the mass. (4)
(*b*) the maximum bending moment due to its own weight. (6)
(*c*) the maximum bending stress. (6)
(Density of steel = 7860 kg/m^3)

65. In a lifting operation, a mass of 8 t is uniformly accelerated vertically upwards, from rest, to a maximum velocity. It is then immediately allowed to retard, under gravity, to come to rest. The total lift is 200 m, and the total time taken is 20 s.
(*a*) Draw the velocity/time graph, indicating important points. (4)
(*b*) Determine the maximum:
(*i*) velocity; (3)
(*ii*) acceleration; (7)
(*iii*) tension in the rope. (2)

66. (*a*)(*i*) Define momentum. (2)
(*ii*) State the law of conversion of momentum. (2)

(*b*) A gun of mass 2 t fires horizontally a shell of mass 25 kg. The gun's horizontal recoil is resisted by a constant force of 8 kN, which brings the gun to rest in 1·5 s. Find the initial velocity of the shell:
(*i*) relative to the ground; (9)
(*ii*) relative to the gun. (3)

67. A ship of mass 4000 t is fitted with engines capable of developing 3 MW. The ship's resistance through water is given by the formula:

$$R = kv^2$$

where: R is the resistance in kN
k is a constant in kN/knot^2
v is the velocity in knots

The value of k for the ship is 1·5 kN/knot^2. Obtain:

(*a*) the ship's maximum velocity when developing full power; (6)
(*b*) the ship's acceleration at a velocity of 10 knots. (10)

Note: 1 nautical mile = 1852 m

68. The torque variation in a shaft is given by the expression:

$$T\ (\text{kNm}) = 140 + 40 \sin A - 20 \cos A$$

where A is any angle between 0° and 360°. The limit of shear stress intensity is 60 MN/m^2. Determine:

(*a*) the maximum torque transmitted by the shaft; (12)
(*b*) the maximum diameter of the shaft necessary to transmit the maximum torque. (4)

69. An observation point is 45 m above the ground and a body projected vertically upwards from the ground is observed to remain above the point for 4 s. Determine, for the projected body:

(*a*) the maximum height above the ground; (6)
(*b*) the initial velocity of the body; (4)
(*c*) the total time of flight. (6)

70. Oil flows from a pipe of diameter 120 mm, smoothly, into a pipe of 80 mm. The velocity in the larger pipe is 4 m/s.

(*a*) Calculate:
(*i*) the velocity in the smaller pipe; (6)
(*ii*) the mass flow rate in the pipes. (5)

(*b*) A manometer containing mercury, is connected across the two pipes, showing a difference in mercury level of 40 mm. Obtain the pressure drop. (5)

Note: Densities: oil = 860 kg/m^3 mercury = 13.6 t/m^3

71. (a) Define momentum, giving its units and stating its quantity (4)
(*b*) Two bodies of masses 3 kg and 5 kg travel in a straight line and in the same direction. The smaller body moves with a velocity of 8 m/s and the larger body at 12 m/s. The bodies

collide and subsequently move together. Calculate:

(*i*) the final velocity of the two bodies; (6)

(*ii*) the energy lost at impact. (6)

72. A steel beam is simply supported at both ends and is loaded at the centre with a 6 kN force, acting vertically downward. The beam, of solid rectangular section, is 2 m long, 200 mm deep and 100 mm wide. Calculate:

(*a*) the mass of the beam; (4)

(*b*) the maximum bending stress in the beam. (12)

Note: Density of steel: = 7800 kg/m^3

73. A point on the rim of a rotating wheel, originally has a tangential velocity of 12 m/s and a centripetal acceleration of 96 m/s^2. The angular speed of the wheel, subsequently is uniformly retarded to 5 rad/s in 4 s. For the wheel, estimate:

(*a*) the diameter; (5)

(*b*) the original angular speed; (5)

(*c*) the subsequent angular retardation; (3)

(*d*) the tangential acceleration of the point of the rim. (3)

74. A single helical reduction gear train, of ratio 20:1, consists of a pinion and a main gear wheel. The helix angle of the gear is 20°, and the pitch circle diameter of the pinion is 120 mm. The pinion runs at 600 rev/min with a power input of 15 kW. Ignoring any losses, obtain:

(*a*) the pitch circle diameter of the main wheel; (3)

(*b*) the tangential force on the gear teeth; (5)

(*c*) the normal contact force between the teeth; (4)

(*d*) the axial load on the thrust pads. (4)

75. (*a*) Define energy and give THREE examples of forms of mechanical energy together with their definitions and equations. (4)

(*b*) A mass of 100 kg, travelling 5 m/s, collides with a mass of 50 kg, travelling at 1 m/s in the opposite direction. After collision, both masses move together as a single body. Find:

(*i*) the velocity of the combined mass; (6)

(*ii*) the loss of energy in the system. (6)

76. (*a*) Distinguish between the work done by a force and the work done by a torque, stating each work expression and identifying all basic units. (4)

(*b*) A gate valve has a mass of 150 kg, and is operated by a vertical screw thread of pitch 20 mm, mean diameter 50 mm and a hand wheel of 600 mm diameter. The coefficient of friction between the screw thread and the nut is 0·2. Determine the total tangential force required on the rim of the hand wheel. (12)

77. A vessel is to be re-engined, and its solid shaft is to be replaced by a hollow shaft with the same safety factor and outer diameter. The details of the shafts are:

Old shaft: ultimate shear stress = 350 MN/m^2; diameter = 500 mm.

New shaft: ultimate shear stress = 440 MN/m^2

The power is increased by 20%, and the speed is increased by 10%. Calculate the maximum inner diameter of the hollow shaft. (16)

78. (*a*) Define both mass flow rate and density, giving their units. (2)

(*b*) Oil of density 850 kg/m^3, flows at 6 m/s in a branch pipe of 75 mm diameter. It joins with a second pipe of 100 mm diameter, containing oil of density 950 kg/m^3, flowing at 10 m/s. These two branch pipes run into a main pipe in which oil flows at 12 m/s. Determine the:

(*i*) diameter of the main pipe; (6)

(*ii*) total mass flow rate; (4)

(*iii*) density of the mixed oil. (4)

79. A man standing on the edge of a cliff, 250 m high, releases a stone. When it has travelled 10 m, he releases a second stone. When the first stone hits the ground,

(*a*) calculate the time for it to reach the ground. (4)

(*b*) When the first stone hits the ground, also calculate:

(*i*) the velocity of impact of the first stone; (4)

(*ii*) the height of the second stone above the ground; (6)

(*iii*) the velocity of the second stone. (2)

80. (*a*) State the beam equation defining the symbols used and giving the units for each part. (7)

(*b*) A pair of radial arm lifeboat davits of solid round section, as shown in the sketch, has the following dimensions:

height = 2 m; overhang = 1·5 m; section = 120 mm diameter.

A lifeboat of mass 2 t is suspended from the davits.

The steel from which the davits are made has a U.T.S. of 347 MN/m^2.

Determine:

(*i*) the magnitude of the maximum bending stress; (7)

(*ii*) the safety factor for the davits. (2)

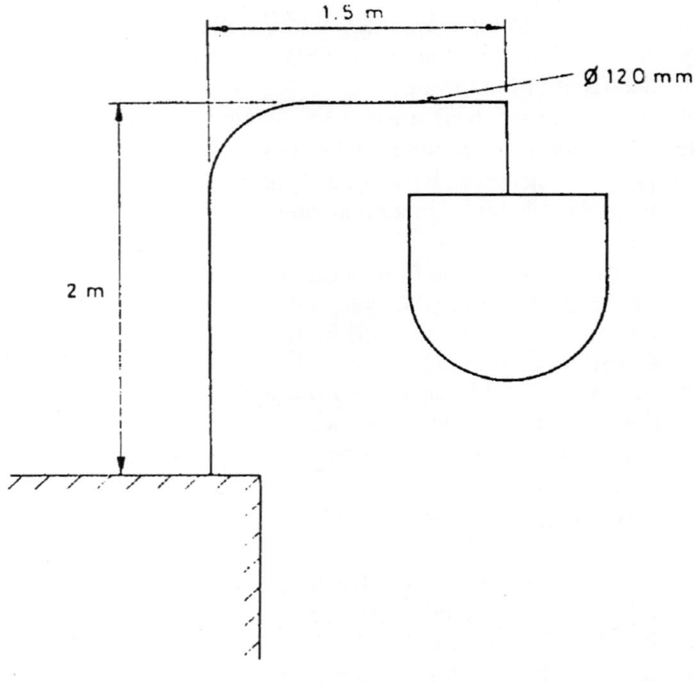

SOLUTIONS TO EXAMINATION QUESTIONS

CLASS TWO

1.

$$\text{Velocity ratio} = \frac{2DN}{d}$$

$$= \frac{2 \times 360 \times 80}{120} = 480 \quad \text{Ans. } (i)$$

$$\text{Ideal effort} = \frac{\text{load}}{\text{velocity ratio}}$$

$$\text{Mass} = 2{\cdot}04 \times 10^3 \text{ kg}$$

$$\text{Force to lift mass (weight)} = 2{\cdot}04 \times 10^3 \times 9{\cdot}81$$

$$= 20{\cdot}01 \times 10^3 \text{ N}$$

$$\text{Ideal effort} = \frac{20{\cdot}01 \times 10^3}{480} = 41{\cdot}7 \text{ N Ans } (ii)$$

$$\text{Mechanical advantage} = \text{velocity ratio} \times \text{efficiency}$$

$$= 480 \times 0{\cdot}45$$

$$\text{Actual effort} = \frac{\text{load}}{\text{mechanical advantage}}$$

$$= \frac{20{\cdot}01 \times 10^3}{480 \times 0{\cdot}45}$$

$$= \frac{41{\cdot}7}{0{\cdot}45} = 92{\cdot}66 \text{ N} \qquad \text{Ans } (iii)$$

2. (a) Power available for pumping water $= 0{\cdot}8 \times 5$
$= 4 \text{ kW}$

Taking into account the K.E. of the water at discharge:
Pump output

$$\begin{aligned}
\text{Power} &= \text{work done/s in lifting water + K.E./s at discharge} \\
&= \text{weight lifted/s} \times \text{height + K.E./s} \\
&= \dot{m}gh + \tfrac{1}{2}\dot{m}v^2 \text{ where } m = \text{mass flow rate (kg/s)} \\
4 \times 10^3 &= (\dot{m} \times 9{\cdot}81 \times 34) + (\tfrac{1}{2} \times \dot{m} \times 20^2) \\
4 \times 10^3 &= 533{\cdot}5 \times \dot{m}
\end{aligned}$$

Mass flow rate, $\dot{m}$ = *7·5 kg/s* *Ans. (a)*

(b) $$\text{Mass flow rate} = \frac{\text{Mass of water pumped}}{\text{Time taken}}$$

$$7{\cdot}5 = \frac{50 \times 10^3}{\text{Time}}$$

$$\begin{aligned}
\text{Discharge time} &= 6{\cdot}667 \times 10^3 \text{ s} \\
&= 1 \text{ h } 51{\cdot}1 \text{ min.}
\end{aligned}$$ Ans. (b)

3. Maximum force required to lift gate will be when the gate is closed, it is then at its greatest depth, the water load acts on one side only, and it has no buoyancy.

Weight of 1 m^3 = $1025 \times 9{\cdot}81$ N

Depth from water level to centre of gravity of gate = $5 - \tfrac{1}{2} \times 0{\cdot}9 = 4{\cdot}55$ m

$$\begin{aligned}
\text{Load on gate} &= HA\rho g \\
&= 4{\cdot}55 \times 0{\cdot}6 \times 0{\cdot}9 \times 1{\cdot}025 \times 10^3 \times 9{\cdot}81 \\
&= 24{\cdot}71 \text{ kN} \\
\text{Friction force} &= \text{coeff. of friction} \times \text{force between surfaces} \\
&= 0{\cdot}28 \times 24{\cdot}71 \\
&= 6{\cdot}918 \text{ kN} \\
\text{Weight of gate} &= 367 \times 9{\cdot}81 = 3{\cdot}6 \text{ kN} \\
\text{Initial lifting force} &= \text{friction force + weight of gate} \\
&= 6{\cdot}918 + 3{\cdot}6 \\
&= 10{\cdot}518 \text{ kN Ans.}
\end{aligned}$$

4. $$\text{Yield stress} = \frac{\text{yield force}}{\text{original area}}$$

$$= \frac{87 \times 10^3}{0{\cdot}7854 \times 20^2 \times 10^{-6}}$$

$$= 2{\cdot}769 \times 10^8 \text{N/m}^2$$

$$= 276.9 \text{ MN/m}^2 \qquad \text{Ans } (i)$$

$$U.T.S. = \frac{\text{breaking force}}{\text{original area}}$$

$$= \frac{145 \times 10^3}{0{\cdot}7854 \times 20^2 \times 10^{-6}}$$

$$= 4{\cdot}615 \times 10^8 \text{N/m}^2$$

$$= 461{\cdot}5 \text{ MN/m}^2 \qquad \text{Ans } (ii)$$

$$\% \text{ Elongation} = \frac{\text{elongation}}{\text{original length}} \times 100$$

$$= \frac{126 - 100}{100} \times 100$$

$$= 26 \qquad \text{Ans. } (iii)$$

$$\% \text{ Reduction of area} = \frac{\text{reduction of area}}{\text{original area}} \times 100$$

$$= \frac{20^2 - 15{\cdot}2^2}{20^2} \times 100$$

$$= 42{\cdot}25 \qquad \text{Ans. } (iv)$$

Fig. 146 shows a typical stress-strain graph.

5. From $\frac{T}{J} = \frac{\tau}{r} = \frac{G\theta}{l}$ $\qquad G = \frac{Tl}{J\theta}$

$$J = \frac{\pi}{32}\,\text{d}^4 = \frac{\pi}{32} \times 0{\cdot}022^4$$

$$\theta = 2° \times \frac{2\pi}{360} \text{ or } \frac{2}{57{\cdot}3} \text{ rad}$$

$$G = \frac{32 \times 343 \times 0{\cdot}2 \times 57{\cdot}3}{\pi \times 0{\cdot}022^4 \times 2}$$

$$= 8{\cdot}543 \times 10^{10}\ \text{N/m}^2 \text{ or } 85{\cdot}43\ \text{GN/m}^2$$ Ans. (*i*)

$$\tau = \frac{G\theta r}{l}$$

$$= \frac{8{\cdot}543 \times 10^{10} \times 2 \times 0{\cdot}011}{0{\cdot}2 \times 57{\cdot}3}$$

$$= 1{\cdot}64 \times 10^8\ \text{N/m}^2$$

Stress in shaft is to be 50% of the above:

$$= 0{\cdot}82 \times 10^8 \text{ or } 8{\cdot}2 \times 10^7\ \text{N/m}^2$$

$$T = \frac{J\tau}{r} = \frac{\pi \times D^4 \times \tau \times 2}{32 \times D} = \frac{\pi}{16} D^3 \tau$$

$$\text{Torque in shaft} = \frac{\pi}{16} \times 0{\cdot}175^3 \times 8{\cdot}2 \times 10^7$$

$$= 8{\cdot}63 \times 10^4\ \text{N m}$$

$$\text{Power} = \text{torque} \times \omega \quad \text{i.e. } P = T\omega$$

$$= 8{\cdot}63 \times 10^4 \times \frac{300 \times 2\pi}{60}$$

$$= 2{\cdot}712 \times 10^6\ \text{W or } 2712\ \text{kW}$$ Ans. (*ii*)

6. Taking moments of area about centre of hole, working in m:

$$\bar{x} = \frac{0{\cdot}7854 \times 0{\cdot}4^2 \times 0{\cdot}1 - 0{\cdot}7854 \times 0{\cdot}125^2 \times 0}{0{\cdot}7854 \times 0{\cdot}4^2 - 0{\cdot}7854 \times 0{\cdot}125^2}$$

$$\bar{x} = \frac{0{\cdot}016 - 0}{0{\cdot}16 - 0{\cdot}01563}$$

$$= 0{\cdot}1108\ \text{m or } 110{\cdot}8\ \text{mm}$$ Ans. (*i*)

As density and thickness is uniform, mass varies as area, which varies as the square of the diameter,

$$\text{Mass of holed disc} = 80 \times \frac{0{\cdot}4^2 - 0{\cdot}125^2}{0{\cdot}4^2}$$

$$= 72{\cdot}19 \text{ kg}$$

$$\frac{120 \text{ rev/min} \times 2\pi}{60} = 4\pi \text{ rad/s}$$

$$\text{Centrifugal force} = m\omega^2 r$$
$$= 72{\cdot}19 \times (4\pi)^2 \times 0{\cdot}1108$$
$$= 1263 \text{ N or } 1{\cdot}263 \text{ kN}$$

Ans. (*ii*)

7.

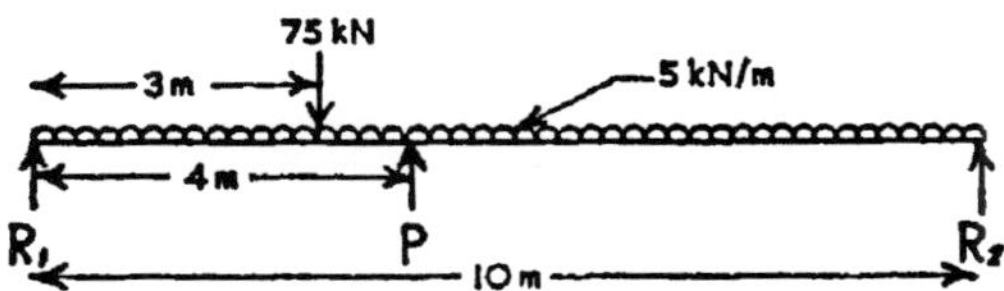

Let P = upward force exerted by the prop at 4 m from left end. Taking moments in kN m, bending moment at mid-span (moments to the right)

and this is to be zero $\therefore 0 = {}^{-}R_2 \times 5 + 5 \times 5 \times 2{\cdot}5$

$$5R_2 = 62{\cdot}5$$
$$R_2 = 12{\cdot}5 \text{ kN}$$

Moments about R_1:

Clockwise moments = Anticlockwise moments

$$75 \times 3 + 5 \times 10 \times 5 = P \times 4 + R_2 \times 10$$
$$225 + 250 = 4P + 12{\cdot}5 \times 10$$
$$4P = 225 + 250 - 125$$
$$4P = 350$$
$$P = 87{\cdot}5 \text{ kN}$$

Ans.

8. Mean depth $= \frac{1}{2}(3{\cdot}5 + 4) = 3{\cdot}75$ m

Theo. vel. through orifice $= \sqrt{2gh}$

Theo. discharge rate $= \sqrt{2gh} \times \text{area}$

Actual volume flow rate $= 0{\cdot}62 \times \sqrt{2gh} \times \text{area}$

Mean flow rate based on mean depth

$$= 0{\cdot}62 \times \sqrt{2 \times 9{\cdot}81 \times 3{\cdot}75} \times 0{\cdot}7854 \times 0{\cdot}3^2$$
$$= 0{\cdot}3759 \text{ m}^3\text{/s}$$

Maximum volume of liquid to be discharged while cock is shut

$$= \text{area} \times \text{change in depth}$$
$$= 25 \times (4 - 3{\cdot}5) = 12{\cdot}5 \text{ m}^3$$

$$\text{Time to discharge} = \frac{\text{volume}}{\text{volume flow rate}}$$

$$= \frac{12{\cdot}5}{0{\cdot}3759} = 33{\cdot}25 \text{ s} \qquad \text{Ans.}$$

9.

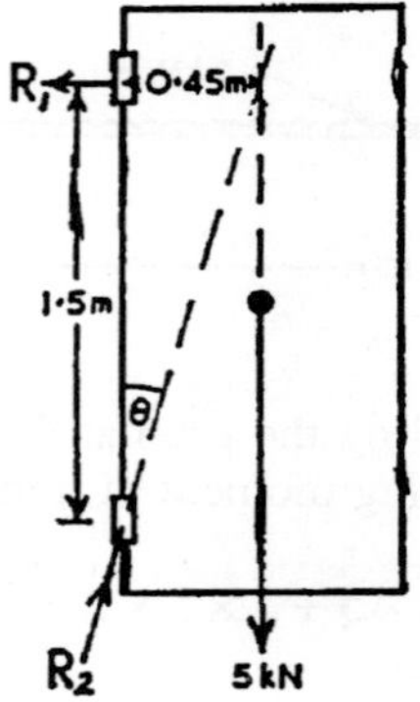

Let R_1 = reaction of top hinge,
R_2 = reaction of bottom hinge
Weight of door = $509{\cdot}7 \times 9{\cdot}81$ N = 5 kN

As the whole weight of the door is carried at the bottom hinge, then the top hinge must exert a horizontal force only, pulling inwards.

Moments about bottom hinge, in kN m,

$$\text{Clockwise moments} = \text{Anticlockwise moments}$$
$$5 \times 0{\cdot}45 = R_1 \times 1{\cdot}5$$
$$R_1 = 1{\cdot}5 \text{ kN}$$

The effects of the reaction of the bottom hinge must be to exert (*i*) a horizontal force of 1·5 kN outwards to balance the inward pull of the top hinge, (*ii*) an upward force of 5 kN to balance the downward weight of the door. The resultant of these is the reaction of the bottom hinge.

$$R_2 = \sqrt{5^2 + 1{\cdot}5^2} = 5.22 \text{ kN}$$

Let θ = angle of R_2 to the vertical,

$$\tan\theta = \frac{1{\cdot}5}{5} = 0{\cdot}3$$

$$\theta = 16^\circ\, 42'$$

Reaction of top hinge = 1·5 kN acting horizontally
Reaction of bottom hinge = 5·22 kN at 16° 42′ to vertical Ans.

10. Force on tank bottom before steel is immersed is the weight of water in the tank.

$$m = V\rho$$
$$= 0{\cdot}9 \times 0{\cdot}9 \times 0{\cdot}6 \times 10^3$$
$$= 486 \text{ kg}$$

Weight of water $= 486 \times 9{\cdot}81 = 4768$ kN

Force on tank bottom = 4·768 kN Ans. (*i*)

Increase of force on tank bottom when steel is immersed is the downward reaction to the upward force of buoyancy on the steel, which is the weight of the water displaced,

$$m = V\rho$$
$$= 0{\cdot}03 \times 10^3 \text{ kg}$$

Weight of water displaced $= 0{\cdot}03 \times 10^3 \times 9{\cdot}81$
$= 294{\cdot}3$ N Ans. (*ii*)

Mass of steel $= V\rho$
$= 0{\cdot}03 \times 7{\cdot}86 \times 10^3$ kg

Weight of steel $= 0{\cdot}03 \times 7{\cdot}86 \times 10^3 \times 9{\cdot}81$
$= 294{\cdot}3 \times 7{\cdot}86 = 2313$ N

Tension in rope = weight of steel – weight of water displaced
$= 2313 - 294{\cdot}3$
$= 2018{\cdot}7 = 2{\cdot}0187$ kN Ans. (*iii*)

11.

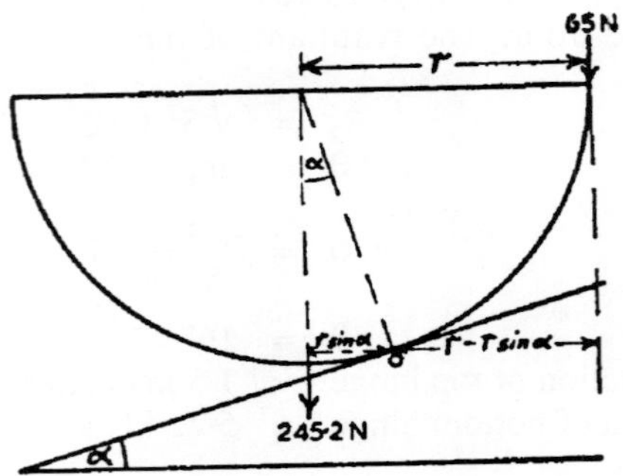

$$\text{Weight of hemisphere} = 25 \times 9{\cdot}81 = 245{\cdot}2 \text{ N}$$

Taking moments about point of contact O:

$$\begin{aligned}
\text{Clockwise moments} &= \text{Anticlockwise moments} \\
65 \times (r - r \sin \alpha) &= 245{\cdot}2 \times r \sin \alpha \\
65r - 65r \sin \alpha &= 245{\cdot}2 \times r \sin \alpha \\
65 - 65 \sin \alpha &= 245{\cdot}2 \sin \alpha \\
245{\cdot}2 \sin \alpha + 65 \sin \alpha &= 65 \\
310{\cdot}2 \sin \alpha &= 65 \\
\sin \alpha &= 0{\cdot}2095 \\
\alpha &= 12^\circ \, 6' \text{ Ans.}
\end{aligned}$$

12.

$$\text{Speed of output shaft} = 100 \times \frac{72}{24} = 300 \text{ rev/min}$$

$$\text{P} = \text{T}\omega \qquad \ldots \ldots \ldots (i)$$

$$J = \frac{\pi}{32} D^4$$

$$T = \frac{\pi}{16} D^3 \tau \qquad \ldots \ldots \ldots \ldots (ii)$$

From (*i*) and (*ii*):

$$\text{Power} = \frac{\pi}{16} D^3 \tau \times \omega$$

$$\text{power output} = 0{\cdot}85 \times \text{power input}$$

$$\frac{\pi}{16} \times 0{\cdot}075^3 \times \tau \times \frac{300 \times 2\pi}{60}$$

$$= 0{\cdot}85 \times \frac{\pi}{16} \times 0{\cdot}1^3 \times 21 \times 10^6 \times \frac{100 \times 2\pi}{60}$$

$$\tau = \frac{0{\cdot}85 \times 0{\cdot}1^3 \times 21 \times 10^6 \times 100}{0{\cdot}075^3 \times 300}$$

$= 1{\cdot}41 \times 10^7$ N/m^2 or $14{\cdot}1$ MN/m^2 Ans. (*i*)

Power $= T\omega$

$$= \frac{\pi}{16} D^3 \tau \times \omega$$

$$= \frac{\pi}{16} \times 0{\cdot}075^3 \times 1{\cdot}41 \times 10^7 \times \frac{300 \times 2\pi}{60}$$

$= 3{\cdot}67 \times 10^4$ W or $36{\cdot}7$ kW Ans. (*ii*)

13. Area of opening = effective circumference × lift

$= \frac{8}{9} \times \pi \times 60 \times 4 = 670{\cdot}4$ mm^2

$= 670{\cdot}4 \times 10^{-6}$ m^2

Volume flow rate = area × velocity

$= 670{\cdot}4 \times 10^{-6} \times 2 \times 3600$

Mass flow rate = volume flow rate × density

$= 670{\cdot}4 \times 10^{-6} \times 2 \times 3600 \times 10^3$

$= 4{\cdot}827 \times 10^3$ kg/h

$= 4{\cdot}827$ Mg/h or $4{\cdot}827$ t/h Ans.

14. Rotational speed of drum

$$= \frac{\text{linear speed}}{\text{circumference}}$$

$$= \frac{1{\cdot}8 \times 60}{\pi \times 1{\cdot}2} = 28{\cdot}64 \text{ rev/min}$$

Speed of motor through 50 to 1 gear ratio

$= 28{\cdot}64 \times 50 = 1432$ rev/min Ans. (*i*)

Mass of body $= 150 \times 10^3$ kg

Weight $= mg = 150 \times 10^3 \times 9{\cdot}81 = 1{\cdot}472 \times 10^6$ N

Force to overcome gravity on incline of 3°

$= W \sin \alpha = 1{\cdot}472 \times 10^6 \times \sin 3° = 7{\cdot}699 \times 10^4$ N

Tractive resistance at 8×10^{-3} N of force per kg of mass

$= 8 \times 10^{-3} \times 150 \times 10^3 = 1200$ N

$$\text{Total force} = 76{\cdot}99 \times 10^3 + 1{\cdot}2 \times 10^3 = 78{\cdot}19 \times 10^3 \text{ N}$$

$$\begin{aligned}\text{Drum torque} &= \text{force} \times \text{radius}\\ &= 78{\cdot}19 \times 10^3 \times 0{\cdot}6\\ &= 4{\cdot}691 \times 10^4 \text{ N m or } 46{\cdot}91 \text{ kN m}\end{aligned}$$ Ans. (*ii*)

Output power of gears

$$\begin{aligned}&= \text{work done/s}\\ &= \text{force} \times \text{velocity i.e. } P = Fv\\ &= 78{\cdot}19 \times 10^3 \times 1{\cdot}8 = 140{\cdot}7 \times 10^3\ W\end{aligned}$$

Input power to gears

$$= \frac{\text{output}}{\text{efficiency}}$$

$$= \frac{140{\cdot}7 \times 10^3}{0{\cdot}75}$$

$$= 187{\cdot}6 \times 10^3\ W \text{ or } 187{\cdot}6 \text{ kW}$$ Ans. (*iii*)

15. Velocity of bullet as it emerges from wood

$$= 480 - 150 = 330 \text{ m/s}$$

Let m = mass of bullet

F = average resisting force of wood

t = original thickness of wood

Kinetic energy $= \frac{1}{2} mv^2$

Loss of $K.E.$ = Ave. res. force × distance

1st case:

$$\tfrac{1}{2}(480^2 - 330^2) = F \times t \qquad \dots\ \dots\ \dots\ (i)$$

2nd case:

$$\tfrac{1}{2} \times 480^2 = F \times \text{new thickness} \qquad \dots\ \dots\ (ii)$$

$$\frac{\text{new thickness}}{t} = \frac{\frac{1}{2} m \times 480^2 \times F}{F \times \frac{1}{2} m\,(480^2 - 330^2)}$$

$$= \frac{480^2}{480^2 - 330^2}$$

$$= 1{\cdot}895$$

thickness of wood should be:
1·895 times original thickness Ans.

Alternatively, note that loss of kinetic energy is proportional to distance, which is thickness of wood; also, kinetic energy is proportional to, and can be represented by, $(\text{velocity})^2$.

thickness of wood should be $\dfrac{480^2}{480^2 - 330^2} \times$ original thickness

16. $$\text{Mechanical advantage} = \frac{\text{load lifted}}{\text{effort applied}}$$

$$= \frac{670 \times 9{\cdot}81}{22{\cdot}24} = 295{\cdot}5 \qquad \text{Ans. } (i)$$

$$\text{Velocity ratio} = \frac{\text{mechanical advantage}}{\text{efficiency}}$$

$$= \frac{295{\cdot}5}{0{\cdot}32} = 923{\cdot}7 \qquad \text{Ans. } (ii)$$

$$\text{Velocity ratio} = \frac{\text{distance moved by effort}}{\text{distance moved by load}}$$

Consider one revolution of the wormwheel:
Distance moved by effort = circ. of effort wheel × no. of revs. of worm
= $\pi \times 72 \times$ no. of teeth in worm wheel
Distance moved by load = pitch of screw = 12 mm

$$\text{Velocity ratio} = \frac{\pi \times 72 \times \text{no. of teeth}}{12} = 923{\cdot}7$$

$$\text{No. of teeth} = \frac{923{\cdot}7 \times 12}{\pi \times 72} = 49 \qquad \text{Ans. } (iii)$$

17. See Fig.
Load on AB = $33 \times 1{\cdot}6 = 52{\cdot}8$ kN
Load on DE = $33 \times 1{\cdot}2 = 39{\cdot}6$ kN

Moments about B,

$$
\begin{aligned}
\text{Clockwise moments} &= \text{Anticlockwise moments} \\
60 \times 1{\cdot}6 + 39{\cdot}6 \times 5{\cdot}4 &= 52{\cdot}8 \times 0{\cdot}8 + R_2 \times 4{\cdot}8 \\
96 + 213{\cdot}8 &= 42{\cdot}24 + R_2 \times 4{\cdot}8 \\
267{\cdot}56 &= R_2 \times 4{\cdot}8 \\
R_2 &= 55{\cdot}74 \text{ kN} \qquad \ldots\ldots\ldots (i)
\end{aligned}
$$

$$
\begin{aligned}
\text{Upward forces} &= \text{Downward forces} \\
R_1 + 55{\cdot}74 &= 52{\cdot}8 + 60 + 39{\cdot}6 \\
R_1 &= 96{\cdot}66 \text{ kN} \qquad \ldots\ldots\ldots (ii)
\end{aligned}
$$

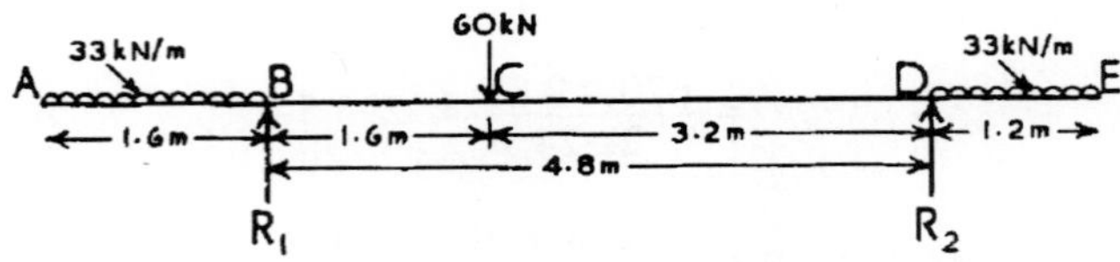

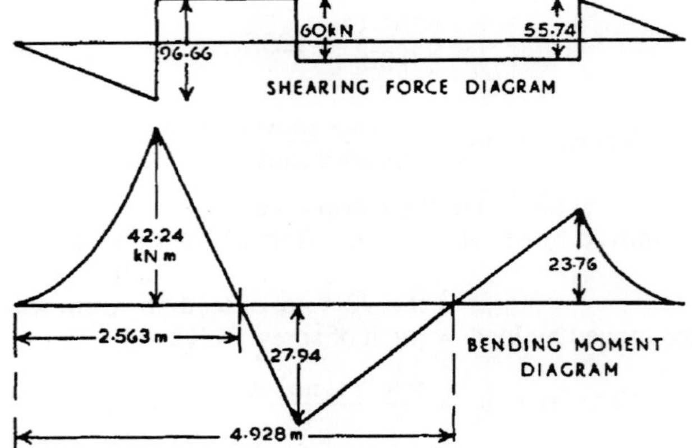

Bending moments at the principal points:

M @ B = $52{\cdot}8 \times 0{\cdot}8 = 42{\cdot}24$ kN m .... (*iii*)

M @ C = $52{\cdot}8 \times 2{\cdot}4 - 96{\cdot}66 \times 1{\cdot}6 = -27{\cdot}94$ kN m ...... (*iv*)

M @ D = $39{\cdot}6 \times 0{\cdot}6 = 23{\cdot}76$ kN m .... (*v*)

Let M = zero at x m from B

$$
\begin{aligned}
M_x &= 52{\cdot}8 \times (0{\cdot}8 + x) - 96{\cdot}66 \times x = 0 \\
42{\cdot}24 + 52{\cdot}8x &= 96{\cdot}66x \\
42{\cdot}24 &= 43{\cdot}86x \\
x &= 0{\cdot}963 \text{ m from B}
\end{aligned}
$$

$$\begin{aligned}
\text{From A, } M &= 0 \text{ at } 1{\cdot}6 + 0{\cdot}963 = 2{\cdot}563 \text{ m}\\
\text{Let } M &= \text{zero at z m from D}\\
&= 39{\cdot}6 \times (0{\cdot}6 + \text{z}) - 55{\cdot}74 \times \text{z} = 0\\
23{\cdot}76 + 39{\cdot}6\,\text{z} &= 55{\cdot}74\,\text{z}\\
23{\cdot}76 &= 16{\cdot}14\,\text{z}\\
\text{z} &= 1{\cdot}472 \text{ m from D}
\end{aligned}$$

From A, $M = 0$ at $6{\cdot}4 - 1{\cdot}472 = 4{\cdot}928$ m

Positions of zero bending moment from A

$= 2{\cdot}563$ m and $4{\cdot}928$ m Ans. (*i*)

Bending equation $\dfrac{M}{I} = \dfrac{\sigma}{y}$

I = 2nd moment of area

M = max. bending moment = $42{\cdot}24 \times 10^3$

σ = max. bending stress = 60×10^6

y = $\frac{1}{2}$ depth

$$I = \frac{My}{\sigma} = \frac{42{\cdot}24 \times 10^3 \times 0{\cdot}2}{60 \times 10^6}$$

$$= 1{\cdot}408 \times 10^{-4}$$ Ans. (*ii*)

18.

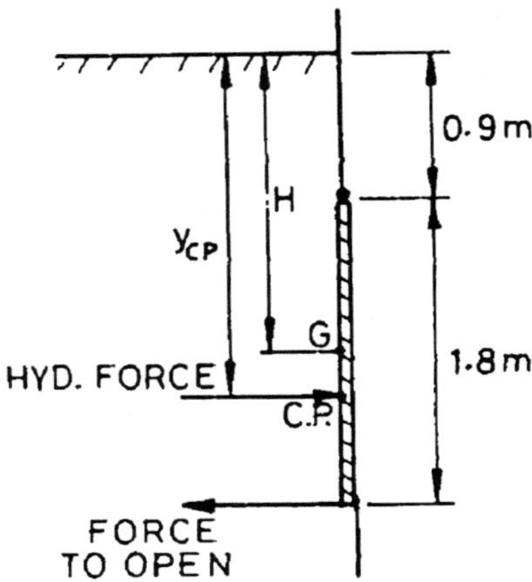

Hydrostatic force = $HA\rho g$

$= 1{\cdot}8 \times 1{\cdot}8^2 \times 1024 \times 9{\cdot}81$
$= 58{\cdot}59$ kN

Note: In determining the position of the centre of pressure, it must be noted that the upper edge of the gates is below the liquid surface. Thus, its position must be calculated as shown below).

$$y_{cp} = \frac{I_G}{AH} + H$$

For a rectangular area, $I_G = \dfrac{BD^3}{12}$

$$y_{cp} = \frac{18 \times 1{\cdot}8^3}{12 \times 1{\cdot}8^2 \times 1{\cdot}8} + 1{\cdot}8$$

$= 1{\cdot}95$ m below the surface
The position of the C. o. P. $= 1{\cdot}95 - 0{\cdot}9$
$= 1{\cdot}05$ m below the hinge.

Take moments about the hinge:
Force to open gate $\times 1{\cdot}8 = 58{\cdot}59 \times 1{\cdot}05$
Force to open $= 34{\cdot}18$ kN. Ans.

19. Cross-sectional area of belt
$= 200 \times 12 \times 10^{-6}$ m^2
Maximum pull in belt = stress × area
$= 400 \times 10^3 \times 200 \times 12 \times 10^{-6}$
$= 960$ N
Tension in slack side $= \frac{1}{3} \times 960 = 320$ N
Effective pull $= 960 - 320 = 640$ N
Effective radius from shaft centre to mid-thickness of belt
$= 250 + 6 = 256$ mm $= 0{\cdot}256$ m
Rotational speed $= 375 \div 60 = 6{\cdot}25$ rev/s
Linear speed of belt $= 6{\cdot}25 \times 2\pi \times 0{\cdot}256$ m/s
Power = work done per second
= effective pull × speed
$= 640 \times 6{\cdot}25 \times 2\pi \times 0{\cdot}256$
$= 6{\cdot}434 \times 10^3$ W
$= 6{\cdot}434$ kW Ans.

20. Fundamental equation $\frac{T}{J} = \frac{\tau}{r} = \frac{G\theta}{l}$

$$J = \frac{\pi}{32}\ (D^4 - d^4) = \frac{\pi}{32}\ (D^2 + d^2)(D^2 - d^2)$$

$$= \frac{\pi}{32}\ (0{\cdot}05^2 + 0.03^2)(0{\cdot}05^2 - 0{\cdot}03^2)$$

$$= \frac{\pi}{32} \times 0{\cdot}0034 \times 0{\cdot}0016$$

$$r = \tfrac{1}{2}D = 25 \text{ mm} = 0{\cdot}025 \text{ m}$$

$$\theta = 1{\cdot}5 \times \frac{2\pi}{360} \text{ or } \frac{1{\cdot}5}{57{\cdot}3} = 0{\cdot}02618 \text{ rad.}$$

$$\tau = \frac{T \times r}{J}$$

$$= \frac{32 \times 1{\cdot}52 \times 10^3 \times 0{\cdot}025}{\pi \times 0{\cdot}0034 \times 0{\cdot}0016}$$

$$= 7{\cdot}114 \times 10^7 \text{ N/m}^2 \text{ or } 71{\cdot}14 \text{ MN/m}^2$$ Ans. (*i*)

$$G = \frac{\tau \times l}{r \times \theta}$$

$$= \frac{7{\cdot}114 \times 10^7 \times 0{\cdot}8}{0{\cdot}025 \times 0{\cdot}02618}$$

$$= 8{\cdot}695 \times 10^{10} \text{ N/m}^2 \text{ or } 86{\cdot}95 \text{ GN/m}^2$$ Ans. (*ii*)

$$\text{P} = T\omega$$

$$= 1{\cdot}5 \times \frac{180 \times 2\pi}{60}$$

$$= 28{\cdot}66 \text{ kW}$$ Ans. (*iii*)

21.

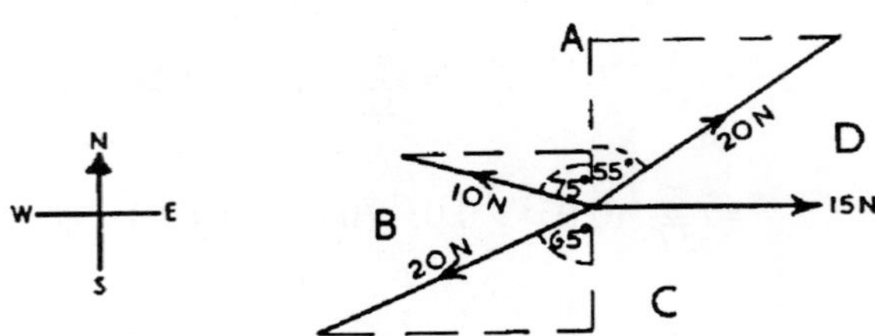

Vertical component of	AB	= 10 cos 75°	= 2·588	North
Horizontal "	"	= 10 sin 75°	= 9·659	West
Vertical "	BC	= 20 cos 65°	= 8·452	South
Horizontal "	"	= 20 sin 65°	= 18·126	West
Vertical "	CD	= 0		
Horizontal "	"	= 15		East
Vertical "	DA	= 20 cos 55°	= 11·472	North
Horizontal "	"	= 20 sin 55°	= 16·384	East

Resultant of vertical components

= 11·472 + 2·588 – 8·452 = 5·608 North

Resultant of horizontal components

= 16·384 + 15 – 18·126 –9·659 = 3·599 East

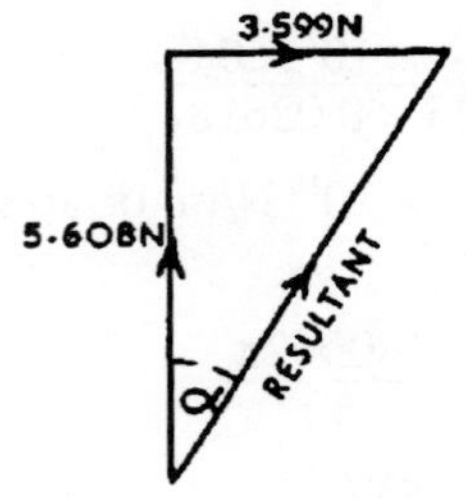

$$\tan \alpha = \frac{3{\cdot}599}{5{\cdot}608} = 0{\cdot}6418 \quad \alpha = 32^\circ\, 42'$$

$$\text{Resultant} = \frac{5{\cdot}608}{\cos 32^\circ\, 42'} = 6{\cdot}664 \text{ N}$$

Resultant = 6·664 N North 32° 42′ East Ans.

22.

$$\text{Velocity ratio} = \frac{\text{distance moved by effort}}{\text{distance moved by load}}$$

$$= \frac{2\pi \times \text{radius of toggle}}{\text{sum of thread pitches}}$$

$$= \frac{2\pi \times 336}{10 + 5} = 140{\cdot}8 \quad \text{Ans. } (i)$$

$$\text{Mechanical advantage} = \text{velocity ratio} \times \text{efficiency}$$

$$= 140{\cdot}8 \times 0{\cdot}25 = 35{\cdot}2$$

$$\text{Effort} = \frac{\text{load}}{\text{mech. advantage}}$$

$$= \frac{10 \times 10^3}{35{\cdot}2} = 284{\cdot}1 \text{ N} \quad \text{Ans. } (ii)$$

23.

(a)

$$\text{Efficiency of joint} = \frac{\text{U.T.S. of weld metal}}{\text{U.T.S. of shell plate}} \times 100$$

$$= \frac{170}{200} \times 100$$

$$= 85\%. \qquad \text{Ans. (a).}$$

(b)

$$\text{Longitudinal seam, stress} = \frac{pd}{2t}$$

$$= \frac{0{\cdot}6 \times 10^6 \times 2}{2 \times 0{\cdot}02}$$

$$= 30 \text{ MN/m}^2 \qquad \text{Ans. (b)}$$

(c)

$$\text{Factor of safety} = \frac{\text{U.T.S.}}{\text{working stress}}$$

$$= \frac{170}{30}$$

$$= 5{\cdot}7. \qquad \text{Ans. (c)}$$

24. Total force on one ram = pressure × area

$$= 75 \times 10^5 \times 0{\cdot}7854 \times 0{\cdot}3^2 \text{ N}$$

$$\text{Total torque} = \text{force of 2 rams} \times \text{leverage}$$

$$= 75 \times 10^5 \times 0{\cdot}7854 \times 0{\cdot}3^2 \times 2 \times 0{\cdot}75 \text{ Nm}$$

$$\text{From } \frac{T}{J} = \frac{\tau}{r} \qquad J = \frac{\pi}{32} d^4 \qquad r = \tfrac{1}{2} d$$

$$T = \frac{\pi}{16} d^3 \tau$$

$$\tau = \frac{16T}{\pi d^3}$$

$$= \frac{16 \times 75 \times 10^5 \times 0{\cdot}7854 \times 0{\cdot}3^2 \times 2 \times 0{\cdot}75}{\pi \times 0{\cdot}45^3}$$

$$= 4{\cdot}444 \times 10^7 \text{ N/m}^2$$

$$= 44{\cdot}44 \text{ MN/m}^2 \quad \text{Ans.}$$

25.

$$W = V\rho g$$

Let d = draught of wood,

For equilibrium of floating wood:

Upward force = Downward force

Weight of oil displayed = Weight of wood

$$0{\cdot}1^2 \times d \times 0{\cdot}84 \times 10^3 \times 9{\cdot}81 = 0{\cdot}1^3 \times 0{\cdot}63 \times 10^3 \times 9{\cdot}81$$

$$d = 0{\cdot}1 \times \frac{0{\cdot}63}{0{\cdot}84} = 0{\cdot}075$$

Volume of immersed part of wood

$$= 0{\cdot}1^2 \times 0{\cdot}075 = 0{\cdot}00075 \text{ m}^3$$

Total volume of oil and wood up to oil level

$$= 0{\cdot}2^2 \times 0{\cdot}1 + 0{\cdot}00075 = 0{\cdot}00475 \text{ m}^3$$

New depth of oil = volume ÷ area

$$= 0{\cdot}00475 \div 0{\cdot}2^2 = 0{\cdot}11875 \text{ m}$$

Hydrostatic Load = $HA\rho g$

$$\text{Load on bottom} = 0{\cdot}11875 \times 0{\cdot}2^2 \times 0{\cdot}84 \times 10^3 \times 9{\cdot}81$$

$$= 0{\cdot}00475 \times 0{\cdot}84 \times 10^3 \times 9{\cdot}81$$

$$= 39{\cdot}14 \text{ N} \qquad \text{Ans. } (i)$$

(alternatively, load on bottom can be obtained by weight of oil + weight of wood)

Load on each side $= \frac{1}{2} \times 0{\cdot}11875 \times 0{\cdot}2 \times 0{\cdot}11875 \times 0{\cdot}84 \times 10^3 \times 9{\cdot}81$

$= 11{\cdot}62$ N Ans. (*ii*)

26.

Kinetic energy $= \frac{1}{2}mv^2 = \frac{1}{2} \times 48 \times 25^2$

$= 15\,000$ J $= 15$ kJ Ans (*i*)

$$\text{Time} = \frac{\text{distance}}{\text{average velocity}}$$

$$= \frac{31{\cdot}25}{\frac{1}{2} \times 25} = 2{\cdot}5 \text{ s}$$ Ans. (*ii*)

$$\text{Retardation} = \frac{\text{change of velocity}}{\text{time}}$$

$$= \frac{25}{2{\cdot}5} = 10 \text{ m/s}^2$$ Ans. (*iii*)

Retarding force = mass × retardation

$= 48 \times 10 = 480$ N Ans. (*iv*)

27.

$\dot{m} = av\rho$

$= 0{\cdot}7854 \times 0{\cdot}125^2 \times 1{\cdot}2 \times 1{\cdot}025 \times 10^3$

$= 15{\cdot}1$ kg/s

Force to lift mass against gravity $= mg$

$= 15{\cdot}1 \times 9{\cdot}81 = 148{\cdot}1$ N/s

Work done/s = weight lifted/s × height

= output power

$= 148{\cdot}1 \times 5{\cdot}8$

$= 858{\cdot}8$ W Ans. (*i*)

$$\text{Input power} = \frac{\text{output power}}{\text{efficiency}}$$

$$= \frac{858{\cdot}8}{0{\cdot}62}$$

$$= 1386 \text{ W} = 1{\cdot}386 \text{ kW} \qquad \text{Ans. } (ii)$$

Note: Flow velocity at discharge is low, hence K.E. of the liquid can be neglected.

If required: Power = K.E./s

$$= \tfrac{1}{2}\,\dot{m}v^2 \text{ W}$$

$$= \tfrac{1}{2} \times 15{\cdot}1 \times 1{\cdot}2^2$$

i.e. Additional output power = 11 W

28.

(a) Resolve 150 N force into vertical and horizontal components:

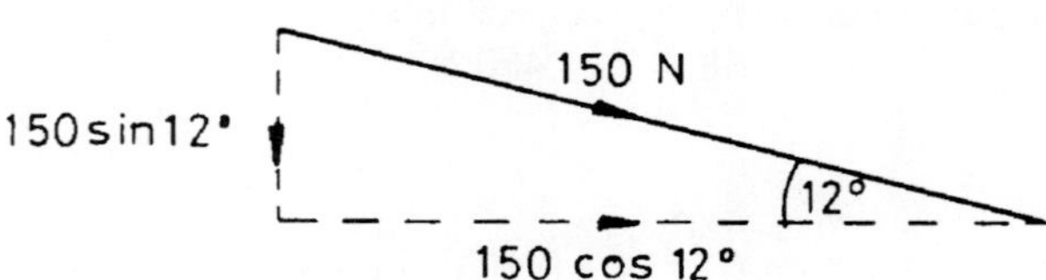

At end A of the rod,

Sum of the downward forces = $250 + 150 \sin 12°$

= 281·2 N

Sum of forces acting to the right = $100 + 150 \cos 12°$

= 246·7 N

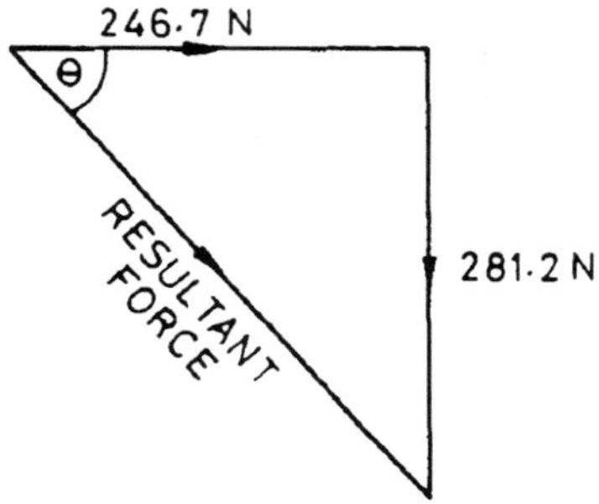

$$\text{From Fig. the resultant force} = \sqrt{281{\cdot}2^2 + 246{\cdot}7^2}$$
$$= 374{\cdot}1 \text{ N}$$

$$\text{Also,}\ \tan\theta = \frac{281{\cdot}2}{246{\cdot}7}$$

$$\text{Angle } \theta = 48{\cdot}7^\circ$$

At A, the resultant force is 374·1 N, acting at 48·7° below horizontal. Ans. (a)

(b)

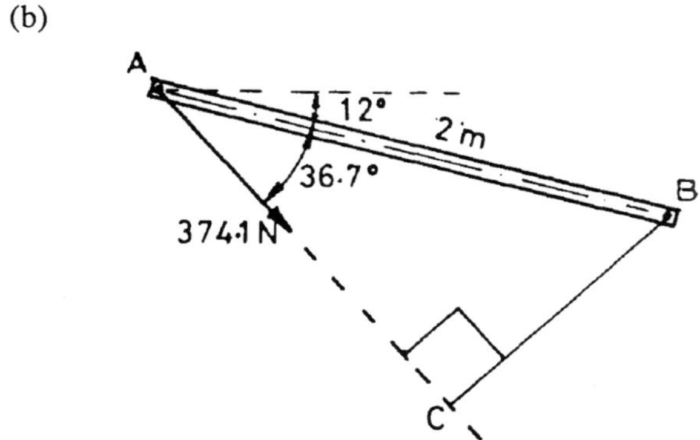

Moment of resultant force (about B)

$$= \text{force} \times \text{perpendicular distance BC}$$
$$= 374{\cdot}1 \times 2 \sin 36{\cdot}7^\circ$$
$$= 447{\cdot}1 \text{ Nm.}$$

Ans. (b)

29. $$V.R. = \frac{\text{effort circle radius}}{\text{load circle radius}} \times \frac{\text{product of teeth in followers}}{\text{product of teeth in drivers}}$$

$$\text{V.R.} = \frac{600}{(150 + 12{\cdot}5)} \times \frac{56 \times 125}{14 \times 20}$$

$$= 92{\cdot}3 \qquad \text{Ans. } (i)$$

$$M.A. = \frac{\text{load lifted}}{\text{effort applied}}$$

$$= \frac{5000 \times 9{\cdot}81}{740} = 66{\cdot}3 \qquad \text{Ans. } (ii)$$

$$\text{Efficiency} = \frac{\text{mechanical advantage}}{\text{velocity ratio}}$$

$$= \frac{66{\cdot}3}{92{\cdot}3} = 0{\cdot}7183 \text{ or } 71{\cdot}83\% \qquad \text{Ans. } (iii)$$

30. Let m = mass of truck
then mg = weight of truck
Tractive resistance on level = $0{\cdot}902\, m$
" " " incline = $0{\cdot}902\, m \cos\alpha$

Force to pull up = force to overcome gravity + tractive resistance
Force to hold = force to overcome gravity – tractive resistance

$$4{\cdot}98 \times 10^3 = mg \sin\alpha + 0{\cdot}902\, m \cos\alpha \quad \ldots\ldots \quad (i)$$

$$3{\cdot}34 \times 10^3 = mg \sin\alpha - 0{\cdot}902\, m \cos\alpha \quad \ldots\ldots \quad (ii)$$

$$8{\cdot}32 \times 10^3 = 2\, mg \sin\alpha \quad \ldots\ldots \quad (iii) \text{ i.e. adding}$$

$$1{\cdot}64 \times 10^3 = 2 \times 0{\cdot}902\, m \cos\alpha \ldots \quad (iv) \text{ i.e. subtracting}$$

$$\frac{8{\cdot}32 \times 10^3}{1{\cdot}64 \times 10^3} = \frac{2\, mg \sin\alpha}{2 \times 0{\cdot}902\, m \cos\alpha} \qquad \text{i.e. dividing}$$

$$\frac{8{\cdot}32 \times 0{\cdot}902}{1{\cdot}64 \times 9{\cdot}81} = \frac{\sin\alpha}{\cos\alpha} = \tan\alpha$$

$$\tan\alpha = 0{\cdot}4665 \qquad \alpha = 25^\circ$$

$$\sin\alpha = 0{\cdot}4226$$

Rise = $0{\cdot}4226$ m/m up the incline
= $422{\cdot}6$ mm/m Ans. (*i*)

From (*iii*)

$$2mg \sin\alpha = 8{\cdot}32 \times 10^3$$

$$m = \frac{8{\cdot}32 \times 10^3}{2 \times 9{\cdot}81 \times 0{\cdot}4226}$$

$$= 1003 \text{ kg or } 1{\cdot}003 \text{ Mg}$$ Ans. (*ii*)

31.

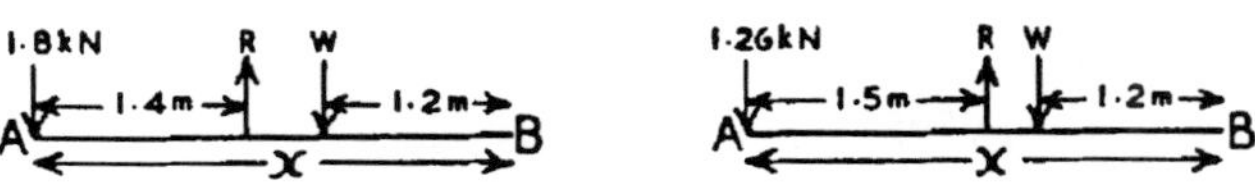

Let length of shaft = x m, weight = W kN
Referring to first condition, moments about R:

$$W \times (x - 1{\cdot}2 - 1{\cdot}4) = 1{\cdot}8 \times 1{\cdot}4$$

$$W = \frac{2{\cdot}52}{x - 2{\cdot}6} \qquad \dots\ \dots\ \dots\ (i)$$

Referring to second condition, moments about R:

$$W \times (x - 1{\cdot}2 - 1{\cdot}5) = 1{\cdot}26 \times 1{\cdot}5$$

$$W = \frac{1{\cdot}89}{x - 2{\cdot}7} \qquad \dots\ \dots\ \dots\ (ii)$$

Equating (*i*) and (*ii*),

$$\frac{2{\cdot}52}{x - 2{\cdot}6} = \frac{1{\cdot}89}{x - 2{\cdot}7}$$

$$2{\cdot}52\,(x - 2{\cdot}7) = 1{\cdot}89\,(x - 2{\cdot}6)$$

$$2{\cdot}52x - 6{\cdot}804 = 1{\cdot}89\,x - 4{\cdot}914$$

$$0{\cdot}63x = 1{\cdot}89$$

$$x = 3 \text{ m}$$ Ans. (*i*)

From (*i*) $W = \dfrac{2{\cdot}52}{3 - 2{\cdot}6} = 6{\cdot}3$ kN

$$\text{Mass} = \frac{W}{g} = \frac{6{\cdot}3 \times 10^3}{9{\cdot}81}$$

$$= 642{\cdot}2 \text{ kg}$$ Ans. (*ii*)

32.

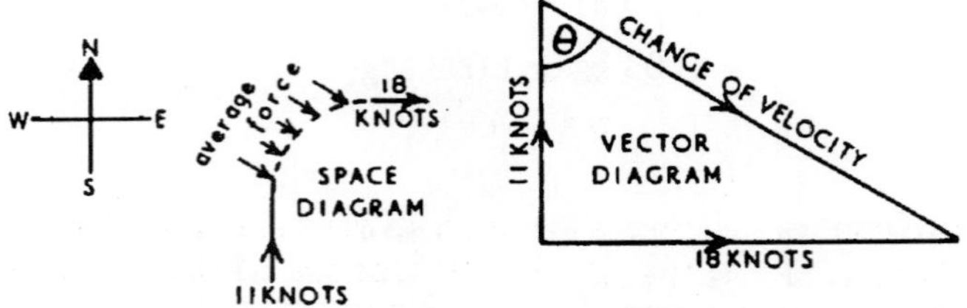

$$\text{Change of velocity} = \sqrt{11^2 + 18^2} = 21{\cdot}1 \text{ knots}$$

$$\text{Acceleration} = \frac{\text{change of velocity}}{\text{time to change}}$$

$$= \frac{21{\cdot}1 \times 1{\cdot}852 \times 10^3}{3600 \times 30} = 0{\cdot}3618 \text{ m/s}^2$$

Ans. (*i*)(a)

$$\tan\theta = \frac{18}{11} = 1{\cdot}6364 \qquad \theta = 58°\ 34'$$

Direction is South 58° 34′ East Ans. (*i*)(b)

$$\text{Accelerating force} = \text{mass} \times \text{acceleration}$$
$$= 4 \times 10^3 \times 0{\cdot}3618$$
$$= 1{\cdot}4472 \times 10^3 \text{ N or } 1{\cdot}4472 \text{ kN}$$

Ans. (*ii*)

33. See Fig. 199, Chapter 11,
Referring to space diagram:

$$\frac{l}{\sin\theta} = \frac{r}{\sin\phi}$$

$$\sin\phi = \frac{0{\cdot}6 \times \sin 30°}{2{\cdot}4} = 0{\cdot}125$$

$$\phi = 7°\ 11'$$

$$\text{Total force on piston} = \text{pressure} \times \text{area}$$
$$= 2{\cdot}3 \times 10^6 \times 0{\cdot}7854 \times 0{\cdot}57^2$$
$$= 5{\cdot}869 \times 10^5 \text{ N}$$

Referring to vector diagram:

$$\text{Thrust in conn. rod} = \frac{\text{piston force}}{\cos \phi}$$

$$= \frac{5{\cdot}869 \times 10^5}{\cos 7^\circ\ 11'} = 5{\cdot}916 \times 10^5\ \text{N}$$

The force on the crank pin is the thrust in the conn. rod, hence

Force on crank pin $= 5{\cdot}916 \times 10^5$ N or $591{\cdot}6$ kN Ans. (*i*)

Referring to space diagram:

Perpendicular distance from line of action of thrust in conn. rod to shaft centre = OP $= r \times \sin(\phi + \theta)$

$= 0{\cdot}6 \times \sin 37^\circ\ 11' = 0{\cdot}3626$ m

Turning moment on crank shaft

= thrust in conn. rod × perp. distance

$= 5{\cdot}916 \times 10^5 \times 0{\cdot}3626$

$= 2{\cdot}145 \times 10^5$ N m or $214{\cdot}5$ kN m Ans (*ii*)

34.

$$\text{Stress} = \frac{\text{load}}{\text{area}}$$

Load on bolt = stress × area

$= 51{\cdot}6 \times 10^6 \times 0{\cdot}7854 \times 0{\cdot}06^2$

$= 1{\cdot}459 \times 10^5$ N or $145{\cdot}9$ kN Ans. (*i*)

$$\frac{\sigma}{\varepsilon} = \text{E}$$

$$\varepsilon = \frac{\sigma}{\text{E}}$$

$$= \frac{51{\cdot}6 \times 10^6}{210 \times 10^9} = 2{\cdot}457 \times 10^{-4}$$

$$\varepsilon = \frac{x}{l}$$

$$x = \varepsilon \times l$$

$$= 2{\cdot}457 \times 10^{-4} \times 560 = 0{\cdot}1376\ \text{mm}$$

and this is half the total deformation.

Total deformation of bolt and other parts

$$= 2 \times 0{\cdot}1376 = 0{\cdot}2752 \text{ mm}$$

4·2 mm axial movement = 360° turn of the nut

$$1\text{mm} \quad " \quad " \quad = \frac{360}{4{\cdot}2} \text{ deg.} = 85{\cdot}72°$$

Total angular movement of nut

$$= 85{\cdot}72 \times 0{\cdot}2752 = 23{\cdot}59°$$ Ans. (*ii*)

35. $$120 \text{ rev/min} \times \frac{2\pi}{60} = 12{\cdot}57 \text{ rad/s}$$

$$\text{P} = \text{T}\omega$$

$$\text{T} = \frac{\text{P}}{\omega} = \frac{2500 \times 10^3}{12{\cdot}57}$$

$$= 1{\cdot}989 \times 10^5 \text{ N m}$$

$\frac{T}{J} = \frac{\tau}{r} = \frac{G\theta}{l}$, for a solid shaft:

$$T = \frac{J\tau}{r} = \frac{\pi}{16} D^3$$

$$\text{Stress in shaft } \tau = \frac{16\,T}{\pi\,D^3}$$

$$= \frac{16 \times 1{\cdot}989 \times 10^5}{\pi \times 0{\cdot}25^3}$$

$$= 6{\cdot}482 \times 10^7 \text{ N/m}^2 \text{ or } 64{\cdot}82 \text{ MN/m}^2$$ Ans. (*i*)

$$\theta = \frac{\tau l}{rG}$$

$$= \frac{6{\cdot}482 \times 10^7 \times 3}{0{\cdot}125 \times 83 \times 10^9} = 0{\cdot}01874 \text{ rad}$$

$$\text{Angle of twist} = 0{\cdot}01874 \times 57{\cdot}3$$

$$= 1{\cdot}074°$$ Ans. (*ii*)

Torque transmitted by coupling bolts

= stress in bolt × area × P.C. radius × no. of bolts

and this is the same as the torque in the shaft.

$$\tau_B \times 0{\cdot}7854 \times 0{\cdot}05^2 \times 0{\cdot}175 \times 6 = 1{\cdot}989 \times 10^5$$

$$\tau_B = \frac{1{\cdot}989 \times 10^5}{0{\cdot}7854 \times 0{\cdot}05^2 \times 0{\cdot}175 \times 6}$$

$$= 9{\cdot}647 \times 10^7 \text{ N/m}^2 \text{ or } 96{\cdot}47 \text{ MN/m}^2$$ Ans. (*iii*)

36. Theoretical velocity $= \sqrt{2gh}$ m/s

Actual velocity of water issuing from hole

$= 0{\cdot}97 \times \sqrt{2 \times 9{\cdot}81 \times 2{\cdot}14}$

$= 6{\cdot}285$ m/s

Area of water jet $= 0{\cdot}64 \times$ area of hole

$= 0{\cdot}64 \times 0{\cdot}7854 \times 16^2$ mm^2

$= 128{\cdot}6 \times 10^{-6}$ m^2

Volume flow rate = area × velocity

$= 128{\cdot}6 \times 10^{-6} \times 6{\cdot}285$

$= 0{\cdot}808 \times 10^{-3}$ m^3/s Ans.

37.

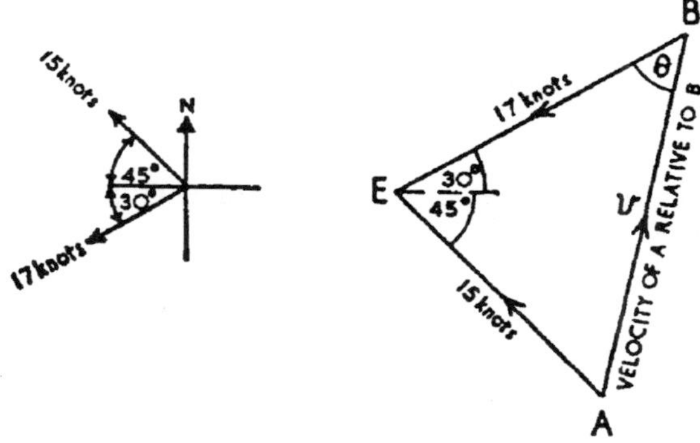

$$v^2 = 15^2 + 17^2 - 2 \times 15 \times 17 \times \cos 75°$$

$$v = \sqrt{382} = 19{\cdot}54 \text{ knots}$$

$$\frac{15}{\sin\theta} = \frac{19{\cdot}54}{\sin 75°}$$

$$\sin\theta = \frac{15 \times 0{\cdot}9659}{19{\cdot}54} = 0{\cdot}7416$$

$$\theta = 47° \, 52'$$

$$\text{Angle East of North} = 90° - 47° \, 52' - 30° = 12° \, 8'$$

$$\left.\begin{array}{r} \text{Velocity of A relative to B} = 19{\cdot}54 \text{ knots} \\ \text{in the direction } 12° \, 8'\text{E. of N.} \end{array}\right\} \quad \text{Ans. } (i)$$

Since the two ships move apart at the rate of 19·54 nautical miles per hour, the time to be 100 nautical miles apart is

$$\frac{100}{19{\cdot}54} = 5{\cdot}119 \text{ h} \qquad \text{Ans. } (ii)$$

38. Circumference of driver = $40 \times 12 = 480$ mm = 0·48 m

$$\text{Work done per revolution} = \text{Force on teeth} \times \text{circumference}$$

$$\text{Work done/s} = \text{Force} \times 0{\cdot}48 \times 7$$

$$6{\cdot}7 \times 10^3 = \text{Force} \times 0{\cdot}48 \times 7$$

$$F = 1995 \text{ N} = 1{\cdot}995 \text{ kN} \qquad \text{Ans. } (i)$$

$$\text{Torque} = \text{Force on teeth} \times \text{radius}$$

As the force on the teeth is the same for both wheels, the torque is in the ratio of the radii of the wheels,

Ratio, torque in driver : torque in driven
= radius of driver : radius of driven

or, number of teeth in driver : no. of teeth in driven
= 40 : 240
= 1 : 6 Ans. (ii)

39.

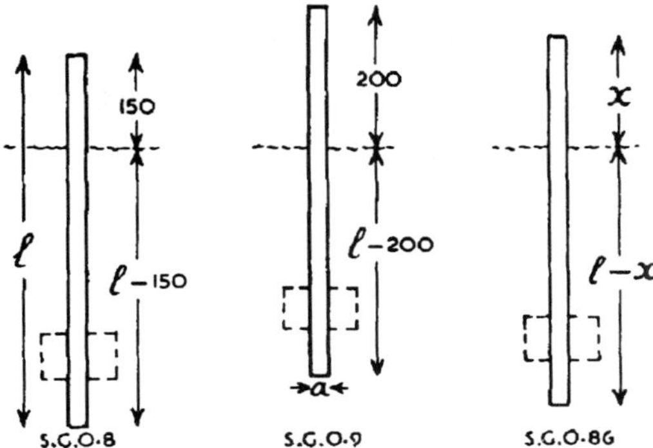

Let the cross-sectional area of the rod be represented by a and the load on the bottom can have the same cross-section so that the 'equivalent length' of rod and load can then be represented by l. Let the length of exposed rod be x when floating in liquid of relative density 0·86.

Weight of liquid displayed displaced = mass × g
= volume × density × g
= length immersed × area × rel. density × 10^3 × g

and this is the same for all three cases because the weight of liquid displaced is equal to the weight of the system which is constant. Equating the two given cases:

$$(l - 0{\cdot}2) \times a \times 0{\cdot}9 \times 10^3 \times g = (l - 0{\cdot}15) \times a \times 0{\cdot}8 \times 10^3 \times g$$
$$(l - 0{\cdot}2) \times 0{\cdot}9 = (l - 0{\cdot}15) \times 0{\cdot}8$$
$$0{\cdot}9l - 0{\cdot}18 = 0{\cdot}8l - 0{\cdot}12$$
$$0{\cdot}1l = 0{\cdot}06$$
$$l = 0{\cdot}6 \text{ m}$$

Equating 2nd and 3rd cases, omitting quantities which obviously cancel and substituting the value of l:

$$(l - 0{\cdot}2) \times 0{\cdot}9 = (l - x) \times 0{\cdot}86$$
$$(0{\cdot}6 - 0{\cdot}2) \times 0{\cdot}9 = (0{\cdot}6 - x) \times 0{\cdot}86$$
$$0{\cdot}4 \times 0{\cdot}9 = 0{\cdot}516 - 0{\cdot}86\,x$$
$$0{\cdot}86\,x = 0{\cdot}516 - 0{\cdot}36$$
$$x = 0{\cdot}1814 \text{ m or } 181{\cdot}4 \text{ mm}$$

Ans.

40. Distributed load of 570 kN acts through its centre of gravity which is at mid-length.

Half length of beam = 5·55 m

80 kN concentrated load is

$$5{\cdot}55 - 4{\cdot}3 = 1{\cdot}25 \text{ m from centre}$$

Let x = distance from centre at which the 20 kN concentrated load is to be placed.

Moments about centre of beam, in kN m,

Clockwise moments = Anticlockwise moments

$$R_1 \times 5{\cdot}55 + 20 \times x = R_2 \times 5{\cdot}55 + 80 \times 1{\cdot}25$$

$R_1 \times 5{\cdot}55$ cancels with $R_2 \times 5{\cdot}55$ because $R_1 = R_2$

$$20x = 80 \times 1{\cdot}25$$

$$x = 5 \text{ m from centre}$$

the 20 kN should be placed at

10·55 m from left end
or, 0·55 m from right end } Ans.

41. (**Note** the entire internal volume of the tank must be subject to the pressure of 40 kN/m^2)

On one side of the tank:

$$\begin{aligned}\text{Force due to internal pressure} &= \text{pressure} \times \text{area}\\ &= 40 \times 10^3 \times 1{\cdot}2 \times 2{\cdot}5\\ &= 120 \text{ kN}\\ \text{Hydrostatic load} &= HA\rho g\\ &= \frac{1{\cdot}5}{2} \times 1{\cdot}5 \times 1{\cdot}2 \times 1000 \times 9{\cdot}81\\ &= 13{\cdot}24 \text{ kN}\\ \text{Total force on side of tank} &= 120 + 13{\cdot}24\\ &= 133{\cdot}24 \text{ kN.}\end{aligned}$$

Ans.

42. Force/m width on the fresh water side $= HA\rho g$

$$= 1{\cdot}5 \times 1 \times 3 \times 10^3 \times 9{\cdot}81$$

$$= 44{\cdot}15 \text{ kN}$$

(a) let y_1 = maximum depth of sea water

$$\begin{aligned}\text{Force due to S.W.} - \text{Force due to F.W.} &= \text{Resultant force}\\ \text{Force due to S.W.} &= 36{\cdot}3 + 44{\cdot}15\\ &= 80{\cdot}45 \text{ kN.}\\ \text{Now, hydrostatic load} &= HA\rho g\end{aligned}$$

$$80{\cdot}45 \times 10^3 = \frac{y_1}{2} \times 1 \times y_1 \times 1025 \times 9{\cdot}81$$

Maximum depth of S.W = 4 m. Ans. (a)

(b)Let y_2 = minimum depth of S.W.

Force due to F.W. – Force due to S.W. = 36·3 kN

Force due to S.W. = 44·15 – 36·3

= 7·85 kN

$$7{\cdot}85 \times 10^3 = \frac{y_2}{2} \times 1 \times y_2 \times 1025 \times 9{\cdot}81$$

Minimum depth of S.W. = 1·25 m. Ans. (b)

43. Moments about lathe centre, for static balance,

$$15 \times 60 = 3 \times r$$

$$r = 300 \text{ mm}$$

Counter-balance should be fixed at a radius of 300 mm, radially opposite to the line on which the centre of gravity of the casting lies.

Ans. (*i*)

$$\frac{150 \text{ rev/min} \times 2\pi}{60} = 5\pi \text{ rad/s}$$

Unbalanced moment set up

= *C.F.* of casting or counter-balance × arm of couple

= $m\omega^2 r \times (120 - 10) \times 10^{-3}$

= $15 \times (5\pi)^2 \times 0{\cdot}06 \times 0{\cdot}11$

= 24·43 N m Ans. (*ii*)

44.

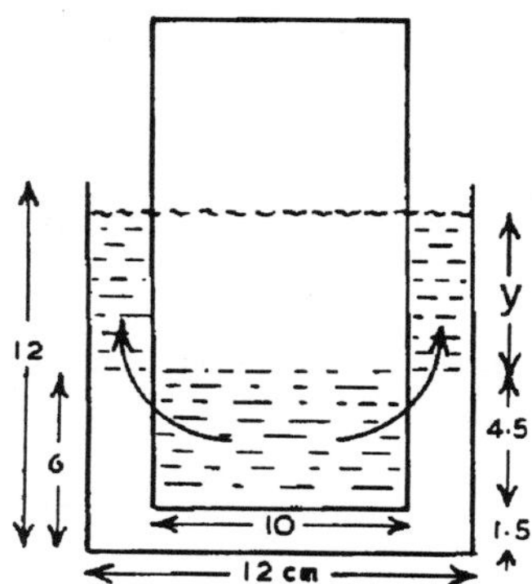

Working in cm

Let y cm = rise of water level.

When bar is immersed, volume of water pushed up to increase the depth is

$$0{\cdot}7854 \times 10^2 \times 4{\cdot}5 = (12^2 - 0{\cdot}7854 \times 10^2) \times y$$
$$353{\cdot}4 = 65{\cdot}46\,y$$
$$y = 5{\cdot}4 \text{ cm}$$

New depth of water = $6 + 5{\cdot}4 = 11{\cdot}4$

Water level from vessel top = $12 - 11{\cdot}4$

= $0{\cdot}6$ cm = 6 mm Ans. (*i*)

Hydrostatic load (thrust) = $HA\rho g$

Thrust on side of vessel:

Final thrust = $\frac{1}{2} \times 0{\cdot}114 \times 0{\cdot}114 \times 0{\cdot}12 \times 9810$

Initial thrust = $\frac{1}{2} \times 0{\cdot}06 \times 0{\cdot}06 \times 0{\cdot}12 \times 9810$

Increased thrust = $0{\cdot}5 \times 0{\cdot}12 \times 9810\,(0{\cdot}114^2 - 0{\cdot}06^2)$

= 5·53 N Ans. (*ii*)

Weight of bar = mass × g

= volume × density × g

= $0{\cdot}7854 \times 0{\cdot}1^2 \times 0{\cdot}16 \times 2{\cdot}56 \times 10^3 \times 9{\cdot}81$

Length of bar in water = $4{\cdot}5 + y$

= $4{\cdot}5 + 5{\cdot}4 = 9{\cdot}9$ cm = 0·099 m

Weight of water displaced = $0{\cdot}7854 \times 0{\cdot}1^2 \times 0{\cdot}099 \times 10^3 \times 9{\cdot}81$

Reading of spring balance = Wt. of bar – wt. of water displaced

= $0{\cdot}7854 \times 0{\cdot}1^2 \times 9810\,(0{\cdot}16 \times 2{\cdot}56 - 0{\cdot}099)$

= $0{\cdot}7854 \times 0{\cdot}1^2 \times 9810 \times 0{\cdot}3106$

= 23·93 N Ans. (*iii*)

45.

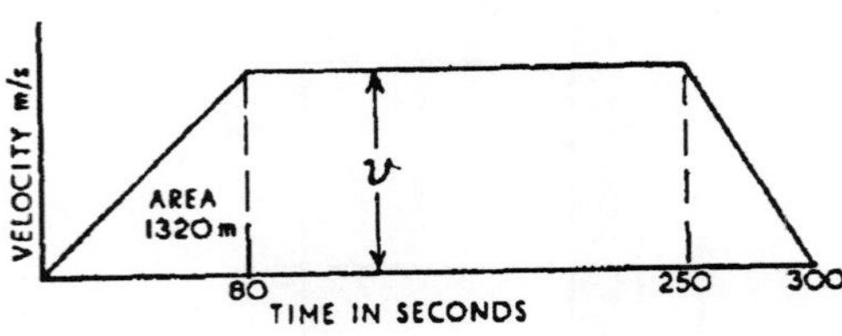

1st period, Distance = average velocity × time

$$1320 = \tfrac{1}{2}(0 + v) \times 80$$

$$v = \frac{2 \times 1320}{80} = 33 \text{ m/s}$$

2nd period, Time = 300 – 80 – 50 = 170 s
Distance = velocity (constant) × time
= 33 × 170 = 5610 = 5·61 km

3rd period, Distance = average velocity × time
$= \frac{1}{2}(33 + 0) \times 50$
= 825 m = 0·825 km
Total distance = 1·32 + 5·61 + 0·825
= 7·755 km Ans.

46. Torque transmitted by plates
= force on each plate × radius × no. of plates
= stress × area × radius × no. of plates

Also,
Torque transmitted by shaft $= \frac{P}{\omega}$

$$\text{Stress} \times 0{\cdot}01 \times 0{\cdot}04 \times 0{\cdot}3 \times 36 = \frac{1{\cdot}5 \times 10^6}{3 \times 2\pi}$$

$$\text{Stress} = \frac{1{\cdot}5 \times 10^6}{0{\cdot}01 \times 0{\cdot}04 \times 0{\cdot}3 \times 36 \times 3 \times 2\pi}$$

$= 1{\cdot}842 \times 10^7$ N/m²
$= 18{\cdot}42$ MN/m² Ans.

47. Expressions used in the calculations:

$$\text{Ideal effort} = \frac{\text{load lifted}}{\text{velocity ratio}}$$

$$\text{Mechanical advantage} = \frac{\text{load lifted}}{\text{effort applied}}$$

$$\%\ \text{Efficiency} = \frac{\text{mechanical advantage} \times 100}{\text{velocity ratio}}$$

Load	30	40	50	60	70	80	90	100
Effort	7	8·5	10·1	11·6	13·1	14·6	16	17·5
Ideal effort	2	2·67	3·33	4	4·67	5·33	6	6·67
Mech. adv.	4·29	4·71	4·95	5·17	5·34	5·48	5·62	5·72
Efficiency %	28·6	31·4	33	34·5	35·6	36·5	37·5	38·1

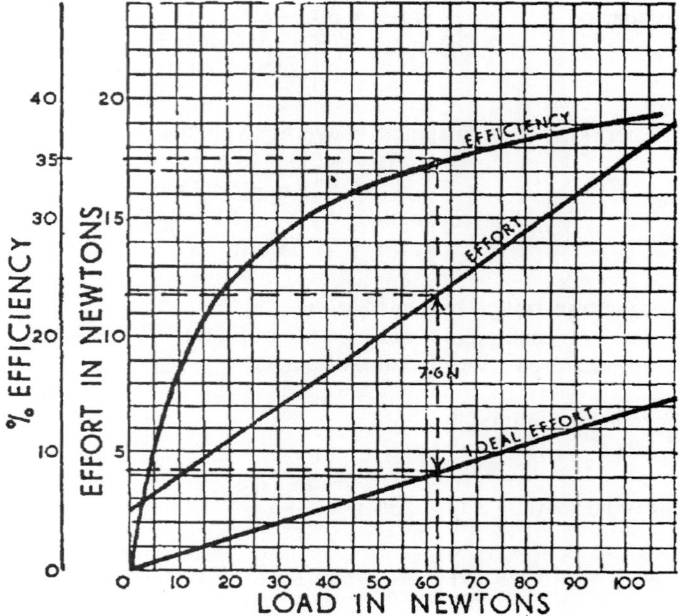

From the curves, when efficiency = 35%

$$\text{Load lifted} = 62 \text{ N}$$
$$\text{Effort applied} = 11{\cdot}8 \text{ N}$$
$$\text{Ideal effort} = 4{\cdot}2\text{N}$$

$$\text{Effort expended in friction} = 11{\cdot}8 - 4{\cdot}2 = 7{\cdot}6 \text{ N} \qquad \text{Ans.}$$

48. When the system is just on the point of moving, the 10 kg is on the point of moving down its own plane but is held by the pull in the cord, the 7·5 kg is on the point of moving up its own plane by the pull in the cord.

Force to pull the 7·5 kg up

$$= W \sin \alpha + \mu\, W \cos \alpha$$
$$= 7{\cdot}5 \times 9{\cdot}81 \times \sin 20° + \mu \times 7{\cdot}5 \times 9.81 \times \cos 20°$$
$$= 9{\cdot}81\,(2{\cdot}565 + 7{\cdot}048\mu) \;\ldots\;\ldots \qquad (i)$$

Force to hold the 10 kg

$$= W \sin \alpha - \mu\, W \cos \alpha$$

$$= 10 \times 9{\cdot}81 \times \sin 60° - 0{\cdot}4 \times 10 \times 9{\cdot}81 \times \cos 60°$$
$$= 9{\cdot}81\,(8{\cdot}66 - 2)$$
$$= 9{\cdot}81 \times 6{\cdot}66 \quad \ldots\ \ldots\ \ldots\ \ldots\ (ii)$$

$$9{\cdot}81\,(2{\cdot}565 + 7{\cdot}048\mu) = 9{\cdot}81 \times 6{\cdot}66$$
$$7{\cdot}048\mu = 6{\cdot}66 - 2{\cdot}565$$
$$7{\cdot}048\mu = 4{\cdot}095$$
$$\mu = 0{\cdot}5809 \quad \text{Ans.}$$

49. From $v = \sqrt{2gh}$, the velocity is proportional to the square root of the head. The quantity of water flowing through is proportional to the velocity. Therefore the quantity is proportional to the square root of the head.

When inlet valve is submerged 4 m, depth of water run up in one min.

$$= \frac{1{\cdot}25}{50} = 0{\cdot}025 \text{ m/min.}$$

When inlet valve is submerged 9 m, depth of water run up in one min.

$$= 0{\cdot}025 \times \sqrt{\frac{9}{4}} = 0{\cdot}025 \times 1{\cdot}5 = 0{\cdot}0375 \text{ m/min.}$$

Rate of pumping water out of tank

$$= \frac{1{\cdot}25}{125} = 0{\cdot}01 \text{ m/min.}$$

With valve open and pump working, rate of filling tank

$$= 0{\cdot}0375 - 0{\cdot}01 = 0{\cdot}0275 \text{ m/min.}$$

$$\text{Depth to fill} = 1{\cdot}25 - 0{\cdot}15 = 1{\cdot}1 \text{ m}$$

$$\text{Time to fill} = \frac{1{\cdot}1}{0{\cdot}0275} = 40 \text{ min.} \quad \text{Ans.}$$

50.

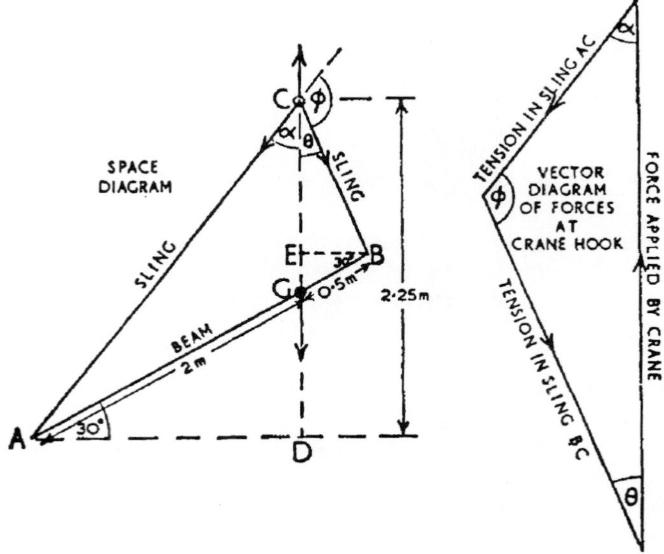

Referring to space diagram,

$$AD = AG \cos 30° = 2 \times 0{\cdot}866 = 1{\cdot}732 \text{ m}$$

$$\tan \alpha = \frac{AD}{CD} = \frac{1{\cdot}732}{2{\cdot}25} = 0{\cdot}7696$$

$$\alpha = 37°\ 35'$$

$$BE = BG \cos 30° = 0{\cdot}5 \times 0{\cdot}866 = 0{\cdot}433$$

$$EG = BG \sin 30° = 0{\cdot}5 \times 0{\cdot}5 = 0{\cdot}25$$

$$DG = AG \sin 30° = 2 \times 0{\cdot}5 = 1$$

$$CE = CD - DG - EG = 2{\cdot}25 - 1 - 0{\cdot}25 = 1$$

$$\tan \theta = \frac{BE}{CE} = \frac{0{\cdot}433}{1} = 0{\cdot}433$$

$$\theta = 23°25'$$

$$\phi = 180° - \alpha - \theta = 180° - 37°\ 35' - 23°\ 25'$$

$$= 119°$$

Force on crane hook = weight of beam

$$= 7{\cdot}5 \times 10^3 \times 9{\cdot}81$$

$$= 7{\cdot}358 \times 10^4 \text{ N} = 73{\cdot}58 \text{ kN}$$

Referring to vector diagram,

$$\frac{\text{Tension in sling AC}}{\sin\theta} = \frac{\text{crane force}}{\sin\phi}$$

$$\text{Tension in sling AC} = \frac{73{\cdot}58 \times 0{\cdot}3974}{0{\cdot}8746}$$

$$= 33{\cdot}44 \text{ kN} \qquad \text{Ans. } (i)$$

$$\frac{\text{Tension in sling BC}}{\sin\alpha} = \frac{\text{crane force}}{\sin\phi}$$

$$\text{Tension in sling BC} = \frac{73{\cdot}58 \times 0{\cdot}61}{0{\cdot}8746}$$

$$= 51{\cdot}33 \text{ kN} \qquad \text{Ans. } (ii)$$

51. The falling body has no initial velocity and an acceleration of 9·81 m/s^2. The rising body has an initial velocity of 36 m/s and a retardation (or negative acceleration) of 9·81 m/s^2. The falling body thus gains velocity at the same rate as the rising body loses velocity, therefore the velocity of approach is constant at 36 m/s.

$$\text{Distance} = 170 \text{ m}$$

$$\text{Time to pass} = \frac{170}{36} = 4{\cdot}722 \text{ s} \qquad \text{Ans. } (i)$$

In this time, falling body:

$$\begin{aligned} \text{gain in velocity} &= \text{accel.} \times \text{time} = 9{\cdot}81 \times 4{\cdot}722 \text{ m/s} \\ \text{average velocity} &= \tfrac{1}{2}(0 + 9{\cdot}81 \times 4{\cdot}722) \text{ m/s} \\ \text{distance fallen} &= \text{average velocity} \times \text{time} \\ &= \tfrac{1}{2} \times 9{\cdot}81 \times 4{\cdot}722 \times 4{\cdot}722 \text{ m} \\ &= 109{\cdot}3 \text{ m} \end{aligned}$$

distance above ground when bodies pass each other

$$= 170 - 109{\cdot}3$$

$$= 60{\cdot}7 \text{ m} \qquad \text{Ans. } (ii)$$

52. Torque transmitted by bolts
= stress × area × radius of P.C. × no. of bolts

$$\text{Torque transmitted by shafts} = \frac{\pi\,(D^4 - d^4)\tau}{16D}$$

Stress is same and cancels, therefore,

$$\frac{\pi}{4} \times 0{\cdot}067^2 \times 0{\cdot}28 \times 8 = \frac{\pi\,(0{\cdot}35^4 - d^4)}{16 \times 0{\cdot}35}$$

$$\frac{\pi \times 0{\cdot}067^2 \times 0{\cdot}28 \times 8 \times 16 \times 0{\cdot}35}{4 \times \pi} = 0{\cdot}35^4 - d^4$$

$$\begin{aligned} 0{\cdot}01408 &= 0{\cdot}01501 - d^4 \\ d &= \sqrt[4]{0{\cdot}00093} \\ &= 0{\cdot}1746 \text{ m} = 174{\cdot}6 \text{ mm} \quad \text{Ans.} \end{aligned}$$

53.

$$\begin{aligned} \text{Water thrust} &= \text{HA}\rho\text{g} \\ &= \frac{6{\cdot}3}{2} \times 9 \times 6{\cdot}3 \times 10^3 \times 9{\cdot}81 \\ &= 1752 \text{ kN} \end{aligned}$$

$$\begin{aligned} \text{Oil thrust} &= \text{HA}\rho\text{g} \\ &= \frac{4{\cdot}5}{2} \times 9 \times 4{\cdot}5 \times 850 \times 9{\cdot}81 \\ &= 759{\cdot}8 \text{ kN} \end{aligned}$$

$$\begin{aligned} \text{Resultant net thrust on bulkhead} &= 1752 - 759{\cdot}8 \\ &= 992{\cdot}2 \text{ kN} \qquad \text{Ans. } (i) \end{aligned}$$

$$\begin{aligned} \text{C.o.P.} &= \tfrac{1}{3} \text{ height above base} \\ \text{of water} &= \tfrac{1}{3} \times 6{\cdot}3 = 2{\cdot}1 \text{ m} \\ \text{of oil} &= \tfrac{1}{3} \times 4{\cdot}5 = 1{\cdot}5 \text{ m} \end{aligned}$$

Moments about base:

$$\begin{aligned} 1752 \times 2{\cdot}1 - 759{\cdot}8 \times 1{\cdot}5 &= 992{\cdot}2 \times y \\ 3679 - 1140 &= 992{\cdot}2 \times y \\ 2539 &= 992{\cdot}2 \times y \\ y &= 2{\cdot}56 \text{ m} \\ \text{Resultant C.o.P.} &= 2{\cdot}56 \text{ m above base} \qquad \text{Ans. } (ii) \end{aligned}$$

54. Referring to space diagram of Fig. 199, Chapter 11,

$$\frac{r}{\sin \phi} = \frac{l}{\sin 45^\circ}$$

$$\sin \phi = \frac{0{\cdot}52 \times 0{\cdot}7071}{1{\cdot}96} = 0{\cdot}1876$$

$$\phi = 10^\circ\ 49'$$

$$OP = r \sin(\phi + \theta)$$

$$= 0{\cdot}52 \sin 55^\circ\ 49' = 0{\cdot}4301 \text{ m}$$

Referring to vector diagram,

Force on guide = piston force × tan ø
= 196 × 0·1911
= 37·45 kN Ans. (*i*)

$$\text{Force in connecting rod} = \frac{\text{piston force}}{\cos \phi}$$

$$\frac{196}{0{\cdot}9822} = 199{\cdot}5 \text{ kN}$$

Turning moment = force in conn. rod × OP
= 199·5 × 0·4301
= 85·82 kN m Ans. (*ii*)

55. Referring to the load diagram of Fig:

Total load = 88 × 3 + 44 × 3 = 396 kN

$R_1 = R_2 = \frac{1}{2} \times 396 = 198$ kN

Moments about R_1

Clockwise moments = anticlockwise moments

$$88 \times 3 \times 1{\cdot}5 + 44 \times 3 \times 4{\cdot}5 = R_2 \times x$$

$$990 = 198 \times x$$

$$x = 5 \text{ m}$$

Right hand reaction should be at 5 m from left end which is 1 m from right end. Ans. (*i*)

Let z = position from left end where *S.F.* is zero and *B.M.* is maximum, adding and subtracting forces,

$$198 - 88 \times z = 0$$

$$z = 2{\cdot}25 \text{ m}$$

$$M_{max} = -198 \times 2{\cdot}25 + 88 \times 2{\cdot}25 \times 1{\cdot}125$$

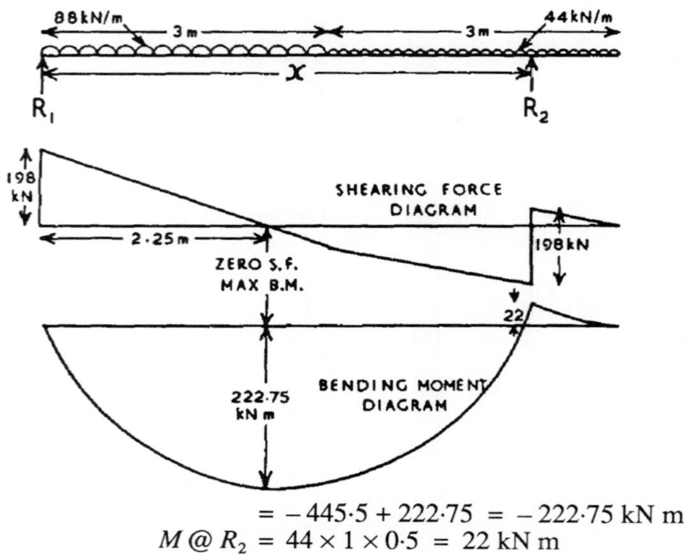

$$= -445{\cdot}5 + 222{\cdot}75 \;=\; -222{\cdot}75 \text{ kN m}$$

$$M @ R_2 \;=\; 44 \times 1 \times 0{\cdot}5 \;=\; 22 \text{ kN m}$$

The shearing force and bending moment diagrams with these principal values are shown.

56.

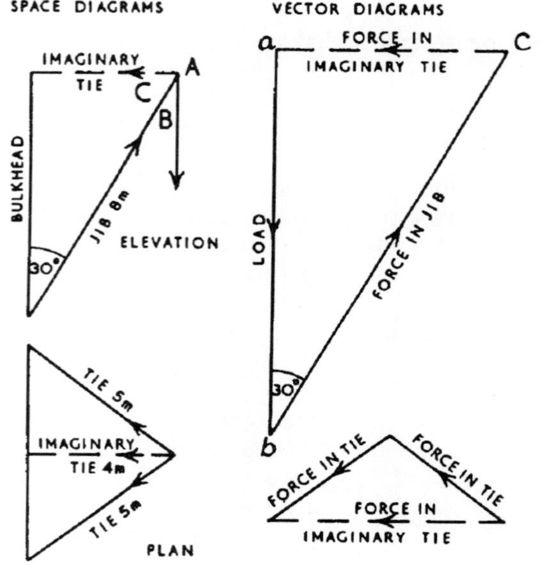

To begin with, insert an imaginary tie in the same plane as the jib and load, in the centre of the two real ties to take their place temporarily, as shown in Fig.

Weight of mass suspended from jib head

$$= mg = 1570 \times 9{\cdot}81 = 15400 \text{ N or } 15{\cdot}4 \text{ kN}$$

Referring to vector diagram,

$$\text{Force in jib} = \frac{\text{load}}{\cos 30^\circ} = \frac{15{\cdot}4}{0{\cdot}866} = 17{\cdot}78 \text{ kN} \qquad \text{Ans. } (i)$$

$$\text{Force in imaginary tie} = \text{load} \times \tan 30^\circ$$
$$= 15{\cdot}4 \times 0{\cdot}5774 = 8{\cdot}892 \text{ kN}$$

Referring to vector diagram of plan view, force in the imaginary tie is the resultant of the two forces in the real ties,

$$\text{Force in each tie} = \tfrac{1}{2} \times \text{force in imaginary tie} \times 5/4$$
$$= \frac{8{\cdot}892 \times 5}{2 \times 4}$$
$$= 5{\cdot}558 \text{ kN} \qquad \text{Ans. } (ii)$$

57.

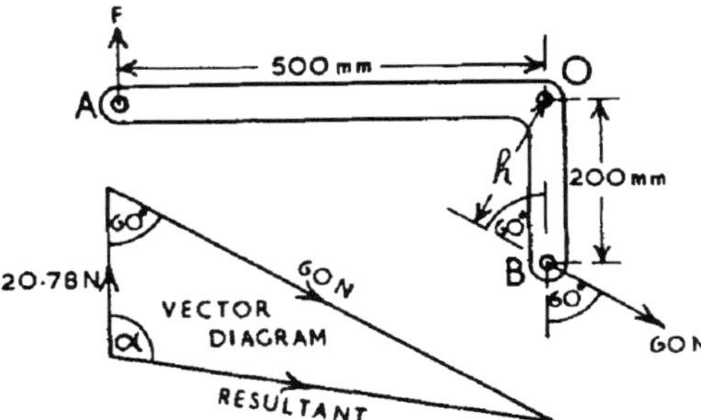

Referring to space diagram, perpendicular distance from line of action of 60 N force to fulcrum 0 =

$$h = 200 \times \sin 60^\circ = 173{\cdot}2 \text{ mm}$$

Moments about 0,

$$F \times 500 = 60 \times 173{\cdot}2$$
$$F = 20{\cdot}78 \text{ N} \qquad \text{Ans. } (i)$$

Referring to vector diagram:

$$(\text{Resultant})^2 = 20{\cdot}78^2 + 60^2 - 2 \times 20{\cdot}78 \times 60 \times \cos 60^\circ$$
$$= 431{\cdot}8 + 3600 - 1247$$

$$\text{Resultant} = \sqrt{2785} = 52{\cdot}78 \text{ N}$$ Ans. (*iia*)

$$\frac{60}{\sin \alpha} = \frac{52{\cdot}78}{\sin 60°}$$

$$\sin \alpha = \frac{60 \times 0{\cdot}866}{52{\cdot}78} = 0{\cdot}9845$$

$\alpha = 79°\ 54'$ or $180° - 79°\ 54' = 100°\ 6'$ to the vertical or, $100°\ 6' - 90° = 10°\ 6'$ to the horizontal. Ans. (*iib*)

58.

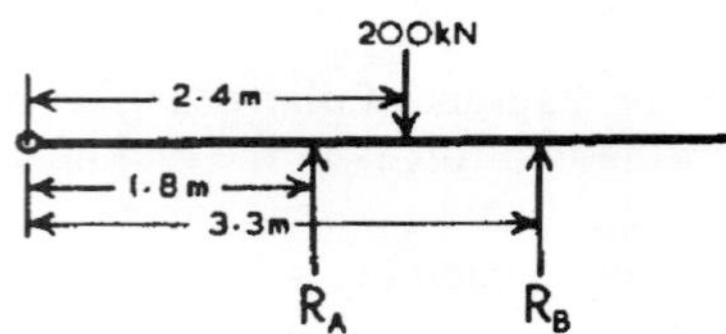

The beam remains rigid and horizontal when the load is placed on it, therefore both props are compressed the same amount and, since their lengths are equal, then the strain in each prop is the same.

Let load on first prop = R_A
and " " second " = R_B

$$\frac{\text{stress}}{\text{strain}} = E \qquad \text{strain} = \frac{\text{stress}}{E}$$

$$\text{stress} = \frac{\text{load}}{\text{area}} \qquad \text{strain} = \frac{\text{load}}{\text{area} \times E}$$

strain in prop A = strain in prop B

$$\frac{\text{load}}{\text{area} \times E} \text{ of prop A} = \frac{\text{load}}{\text{area} \times E} \text{ of prop B}$$

$$\frac{R_A}{11{\cdot}6 \times 10^{-4} \times 180 \times 10^9} = \frac{R_B}{14{\cdot}8 \times 10^{-4} \times 150 \times 10^9}$$

$$R_B = R_A \times \frac{14{\cdot}8 \times 150}{11{\cdot}6 \times 180}$$

$$R_B = 1{\cdot}063 \times R_A \qquad \ldots \ldots (i)$$

Moments about hinge,

$$200 \times 10^3 \times 2{\cdot}4 = R_A \times 1{\cdot}8 + R_B \times 3{\cdot}3$$

$$200 \times 10^3 \times 2{\cdot}4 = R_A \times 1{\cdot}8 + 1{\cdot}063 \times R_A \times 3{\cdot}3$$

$$200 \times 10^3 \times 2{\cdot}4 = 5{\cdot}308 \times R_A$$

$$R_A = 9{\cdot}043 \times 10^4 \text{ N or } 90{\cdot}43 \text{ kN}$$ Ans. (*i*)

From (*i*) $$R_B = 1{\cdot}063 \times 90{\cdot}43 = 96{\cdot}13 \text{ kN}$$ Ans. (*ii*)

Total forces up = Total forces down

Hinge force + R_A + R_B = 200

Hinge force = 200 – 90·43 – 96·43

= 13·44 kN Ans. (*iii*)

59. Volume flow rate through orifice

$= C_D \times$ velocity $\times$ area of orifice

and for a steady level this is equal to the volume flow rate into the tank which is $1{\cdot}2 \div 60 = 0{\cdot}02 \text{m}^3/\text{s}$, therefore,

$$0{\cdot}7 \times \sqrt{2gh} \times 0{\cdot}7854 \times 65^2 \times 10^{-6} = 0{\cdot}02$$

$$\sqrt{2 \times 9{\cdot}81 \times h} = \frac{0{\cdot}02 \times 10^6}{0{\cdot}7 \times 0{\cdot}7854 \times 65^2}$$

$$= 8{\cdot}61$$

squaring both sides,

$$2 \times 9{\cdot}81 \times h = 8{\cdot}61^2$$

$$h = \frac{8{\cdot}61^2}{2 \times 9{\cdot}81} = 3{\cdot}778 \text{ m}$$ Ans. (*i*)

Volume in tank = area × depth

$= 0{\cdot}7854 \times 1{\cdot}5^2 \times 3{\cdot}778$

$= 6{\cdot}676 \text{ m}^3$

Initial volume flow = $1{\cdot}2 \text{ m}^3/\text{min}$.

mean " " = $\frac{1}{3} \times 1{\cdot}2 = 0{\cdot}4 \text{ m}^3/\text{min}$.

$$\text{Time to empty} = \frac{\text{volume}}{\text{mean vol. flow}}$$

$$= \frac{6{\cdot}676}{0{\cdot}4} = 16{\cdot}69 \text{ min.}$$ Ans. (*ii*)

60. Tensile (hoop) stress set up in the rim of a flywheel due to centrifugal force

$$= \rho\omega^2 r^2$$

$$\omega = \frac{250 \times 2\pi}{60} = 26{\cdot}18 \text{ rad/s}$$

$$\begin{aligned} \text{stress} &= \rho\omega^2 r^2 \\ &= 7{\cdot}2 \times 10^3 \times 26{\cdot}18^2 \times 1{\cdot}8^2 \\ &= 1{\cdot}599 \times 10^7 \text{ N/m}^2 \text{ or } 15{\cdot}99 \text{ MN/m}^2 \quad \text{Ans. } (i) \\ K.E. \text{ of rotation} &= \tfrac{1}{2} mk^2\omega^2 \end{aligned}$$

For a thin rim, the volume may be taken as the product of the mean circumference and cross-sectional area, and the radius of gyration as the mean radius.

$$\begin{aligned} m &= V\rho \\ &= 2\pi \times 1{\cdot}8 \times 0{\cdot}1 \times 0{\cdot}15 \times 7{\cdot}2 \times 10^3 \\ &= 1{\cdot}222 \times 10^3 \text{ kg} \end{aligned}$$

$$\begin{aligned} K.E. \text{ stored up} &= \tfrac{1}{2} mk^2 \omega^2 \\ &= 0{\cdot}5 \times 1{\cdot}222 \times 10^3 \times 1{\cdot}8^2 \times 26{\cdot}18^2 \\ &= 1{\cdot}357 \times 10^6 \text{ J or } 1357 \text{ kJ} \qquad \text{Ans. } (ii) \end{aligned}$$

$K.E.$ remaining at reduced speed

$$\begin{aligned} &= 1357 - 81 = 1276 \text{ kJ} \\ K.E. &= \tfrac{1}{2} mk^2 \omega^2 \end{aligned}$$

$$\omega = \sqrt{\frac{1276 \times 10^3 \times 2}{1{\cdot}222 \times 10^3 \times 1{\cdot}8^2}} = 25{\cdot}39 \text{ rad/s}$$

$$\text{speed} = \frac{25{\cdot}39 \times 60}{2\pi} = 242{\cdot}4 \text{ rev/min}$$

$$\begin{aligned} \text{Reduction in speed} &= 250 - 242{\cdot}4 \\ &= 7{\cdot}6 \text{ rev/min} \qquad \text{Ans. } (iii) \end{aligned}$$

61. Cross-sectional area of tube

$$\begin{aligned} &= 0{\cdot}7854 \times (0{\cdot}04^2 - 0{\cdot}03^2) \\ &= 5{\cdot}498 \times 10^{-4} \text{ m}^2 \end{aligned}$$

Cross-sectional area of bar

$$= 0{\cdot}7854 \times 0{\cdot}02^2 = 3{\cdot}142 \times 10^{-4} \text{ m}^2$$

Tensile force in bar

$$= \text{compressive force in tube}$$
$$= 12{\cdot}5 \times 10^3 \text{N}$$

$$\text{Stress in tube} = \frac{\text{load}}{\text{area}} = \frac{12{\cdot}5 \times 10^3}{5{\cdot}498 \times 10^{-4}}$$

$$= 2{\cdot}274 \times 10^7 \text{ N/m}^2 \text{ or } 22{\cdot}74 \text{ MN/m}^2 \text{ Ans. } (a)$$

$$\text{Stress in bar} = \frac{12{\cdot}5 \times 10^3}{3{\cdot}142 \times 10^{-4}}$$

$$= 3{\cdot}978 \times 10^7 \text{ N/m}^2 \text{ or } 39{\cdot}78 \text{ MN/m}^2 \text{ Ans. } (b)$$

Result of further tightening:

$$\text{Load on bolt} = \text{Load on tube}$$
$$\text{stress}_B \times \text{area}_B = \text{stress}_T \times \text{area}_T$$

$$\text{stress}_B = \text{stress}_T \times \frac{\text{area}_T}{\text{area}_B}$$

$$\text{stress}_B = \text{stress}_T \times \frac{5{\cdot}498 \times 10^{-4}}{3{\cdot}142 \times 10^{-4}}$$

$$= 1{\cdot}75 \times \text{stress}_T \qquad \dots \dots (i)$$

When nut is turned one revolution, total deformation is 1 mm = 10^{-3} m which is shared between tube and bolt.

$$\text{Compression of tube + stretch of bar} = 10^{-3} \text{ m}$$
$$\text{strain}_T \times \text{length}_T + \text{strain}_B \times \text{length}_B = 10^{-3}$$

$$\frac{\text{stress}_T}{E_T} \times \text{length}_T + \frac{\text{stress}_B}{E_B} \times \text{length}_B = 10^{-3}$$

$$\text{Stress}_T + \text{stress}_B = \frac{10^{-3} \times 210 \times 10^9}{12}$$

$$= 1{\cdot}75 \times 10^7 \qquad \dots \dots (ii)$$

Substituting for stress_B from (*i*) into (*ii*),

$$\text{stress}_T + 1{\cdot}75 \text{ stress}_T = 1{\cdot}75 \times 10^7$$

$$\text{Stress}_T = \frac{1{\cdot}75 \times 10^7}{2{\cdot}75}$$

$$= 6{\cdot}364 \times 10^6 \text{ N/m}^2$$

Increase in stress in tube $= 6{\cdot}364 \text{ MN/m}^2$ Ans. (*c*)

Increase in stress in bar from (*i*) $= 1{\cdot}75 \times 6{\cdot}364 = 11{\cdot}14 \text{ MN/m}^2$

Ans. (*d*)

62.

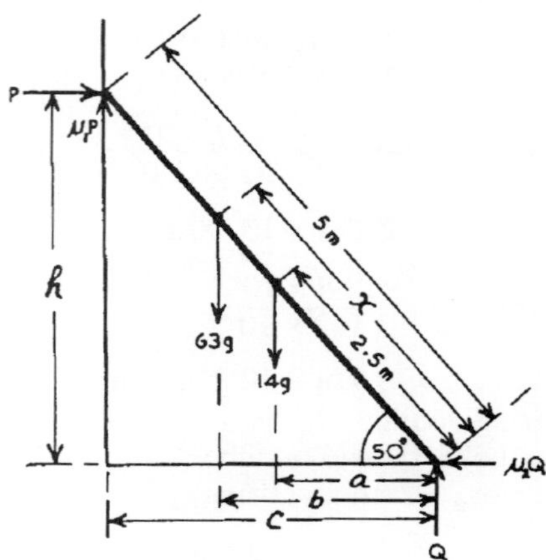

The ladder being uniform, its centre of gravity through which its weight acts is at mid-length.

Referring to Fig.

Let normal reaction at wall = P

and " " ground = Q

Friction force opposing motion at wall

$$= \text{coeff. of friction} \times \text{force between surfaces}$$

$$= \mu_1 P = 0{\cdot}2P$$

Friction force opposing motion along ground

$$= \mu_2 Q = 0{\cdot}5Q$$

$$\text{Weight of man} = 63 \times g$$

$$\text{" " ladder} = 14 \times g$$

$$\text{Forces pushing right} = \text{Forces pushing left}$$

$$P = 0{\cdot}5Q$$

$$Q = 2P$$

$$\text{Upward forces} = \text{Downward forces}$$

$$Q + 0{\cdot}2P = 63g + 14g$$

$$2P + 0{\cdot}2P = 77g$$

$$2{\cdot}2P = 77g$$

$$P = 35g$$

Let x = distance up ladder for position of man. To take moments about foot of ladder, distances required:

$$h = 5 \sin 50^\circ \quad = 3{\cdot}83$$

$$a = 2{\cdot}5 \cos 50° = 1{\cdot}607$$
$$b = x \cos 50° = 0{\cdot}6428x$$
$$c = 5 \cos 50° = 3{\cdot}214$$

Clockwise moments = Anticlockwise moments

$$P \times h + \mu_1 P \times c = 63g \times b + 14g \times a$$
$$35g \times 3{\cdot}83 + 0{\cdot}2 \times 35g \times 3{\cdot}214 = 63g \times 0{\cdot}6428x + 14g \times 1{\cdot}607$$
$$134{\cdot}1 + 22{\cdot}5 = 40{\cdot}5x + 22{\cdot}5$$
$$134{\cdot}1 = 40{\cdot}5x$$
$$x = 3{\cdot}31 \text{ m}$$ Ans.

63.

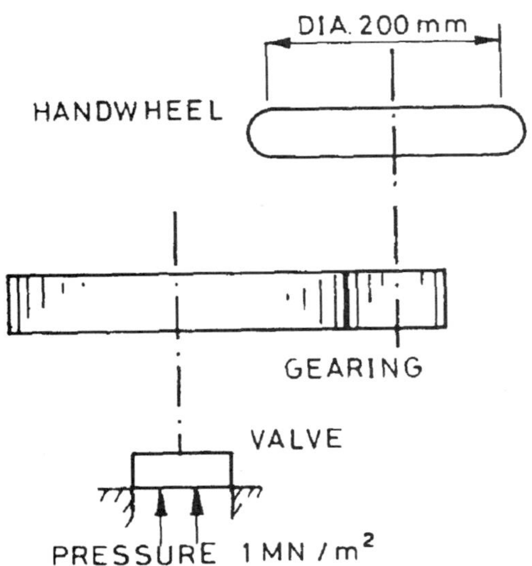

$$\text{Force on valve} = \text{pressure} \times \text{area}$$
$$= 1 \times 10^6 \times 27\,000 \times 10^{-6}$$
$$= 27\,000 \text{ N.}$$

Consider one revolution of the valve gearwheel:

$$\text{Distance moved by effort} = 4 \times 200 \times \pi \text{ mm}$$
$$\text{Distance moved by load} = 4 \text{ mm}$$

$$\text{V.R.} = \frac{4 \times 200 \times \pi}{4}$$
$$= 200\pi$$

$$\text{Efficiency} = \frac{\text{M.A.}}{\text{V.R.}}$$

$$\text{M.A.} = 0{\cdot}43 \times 200\pi$$
$$= 270$$

$$\text{Also, M.A.} = \frac{\text{load}}{\text{effort}}$$

$$\text{Effort at handwheel} = \frac{27\,000}{270}$$

$$= 100 \text{ N}$$ Ans.

64.

$$m = V\rho$$

$$= \frac{\pi}{4}\,(0{\cdot}5^2 - 0{\cdot}4^2) \times 9 \times 7860$$

$$= 5t$$ Ans. (*a*)

Max. bending will occur at mid-point

$$\text{and } R_A = R_B = \frac{5 \times 9{\cdot}81}{2}$$

$$M = -\left[\left(\frac{5\times 9{\cdot}81}{2}\right) \times 4{\cdot}5\right] + \left[\frac{5\times 9{\cdot}81}{2} \times 4{\cdot}5 \times \frac{4{\cdot}5}{2}\right]$$

$$= -55{\cdot}2 \text{ kNm (i.e. sagging)}$$ Ans. (*b*)

$$\sigma = \frac{My}{I} = \frac{55{\cdot}2 \times 10^3 \times 0{\cdot}25 \times 64}{\pi\,(0{\cdot}5^4 - 0{\cdot}4^4)}$$

$$= 7{\cdot}62 \text{ MN/m}^2$$ Ans. (*c*)

65.

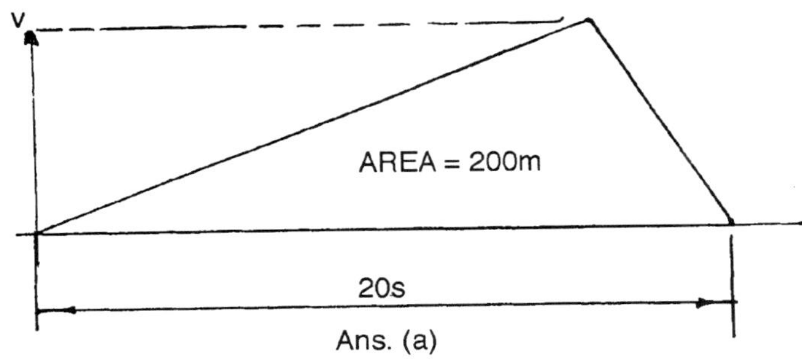

$$s = \frac{v}{2} \times 20 = 10v$$

$$200 = 10v$$

$$v = 20 \text{ m/s}$$ Ans. $(b)(i)$

$$0 = u + at \;\therefore\; 0 = 20 - 9{\cdot}81t.$$

$$t = \frac{20}{9{\cdot}81} = 20{\cdot}04 \text{ s}$$

Time to accelerate = $20 - 2{\cdot}04 = 17{\cdot}96$ s Ans $(b)(ii)$

$$v = u + at$$

$$20 = a \times 17{\cdot}96$$

$$a = 1{\cdot}113 \text{ m/s}^2$$

Acceleration = $1{\cdot}113$ m/s^2

Tension in rope = $ma + mg$

$= 8000 \times 1{\cdot}113 + 8000 \times 9{\cdot}81$

$= 87384$ N or $87{\cdot}384$ kN

Tension in rope = $87{\cdot}384$ kN Ans. $(b)(iii)$

66.

$$F = ma$$

$$a = \frac{8 \times 10^3}{2 \times 10^3}$$

$$= 4 \text{ m/s}^2$$

Velocity of gun = $a \times t$

$= 4 \times 1{\cdot}5$

$= 6$ m/s

$$\text{Momentum of gun} = \text{Momentum of shell}$$
$$2 \times 10^3 \times 6 = 25\text{v}$$
$$\text{Velocity of shell} = \frac{2 \times 10^2 \times 6}{25}$$
$$= 480 \text{ m/s Relative to ground} \quad \text{Ans. } (i)$$
$$= 486 \text{ m/s Relative to gun} \quad \text{Ans. } (ii)$$

67.

$$P = F\text{v} \qquad \text{Velocity (m/s)} = \frac{\text{v} \times 1852}{3600} \quad (\text{v} = \text{knots})$$

$$R = \text{k}v^2 \qquad F = \frac{P}{\text{v}.} \qquad \text{kv}^2 = \frac{P}{\text{v}}$$

$$P = \text{kv}^2 \times \frac{\text{v} \times 1852}{3600}$$

$$3 \times 10^6 = 1{\cdot}5 \times 10^3 \times \text{v}^3 \times \frac{1853}{3600}$$

$$\text{v} = 15{\cdot}79 \text{ knots} \qquad \text{Ans. } (a)$$
$$R = \text{kv}^2 = 1{\cdot}5 \times 10^2 = 150 \text{ kN}$$

$$P = F\text{v} = 150 \times 10^3 \times 10 \times \frac{1852}{3600} = 771{\cdot}63 \text{ kW}$$

$$\text{Accel. power} = 3 \times 10^6 - 771{\cdot}63^3 = 2{\cdot}28 \text{ MW}$$

$$F = \frac{3600 \times 2{\cdot}28 \times 10^6}{10 \times 18{\cdot}52}$$
$$= 433{\cdot}15 \text{ kN}$$
$$F = ma$$
$$a = \frac{433{\cdot}15 \times 10^3}{4000 \times 10^3}$$
$$\text{Acceleration} = 0{\cdot}108 \text{ m/s}^2 \text{ at 10 knots} \qquad \text{Ans. } (b)$$

68.

$$T = 140 + 40 \sin A - 20 \cos A$$

$$\frac{dT}{dA} = 40 \cos A - (-20 \sin A)$$

$$\frac{dT}{dA} = 0 \text{ at turning points}$$

$$-40\cos A = 20\sin A$$
$$\tan A = -2 \quad A = -63{\cdot}44^\circ$$
$$A = 116{\cdot}56^\circ \text{ or } 296{\cdot}56^\circ$$

$$\frac{d^2T}{dA^2} = -40\sin A + 20\cos A$$

$$A = 116{\cdot}56^\circ \quad \frac{d^2T}{dA^2} = -44{\cdot}72 \text{ (Max)}$$

$$A = 296{\cdot}56^\circ \quad \frac{d^2T}{dA^2} = +44{\cdot}72 \text{ (Min)}$$

$$T_{Max} = 140 + 40\sin 116{\cdot}56^\circ - 20\cos 116{\cdot}56^\circ$$
$$T_{Max} = 184{\cdot}72 \text{ kN m}$$ Ans. (*a*)

$$\frac{T}{J} = \frac{\tau}{r} \qquad J = \frac{\pi D^4}{32}$$

$$\frac{184{\cdot}72 \times 10^3 \times 32}{\pi D^4} = \frac{60 \times 10^6 \times 2}{D}$$

$$\frac{184{\cdot}72 \times 10^3 \times 32}{\pi \times 60 \times 10^6 \times 2} = \frac{D^4}{D} = D^3$$

$$D = 0{\cdot}250\,m$$
$$\text{Minimum diameter} = 250\,mm$$ Ans. (*b*)

69.

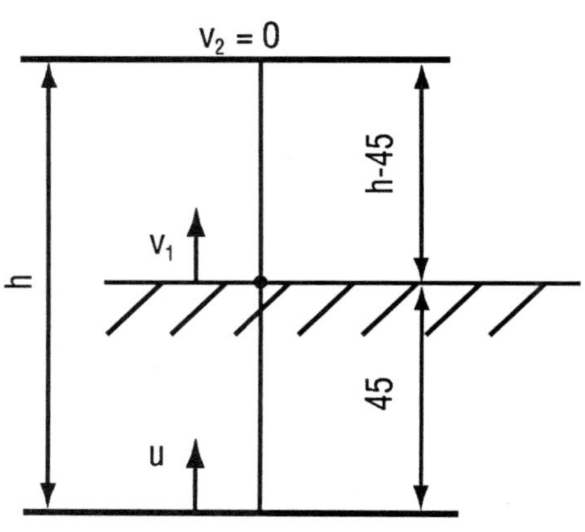

$$\begin{aligned}
v_2 &= v_1 - 9{\cdot}81t \\
0 &= v_1 - 9{\cdot}81 \times 2 \\
v_1 &= 19{\cdot}62 \text{ m/s} \\
h &= v_1 t - \tfrac{1}{2} g t^2 \\
h - 45 &= 2 \times 19{\cdot}62 - \tfrac{1}{2} \times 9{\cdot}81 \times 2^2 \\
\text{Maximum height} &= 39{\cdot}24 - 19{\cdot}62 + 45 = 64{\cdot}62 \text{ m} \quad \text{Ans. } (a) \\
v_1^{\ 2} &= u^2 - 2 \times 9{\cdot}81 \times 45 \\
19{\cdot}62^2 &= u^2 - 882{\cdot}9 \\
u^2 &= 384{\cdot}9 + 882{\cdot}9 \\
u^2 &= 1267{\cdot}8 \\
\text{Initial velocity} &= 35{\cdot}6 \text{ m/s} \quad \text{Ans. } (b) \\
19{\cdot}62 &= 35{\cdot}6 - 9{\cdot}81\, t \\
t &= \frac{19{\cdot}62 - 35{\cdot}6}{9{\cdot}81} = 1{\cdot}63 \text{ s} \\
\text{Total time} &= 2(1{\cdot}63 + 2) = 7{\cdot}26 \text{ s} \quad \text{Ans. } (c)
\end{aligned}$$

70.

$$\begin{aligned}
v_1 a_1 &= v_2 a_2 \\
0{\cdot}7854 \times 0{\cdot}12^2 \times 4 &= 0{\cdot}7854 \times 0{\cdot}08^2 \times v_2 \\
v_2 &= 9 \text{ m/s} \\
\text{Velocity in smaller pipe} &= 9 \text{ m/s} \quad \text{Ans.}(a)(i) \\
\dot{m} &= \rho A v \\
&= 860 \times 0{\cdot}7854 \times 0{\cdot}12^2 \times 4 \\
\text{Mass flow rate} &= 38{\cdot}9 \text{ kg/s} \quad \text{Ans. } (a)(ii)
\end{aligned}$$

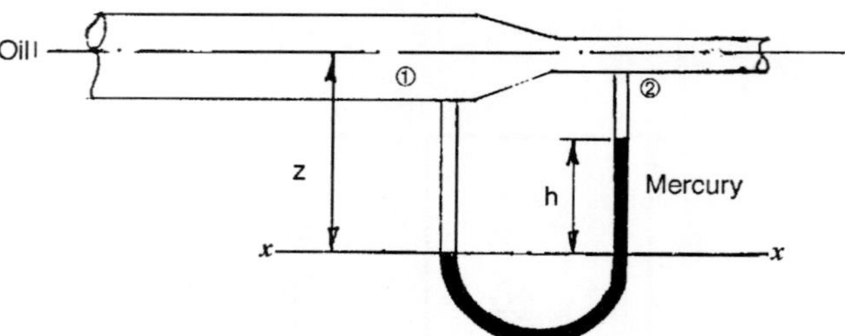

$$\begin{aligned}
\text{Pressure at } xx &= \text{constant} \\
p_1 + \rho_0 g z &= \rho_2 + \rho_0 g(z - h) + \rho_M g h \\
p_1 - p_2 &= \rho_M g h - \rho_o g h \\
&= 9{\cdot}81 \times 0{\cdot}04 \times (13600 - 1000) \\
\text{Pressure drop} &= 4944 \text{ N/m}^2 \quad \text{Ans. } (b)
\end{aligned}$$

71. See definition (Chapter 3), etc. Ans. (*a*)

$$\text{Momentum before impact} = \text{Momentum after impact}$$
$$m_1 \text{v}_1 + m_2 \text{v}_2 = (m_1 + m_2)\,\text{v}_3$$
$$(3 \times 8) + (5 \times 12) = (3 + 5)\,\text{v}_3$$
$$\text{v}_3 = \frac{24 + 60}{8}$$

Final velocity of two bodies $= 10{\cdot}5$ m/s Ans. (*b*)(*i*)

$$K.E. = \tfrac{1}{2} m\text{v}^2$$
$$K.E. \text{ before impact} = \tfrac{1}{2} m_1 \text{v}_1{}^2 + \tfrac{1}{2} m\text{v}_2{}^2$$
$$= \tfrac{1}{2} \times 3 \times 8^2 + \tfrac{1}{2} \times 5 \times 12^2$$
$$= 96 + 360$$
$$= 456\,J$$

$$K.E. \text{ after impact} = \tfrac{1}{2} m_3 \text{v}_3{}^2$$
$$= \tfrac{1}{2} \times 8 \times 10{\cdot}5^2$$
$$= 441J$$

$$\text{Loss of } K.E. = \text{before} - \text{after}$$
$$= 456 - 441$$

Energy lost $= 15J$ Ans. (*b*)(*ii*)

72. Mass of beam = volume × density

$$= 2 \times 0{\cdot}1 \times 0{\cdot}2 \times 7800$$

$= 312$ kg Ans. (*a*)

$$R_A = R_B = \frac{312 \times 9{\cdot}81 + 6 \times 10^3}{2}$$
$$= 4{\cdot}53 \text{ kN}$$

$$B.M. = -(4{\cdot}53 \times 1) + (1{\cdot}53 \times 1 \times 0{\cdot}5)$$
$$= 3{\cdot}765 \text{ kNm i.e. sagging}$$

$$\sigma = \frac{My}{I} = \frac{3{\cdot}765 \times 10^3 \times 12 \times 0{\cdot}1}{0{\cdot}1 \times 0{\cdot}2^3}$$

$= 5{\cdot}648$ MN/m^2 Ans. (*b*)

73.

$$\text{v} = \omega r$$
$$\alpha = \omega^2 r$$
$$12 = \omega r$$
$$96 = \omega^2 r$$

$$\omega = \frac{12}{r}$$

$$96 = \left(\frac{12}{r}\right)^2 r$$

$$r = \frac{144}{96}$$

$$= 1{\cdot}5 \text{ m}$$

$$\text{Diameter} = 3 \text{ m} \qquad \text{Ans. } (a)$$

$$\text{v} = \omega r$$

$$\omega = \frac{1{\cdot}2}{1{\cdot}5}$$

$$\text{Original angular speed} = 8 \text{ rad/s} \qquad \text{Ans. } (b)$$

$$\alpha = \frac{\text{change in velocity}}{\text{time}}$$

$$= \frac{8-6}{4}$$

$$\text{Angular acceleration} = 0{\cdot}75 \text{ rad/s}^2 \qquad \text{Ans. } (c)$$

$$a = \alpha\, r$$

$$= 0{\cdot}75 \times 1{\cdot}5$$

$$\text{Tangential acceleration} = 1{\cdot}125 \text{ m/s}^2 \qquad \text{Ans. } (d)$$

74.

$$20 = \frac{\text{main wheel dia.}}{\text{pinion dia.}}$$

$$\text{Dia. main wheel} = 20 \times 120 = 2400\text{mm}$$

$$\text{Pitch circle diameter (main wheel)} = 2{\cdot}4 \text{ m} \qquad \text{Ans. } (a)$$

$$\text{P} = \omega\, T$$

$$15 \times 10^3 = 2\pi \times \frac{600}{60} T \qquad T = \frac{15 \times 10^3}{2\pi \times 10} = 238{\cdot}7 \text{ Nm}$$

$$\text{Tangential force} = \frac{\text{torque}}{\text{radius}} = \frac{238{\cdot}7}{0{\cdot}06} = 3979 \text{ N} \qquad \text{Ans. } (b)$$

$$\text{N}_\text{F} = \frac{T_\text{F}}{\cos 20°} = \frac{3979}{\cos 20°}$$

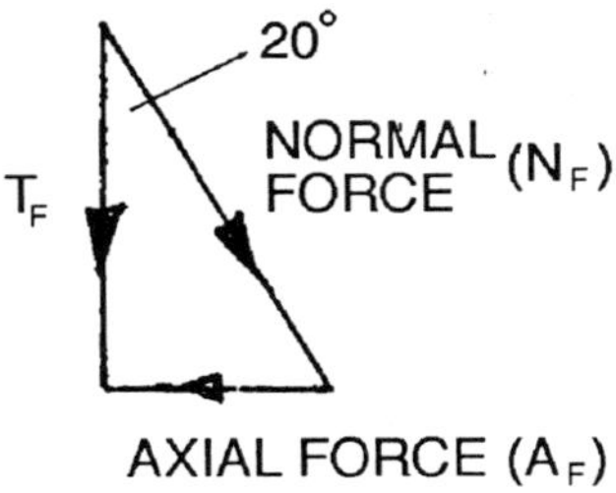

$N_F = 4234$ N Ans. (*c*)

$A_F = T_F \times \tan 20° = 3979 \times \tan 20°$

$A_F = 1448$N Ans. (*d*)

75. The energy of a body or system is its capacity for doing work

Kinetic energy – energy of movement ($\frac{1}{2}mv^2$)
Potential energy – energy of position (*m.g.h.*)
Internal energy – energy of state ($mc_v\ \delta T$)
} Ans. (*a*)

Momentum of *A* before impact (say left) = – 100 × 5
= – 500 Kgm/s

Momentum of *B* before impact (say right) = 50 × 1
= 50 Kgm/s

Momentum of *A* and *B* after impact = (100 +50) × –v
= – 150 v (left)

Momentum before = momentum after

– 500 + 50 = – 150v

– 450 = – 150v

Velocity of combined mass = 3 m/s Ans. (*b*)(*i*)

K.E. before = $\frac{1}{2} \times 100 \times 5^2 + \frac{1}{2} \times 50 \times 1^2$
= 1275 Nm

K.E. after = $\frac{1}{2} \times 150 \times 3^2$ = 675 Nm

System loss of energy = 1275 – 675 = 600 Nm Ans.(*b*)(*ii*)

76. Work done by force $= F \times s \quad J$
Work done by torque $= T \times \theta \quad J$ } Ans. (*a*)

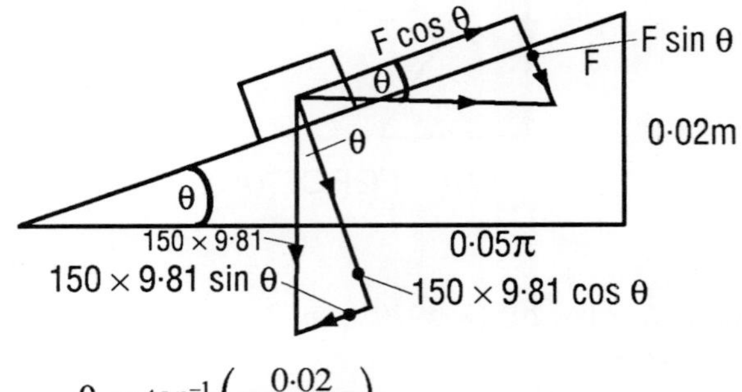

$$\theta = \tan^{-1}\left(\frac{0{\cdot}02}{\pi \times 0{\cdot}05}\right)$$

$$\theta = 7{\cdot}256°$$

Force up plane $= F \cos \theta$

$$F \cos \theta = \mu\,(150 \times 9{\cdot}81 \cos \theta + F \sin \theta) + (150 \times 9{\cdot}81 \sin \theta)$$
$$= 0{\cdot}2 \times 150 \times 9{\cdot}81 \times \cos 7{\cdot}256° + 0{\cdot}2 \times F \times \sin 7{\cdot}256° + 150 \times 9{\cdot}81 \times \sin 7{\cdot}256°$$
$$0{\cdot}9919F = 291{\cdot}94 + 0{\cdot}0253\,F$$
$$0{\cdot}9919F - 0{\cdot}0253F = 291{\cdot}94$$
$$0{\cdot}966F = 291{\cdot}94$$
$$F = 302$$

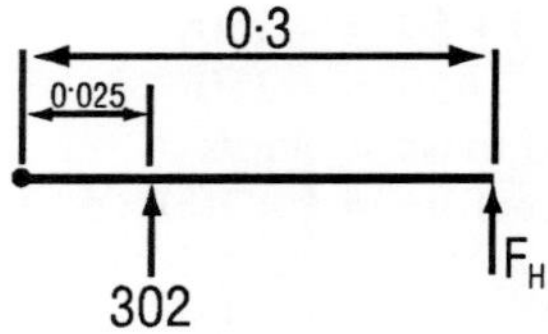

$$F_H \times 0{\cdot}3 = 0{\cdot}025 \times 302$$

$$F_H = \frac{0{\cdot}025 \times 302}{0{\cdot}3} = 25{\cdot}17 \text{ N}$$

Force on handwheel $= 25{\cdot}17$ N Ans. (*b*)

77. $\frac{T}{J} = \frac{\tau}{r}$ $\qquad \tau_N = \frac{\tau_0 \times 440}{350} \qquad = 1{\cdot}257\ \tau_0$

$$T_0 = \frac{\text{Power}_0}{2\pi \times n_0} \quad \therefore \quad T_N = \frac{1{\cdot}2\ \text{Power}_0}{2\pi \times 1{\cdot}1 \times n_0} \qquad = 1{\cdot}091\ T_0$$

$$J_0 = \frac{\pi D^4}{32} \qquad = \frac{\pi \times 0{\cdot}5^4}{32} \qquad = 6{\cdot}136 \times 10^{-3}$$

$$\frac{T_0}{J_0} = \frac{\tau_0}{r} \qquad \tau_0 = \frac{T_0 \times 0{\cdot}25}{6{\cdot}136 \times 10^{-3}} \qquad = 40{\cdot}74\ T_0$$

$$J_N = \frac{\pi \times 0{\cdot}5^4}{32} - \frac{\pi\, d^4}{32} = 0{\cdot}006136 - 0{\cdot}098\ d^4$$

$$\frac{T_N}{J_N} = \frac{\tau_N}{0{\cdot}25} \qquad = \frac{1{\cdot}257\ \tau_0}{0{\cdot}25} \qquad = 5{\cdot}028\ \tau_0$$

$$\frac{1{\cdot}091\ T_0}{J_N} \qquad = 5{\cdot}028 \times 40{\cdot}74\ T_0 = 204{\cdot}84\ T_0$$

$$\frac{1{\cdot}091}{204{\cdot}84} = J_N = 0{\cdot}006136 - 0{\cdot}098 d^4$$

$$5{\cdot}326 \times 10^{-3} - 0{\cdot}006136 = -0{\cdot}098 d^4$$

$$0{\cdot}098 d^4 = 8{\cdot}08 \times 10^{-4}$$

$$d^4 = \frac{8{\cdot}08 \times 10^{-4}}{0{\cdot}098}$$

$$d = 0{\cdot}301 \text{ m}$$

Maximum inner diameter = 301 mm Ans.

78. Mass flow rate is mass of fluid flowing per second
$\dot{m}$ is kg/s
Density is mass per unit volume
ρ is kg/m³ Ans.(*a*)

$$\dot{m}_1 = \rho_1 a_1 v_1$$
$$= 850 \times 0{\cdot}7854 \times 0{\cdot}075^2 \times 6$$
$$= 22{\cdot}53 \text{ kg/s}$$
$$\dot{m}_2 = \rho_2 a_2 v_2$$

$$= 950 \times 0{\cdot}7854 \times 0{\cdot}1^2 \times 10$$
$$= 74{\cdot}61 \text{ kg/s}$$
$$a_3\, v_3 = a_1\, v_1 + a_2\, v_2$$
$$d^2 \times 12 = 0{\cdot}075^2 \times 6 + 0{\cdot}1^2 \times 10$$
$$d = 0{\cdot}1056 \text{ m}$$

Diameter of main pipe is 105·6 mm Ans. (*b*)(*i*)

Total mass flow rate is 97·14 kg/s Ans. (*b*)(*ii*)

$$\rho_3\, a_3\, v_3 = \rho_1\, a_1\, v_1 + \rho_2\, a_2\, v_2$$
$$\rho_3 \times 0{\cdot}1056^2 \times 12 = 850 \times 0{\cdot}075^2 \times 6 + 950 \times 0{\cdot}1^2 \times 10$$
$$\rho_3 = 924 \text{ kg/m}^3$$

Density of mixed oil is 924 kg/m³ Ans. (*b*)(*iii*)

79.
$$s = ut + \tfrac{1}{2} gt^2$$
$$250 = \tfrac{1}{2} \times 9{\cdot}81 \times t^2$$
$$t = \sqrt{\frac{2 \times 250}{9{\cdot}81}} = 7{\cdot}14$$

Time $= 7{\cdot}14$ s Ans. (*a*)

$v = 0 + gt = 9{\cdot}81 \times 7{\cdot}14 = 70$ m/s Ans. (*b*)(*i*)

$$s = ut + \tfrac{1}{2} gt^2 \quad \therefore \quad t = \sqrt{\frac{2 \times 10}{9{\cdot}81}} = 1{\cdot}43 \text{ s}$$

Time of flight $= 7{\cdot}14 - 1{\cdot}43 = 5{\cdot}71$ s

Distance (s) $= 0 \times 5{\cdot}71 + \tfrac{1}{2} \times 9{\cdot}81 \times 5{\cdot}71^2$

$= 160$ m

Height of second stone above ground $= 250 - 160 = 90$ m

Ans. (*b*)(*ii*)

Velocity of second stone
$$v^2 = u^2 + 2gs$$
$$v^2 = 0{\cdot}2 \times 9{\cdot}81 \times 160$$
$$v = 56 \text{ m/s}$$
Ans.(*b*)(*iii*)

80. See text (Chapter 10):

Using $\dfrac{M}{I} = \dfrac{\sigma}{y} = \dfrac{E}{R}$ Ans. (*a*)

Two davits; Weight on one $= 1000 \times 9{\cdot}81$

$M_{MAX} = 1000 \times 9{\cdot}81 \times 1{\cdot}5$

$= 14715$ kNm

$$\sigma_{MAX} = \frac{My}{I} = \frac{14{\cdot}715 \times 10^3 \times 0{\cdot}06 \times 64}{\pi \times 0{\cdot}12^4}$$

Maximum bending stress = 86·74 MN/m² Ans. (*b*)(*i*)

$$\text{Factor of safety} = \frac{UTS}{WS} = \frac{347}{86{\cdot}74}$$

$$= 4$$ Ans. (*b*)(*ii*)

SELECTION OF EXAMINATION QUESTIONS

CLASS ONE

1. A short hollow cylindrical pillar is 250 mm outside diameter and 25 mm thick, and supports an axial compressive load of 400 kN, the line of action of which is 20 mm from the centre of the pillar. Find the maximum and minimum stresses in the pillar. Find also the maximum permissible eccentricity of the load so that one side is just about to be in tension.

2. A motor produces a constant output torque of 0·5 kN m during its run up from rest to an operational speed of 330 rev/min. If the rotational parts have an equivalent mass of 425 kg with a radius of gyration of 750 mm, determine (*i*) the angular acceleration during run up, (*ii*) time taken to reach operational speed, (*iii*) the change in kinetic energy during the last 5 s of the the run up to operational speed.

3. A mass of 4000 kg hangs on the end of a lifting cable. Find the shortest time the mass can be lifted through a height of 27 m, starting from rest and coming to rest, if the tension in the cable is not to exceed 50 kN.

4. The pressure in a sea water main is 7 bar (= 7×10^5 N/m^2) and to this is connected a hose 50 mm diameter and 12 m long with a nozzle at the end which is 20 mm diameter and 6 m high above the main. Find the velocity of the water at discharge from the nozzle, neglecting effects of bends in the hose. Take the density of sea water as 7860 kg/m^3, and the friction in the hose. Take the density of sea water as 1025 kg/m^3, and the friction in the hose as equivalent to a loss of head of:

$$\frac{4fl}{d} \times \frac{v^2}{2g} \quad \text{(m)}$$

where f = 0·01
l = length (m)

v = velocity through hose (m/s)
d = diameter (m)

5. A beam of uniform section is 8 m long, its mass is 3·262 t and it is simply supported at each end. The beam carries concentrated loads of 20 kN at 2·5 m from the left end and 10 kN at 5 m from the left end. Sketch and dimension the shearing force and bending moment diagrams and calculate the position and magnitude of the maximum bending moment.

6. A mild steel bar 50 mm diameter and 430 mm long is reduced in diameter to 40 mm along the centre part of the bar for a length of 180 mm. When subjected to an axial pull the total elongation is 0·5 mm. Taking E for steel as 210 GN/m^2, calculate the tensile load, the resilience in each part of the bar, and the total resilience.

7. The mass of a flywheel and its shaft is 2000 kg and the effective radius of gyration is 375 mm. The shaft runs in bearings 125 mm diameter and the coefficient of friction between shaft and bearings in 0·12 which may be taken as a constant at all speeds. Calculate (*i*) the kinetic energy stored in the wheel and shaft when rotating at 480 rev/min, (*ii*) energy absorbed in friction at the bearings per revolution, (*iii*) time taken to come to rest from its running speed of 480 rev/min after the driving power is cut off.

8. A casting of mass 50 kg is placed on a plane inclined at 30° to the horizontal. If the coefficient of friction is 0·4, find the time taken for the casting to slide down the plane a distance of 10m without any external force being applied. Find also, the force required to give the casting an acceleration of 2 m/s^2 up the plane.

9. A vertical pipe tapers from 150 mm diameter at the top to 100 mm diameter at the bottom. Water (density 1000 kg/m^3) flows upward through the pipe at the rate of 360 m^3/h, the outlet pressure being 80 kN/m^2 less than the inlet pressure. If frictional loss per unit mass of water flow is 10 J, determine the length of the pipe.

10. One ship is steaming due South at 12·5 knots and another ship is steaming 20° North of West at 11 knots. When the first ship is 2 nautical miles from the point of intersection of their courses, the other is 5 nautical miles from this point. Find the distance between the ships at their closest approach to each other and the time to

reach these positions. Graphical solution is acceptable.

11. A solid copper bar 25 mm diameter is sheathed with a steel tube of equal length 40 mm outside diameter and 25 mm inside diameter, the tube being a sliding fit over the bar. They are secured together by a 10 mm diameter pin through the tube and bar at each end. Calculate the stress in the bar, tube and pins if the combination is raised in temperature through 100°C. Take the moduli of elasticity as 206 GN/m^2 for steel and 103 GN/m^2 for copper and the coefficients of linear expansion per degree C as 12×10^{-6} for steel and 17×10^{-6} for copper.

12. The outside diameter of a hollow shaft is 1·6 times the inside diameter. When transmitting 420 kW of power at 180 rev/min the angle of twist is one degree over a length of 40 times the inside diameter. Taking G for the material as 85 GN/m^2, calculate (*i*) the outside and inside diameters, (*ii*) the maximum shear stress, and (*iii*) the shear strain at the inner fibres.

13. In a Porter governor the four arms are each 300 mm long, the mass of each ball is 2·25 kg and the mass of the central sleeve is 38·25 kg. The arms incline at 45° and 30° to the vertical under the limiting conditions of the speed range, and the effect of friction in the mechanism is equivalent to a frictional force of 9 N between the spindle and cental sleeve. If the governor runs at twice the speed of the engine, calculate the maximum and minimum engine speeds.

14. The difference in pressure between the inlet and throat of a horizontal venturi meter is measured by a submerged manometer containing mercury. The inlet diameter of the meter is 300 mm. When the volume flow of fresh water is 1764 l/min the manometer reads 56 mm difference in mercury level. Take the relative density of mercury as 13·6 and calculate the throat diameter of the meter.

15. A beam is 20 m long and of uniform section. It is to be lifted by two vertical slings keeping the beam horizontal. Find the positions to secure the slings on the beam so that the stress in the beam due to its own weight is a minimum. Sketch the shearing force and bending moment diagrams.

16. The speed reduction ratio of a single reduction gearing consisting of a pinion driving a gear wheel is three to one. The

masses of the pinion and gear wheel are 45 kg and 420 kg respectively, and their radii of gyration are 150 mm and 440 mm respectively. On no-load the pinion attains its maximum rotational speed of 15 rev/s from rest while turning through 150 rev. Find (*i*) the time to attain maximum speed, (*ii*) the total torque in the pinion shaft to accelerate the whole gearing, and (*iii*) the total kinetic energy in the gears when running at maximum speed.

17. The stiffness of a close-coiled helical spring is such that it deflects 36 mm when an axial force of 10 N is applied on the end hook. Calculate the mass to be hung on the hook so that, when set vibrating, it will make one complete oscillation per second. The mass of the spring is 0·6 kg and the effect of this when vibrating is to be taken as equivalent to a mass on the end hook equal to one-third of the mass of the spring. Calculate also the maximum velocity and maximum acceleration of the vibrating mass when initially displaced 12 mm from the equilibrium.

18. A horizontal jet of fresh water 25 mm diameter with a velocity of 15 m/s strikes a flat plate which is inclined at 45° to the direction of the jet. Assuming no splash back of the water, calculate the normal force exerted on the plate when the plate is (*i*) stationary, (*ii*) moving at 10 m/s in the direction of the jet.

19. A solid steel shaft 240 mm diameter has a bronze liner shrunk on it over its entire length, the outer diameter of the liner being 290 mm. Calculate (*i*) the ratio of the stresses in the shaft and liner, (*ii*)the maximum stresses in the shaft and liner, and (*iii*) the angle of twist over a length of 4 m, when the torque transmitted is 100 kN m. Take the moduli of rigidity for steel and bronze as 90 and 42 GN/m^2 respectively and assume no slipping between shaft and liner.

20. A beam 6 m long is simply supported at each end and carries a uniformly distributed load of 1·2 kN/m for the first 3 m from the left end. The beam is of symmetrical cross-section, its depth is 300 mm, and the second moment of the section about its neutral axis is $1{\cdot}33 \times 10^{-4}$ m^4. Calculate the position and magnitude of the maximum stress in the beam and sketch the shearing force and bending moment diagrams.

21. Mass of the flywheel of a shearing machine is 3 t and its radius of gyration is 0·55 m. At the beginning of its cutting stroke the

speed of the flywheel is 110 rev/min. If 12 kJ of work is done in shearing a plate in 3 s, calculate the rotational speed of the flywheel at the end of the cutting stroke in rev/min, and the angular retardation in m/s^2.

22. The power transmitted by a belt driven pulley is 8·2 kW when the tension in the tight side of the belt is 1·855 kN, the linear speed of the belt is 6·6 m/s and the coefficient of friction between belt and pulley is 0·3. Calculate the angle of lap of the belt around the pulley, using the expression:

$$\frac{F_1}{F_2} = \varepsilon^{\mu\theta}$$

where F_1 = force in tight side of belt
F_2 = force in slack side of belt
μ = coefficient of friction
θ = angle of lap in radians

23. A pump is to draw water from a depth of 3 m and in order to prevent the water breaking up a pressure not less than an equivalent head of 2 m is to be maintained in the pipe line. Taking the atmospheric pressure as 760 mm Hg find the minimum diameter of the suction pipe to deal with 300m^3/h.
Density of mercury = $13{\cdot}6 \times 10^3$ kg/m^3.

24. A close coiled helical spring of mean coil diameter 40 mm is made of wire 5 mm diameter, the modulus of rigidity of the wire material being 88 kN/mm^2. If the stress in the wire is not to exceed 250 N/mm^2 when the spring is at its maximum deflection of 20 mm, find (*i*) the number of coils, (*ii*) the maximum load, and (*iii*) the energy stored in the spring at maximum deflection.

25. A projectile leaves a gun at a velocity of v at an angle of α to the horizontal. If R represents the horizontal range and h the maximum height attained, prove the following expression and use it to find the maximum height when the range is 1000 m and the angle of elevation 20°:

$$R = \frac{4h}{\tan\alpha}$$

26. A body of mass 150 kg rests on a plane inclined at 18° to the horizontal and the coefficient of friction between body and plane is 0·33. The body is to be pulled by a rope running parallel to the plane and wound on to a drum, the diameter of the drum is 800 mm and its moment of inertia is 1·6 kg m^2. If a torque of 373 N m is applied to the drum, find the acceleration of the body up the plane, the distance travelled in 12 s, and the pull in the rope.

27. A beam is 8 m long and is simply supported at points 2 m from each end. It carries a concentrated load of 5 kN at each end and a distributed load uniformly spread along the beam over the length between the supports. If the bending moment is zero at mid-span, find the magnitude of the distributed load and sketch the shearing force and bending moment diagrams.

28. A venturi meter having an inlet diameter of 50 mm and throat diameter of 25 mm is used to measure the upward flow of water in a pipeline. The throat is 150 mm above the inlet and a differential pressure gauge connected to these points gives a reading of 250 mm head of water (density 1000 kg/m^3). Taking the meter coefficient as 0·95 determine the volumetric flow rate of water.

29. A vertical spring-loaded governor is operated by a pair of bell-crank levers, their fulcrums being at 40 mm from the centre-line of the spindle. The horizontal arms engage with the sleeve which compresses the spring on the spindle and the vertical arms each carry a mass of 0·5 kg on their ends. The bell-crank leverage is 1·5:1. Find (*i*) the force on the spring when the rotational speed of the governor is 25 rev/s. If the spring is further compressed by 5 mm when the speed is increased by 2%, find (*ii*) the stiffness of the spring in N/mm.

30. A wheel and shaft of total mass 14 kg runs in horizontal bearings. The shaft is 50 mm diameter and has a cord wound around it with a hook attached to its free end. It is found that a mass of 0·3 kg hanging on the hook is just sufficient to overcome friction. With the 0·3 kg mass removed and replaced with a mass of 2 kg, it falls 1·05m in 8 s from rest. Calculate the radius of gyration of the wheel and shaft.

31. A mass of 250 kg is lifted out of a ship's hold by a rope from a winch drum with an acceleration of 0·2 m/s^2. The diameter of the

drum is 600 mm and the diameter of the rope is 30 mm. The equivalent mass and radius of gyration of the rotating parts are 275 kg and 240 mm respectively. Find the tension in the rope and the torque in the drum shaft.

32. The open area *A* of a valve in an air flow system is governed by the differential pressure *h* across a section of the ducting and is related by:

$$A = C_1h + C_2h^2$$

Where C_1 and C_2 are constants.
The following readings were taken during a test:

A (mm^2)	h (mm of water)
1000	3·75
1670	5

Estimate the value of *h* when *A* is 1870 mm^2.

33. A metal disc of diameter 0·5 m and mass 24 kg takes 5·25 s to roll 6 m down an incline of 1 in 15, starting from rest. Calculate the moment of inertia of the disc.

34. A hydraulic steering gear has 4 rams each 300 mm diameter, the distance between centre-line of rams and centre of rudder stock is 750 mm, the maximum rudder angle is 35°, and the by-pass valves lift when the pressure in the cylinders is 75 bar (= 75×10^5 N/m^2). (*i*) If the stress in the rudder stock is not to exceed 70 MN/m^2, find the diameter of the rudder stock. (*ii*) If the stress in the tiller arms at 600 mm from the centre of the rudder stock is not to exceed 100 MN/m^2, find the diameter of the tiller arms at this section.

35. Fresh water flows through a horizontal venturi meter which has diameters at entrance and throat of 150 mm and 50 mm respectively. The difference in pressure between these two points is measured by a submerged U-tube containing mercury and water. Taking the density of mercury as $13{\cdot}6 \times 10^3$ kg/m^3 and the discharge coefficient for the meter as 0·9, calculate the mass flow of water per hour through the meter when the recording on the U-tube is 50 mm of mercury.

36. The two axles of a four-wheeled truck are 3 m apart and the

centre of gravity of the truck lies mid-way between the axles and at a height of 1·5 m above the ground. Calculate the inclination of a gradient upon which, with its rear wheels locked, the truck would be at rest but just on the point of slipping, taking the coefficient of sliding friction between the locked wheels and the ground as 0·45.

37. A solid steel beam 2 m long and of uniform square section of 150 mm sides is suspended from a crane hook by a single rope, the point of attachment being at 300 mm from mid-length of the beam, and a downward force is applied at one end to keep it horizontal. Find (*i*) the magnitude of the end force, (*ii*) the tension in the lifting rope, and (*iii*) the stress induced in the beam at the point of attachment. Take the density of steel as 7860 kg/m^3.

38. A tube 5 m long, 13 mm outside diameter and 1·5 mm thick, is coiled into a helical spring of mean coil diameter 150 mm. The spring is placed in a cylinder to oppose the motion of a piston in the cylinder. The diameter of the piston is 200 mm and the effective pressure on the piston is 0·105 bar (= $0{\cdot}105 \times 10^5$ N/m^2). Calculate (*i*) the twisting moment applied to the spring, (*ii*) the maximum stress in the spring, and (*iii*) the energy stored in the spring. The modulus of rigidity of the spring material is 103 GN/m^2.

39. A mass of 1000 kg is being lowered at a velocity of 0·6 m/s by a cable which has an effective cross-sectional area of 645 mm^2. When 12 m length of cable is unwound, a brake suddenly stops the lowering. Taking the modulus of elasticity of the cable material as 207 GN/m^2, calculate the maximum stress set up in the cable due to the sudden braking, and the maximum extension of the cable.

40. A cantilever beam is fabricated by steel plates welded together in the form of a hollow rectangular box section, uniform over its entire length of 2 m. The overall depth is 160 mm, overall width 120 mm, the side plates are 10 mm thick and the top and bottom plates 20 mm thick. Calculate the maximum concentrated mass that can be carried on the free end of the cantilever so that the maximum stress will not exceed 40 MN/m^2. Density of steel = 7860 kg/m^3.

41. A 2 m diameter circular flap valve closes an outlet in a reservoir. The flap is sloped at 30° to the horizontal and it is hinged at its top edge, which is 1·25 m vertically below the free surface of the water (density 1000 kg/m^3).

Calculate:
(a) the total hydrostatic thrust on the valve
(b) the depth of the centre of pressure
(c) the magnitude of the minimum force required, on a vertical chain, attached to the centre of the flap, in order to open it.

42. A bar 1·5 m long is hung horizontally by a vertical wire at each end, one wire is steel and the other is brass, and the diameter of the brass wire is twice the diameter of the steel wire. A mass is to be hung from the bar, find the position along the bar from which the mass should be hung so that the bar will remain horizontal and state what fraction of the mass will be carried by each wire. Take the modulus of elasticity of steel as being twice that of brass.

43. A pump delivers 8 kg/s of water to a fire hose fitted with a nozzle held at a height of 20 m above the pump. Water velocity in the hose is 7 m/s and the velocity of the jet leaving the nozzle is 37 m/s. Energy loss in the nozzle is 12% of the kinetic energy gained in the nozzle, and frictional loss in the hose is 0·5 kW.

The mechanical efficiency of the pump is 80%

Estimate the power required to drive the pump.

44. The volumetric flow rate Q(m^3/h) of air through an orifice plate in a ventilation trunking is given by:

$$Q = ACd^2 \left\{\frac{h}{\rho}\right\}^{\frac{1}{2}}$$

where $A = 125{\cdot}2 \times 10^{-4}$, $C = 0{\cdot}596$, d = diameter (mm) of the orifice = 180, h = differential pressure (mm water) across the orifice plate, and ρ = density (kg/m^3) of air = 1·22.

The output signal S(N/m^2) from the differential pressure measuring instrument is given by:

$$S = (2{\cdot}1 \times 10^4) + kh$$

where k is a constant.

If the flow rate is 3340 m^3/h when the output signal is $10{\cdot}3 \times 10^4$ N/m^2, calculate the flow rate when the output signal is $4{\cdot}2 \times 10^4$ N/m^2.

45. The motion of the bucket of a vertical reciprocating pump may be taken as simple harmonic motion. The bucket diameter is 150

mm, stroke 200 mm, and it does 70 double strokes per minute. Calculate the force to drive the bucket and the power required, at the instant the bucket is 15 mm from the bottom of its stroke and moving upwards. Take the combined mass of the reciprocating parts and water above the bucket as 30 kg, and discharge pressure 80 kN/m^2 at that instant.

46. A bulkhead watertight door 2 m high and 800 mm wide has a pair of hinges on one of its vertical edges symmetrically placed at 850 mm above and below the centre, and a draw-bolt at the centre of its other vertical edge. On one side, tending to open the door, there is sea water of density 1024 kg/m^3 to a depth of 2·5 m above the sill. Calculate the forces on the bolt and hinges.

47. A wire rope is constructed of two steel wires, each 2·8 mm diameter, and six bronze wires, each 2·5 mm diameter. Calculate (*i*) the stresses in the wires when the rope carries a load of 4·5 kN, and (*ii*) the equivalent modulus of elasticity for the composite wire rope. Take E for steel = 210 GN/m^2 and E for bronze = 84 GN/m^2.

48. A rigid steel beam 5 m long is hinged at one end and supported horizontally by two vertical steel tie rods. One tie rod is 3 m long, 10 mm diameter, and attached to the beam at 1 m from the hinge. The other tie rod is 6 m long, 20 mm diameter, and attached to the free end of the beam. A load of 102 kN is now suspended from a point on the beam at 3·5 m from the hinge. Calculate the stress in each tie rod and the maximum deflection of the beam. Take E for steel = 206 GN/m^2.

49. The nozzle of a hose is 500 mm long and tapers from 75 mm diameter at the inlet to 25 mm diameter at the exit. Calculate (*a*) the velocity of the jet at exit and (*b*) the mass flow rate, when the nozzle is inclined upwards at 45° to the horizontal and fresh water is supplied to it at a pressure of 140 kN/m^2.

50. The outer and inner radii of a crane hook at the section where maximum bending moment occurs are 108 and 44 mm respectively. The cross-section there is elliptical, major axis 64 mm and minor axis 38 mm. Calculate the maximum tensile and comprehensive stresses when a load of 20 kN is suspended on the hook. Take I for an ellipse about the minor axis as $\frac{\pi}{64}bd^3$ where b = minor axis and d = major axis.

51. A bar of bronze, 2·5 m long and 75 mm diameter is rigidly connected end on to a bar of steel 1·5 m long and 50 mm diameter so as to form one long bar. The extreme ends of the compound bar are securely fixed and a torque of 6 kN m is now applied at the junction of the bars. Calculate for each bar (*a*) the torque carried, (*b*) the maximum stress, and (*c*) the maximum angle of twist. Take the modulii of rigidity for bronze and steel as 42 and 84 GN/m^2 respectively.

52. In a pneumatically operated valve which includes a diaphragm and spring, the following expresses the relationship between the control pressure p in kN/m^2, the area A of the diaphragm in mm^2 and the valve displacement x from the closed position in mm, constants are represented by h_1, h_2 and h_3,

$$p = h_1 - \frac{xh_2}{A} - \frac{x^2h_3}{A}$$

When the valve is closed the control pressure is 20·7 kN/m^2, when $x = 6$ mm $p = 48$ kN/m^2, and when $x = 25$ mm $p = 103{\cdot}5$ kN/m^2. The area of the diaphragm is 65·4 cm^2. Calculate the valve lift when the control pressure is 76 kN/m^2.

53. The inlet and outlet diameters of the impeller of a centrifugal pump are 100 mm and 300 mm respectively. When rotating at 600 rev/min the radial velocity of the water is constant at 3 m/s, and 4·5 t of sea water of density 1024 kg/m^3 are discharged per minute. The exit angle of the vanes is 30°. Calculate (*i*) the width of the impeller at entrance and exit, (*ii*) the angle of the vanes at entrance if the water enters without shock, (*iii*) the kinetic energy of the water per kilogramme as it leaves the impeller.

54. Prove the following expressions for a closely coiled helical spring of N coils, mean coil diameter D, wire diameter d, and axial load W:

$$\text{Shear stress in the wire} = \frac{8WD}{\pi d^3}$$

$$\text{Axial deflection of the spring} = \frac{8WD^3N}{Gd^4}$$

The free length of the spring of a relief valve is 210 mm, length when valve is open 170 mm, diameter of valve seat 150 mm, pressure on valve when fully open 70 kN/m^2. The spring has 8 coils and the diameter of the wire is 18 mm. Calculate the mean coil diameter and the shear stress in the wire when fully open. Take G = 90 GN/m^2.

55. A thin cylindrical vessel is 1·2 m diameter and the plate thickness is 50 mm. The circumferential seams are Vee welded at an angle of 60°. Calculate the normal and shear stresses at the junction of the plate and deposited metal when the pressure inside the vessel is 35 bar (= $3{\cdot}5 \times 10^6$N/m^2).

56. Fresh water emerges as a horizontal jet from a sharp edged orifice 25 mm diameter in the side of a tank under a constant head of 1·22 m. The diameter of the vena contracta of the jet is 20 mm and the rate of discharge is 89·7 l/m. Calculate the coefficients of velocity, contraction of area and discharge. If the jet leaving the orifice impinges on a stationary flat plate inclined at 45° to the jet, calculate the force on the plate.

57. Five masses, A, B, C, D and E are fixed on a faceplate in the same plane and they are to be dynamically balanced when the system is rotated. The magnitude of each mass, radius from centre, and clockwise angular displacement from mass A are respectively: A = 2·8 kg, 125 mm, 0°; B = 3·5 kg, 100 mm, 45°; C = 4·2 kg, 75 mm, 120°; D = 4·8 kg, 125 mm, 210°; E is to be placed at a radius of 200 mm, calculate the mass of E and its angular displacement from A. Sketch the vector diagram of forces.

58. A solid steel vertical column 120 mm diameter is subjected to an eccentric compressive load. Calculate the magnitude of the load and its eccentricity if the stresses are not to exceed 15 MN/m^2 compressive and 3 MN/m^2 tensile. Express the load as a force (N) and also as a mass (t).

59. A hollow shaft is coupled to a solid shaft as shown. The ends are secured against movement and a torque of 6 kN m is applied at the junction of shafts. Calculate:
(a) the torque transmitted by each shaft
(b) the angular movement of the coupling.

The modulus of rigidity for both shafts is 83 GN/m^2.

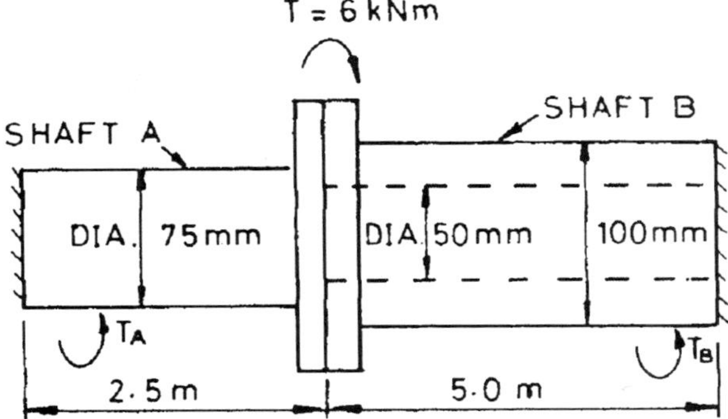

60. A cast iron beam of I section is simply supported at each end and carries a uniformly distributed load over its entire length of 3 m. The compression flange is 100 mm wide, tension flange 250 mm wide, height of centre web between flanges 200 mm, and uniformly 50 mm thick throughout. The second moment of area about its neutral axis is 2·785 × 10^8 mm^4 and the density of the material is 7210 kg/m^3. If the maximum permissible tension in the cast iron is 15·4 MN/m^2, calculate (*i*) the maximum uniformly distributed load that can be carried in addition to the weight of the beam and (*ii*) the compressive stress in the cast iron when carrying this load.

61. A wood chest is secured by its base to the horizontal floor of a ship's compartment. Sea water of density 1024 kg/m^3 enters the compartment and when the water level is 912 mm above the hinges the lid floats at an angle of 36° 52′ to the vertical. The dimensions of the lid are 1·52 m by 1·52 m by 25 mm thick. Find the relative density of the lid material.

62.

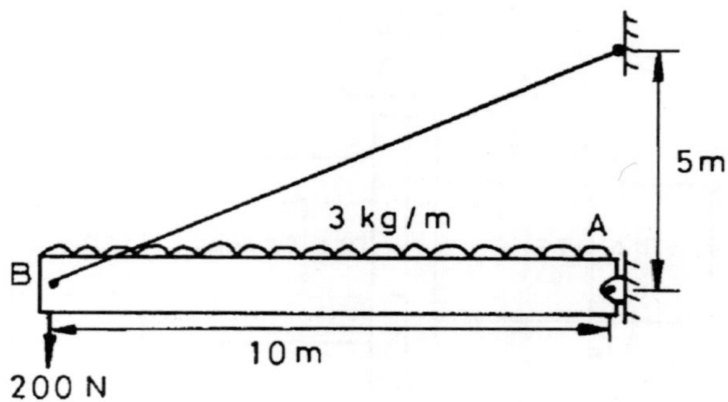

The derrick arm shown is 10 m long of uniform cross section and has a mass of 3 kg/m length. It is supported by a pin at *A* and has a cable at *B*. A vertical force of 200 N is applied at B.

Determine the forces and directions of the reaction in the cable and in the pin.

Note: There are 9 examination questions in each examination paper, each question indicates the mark allocation. The remaining 18 questions here follow this practice.

63. A water duct of rectangular section is 900 mm high, 1200 mm wide and is closed at the end by a rectangular sluice gate AB of uniform thickness, mass 1·6 t, which is inclined at 60° to the horizontal, as shown in the sketch.

The gate is hinged at the top edge and the head of water is 150 mm above the hinge.

Determine the least distance from the hinge at which a 2 kN weight should be attached to keep the gate closed. (16)

Density of water = 1000 kg/m^3.

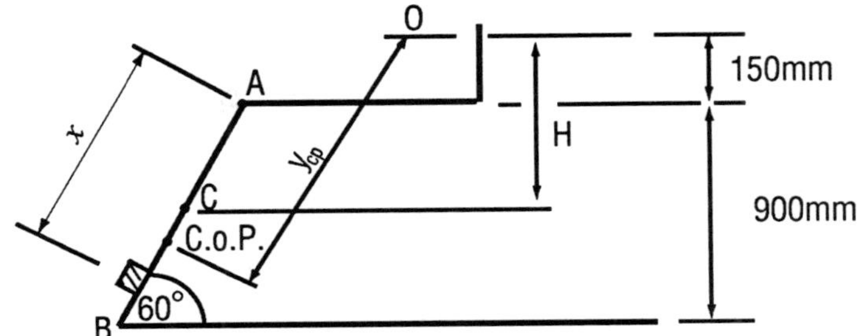

64. A cam operated exhaust valve opens vertically with a lift of 50 mm, its motion being simple harmonic. The valve is opened and closed during 95° of camshaft movement. The mass of the valve is 1·1 kg and there is a spring force at the highest and lowest positions of the valve of 1000 N and 150 N respectively.

The camshaft speed is 300 rev/min, calculate:

(*a*) the maximum velocity of the valve; (8)

(*b*) the maximum acceleration of the valve; (2)

(*c*) the force between the cam and the valve when it is at the top and when it is at the bottom of its travel. (6)

65. (*a*) A vertical metal rod 32 mm diameter and 3 m long is fixed at its upper end. When a steady pull of 81 kN is applied at the lower end, the rod extends by 1·67 mm.

Calculate the Modulus of Elasticity for the metal. (4)

(*b*) The same rod has a collar fixed at its lower end, and when a mass m kg is dropped on to the collar from a height of 19 mm, the instantaneous extension due to the impact is observed to be 3·34 mm.

Determine the magnitude of the mass m. (12)

66. A landing ramp of mass 2 t, lies at an angle of 10° below

horizontal. It is hinged at one end and is to be lifted by two cables, connected at 4 m from the hinge as shown in the sketch.

(*a*) Calculate the minimum force required in each cable when lifting commences. (7)

(*b*) Determine the magnitude and direction of the reaction force at the hinge when lifting commences. (9)

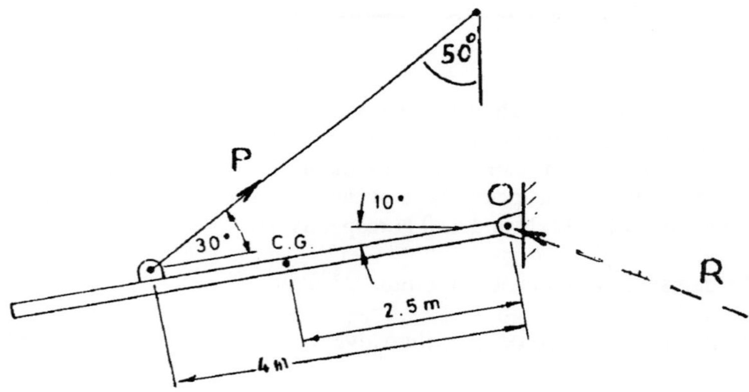

67. A circular flap door is fitted in the vertical side of a tank which contains liquid to some height above the top of the door. The door is able to pivot in horizontal bearings as shown in the sketch.

Show that the torque T required at the spindle to keep the door closed is independent of the head of liquid and is given by:

$$T = I_G \rho g$$

Where I_G is the second moment of the door about an axis through a diameter and ρ is the density of the liquid. (16)

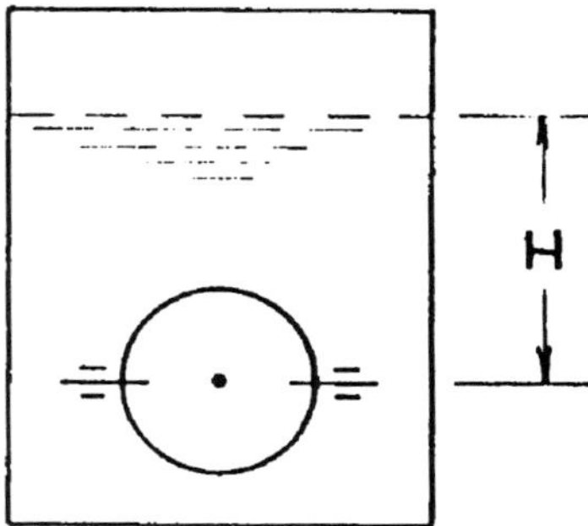

68. (*a*) Derive an expression for the torque required to overcome friction at a single collar thrust bearing where:

r_1 = inner radius of thrust pads, r_2 = outer radius of pads
μ = coefficient of friction between collar and pads
P = axial thrust in the shaft
(The thrust may be considered to act at the mean radius of the pads). (8)

(*b*) A thrust block transmits a thrust of 160 kN when the shaft power is 2 MW at 2·5 rev/s. Inner and outer radii of the thrust pads are 150 mm and 300 mm respectively. Coefficient of friction is 0·02 between pads and collar. Calculate the power loss due to friction, expressed as a percentage of the power transmitted. (8)

69. (*a*) Define *point of contraflexure*, with reference to beams. (2)

(*b*) A container of total mass 40 t is suspended from a simply supported steel beam, 10 m long, as shown in the sketch. The beam is of uniform section and has a mass of 2·6 t. Calculate the position of the point of contraflexure in the beam, taking the mass of the beam into account. (14)

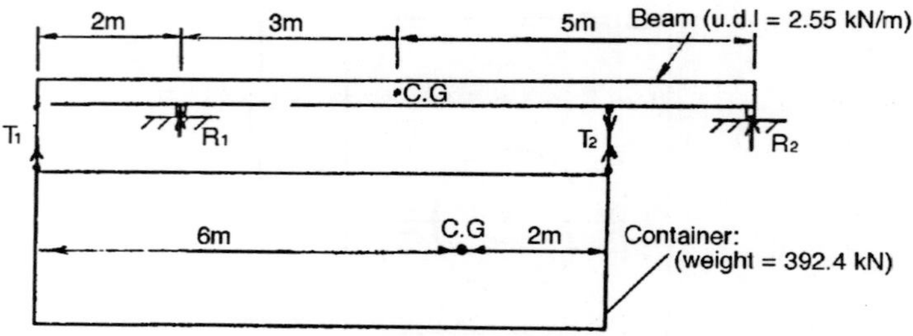

70. The flow of oil in a pipeline is automatically controlled by varying the diameter of the opening in an orifice plate in the pipe.

Flow rate through the orifice is given by the equation:

$\dot{m} = d\,(1{\cdot}1d - 40)\sqrt{h\rho}$

where d is the orifice diameter in mm and is always greater than zero.
h is the differential pressure across the orifice in metres of water (constant at 4 m).
ρ is the density of the oil (constant at 920 kg/m³).

(*a*) Calculate the required diameter d for the orifice:

(*i*) when the flow rate is zero. (5)

(*ii*) when the flow rate is a maximum at 30 000 kg/h (6)

(*b*) Determine the flow rate when d is half-way between the two diameters calculated for part (*a*). (5)

71. A pressurised spherical tank 10 m diameter is partly filled with liquified gas. Gauge pressure measured in the liquid, at the bottom of the tank, is 816 kN/m² and in the gas space above the liquid it is 770 kN/m². Relative density of the liquified gas is 0·518.

(*a*) Calculate the depth of the liquid in the tank. (7)

(*b*) Determine the weight of the liquid in the tank. (9)

Volume of segment of a sphere = $\frac{\pi h^2}{3}(3r - h)$

Where r = *radius of the sphere*
h = *depth of the segment*

72. A sea water centrifugal pump impeller is 300 mm outside diameter and 100 mm inside diameter. Outlet width is 30 mm and speed is 900 rev/min. At exit from the vanes, the water has an absolute velocity of 12 m/s at 12° to the direction of motion of the vane tip.

(*a*) Determine the vane exit angle for shockless flow. (8)

(*b*) Calculate the mass flow rate of water through the pump. (5)

(*C*) The radial velocity of flow at inlet is 2 m/s. Calculate the impeller width at inlet. (3)

Relative density of sea water = *1·025.*

73. An engine has two connecting rods attached to a single crankpin at C, as shown. The longitudinal axis of each cylinder is inclined at 30° to the vertical. The stroke length for each piston is 600 mm and the crankshaft speed is 600 rev/min. Connecting rod length is 700 mm. Determine the instantaneous linear velocity of the pistons A and B when the crankpin is 40° before its top position moving clockwise. (16)

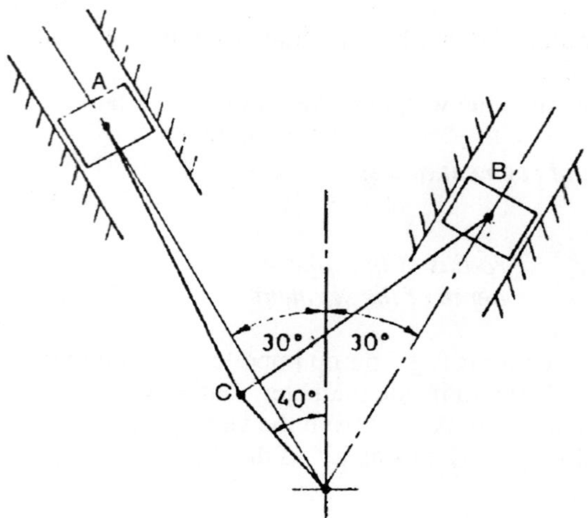

74. Oil of relative density 0·86 flows around a 90° bend in a horizontal pipeline 0·6 m diameter. Oil pressure head in the pipe is 20 m and the flow rate is $1 m^3/s$.

Calculate the magnitude, and state the direction, of the force exerted by the oil on the bend. (16)

75. A box barge 18 m long and 3 m wide floats on an even keel. Two point loads, of 108 kN each are placed on the barge, one at 4 m from each end.

(*a*) Neglecting the weight of the barge, draw the bending moment diagram for the barge, showing relevant values of bending moment. (The hydrostatic upthrust may be considered as a uniformly distributed upward force). (7)

(*b*) Determine the positions of the points of contraflexure in the barge. (5)

(*c*) Calculate the increase in the draught of the barge due to the two loads. (4)

Density of water = 1000 kg/m³.

76. A canal lock gate is 3 m long. The gate is closed and has fresh water at a depth of 4·5 m on one side and 1·8 m on the other side as shown. Calculate:

(*a*) the resultant hydrostatic force on the gate, (6)

(*b*) the height from the bottom of the gate to the resultant centre of pressure, (6)

(*c*) the reaction force R between the gate and the canal side. (4)

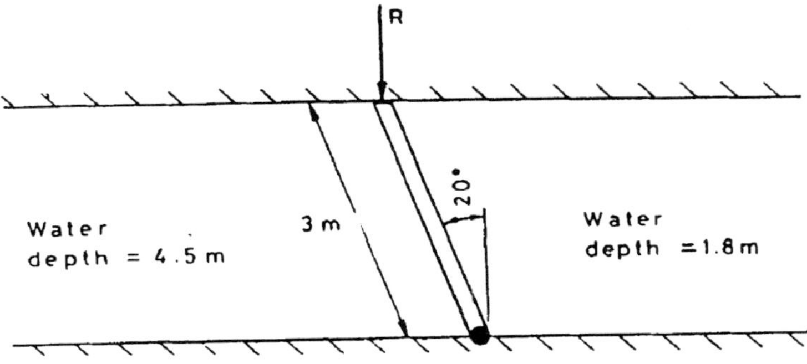

77. (*a*) Define the terms frequency and periodic time for a body moving with simple harmonic motion. (2)

(*b*) The body of a truck is supported by four springs of equal stiffness. When a 24 t load is placed in the truck, it is observed that each of the springs is compressed by 75 mm due to the load.

Determine the stiffness of each spring. (5)

(*c*) Given that the mass of the truck body is 5 t, calculate the frequency of free vertical oscillations of the truck:

(*i*) when it is empty; (6)

(*ii*) when it is carrying the 24 t load. (3)

78. A centrifugal clutch has four identical shoes on the driving shaft, each of mass 0·75 kg, one of which is shown. The C.G. of each shoe will be at 150 mm from the centre of rotation when the shoes engage with the driven shaft.

At rest, each shoe is held by a spring of stiffness 100 kN/m which is extended by 10 mm when the shoe moves from the rest position. The spring force is negligible at the rest position.

(*a*) Calculate the speed at which the clutch will just engage. (6)

(*b*) Calculate the force between the shoe and the rim when the speed is 1500 rev/min. (5)

(*c*) Determine the total torque which can be transmitted at 1500 rev/min, given that $\mu = 0{\cdot}3$ between shoes and rim, and the rim diameter is 360 mm. (5)

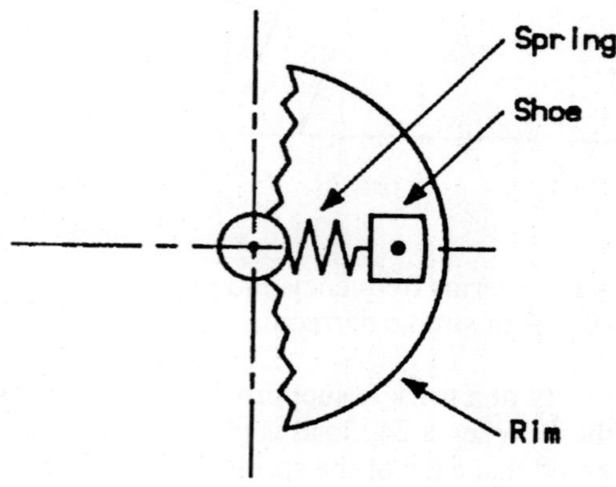

79. Oil (density 900 kg/m^3) flows at 15 000 kg/h in a pipe which branches into two pipes. One branch has a fixed orifice 50 mm diameter, the other has a variable orifice. The differential pressure across each orifice is equal, and each is measured by a manometer.

For the fixed orifice, flow rate is given by:

$\dot{m}$ (kg/h) $= 30\sqrt{h\rho}$

where h is the manometer reading in mm
ρ is the oil density.

For the variable orifice:

$\dot{m}$ (kg/h) = $Ch^{1\cdot3}$

Where C is a constant which depends upon the mass flow rate in the branches.

Find the value of C:

(*a*) when the mass flow rates are equal in the branch pipes; (8)

(*b*) when the flow rate at the fixed orifice is four times the flow rate at the variable orifice. (8)

80. A circular door 1·2 m diameter is fitted in a vertical division plate of a tank. One side of the plate has sea water to a height of 1·8 m above the top edge of the door, the other side is subject to an air pressure of 12 kN/m^2.

Calculate.

(*a*) the resultant force on the door. (7)

(*b*) the position of the resultant centre of pressure. (9)

Density of sea water = 1025 kg/m^3.

SOLUTIONS TO EXAMINATION QUESTIONS
CLASS ONE

1. $$\text{Direct stress} = \frac{\text{load}}{\text{area}}$$

$$= \frac{400 \times 10^3}{0{\cdot}7854\,(0.25^2 - 0{\cdot}2^2)}$$

$$= 2{\cdot}263 \times 10^7 \text{ N/M}^2 = 22{\cdot}63 \text{ MN/m}^2$$

$$\frac{M}{I} = \frac{\sigma}{y} \qquad \sigma = \frac{My}{I}$$

Where M = bending moment due to eccentricity of load

$= 400 \times 0{\cdot}02 = 8$ kN m

$$I = \frac{\pi}{64}\,(D^4 - d^4)$$

$$= \frac{\pi}{64}\,(0{\cdot}25^4 - 0{\cdot}2^4) = 1{\cdot}132 \times 10^{-4} \text{ m}^4$$

$$y = \tfrac{1}{2} \times 0{\cdot}25 = 0{\cdot}125 \text{ m}$$

$$\sigma = \frac{My}{I} = \frac{8 \times 10^3 \times 0{\cdot}125}{1{\cdot}132 \times 10^{-4}}$$

$$= 8{\cdot}833 \times 10^6 \text{ N/m}^2 = 8{\cdot}833 \text{ MN/m}^2$$

The effect of the bending stress is to cause compression on one side and tension on the other.

Maximum stress $= 22{\cdot}63 + 8{\cdot}833 = 31{\cdot}463$ MN/m^2 (compression) Ans.

Minimum stress $= 22{\cdot}63 - 8{\cdot}833 = 13{\cdot}797$ MN/m^2 (compression) Ans.

Minimum stress is to be zero for maximum permissible eccentricity, direct stress – bending stress = 0.

bending stress is to be 22·63 MN/m^2

$$M = \frac{I\sigma}{y}$$

$$400 \times 10^3 \times x = \frac{1{\cdot}132 \times 10^{-4} \times 22{\cdot}63 \times 10^6}{0{\cdot}125}$$

$$x(\text{max. eccentricity}) = \frac{1{\cdot}132 \times 10^{-4} \times 22{\cdot}63 \times 10^6}{400 \times 10^3 \times 0{\cdot}125}$$

$$= 0{\cdot}05124 \text{ m} = 51{\cdot}24 \text{ mm}$$ Ans.

2. $$\frac{330 \text{ rev/min} \times 2\pi}{60} = 34{\cdot}56 \text{ rad/s}$$

$$\text{Moment of inertia } I = mk^2$$

$$= 425 \times 0{\cdot}75^2 = 239{\cdot}1 \text{ kg m}^2$$

$$\text{Accel. torque} = \text{moment of inertia} \times \text{ang. accel.}$$

$$T = I\alpha$$

$$\alpha = \frac{0{\cdot}5 \times 10^3}{239{\cdot}1} = 2{\cdot}09 \text{ rad/s}^2$$ Ans. (*i*)

$$\text{Time} = \frac{\text{change in angular velocity}}{\text{acceleration}}$$

$$= \frac{34{\cdot}56 - 0}{2{\cdot}09} = 16{\cdot}54 \text{ s}$$ Ans. (*ii*)

At 5 s before operational speed:

$$\text{time from rest} = 16{\cdot}54 - 5 = 11{\cdot}54 \text{ s}$$

$$\omega = 2{\cdot}09 \times 11{\cdot}54 = 24{\cdot}12 \text{ rad/s}$$

$$\text{Change in K.E.} = \tfrac{1}{2} I (\omega_2{}^2 - \omega_1{}^2)$$

$$\tfrac{1}{2} \times 239{\cdot}1 (34{\cdot}56^2 - 14{\cdot}12^2)$$

$$= 7{\cdot}324 \times 10^4 \text{ Nm}$$ Ans. (*iii*)

3. The shortest time will be when the acceleration and retardation are at their maximum values. The maximum acceleration depends upon the maximum accelerating force permissible without exceeding the allowable total tension in the cable. The maximum retardation is when the tension in the cable is zero, that is, when the retarding force is equal to the supporting force and the cable is slack, this is when the retardation is g.

$$\text{Supporting force} = \text{weight of mass} = 4\times 10^3 \times 9{\cdot}81\ \text{N}$$

$$\text{Accelerating force} = ma$$

For maximum acceleration:

$$\text{Total tension} = \text{supporting force} + \text{accelerating force}$$

$$50 \times 10^3 = 4\times 10^3 \times 9{\cdot}81 + 4 \times 10^3 \times a$$

$$a = 12{\cdot}5 - 9{\cdot}81$$

$$= 2{\cdot}69\ \text{m/s}^2$$

For maximum retardation:

$$\text{Total tension} = \text{supporting force} - \text{retarding force}$$

$$0 = 4 \times 10^2 \times 9{\cdot}81 - 4 \times 10^3 \times \text{retardation}$$

$$\text{retardation} = 9{\cdot}81\ \text{m/s}^2$$

$$\text{Let } v = \text{maximum velocity in m/s}$$

$$t = \text{total time in s}$$

$$\text{Time to accelerate} = \frac{v}{2{\cdot}69} = 0{\cdot}3717\ v$$

$$\text{Time to retard} = \frac{v}{9{\cdot}81} = 0{\cdot}102v$$

$$\text{Total time} = 0{\cdot}3717v + 0{\cdot}102v = 0{\cdot}4737v$$

$$v = \frac{t}{0{\cdot}4737} \qquad \dots\ \dots\ \dots\ \dots\ (i)$$

$$\text{Also, total time} = \frac{\text{total distance}}{\text{average velocity}}$$

$$t = \frac{27}{\frac{1}{2}v} = \frac{54}{v} \qquad \dots\ \dots\ \dots\ \dots\ (ii)$$

Substituting value of v from (i) into (ii),

$$t = \frac{54 \times 0{\cdot}4737}{t}$$

$$t = \sqrt{54 \times 0{\cdot}4737} = 5{\cdot}058 \text{ s} \qquad \text{Ans.}$$

4.

$$z_2 = 6\text{m}$$
$$p_2 = 0 \text{ (i.e. atmos. pres.)}$$
$$v_2 = \text{discharge vel. m/s}$$

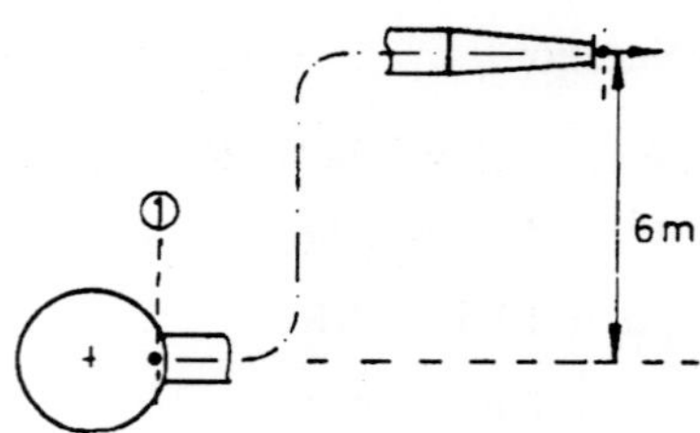

$$z_1 = 0$$
$$p_1 = 7 \text{ bar}$$
$$v_1 = 0$$

Let v_H = the velocity in the hose

v_2 = the velocity at discharge from the nozzle

Volumetric flow rate is constant.

$$\text{Velocity in hose} \times \text{C.S.A. of hose} = \text{Velocity at nozzle} \times \text{C.S.A. of nozzle}$$

$$v \times 0{\cdot}7854 \times 0{\cdot}05^2 = v_2 \times 0{\cdot}7854 \times 0{\cdot}02^2$$

$$v_H = 0{\cdot}16_2$$

$$z_2 + \frac{p_1}{w} + \frac{v_1^2}{2g} = z_2 + \frac{p_2}{w} + \frac{v_2^2}{2g} + \text{friction losses}$$

$$\text{Friction losses} = \frac{4 \times 0{\cdot}01 \times 12 \times (0{\cdot}16\, v_2)^2}{2 \times g \times 0{\cdot}05}$$

$$= 0{\cdot}0125\, v_2^2 \text{ m head}$$

$$0 + \frac{7 \times 10^5}{1025 \times g} + 0 = 6 + 0 + \frac{v_2^2}{2g} + 0{\cdot}0125\ v_2^2$$

$$0{\cdot}0635\ v_2^2 = 63{\cdot}5$$

$$v_2 = 31{\cdot}6$$

Discharge velocity at nozzle $= 31{\cdot}6$ m/s Ans.

5.Weight of beam (which is uniformly distributed)

$$= 3{\cdot}262 \times 10^3 \times 9{\cdot}81 = 32 \times 10^3 \text{ N} = 32 \text{ kN total}$$

$$32 \div 8 = 4 \text{ kN/m}$$

Moments about R_1

$$20 \times 2{\cdot}5 + 10 \times 5 + 32 \times 4 = R_2 \times 8$$

$$R_2 = 28{\cdot}5 \text{ kN}$$

$$R_1 + R_2 = 20 + 10 + 32$$

$$R_1 = 62 - 28{\cdot}5 = 33{\cdot}5 \text{ kN}$$

Neglecting concentrated loads and considering only the weight of the beam, each reaction would be $\frac{1}{2} \times 32 = 16$ kN,

$$M_{centre} = -16 \times 4 + 16 \times 2 = -32 \text{ kN m (i.e.}$$

sagging M)

Neglecting weight of the beam and considering concentrated loads only,

R_1 would be $33{\cdot}5 - 16 = 17{\cdot}5$ kN

R_2 would be $28{\cdot}5 - 16 = 12{\cdot}5$ kN

M under 10 kN load $= 12{\cdot}5 \times 3 = 37{\cdot}5$ kN m

M under 20 kN load $= 17{\cdot}5 \times 2{\cdot}5 = 43{\cdot}75$ kN m

Strictly these values would be negative (sagging M), however the combined M diagram obviates against such sign consideration.

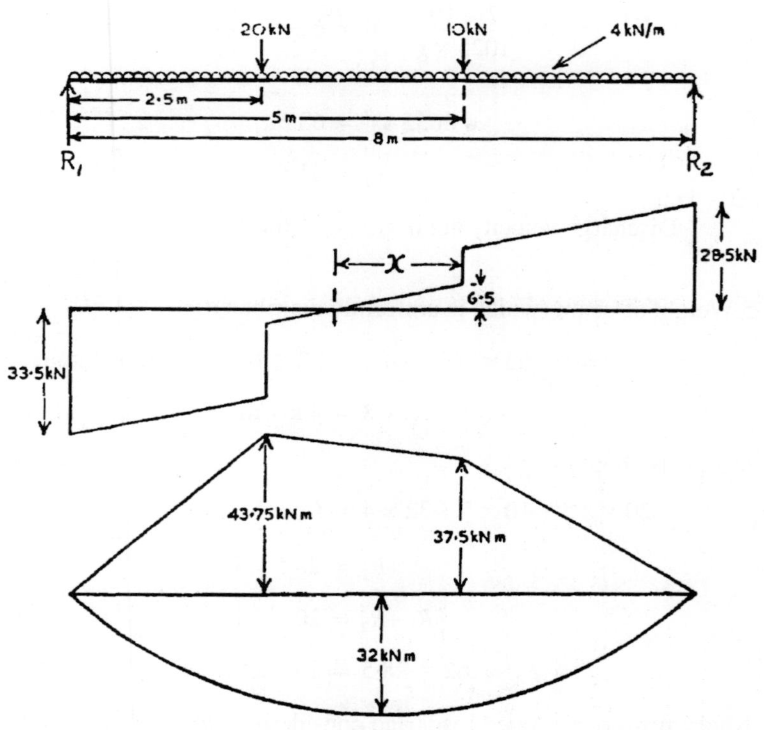

Maximum bending moment occurs where the shearing force is zero, let this position be x m from the 10 kN load as shown in the shearing force diagram.

$$F \text{ at } x = -4 \times 3 - 10 - 4x = 0$$

$$4x = 28{\cdot}5 - 12 - 10$$

$$4x = 6{\cdot}5$$

$$x = 1{\cdot}625 \text{ m}$$

Maximum bending moment occurs at

$$3 + 1{\cdot}625 = 4{\cdot}625 \text{ m from right end} \qquad \text{Ans. } (i)$$

$$M_{max} = -28{\cdot}5 \times 4{\cdot}625 + 4 \times 4{\cdot}625 \times (\tfrac{1}{2} \times 4{\cdot}625) + 10 \times 1{\cdot}625$$

$$= -131{\cdot}8 + 42{\cdot}78 + 16{\cdot}25$$

$$= -72{\cdot}77 \text{ kN m (i.e. sagging)} \qquad \text{Ans. } (ii)$$

6.

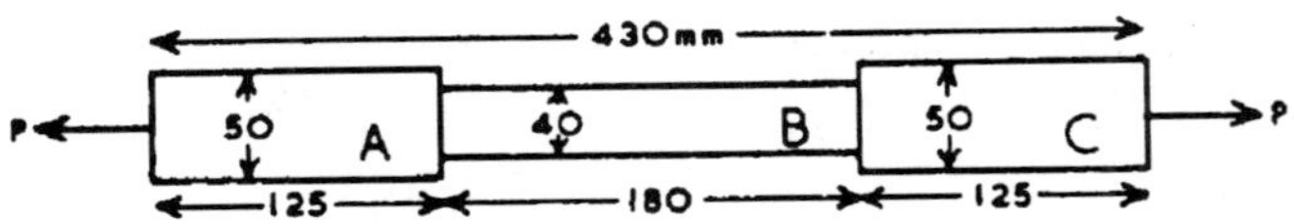

$$\text{Extension} = \text{strain} \times \text{length}$$

$$= \frac{\text{stress}}{E} \times \text{length}$$

$$= \frac{\text{load} \times \text{length}}{\text{area} \times E} = \frac{Pl}{AE}$$

$$\text{Sum of extension of each part} = \text{total extension}$$

$$\frac{P_A\, l_A}{A_A E_A} + \frac{P_B l_B}{A_B E_B} + \frac{P_C l_C}{A_C E_C} = \frac{0{\cdot}5}{10^3}$$

P, E, and 0·7854 are common to the three parts, therefore multiply throughout by E and 0·7854, insert values and solve for P:

$$\frac{P \times 0{\cdot}125}{0{\cdot}05^2} + \frac{P \times 0{\cdot}18}{0{\cdot}04^2} + \frac{P \times 0{\cdot}125}{0{\cdot}05^2} = \frac{0{\cdot}5 \times 0{\cdot}7854 \times 210 \times 10^9}{10^3}$$

$$50P + 112{\cdot}5P + 50P = 8{\cdot}247 \times 10^7$$

$$212{\cdot}5P = 8{\cdot}247 \times 10^7$$

$$P = 3{\cdot}881 \times 10^5 \text{ N or } 388{\cdot}1 \text{ kN}$$

Ans. (*i*)

$$\text{Resilience} = \frac{\sigma^2\, V}{2E}$$

$$= \frac{p^2 \times A \times l}{A^2 \times 2 \times E} = \frac{p^2\, l}{2AE}$$

$$\text{Resilience in each end part} = \frac{(388{\cdot}1 \times 10^3)^2 \times 0{\cdot}125}{2 \times 0{\cdot}7854 \times 0{\cdot}05^2 \times 210 \times 10^9}$$

$$= 22{\cdot}83 \text{ J}$$

Ans.(*ii*)

7. $$\frac{480 \text{ rev/min} \times 2\pi}{60} = 16\pi \text{ rad/s}$$

$$\begin{aligned}\text{Kinetic energy} &= \tfrac{1}{2} mk^2 \omega^2 \\ &= \tfrac{1}{2} \times 2000 \times 0{\cdot}375^2 \times (16\pi)^2 \\ &= 3{\cdot}553 \times 10^5 \text{ J} \\ &= 355{\cdot}3 \text{ kJ} \qquad \text{Ans. } (i)\end{aligned}$$

$$\begin{aligned}\text{Friction force at skin of shaft} &= \mu \times \text{force (weight) on bearing} \\ &= 0{\cdot}12 \times 2000 \times 9{\cdot}81 \text{ N}\end{aligned}$$

$$\begin{aligned}\text{Energy absorbed per revolution} &= \text{force} \times \text{distance} \\ &= \text{force} \times \text{circumference} \\ &= 0{\cdot}12 \times 2000 \times 9{\cdot}81 \times \pi \times 0{\cdot}125 \\ &= 924{\cdot}7 \text{ J} \qquad \text{Ans. } (ii)\end{aligned}$$

Number of revolutions turned while coming to rest

$$\begin{aligned}&= \frac{\text{total energy absorbed}}{\text{energy absorbed per revolution}} \\ &= \frac{355{\cdot}3 \times 10^3}{924{\cdot}7} = 384{\cdot}2 \text{ rev}\end{aligned}$$

Average speed in coming to rest

$$= \tfrac{1}{2}(480 + 0) = 240 \text{ rev/min}$$

$$\begin{aligned}\text{Time} &= \frac{\text{distance}}{\text{average speed}} \\ &= \frac{384{\cdot}2}{240} \\ &= 1{\cdot}601 \text{ min.} = 1 \text{ min } 36{\cdot}06\text{s}\end{aligned}$$

Ans. (*iii*)

8. $$\begin{aligned}\text{Weight of casting} &= 50 \times 9{\cdot}81 = 490{\cdot}5 \text{ N} \\ \text{Component force down the plane} &= \text{weight} \times \sin \alpha \\ &= 490{\cdot}5 \times \sin 30°\end{aligned}$$

$$= 245{\cdot}25 \text{ N}$$

$$\text{Friction force opposing motion} = \mu \times \text{force between surfaces}$$

$$= 0{\cdot}4 \times 490{\cdot}5 \times \cos 30°$$

$$= 169{\cdot}9 \text{ N}$$

Force causing acceleration down the plane

$$= 245{\cdot}25 - 169{\cdot}9$$

$$= 75{\cdot}35 \text{ N}$$

$$\text{Acceleration} = \frac{\text{accelerating force}}{\text{mass}}$$

$$= \frac{75{\cdot}35}{50} = 1{\cdot}507 \text{ m/s}^2$$

$$\text{Velocity after } t \text{ s} = 1{\cdot}507\, t \text{ m/s}$$

$$\text{Average velocity} = \tfrac{1}{2}(0 + 1{\cdot}507\, t) \quad = 0{\cdot}7535t$$

$$\text{Distance} = \text{average velocity} \times \text{time}$$

$$10 = 0{\cdot}7535t \times t$$

$$t = \sqrt{\frac{10}{0{\cdot}7535}} = 3{\cdot}642 \text{ s} \qquad \text{Ans. } (i)$$

Total force to pull up with acceleration

$$= \text{force of gravity} + \text{friction} + \text{accel. force}$$

$$= W \sin \alpha + \mu\, W \cos \alpha + \text{ma}$$

$$= 245{\cdot}25 + 169{\cdot}9 + 50 \times 2$$

$$= 515{\cdot}15 \text{ N} \qquad \text{Ans. } (ii)$$

9.

$$z_1 + \frac{p_1}{w} + \frac{v_1^2}{2g} = z_2 + \frac{p_2}{w} + \frac{v_2^2}{2g} + \text{Friction loss}$$

Note: The energy loss, due to friction , must be expressed as *metres head* of the liquid,

i.e. Friction loss is 10 *J/kg*

$$\text{Friction loss} = \frac{10}{g}\text{ J/N}$$

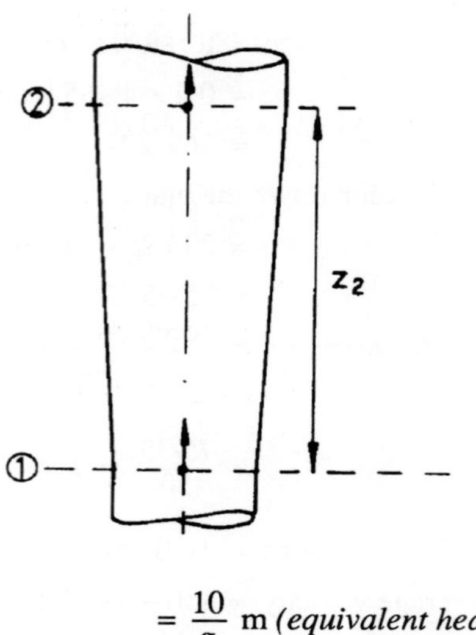

$$= \frac{10}{g} \text{ m } \textit{(equivalent head)}$$

$$\textit{Inlet velocity } v_1 = \frac{360}{3600 \times 0{\cdot}7854 \times 0{\cdot}1^2}$$

$$v_1 = 12{\cdot}73 \text{ m/s.}$$

Volumetric flow rate is constant,

$$v_1 \times a_1 = v_2 \times a_2$$

$$12{\cdot}73 \times \frac{\pi}{4} \times 0{\cdot}1^2 = v^2 \times \frac{\pi}{4} \times 0{\cdot}15^2$$

$$v_2 = 5{\cdot}658 \text{ m/s}$$

Inserting these values in Bernoulli's equation,

$$0 + \frac{p_1}{1000 \times g} + \frac{12{\cdot}73^2}{2 \times g} = z_2 + \frac{p_2}{1000 \times g} + \frac{5{\cdot}658^2}{2 \times g} + \frac{10}{g}$$

$$\frac{p_1}{1000 \times g} + 8{\cdot}26 = z_2 + \frac{p_2}{1000 \times g} + 1{\cdot}632 + 1{\cdot}02$$

$$z_2 = \frac{p_1 - p_2}{1000 \times g} + 5{\cdot}61$$

$$\text{But, } p_1 - p_2 = 80 \times 10^3 \ \text{N/m}^2$$

$$z_2 = \frac{80 \times 10^3}{1000 \times g} + 5{\cdot}61$$

$$z_2 = 13{\cdot}76$$

$$\text{Length of pipe} = 13{\cdot}76 \text{ m}$$ Ans.

10.

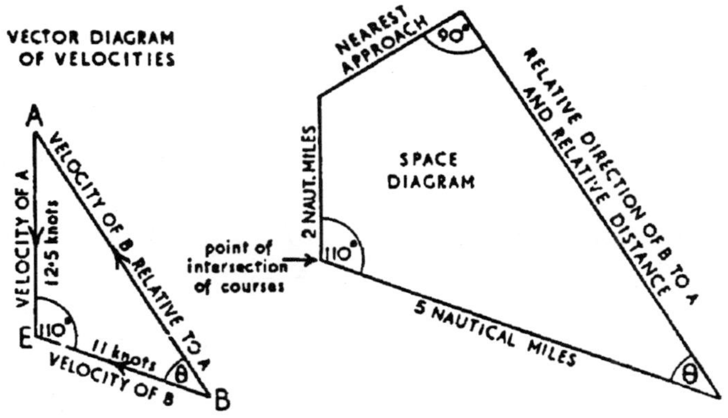

From Fig.

Nearest approach measures 1·97 naut. miles Ans. (*i*)

Relative distance measures 5·65 naut. miles

Relative velocity measures 19·27 knots

Time to be at position of nearest approach

$$= \frac{5{\cdot}65}{19{\cdot}27} \times 60$$

$$= 17{\cdot}6 \text{ min.} \qquad \text{Ans. } (ii)$$

11. When compound bar is heated, copper tends to expand more than the steel, the steel tends to expand less than the copper. Since they are secured together, one pulls against the other but their forces must be equal.

$$\text{Force stretching steel} = \text{Force compressing copper}$$

$$\text{stress}_S \times \text{area}_S = \text{stress}_C \times \text{area}_C$$

$$\text{stress}_S \times 0{\cdot}7854\,(40^2 - 25^2) = \text{stress}_C \times 0{\cdot}7854 \times 25^2$$

$$\text{stress}_S \times 975 = \text{stress}_C \times 625$$

$$\text{stress}_C = 1{\cdot}56 \times \text{stress}_S \qquad \ldots\ \ldots\ (i)$$

Also,

Sum of the strains = Difference in free expansion per unit length

$$\frac{\text{Stress}_S}{E_C} + \frac{\text{stress}_C}{E_C} = \alpha_C\theta - \alpha_S\theta$$

$$\frac{\text{stress}_S}{206 \times 10^9} + \frac{\text{stress}_C}{103 \times 10^9} = 100 \times 10^{-6}\,(17 - 12)$$

$$\text{stress}_S + 2 \times \text{stress}_C = 206 \times 10^9 \times 100 \times 10^{-6} \times 5$$

Substituting stress_C from (i) and simplifying,

$$\text{stress}_S + 2 \times 1{\cdot}56 \times \text{stress}_S = 206 \times 5 \times 10^5$$

$$4{\cdot}12\ \text{stress}_S = 206 \times 5 \times 10^2$$

$$\text{stress}_S = 2{\cdot}5 \times 10^7 \text{ N/m}^2$$

$$\text{stress in steel tube} = 25 \text{ MN/m}^2 \qquad \text{Ans.}$$

$$\text{stress in copper bar} = 1{\cdot}56 \times \text{stress}_S \text{ (from } (i))$$

$$= 39 \text{ MN/m}^2 \qquad \text{Ans.}$$

$$\text{Cross-sect. area of pin} = 0{\cdot}7854 \times 10^2 \text{ mm}^2$$

Being in double shear, area to be sheared through

$$= 2 \times 0{\cdot}7854 \times 10^2 \times 10^{-6} \text{ m}^2$$

$$\text{Force on pin} = \text{force in bar} = \text{stress}_C \times \text{area}_C$$

$$= 39 \times 10^6 \times 0{\cdot}7854 \times 25^2 \times 10^{-6}$$

$$= 39 \times 0{\cdot}7854 \times 25^2 \text{ N}$$

$$\text{Stress in pin} = \frac{\text{force}}{\text{area}}$$

$$= \frac{39 \times 0{\cdot}7854 \times 25^2}{2 \times 0{\cdot}7854 \times 10^2 \times 10^{-6}}$$

$$= 1{\cdot}219 \times 10^8 \text{ N/m}^2$$

$$\text{Stress in pins} = 121{\cdot}9 \text{ MN/m}^2$$ Ans.

12. $$P = T\omega$$

$$\text{Torque in shaft} = \frac{420 \times 60}{180 \times 2\pi} = 22{\cdot}28 \text{ kN m}$$

$$J \text{ for shaft} = \frac{\pi}{32}(D^4 - d^4) = \frac{\pi}{32}\left\{(1{\cdot}6d)^4 - d^4\right\}$$

$$= \frac{\pi}{32}(6{\cdot}554d^4 - d^4)$$

$$= \frac{\pi}{32} \times 5{\cdot}554d^4$$

$$\frac{T}{J} = \frac{G\theta}{l}$$

$$J \times G \times \theta = T \times l$$

$$\pi \times 5{\cdot}554d^4 \times 85 \times 10^9 \times \frac{1 \times 2\pi}{360}$$

$$= 22{\cdot}28 \times 10^3 \times 40d$$

$$d^3 = \frac{32 \times 22{\cdot}28 \times 40 \times 360}{\pi \times 5{\cdot}554 \times 85 \times 10^6 \times 2\pi}$$

$$d = 0{\cdot}1032 \text{ m}$$

$$D = 1{\cdot}6 \times 0{\cdot}1032 = 0{\cdot}1651 \text{ m}$$

At entrance:

$$\frac{1764 \times 10^3}{60} = 0{\cdot}7854 \times 0{\cdot}3^2 \times v_1$$

$$0{\cdot}0294 = 0{\cdot}7854 \times 0{\cdot}3^2 \times v_1$$

$$v_1 = 0{\cdot}4159 \text{m/s}$$

From (i),

$$v_2^2 - v_1^2 = 13{\cdot}85$$

$$v_2 = \sqrt{13{\cdot}85 + 0\ 4159^2}$$

$$= 3{\cdot}744 \text{ m/s}$$

Let d = diameter at throat,

$$\dot{v} = av$$

$$0{\cdot}0294 = 0{\cdot}7854 \times d^2 \times 3{\cdot}744$$

$$d = 0{\cdot}09997 \text{ m}$$

$$= 100 \text{ mm}$$ Ans.

15.

Let w = weight N/m length of beam

Total weight = $20\,w$ N

Each reaction = $10\,w$ N

Let each reaction be at x m from the ends.

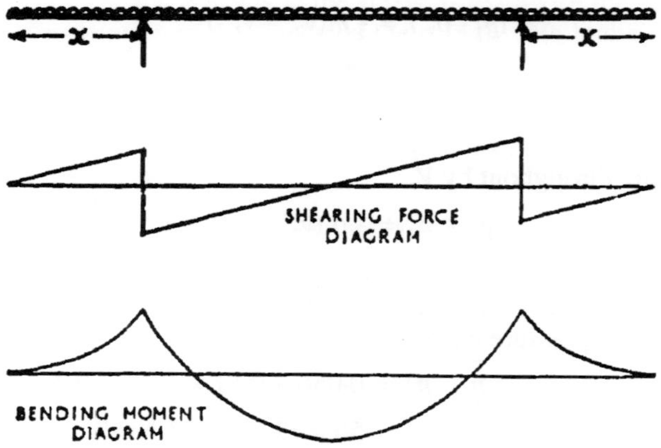

Maximum bending moment (and therefore maximum stress) will be least when the bending moment at the centre is numerically equal to the bending moment at each reaction.

$$M \text{ at reactions} = M \text{ at centre}$$
$$wx \times \tfrac{1}{2}x = 10w\,(10 - x) - 10w \times 5$$
$$\tfrac{1}{2}\,wx^2 = 100w - 10wx - 50w$$
$$x^2 = 200 - 20x - 100$$
$$x^3 + 20x - 100 = 0$$

From this quadratic $x = 4{\cdot}14$

The two slings should be attached at 4·14 m from each end. Ans.

16. Average speed of pinion $= \tfrac{1}{2}\,(0 + 15) = 7{\cdot}5$ rev/s

$$\text{time} = \frac{\text{distance}}{\text{average speed}}$$

$$= \frac{150}{7{\cdot}5} = 20 \text{ s}$$ Ans. (*i*)

Maximum angular velocity of pinion

$$= 15 \times 2\pi = 94{\cdot}26 \text{ rad/s}$$

Maximum angular velocity of wheel

$$= \tfrac{1}{3} \times 94{\cdot}26 = 31{\cdot}42 \text{ rad/s}$$

Angular acceleration of pinion

$$= 94{\cdot}26 \div 20 = 4{\cdot}713 \text{ rad/s}^2$$

Angular acceleration of wheel

$$= 31{\cdot}42 \div 20 = 1{\cdot}571 \text{ rad/s}^2$$
$$I = \text{mass} \times (\text{radius of gyration})^2$$
$$I \text{ of pinion} = 45 \times 0{\cdot}15^2 = 1{\cdot}013 \text{ kg m}^2$$
$$I \text{ of wheel} = 420 \times 0{\cdot}44^2 = 81{\cdot}32 \text{ kg m}^2$$
$$\text{Accelerating torque} = I\alpha$$

Torque in pinion shaft to accelerate pinion

$$= 1{\cdot}013 \times 4{\cdot}713 = 4{\cdot}772$$

Torque in wheel shaft to accelerate wheel

$$= 81{\cdot}32 \times 1{\cdot}571 = 127{\cdot}7 \text{ N m}$$

Force on teeth is common:

$$\frac{\text{torque in pinion shaft}}{\text{radius of pinion}} = \frac{\text{torque in wheel shaft}}{\text{radius of wheel}}$$

Torque in pinion shaft to accelerate wheel

$$= \text{torque in wheel shaft} \times \frac{\text{radius of pinion}}{\text{radius of wheel}}$$

$$= 127{\cdot}7 \times \tfrac{1}{3} = 42{\cdot}57 \text{ N m}$$

Total torque in pinion shaft

$$= 4{\cdot}772 + 42{\cdot}57 = 47{\cdot}342 \text{ N m} \qquad \text{Ans. } (ii)$$

$$\begin{aligned} \text{Energy stored at max. speed} &= \text{Work done to attain speed} \\ \text{Work done/rev} &= \text{force} \times 2\pi \times r \\ &= \text{torque} \times 2\pi \\ \text{Work done in 150 rev} &= 47{\cdot}342 \times 2\pi \times 150 \\ &= 4{\cdot}463 \times 10^4 \text{ J} \\ &= 44{\cdot}63 \text{ kJ} \end{aligned} \qquad \text{Ans. } (iii)$$

17. $$\text{Spring stiffness} = \frac{\text{Force}}{\text{corresponding deflection}}$$

$$= \frac{10}{0{\cdot}036}$$

$$= 277{\cdot}8 \text{ N/m}$$

If frequency = 1 osc/s

then, periodic time t = 1 s

Let m = mass to be attached to the end of the spring

$$t = 2\pi \sqrt{\frac{m + \frac{m_3}{3}}{S}}$$

$$1 = 2\pi \sqrt{\frac{m + \frac{0{\cdot}6}{3}}{277{\cdot}8}}$$

$$m + 0{\cdot}2 = 277{\cdot}8 \times \left(\frac{1}{2\pi}\right)^2$$

Mass to be attached = 6·84 kg. Ans. (i)

$$v = \omega\sqrt{r^2 - x^2}$$

$$= \omega^2 x$$

where r = amplitude

x = displacement from mid-travel

$$\omega = \frac{2\pi}{\text{periodic time}}$$

When x = 0, velocity is maximum, therefore:

max. velocity $= \omega r$

$$= \frac{2\pi}{1} \times 12 \times 10^{-3}$$

$= 0{\cdot}0754$ m/s Ans. (*ii*)

When $x = r$, acceleration is maximum, therefore:

Max. acceleration $= \omega^2 r$

$$= \left\{\frac{2\pi}{1}\right\}^2 \times 12 \times 10^{-3}$$

$= 0{\cdot}4739$ m/s^2 Ans. (*iii*)

18.

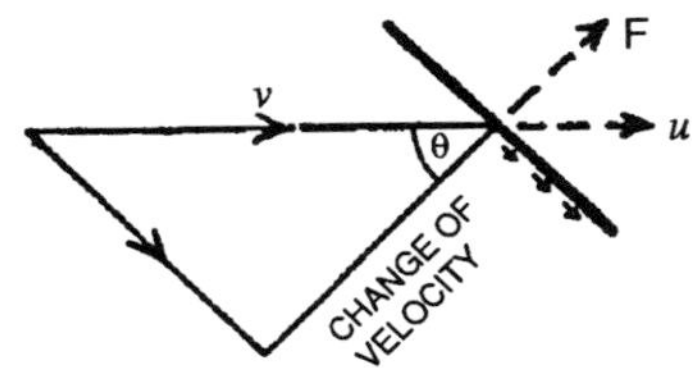

(*i*) Plate stationary:

Mass of water striking plate per second

= area × velocity × density

$= 0{\cdot}7854 \times 0{\cdot}025^2 \times 15 \times 10^3$

$= 7{\cdot}362$ kg/s

Change of velocity $= v \cos\theta$

$= 15 \times \cos 45° = 10{\cdot}6$ m/s

Force = mass × change of velocity per second

$= 7{\cdot}362 \times 10{\cdot}6 = 78{\cdot}09$ N Ans. (*i*)

(*ii*) Plate moving:

Velocity of jet relative to plate

$= v - u = 15 - 10 = 5$ m/s

Mass of water striking plate per second

$$= 7{\cdot}362 \times \frac{5}{15} = 2{\cdot}454 \text{ kg/s}$$

$$\text{Change of velocity} = (v - u)\cos\theta$$
$$= 5\cos 45° = 3{\cdot}535 \text{ m/s}$$
$$\text{Force} = 2{\cdot}454 \times 3{\cdot}535 = 8{\cdot}676 \text{ N}$$ Ans. (*ii*)

19. From $\frac{T}{J} = \frac{\tau}{r} = \frac{G\theta}{l}$ $\qquad \theta = \frac{\tau l}{rG}$

Shaft and liner have same angle of twist,

$$\theta \text{ for shaft}_1 = \theta \text{ for liner}_2$$

$$\frac{\tau_1 l_1}{r_1 G_1} = \frac{\tau_2 l_2}{r_2 G_2}$$ $l_1 = l_2$ and cancels

$$\frac{\tau_1}{0{\cdot}12 \times 90 \times 10^9} = \frac{\tau_2}{0{\cdot}145 \times 42 \times 10^9}$$

$$\frac{\tau_1}{\tau_2} = \frac{0{\cdot}12 \times 90}{0{\cdot}145 \times 42} = 1{\cdot}773$$

Ratio of stress in shaft to stress in liner

$= 1{\cdot}773 : 1$ Ans (*i*)

$$\frac{T}{J} = \frac{\tau}{r}$$

For shaft, $J = \frac{\pi}{32} D^4$ $\qquad r = \frac{1}{2} D$

$$T = \frac{\pi}{16} D^3 \tau$$

$$= \frac{\pi}{16} \times 0{\cdot}24^3\, \tau_1 = 0{\cdot}002714\tau_1$$

For liner, $J = \frac{\pi}{32}(D^4 - d^4)$ $\qquad r = \frac{1}{2} D$

$$T = \frac{\pi(D^4 - d^4)\tau}{16D}$$

$$= \frac{\pi(0{\cdot}29^4 - 0{\cdot}24^4) \times \tau_1}{16 \times 0{\cdot}29} = 0{\cdot}002543\tau_2$$

Total torque

$$= 0{\cdot}002714\tau_1 + 0{\cdot}002543\tau_2 = 100 \times 10^3$$

$$\text{Substituting } \tau_2 = 1{\cdot}773\ \tau_2$$

$$0{\cdot}002714 \times 1{\cdot}773\ \tau_2 + 0{\cdot}002543\tau_2 = 100 \times 10^3$$
$$0{\cdot}004812\ \tau_2 + 0{\cdot}002543\ \tau_2 = 10^5$$
$$0{\cdot}007355\ \tau_2 = 10^5$$
$$\tau_2 = 1{\cdot}359 \times 10^7 \text{ N/m}^2$$

$$\tau_1 = 1{\cdot}773 \times 1{\cdot}359 \times 10^7 = 2{\cdot}411 \times 10^7 \text{ N/m}^2$$

$$\left.\begin{array}{r}\text{Stress in shaft} = 24{\cdot}11 \text{ MN/m}^2\\ \text{Stress in liner} = 13{\cdot}59 \text{ MN/m}^2\end{array}\right\}$$ Ans. (*ii*)

$$\theta = \frac{\tau l}{rG} \quad \text{rad}$$

$$\text{Angle of twist} = \frac{2{\cdot}411 \times 10^7 \times 4}{0{\cdot}12 \times 90 \times 10^9} \times 57{\cdot}3 \text{ deg}$$

$$= 0{\cdot}5118°$$ Ans. (*iii*)

20.

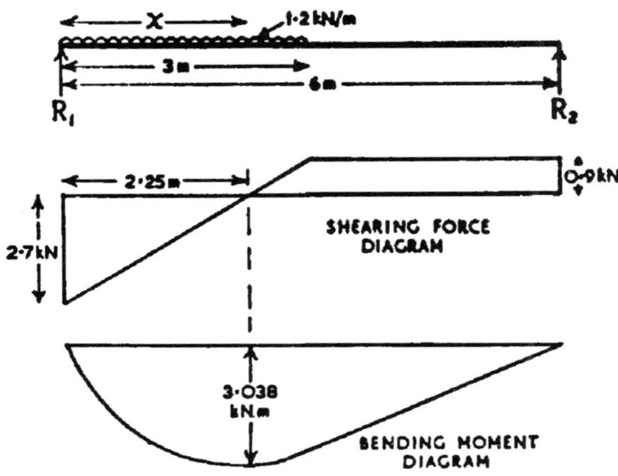

Moments about R_1

$$1{\cdot}2 \times 3 \times 1{\cdot}5 = R_2 \times 6$$
$$R_2 = 0{\cdot}9 \text{ kN}$$
$$R_1 = 1{\cdot}2 \times 3 - 0{\cdot}9 = 2{\cdot}7 \text{ kN}$$

Shearing force at x from left end = $2{\cdot}7 - 1{\cdot}2x$ this is to be zero for position of maximum bending moment and therefore maximum stress:

$$0 = 2{\cdot}7 - 1{\cdot}2x$$
$$x = 2{\cdot}25 \text{ m}$$
$$M_x = -R_1 \times x + 1{\cdot}2 \times x \times \tfrac{1}{2}x$$
$$= -2{\cdot}7 \times 2{\cdot}25 + 0{\cdot}6 \times 2{\cdot}25^2$$
$$= -3{\cdot}038 \text{ kN m}$$

$$\frac{M}{I} = \frac{\sigma}{y} \qquad \sigma = \frac{My}{I}$$

$$\sigma_{max} = \frac{3{\cdot}038 \times 10^3 \times 0{\cdot}15}{1{\cdot}33 \times 10^{-4}}$$
$$= 3{\cdot}426 \times 10^6 \text{ N/m}^2$$

The position of maximum stress is 2·25 m from left end and its magnitude is 3·426 MN/m^2

21.
$$\frac{110 \text{ rev/min} \times 2\pi}{60} = 11{\cdot}52 \text{ rad/s}$$

$$\text{Mass of flywheel} = 3 \times 10^3 \text{ kg}$$
$$\text{Kinetic energy} = \tfrac{1}{2} mk^2 \omega^2$$
$$\text{Internal } K.E. - \text{Final } K.E. = \tfrac{1}{2} mk^2 \omega_1^2 - \tfrac{1}{2} mk^2 \omega_2^2$$
$$= \tfrac{1}{2} mk^2 (\omega_1^2 - \omega_2^2)$$

and this is the energy to do the work in cutting.

$$\tfrac{1}{2} \times 3 \times 10^3 \times 0{\cdot}55^2 (11{\cdot}52^2 - \omega_2^2) = 12 \times 10^3$$

$$11{\cdot}52^2 - \omega_2^2 = \frac{2 \times 12}{3 \times 0{\cdot}55^2}$$

$$\omega_2 = 10{\cdot}31 \text{ rad/s}$$

$$\frac{10{\cdot}31 \times 60}{2\pi} = 98{\cdot}45 \text{ rev/min}$$ Ans. (*i*)

$$\alpha = \frac{\text{change of angular velocity}}{\text{time}}$$

$$= \frac{11{\cdot}52 - 10{\cdot}31}{3}$$

$$= 0{\cdot}4033 \text{ rad/s}^2$$ Ans. (*ii*)

22.

$$\text{Power} = \text{effective pull in belt} \times \text{speed of belt}$$

$$8{\cdot}2 \times 10^3 = (F_1 - F_2) \times 6{\cdot}6$$

$$F_1 - F_2 = 1243$$

$$1855 - F_2 = 1243$$

$$F_2 = 612 \text{ N}$$

$$\frac{F_1}{F_2} = \frac{1855}{612} = 3{\cdot}031$$

$$\frac{F_1}{F_2} = \varepsilon\mu\theta$$

$$3{\cdot}031 = 2{\cdot}718^{\mu\theta}$$

$$\log 3{\cdot}031 = \log 2{\cdot}718 \times 0{\cdot}3 \times \theta$$

$$0{\cdot}4815 = 0{\cdot}4343 \times \theta$$

$$\theta = \frac{0{\cdot}4815}{0{\cdot}4343 \times 0{\cdot}3} = 3{\cdot}695 \text{ rad}$$

$$\text{Angle of lap} = 3{\cdot}695 \times \frac{360}{2\pi} = 211{\cdot}6^\circ$$ Ans.

Alternatively, for second part:

$$\frac{F_1}{F_2} = \varepsilon\mu\theta$$

$$\log_e \frac{F_1}{F_2} = \mu\theta$$

$$\log_e 3{\cdot}031 = 0{\cdot}3 \times \theta$$

$$1{\cdot}1089 = 0{\cdot}3 \times \theta$$

$$\theta = 3{\cdot}696 \text{ rad (as previous)}$$

23.

At the bottom of the pipe, pressure $p_1 = \rho g h$

$$= 13{\cdot}6 \times 10^3 \times 9{\cdot}81 \times 0{\cdot}76$$

$$= 101{\cdot}4 \text{ kN/m}^2$$

At the top of the pipe, pressure $p_2 = 10^3 \times 9{\cdot}81 \times 2$

$$= 19{\cdot}62 \text{ kN/m}^2$$

$$z_1 + \frac{p_1}{w} + \frac{v_1^2}{2g} = z_2 + \frac{p_2}{w} + \frac{v_2^2}{2g}$$

$$0 + \frac{101{\cdot}4 \times 10^3}{10^3 \times g} + 0 = 3 + \frac{19{\cdot}62 \times 10^3}{10^3 \times g} + \frac{v_2^2}{2g}$$

$$101{\cdot}4 = 3 \times g + 19{\cdot}62 + \frac{v_2^2}{2}$$

υ_2 velocity in pipe $= 10{\cdot}23$ m/s Ans.

$$\frac{300}{3600} = 0{\cdot}7854 \times d^2 \times 10{\cdot}23$$

$$d = \sqrt{\frac{300}{3600 \times 0{\cdot}7854 \times 10{\cdot}23}}$$

$$= 0{\cdot}1018 \text{ m} = 101{\cdot}8 \text{ mm}$$ Ans.

24. Working in mm

$$\frac{\tau}{r} = \frac{G\theta}{l}$$

$$\frac{250}{2{\cdot}5} = \frac{88 \times 10^3 \times 1}{\pi \times 40 \times N}$$

$$N = \frac{2{\cdot}5 \times 88 \times 10^3 \times 1}{250 \times \pi \times 40} = 7$$ Ans. (i)

$$\frac{T}{J} = \frac{\tau}{r}$$

T = twisting moment applied to wire, in N mm
= load × mean coil radius
= load × 20

$$J = \frac{\pi}{32} d^4 = \frac{\pi}{32} \times 5^4 \text{ mm}^4$$

$$\frac{32 \times \text{load} \times 20}{\pi \times 5^4} = \frac{250}{2{\cdot}5}$$

$$\text{load} = \frac{\pi \times 5^4 \times 250}{32 \times 20 \times 2{\cdot}5}$$

$= 306{\cdot}8$ N   Ans. (*ii*)

Energy stored = work done to deflect
= average force × deflection
$= \frac{1}{2}(0 + 306{\cdot}8) \times 20 \times 10^{-3}$
$= 3{\cdot}068$ J   Ans. (*iii*)

25. Initial vertical velocity = $v \sin \alpha$

At max. height, vertical velocity = 0

Time to reach max. height

$$= \frac{\text{change of vertical velocity}}{\text{vertical retardation}} = \frac{v \sin \alpha}{g}$$

Vertical distance = average vertical velocity × time

$$= \tfrac{1}{2}(v \sin \alpha + 0) \times \frac{v \sin \alpha}{g}$$

$$h = \frac{v^2 \sin^2 \alpha}{2g} \qquad \dots \dots \dots \dots (i)$$

Total time in the air = time up + time down

$$= \frac{2v \sin \alpha}{g}$$

$$\text{Horizontal distance} = \text{horizontal velocity} \times \text{time}$$

$$= v \cos \alpha \times \frac{2\,v \sin \alpha}{g}$$

$$R = \frac{2v^2 \sin \alpha \cos \alpha}{g} \qquad \dots \dots \dots \dots (ii)$$

From (i), $\frac{v^2 \sin \alpha}{g} = \frac{2h}{\sin \alpha}$ substituting for this into (ii)

$$R = \frac{2 \times 2\,h \times \cos \alpha}{\sin \alpha}$$

$$R = \frac{4h}{\tan \alpha} \qquad \text{Ans } (i)$$

When $R = 1000$ m and $x = 20°$,

$$h = \frac{R \tan \alpha}{4}$$

$$= \frac{1000 \times 0{\cdot}364}{4}$$

$$= 91 \text{ m} \qquad \text{Ans. } (ii)$$

26. Let a = acceleration up the plane in m/s^2

Force to overcome gravity $= mg \sin \alpha$

$= 150 \times 9{\cdot}81 \times \sin 18° = 454{\cdot}7$ N

Force to overcome friction $= \mu mg \cos \alpha$

$= 0{\cdot}33 \times 150 \times 9{\cdot}81 \times \cos 18° = 461{\cdot}8$ N

Force to cause acceleration = mass × acceleration

$= 150 \times a$ N

$$\begin{aligned}
\text{Total force} &= 454{\cdot}7 + 461{\cdot}8 + 150a \\
&= 916{\cdot}5 + 150a \\
\text{Torque at drum} &= \text{force} \times \text{radius} \\
&= (916{\cdot}5 + 150a) \times 0{\cdot}4 \\
&= 366{\cdot}6 + 60a
\end{aligned}$$

$$\begin{aligned}
\alpha &= \frac{a}{r} \\
&= \frac{a}{0.4}\ \text{rad/s}^2
\end{aligned}$$

$$\text{Torque to accelerate drum} = I\alpha$$

$$1{\cdot}6 \times \frac{a}{0{\cdot}4} = 4a \quad \text{N m}$$

$$\begin{aligned}
\text{Total torque} &= \text{torque to accelerate body and drum} \\
373 &= 366{\cdot}6 + 60a + 4a \\
64a &= 6{\cdot}4 \\
a &= 0{\cdot}1\ \text{m/s}^2 && \text{Ans. } (i) \\
\text{Velocity after 12 s} &= 0{\cdot}1 \times 12 = 1{\cdot}2\ \text{m/s} \\
\text{Average velocity} &= \tfrac{1}{2}(0 + 1{\cdot}2) = 0{\cdot}6\ \text{m/s} \\
\text{Displacement} &= \text{average velocity} \times \text{time} \\
&= 0{\cdot}6 \times 12 = 7{\cdot}2\ \text{m} && \text{Ans. } (ii) \\
\text{Pull in rope} &= \text{total force} \\
&= 916{\cdot}5 + 150a \\
&= 916{\cdot}5 + 150 \times 0{\cdot}1 \\
&= 931{\cdot}5\ \text{N} && \text{Ans. } (iii)
\end{aligned}$$

27.

$$\begin{aligned}
\text{let } w &= \text{kN/m of distributed load} \\
\text{Total distributed load} &= 4w \\
\text{Total load on beam} &= 4w + 5 + 5 = 4w + 10\ \text{kN}
\end{aligned}$$

Each reaction carries half of total load

$$R_1 = R_2 = 2w + 5$$

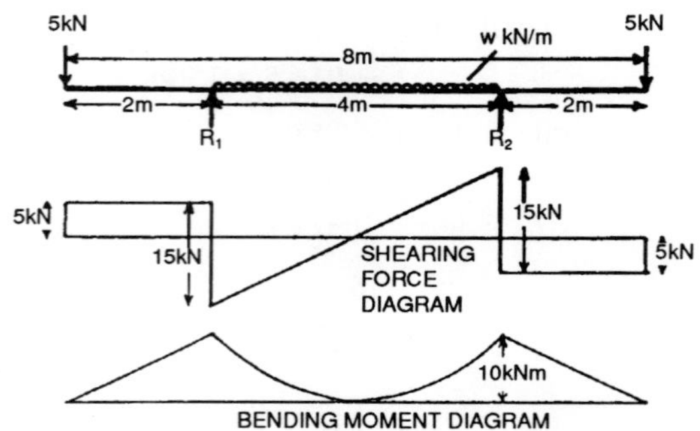

$$M \text{ @ mid-span} = -(2w+5)\times 2 + 2w \times 1 + 5 \times 4$$

and this is to be zero

$$-4w - 10 + 2w + 20 = 0$$

$$2w = 10$$

$$w = 5 \text{ kN/m}$$

$$\text{Total distributed load} = 4\times 5 = 20 \text{ kN}$$ Ans.

Measurements for shearing force and bending moment diagrams:

$$R_1 = R_2 = 2w + 5 = 2\times 5 + 5 = 15 \text{ kn}$$

$$M \text{ @ reactions} = 5 \times 2 = 10 \text{ kN m}$$

28.

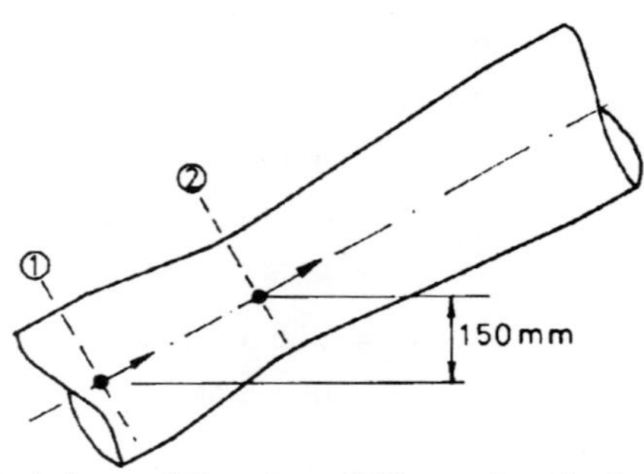

Pressure difference, inlet and throat = 250 mm head of water

$$\text{Pressure difference} = \rho g h$$

$$= 10^3 \times 9{\cdot}81 \times 0{\cdot}25$$

$$p_1 - p_2 = 2{\cdot}452 \times 10^3 \text{ N/m}^2$$

Volumetric flow rate is constant,

$$v_1\, a_1 = v_2\, a_2$$

$$v_1 \times \frac{\pi}{4} \times 0{\cdot}05^2 = v_2 \times \frac{\pi}{4} \times 0{\cdot}025^2$$

$$v_2 = 4\, v_1$$

$$z_1 + \frac{p_1}{w} + \frac{v_1^2}{2g} = z_2 + \frac{p_2}{w} + \frac{v_2^2}{2g}$$

$$0 + \frac{p_1}{10^3 \times g} + \frac{v_1^2}{2g} = 0{\cdot}15 + \frac{p_2}{10^3 \times g} + \frac{(4v_1)^2}{2g}$$

$$\frac{p_1 - p_2}{10^3 \times g} - 0{\cdot}15 = \frac{16v_1^2 - v_1^2}{2g}$$

$$\frac{2{\cdot}452 \times 10^3}{10^3 \times g} - 0{\cdot}15 = \frac{15v_1^2}{2g}$$

$$v_1^2 = (0{\cdot}25 - 0{\cdot}0{\cdot}15) \times \frac{2g}{15}$$

$$v_1 = 0{\cdot}3617 \text{ m/s}$$

$$\text{Volume flow rate} = v_1 \times a_1 \times \text{meter coeff.}$$

$$= 0{\cdot}3617 \times \frac{\pi}{4} \times 0{\cdot}05^2 \times 0{\cdot}95$$

$$= 0{\cdot}000675 \text{ m}^3\text{/s}$$ Ans.

29. $25 \text{ rev/s} \times 2\pi = 50\pi$ rad/s

Centrifugal force of each mass $= m\omega^2 r$

$= 0{\cdot}5 \times (50\pi)^2 \times 0{\cdot}04$

$= 493{\cdot}7$ N

Force of 2 arms on spring (leverage 1·5 : 1),

$= 2 \times 493{\cdot}7 \times 1{\cdot}5$

$= 1481$ N Ans. (*i*)

When speed is increased by 2%,

$$\text{New rotational speed} = 25 \times 1{\cdot}02 = 25{\cdot}5 \text{ rev/s}$$

Centrifugal force varies as square of velocity,

$$\text{New centrifugal force} = 493{\cdot}7 \times \frac{25{\cdot}52^2}{25^2}$$

$$= 513{\cdot}5 \text{ N}$$

$$\text{New force on spring} = 2 \times 513{\cdot}5 \times 1{\cdot}5$$

$$= 1541 \text{ N}$$

$$\text{Stiffness} = \frac{\text{increased force}}{\text{increased deflection}}$$

$$= \frac{1541 - 1481}{5}$$

$$= 12 \text{ N/mm} \qquad \text{Ans. } (ii)$$

30. Mass falls 1·05 in 8 s from rest,

$$\text{Average linear velocity} = 1{\cdot}05 \div 8 \text{ m/s}$$

$$\text{Final linear velocity} = 2\,(1{\cdot}05 \div 8) = 0{\cdot}2625 \text{ m/s}$$

$$\omega = \frac{v}{r}$$

$$= \frac{0{\cdot}2625}{0{\cdot}025} = 10{\cdot}5 \text{ rad/s}$$

Potential energy converted into kinetic energy

$$= (2 - 0{\cdot}3) \times 9{\cdot}81 \times 1{\cdot}05$$

$$= 17{\cdot}51 \text{ J} \qquad \dots\ \dots\ \dots\ \dots\ (i)$$

Kinetic energy gained by system

= $K.E.$ of falling mass + $K.E.$ of rotating parts

$$= \tfrac{1}{2} m_1 v^2 + \tfrac{1}{2} m_2 k^2 \omega^2$$

$$= \tfrac{1}{2} \times 2 \times 0{\cdot}2625^2 + \tfrac{1}{2} \times 14 \times k^2 \times 10{\cdot}5^2$$

$$= 0{\cdot}0689 + 771{\cdot}8k^2 \qquad \ldots \ldots \ldots (ii)$$

$$0{\cdot}0689 + 771{\cdot}8k^2 = 17{\cdot}51$$

$$771{\cdot}8k^2 = 17{\cdot}4411$$

$$k = 0{\cdot}1503 \text{ m}$$

$$= 150{\cdot}3 \text{ mm}$$ Ans.

31. Pull in rope to support weight of mass

$$= 250 \times 9{\cdot}81 = 2453 \text{ N}$$

Pull in rope to accelerate mass = mass × acceleration

$$= 250 \times 0{\cdot}2 = 50 \text{ N}$$

Total tension in rope $= 2453 + 50 = 2503$ N Ans. (*i*)

Effective radius of hoist = drum radius + rope radius

$$= 0{\cdot}3 + 0{\cdot}015 = 0{\cdot}315 \text{ m}$$

Torque to lift and accelerate mass

$$= 2503 \times 0{\cdot}315 = 788{\cdot}3 \text{ N m}$$

Angular acceleration of drum $\alpha = \dfrac{a}{r}$

$$= \frac{0{\cdot}2}{0{\cdot}315} = 0{\cdot}6349 \text{ rad/s}^2$$

Torque to accelerate rotating parts $= mk^2\alpha$

$$= 275 \times 0{\cdot}24^2 \times 0{\cdot}6349 = 10{\cdot}06 \text{ N m}$$

Total torque in winch drum shaft

$$= 788{\cdot}3 + 10{\cdot}06 = 798{\cdot}36 \text{ N m}$$

Ans.(*ii*)

32.

$$1000 = 3{\cdot}75\, C_1 + 14{\cdot}062\, C_2 \qquad \text{(i)}$$

$$1670 = 5\, C_1 + 25\, C_2 \qquad \text{(ii)}$$

Multiply (i) by 5, (ii) by 3·75:

$$5000 = 18{\cdot}75\, C_1 + 70{\cdot}312\, C_2 \qquad \text{(iii)}$$

$$6262{\cdot}5 = 18{\cdot}75\, C_1 + 93{\cdot}75\, C_2 \qquad \text{(iv)}$$

Subtract (iii) from (iv):

$$1262{\cdot}5 = 23{\cdot}438\ C_2$$
$$C_2 = 53{\cdot}865$$

$$\text{Substituting } C_2 = 53{\cdot}865 \text{ in equation (ii):}$$
$$1670 = 5\ C_1 + (25 \times 53{\cdot}865)$$
$$C_1 = 64{\cdot}675$$

When A is 1870 mm^2:

$$1870 = 64{\cdot}675\ h + 53{\cdot}865\ h^2$$

Simplify and re-arrange:

$$h^2 + 1{\cdot}2h - 34{\cdot}716 = 0$$

Solving this quadratic:

$$h = 5{\cdot}32 \text{ mm}$$ Ans.

33. By the principle of Conservation of Energy,

Kinetic energy gained = Potential energy lost

The total kinetic energy gained by the disc consists of kinetic energy of translation due to linear velocity, and kinetic energy of rotation due to angular velocity.

The potential energy lost is due to loss of height.

$$\text{Loss of height} = 6 \times \text{sin of angle of incline}$$
$$= 6 \times \tfrac{1}{15} = 0{\cdot}4 \text{ m}$$
$$\text{Weight of mass} = 24 \times 9{\cdot}81 \text{ N}$$
$$\text{Potential energy lost} = 24 \times 9{\cdot}81 \times 0{\cdot}4$$
$$= 94{\cdot}18 \text{ J}$$

Average linear velocity down plane

$$= \frac{\text{distance}}{\text{time}} = \frac{6}{5{\cdot}25} \text{ m/s}$$

Since it started from rest,

$$\text{final linear velocity} = 2 \times \text{average velocity}$$

$$= \frac{2 \times 6}{5{\cdot}25} = \frac{16}{7} \text{ m/s}$$

$$\text{final angular velocity} = \frac{\text{linear velocity}}{\text{radius}}$$

$$= \frac{16}{7 \times 0{\cdot}25} = \frac{64}{7} \text{ rad/s}$$

Kinetic energy of translation $= \frac{1}{2} mv^2$
Kinetic energy of rotation $= \frac{1}{2} I\omega^2$
Kinetic energy gained = Potential energy lost

$$\tfrac{1}{2} mv^2 + \tfrac{1}{2} I\omega^2 = 94{\cdot}18$$

$$\tfrac{1}{2} \times 24 \times \frac{16^2}{7^2} + \tfrac{1}{2} \times I \times \frac{64^2}{7^2} = 94{\cdot}18$$

$$24 + I \times 4^2 = \frac{94{\cdot}18 \times 2 \times 7^2}{16^2}$$

$$24 + 16I = 36{\cdot}06$$

$$I = 0{\cdot}7537 \text{ kgm}^2$$ Ans.

34. Force of each ram = pressure × area

$$= 75 \times 10^5 \times 0{\cdot}7854 \times 0{\cdot}3^2$$

Force applied at right angles to tiller

$$= \frac{75 \times 10^5 \times 0{\cdot}7854 \times 0{\cdot}3^2}{\cos 35^\circ}$$

$$= 6{\cdot}471 \times 10^5 \text{ N}$$

Leverage from block to rudder stock centre

$$= \frac{750}{\cos 35^\circ} = 915{\cdot}6 \text{ mm} = 0{\cdot}9156 \text{ m}$$

Torque in rudder stock due to 2 rams

$$= 6{\cdot}471 \times 10^5 \times 0{\cdot}9156 \times 2$$

$$= 1{\cdot}185 \times 10^6 \text{ N m}$$

For a solid round section subject to twisting

$$T = \frac{\pi}{16} D^3 \tau$$

$$\text{Diameter of stock} = \sqrt[3]{\frac{16 \times 1{\cdot}185 \times 10^6}{\pi \times 70 \times 10^2}}$$

$$= 0{\cdot}4418 \text{ m} = 441{\cdot}8 \text{ mm} \qquad \text{Ans. } (i)$$

Length from block to section of tiller arm considered

$$= 915{\cdot}6 - 600$$

$$= 315{\cdot}6 \text{ mm} = 0{\cdot}3156 \text{ m}$$

Bending moment at this section

$$= 6{\cdot}471 \times 10^5 \times 0{\cdot}3156$$

$$= 2{\cdot}042 \times 10^5 \text{ N m}$$

For a solid round section subject to bending

$$\sigma = \frac{32M}{\pi D^3}$$

$$\text{Diameter of tiller arms} = \sqrt[3]{\frac{32 \times 2{\cdot}042 \times 10^5}{\pi \times 100 \times 10^6}}$$

$$= 0{\cdot}275\text{m} = 275 \text{ mm} \qquad \text{Ans } (ii)$$

35. Since meter is horizontal, when the water flows from entrance to throat:

$$\text{Loss of pressure energy} = \text{gain in kinetic energy}$$

$$p_1V - p_2V = \tfrac{1}{2} mv_2^2 - \tfrac{1}{2} mv_1^2$$

$$V(p_1 - p_2) = \tfrac{1}{2} m\,(v_2^2 - v_1^2)$$

mass $m \div$ volume V = density ρ therefore, dividing throughout by V:

$$p_1 - p_2 = \tfrac{1}{2} \rho\,(v_2^2 - v_1^2) \qquad \dots\ \dots\ (i)$$

Volume flow rate through the meter is constant, hence

$$a_1 v_1 = a_2 v_2$$

$$v_2 = \frac{0{\cdot}7854 \times 0{\cdot}15^2}{0{\cdot}754 \times 0{\cdot}05^2} \times v_1$$

$$= 3^2 \times v_1 = 9v_1$$

$$\text{From } p = \rho g h$$

$$\text{Pressure difference } p_1 - p_2 = h \times 9{\cdot}81 \times 12{\cdot}6 \times 10^3$$

$$= 0{\cdot}05 \times 9{\cdot}81 \times 12{\cdot}6 \times 10^3$$

From (i)

$$p_1 - p_2 = \tfrac{1}{2}\rho\,(v_2{}^2 - v_1{}^2)$$

$$50 \times 12{\cdot}6 \times 9{\cdot}81 = \tfrac{1}{2} \times 10^3\,(v_2{}^2 - v_1{}^2)$$

$$12{\cdot}36 = v_2{}^2 - v_1{}^2$$

$$12{\cdot}36 = (9v_1)^2 - v_1{}^2$$

$$12{\cdot}36 = 80v_1{}^2$$

$$v_1 = 0{\cdot}3931 \text{ m/s}$$

$$\dot{V} = a\text{v}$$

$$\dot{m} = \dot{V}\rho$$

$$\text{Actual mass flow} = 0{\cdot}9 \times \text{theoretical mass flow}$$

$$= 0{\cdot}9 \times 0{\cdot}7854 \times 0{\cdot}15^2 \times 0{\cdot}3931 \times 3600 \times 10^3$$

$$= 2{\cdot}251 \times 10^4 \text{ kg/h}$$

$$= 22{\cdot}51 \text{ t/h}$$

Ans.

36.

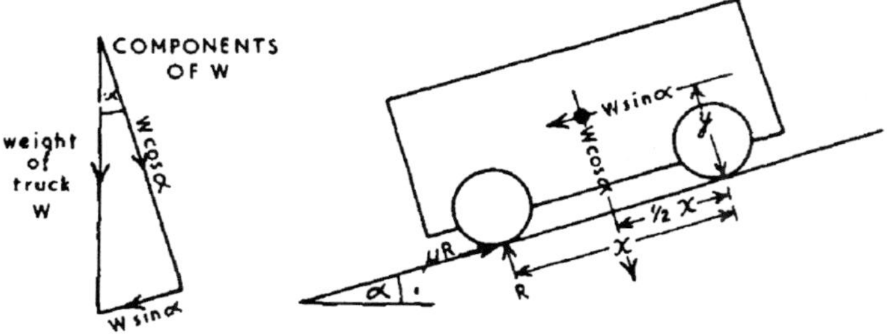

Components of weight of truck W, parallel, and normal, to the plane, are $W \sin \alpha$ and $W \cos \alpha$ respectively, acting through the centre of gravity as shown in Fig.

Let R = normal reaction of plane at point of contact of rear wheels.

Friction force opposing sliding $= \mu R$

Equating forces parallel to plane:

$$\text{Forces up plane} = \text{Forces down plane}$$

$$\mu R = W \sin \alpha \qquad \dots \dots \dots (i)$$

Taking moments about point of contact of front wheels:

$$R \times x = W \sin \alpha \times y + W \cos \alpha \times \tfrac{1}{2} x$$

$$R \times 3 = W \sin \alpha \times 1{\cdot}5 + W \cos \alpha \times 1{\cdot}5 \quad (ii)$$

Values of R from (*i*) and (*ii*) can now be equated to eliminate this unwanted unknown:

From (*i*), $$R = \frac{W \sin \alpha}{\mu} = \frac{W \sin \alpha}{0{\cdot}45}$$

From (*ii*) $$R = 0{\cdot}5\, W \sin \alpha + 0{\cdot}5\, W \cos \alpha$$

Equating, cancelling W, and evaluating α:

$$\frac{W \sin \alpha}{0{\cdot}45} = 0{\cdot}5\, W \sin \alpha + 0{\cdot}5\, W \cos \alpha$$

$$\sin \alpha = 0{\cdot}225 \sin \alpha + 0{\cdot}225 \cos \alpha$$

$$0{\cdot}775 \sin \alpha = 0{\cdot}225 \cos \alpha$$

$$\frac{\sin \alpha}{\cos \alpha} = \frac{0{\cdot}225}{0{\cdot}775}$$

$$\tan \alpha = 0{\cdot}2903$$

$$\alpha = 16^\circ 11' \qquad \text{Ans.}$$

37.

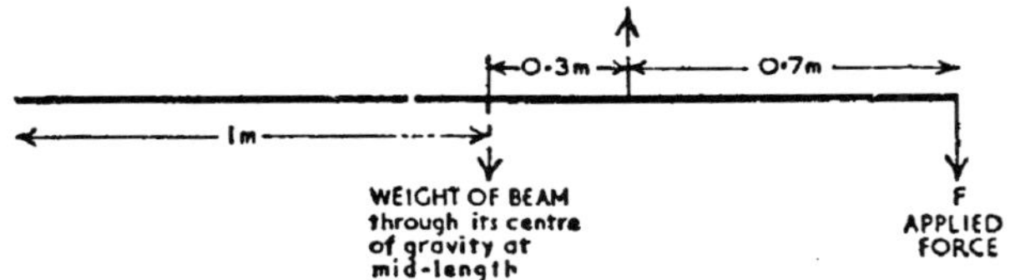

$$\text{Mass of beam} = \text{volume} \times \text{density}$$
$$= 0{\cdot}15^2 \times 2 \times 7{\cdot}86 \times 10^3$$
$$= 353{\cdot}7 \text{ kg}$$

$$\text{Weight of beam} = 353{\cdot}7 \times 9{\cdot}81$$
$$= 3470 \text{ N} = 3{\cdot}47 \text{ kN}$$

Moments about point of suspension

$$F \times 0{\cdot}7 = 3{\cdot}47 \times 0{\cdot}3$$
$$F = 1{\cdot}487 \text{ kN}$$ Ans. (*i*)

$$\text{Tension in rope} = \text{total downward forces}$$
$$= 3{\cdot}47 + 1{\cdot}487$$
$$= 4{\cdot}957 \text{ kN}$$ Ans. (*ii*)

Bending moment at point of suspension (taking moments to the left)

$$= \text{weight of } 1{\cdot}3 \text{ m of beam} \times \text{distance of its C.G.}$$

$$= \frac{1{\cdot}3}{2} \times 3{\cdot}47 \times \frac{1{\cdot}3}{2}$$

$$= 1{\cdot}466 \text{ kN m}$$

$$\frac{M}{I} = \frac{\sigma}{y} \qquad \sigma = \frac{my}{I}$$

For a solid rectangular beam $I = \dfrac{BD^3}{12}$ $\qquad y = \dfrac{D}{2}$

$$= \frac{6M}{BD^2}$$

$$= \frac{6 \times 1{\cdot}466 \times 10^3}{0{\cdot}15 \times 0{\cdot}15^2}$$

$$= 2{\cdot}606 \times 10^6 \text{ N/m}^2$$

or $2{\cdot}606$ MN/m^2 Ans. (*iii*)

38. Force on spring = area of piston × pressure

$$= 0{\cdot}7854 \times 0{\cdot}2^2 \times 0{\cdot}105 \times 10^5$$

$$= 329{\cdot}9 \text{ N}$$

Torque applied to spring = force × mean radius of coils

$$= 329{\cdot}9 \times 0{\cdot}075$$

$$= 24{\cdot}75 \text{ N m}$$

Ans. (*i*)

From $\frac{T}{J} = \frac{\tau}{r} = \frac{G\theta}{l}$ $\qquad \tau = \frac{Tr}{J}$

Working in mm:

$$T = 24{\cdot}75 \times 10^3 \text{ N mm}$$

$$r = \text{outer radius of material} = 6{\cdot}5 \text{ mm}$$

$$J = \frac{\pi}{32}(D^4 - d^4) = \frac{\pi}{32}(D^2 + d^2)(D^2 + d^2)$$

$$= \frac{\pi}{32}(13^2 + 10^2)(13^2 - 10^2)$$

$$= \frac{\pi}{32} \times 269 \times 69 \text{ mm}^4$$

$$\tau = \frac{Tr}{J}$$

$$= \frac{32 \times 24{\cdot}75 \times 10^3 \times 6{\cdot}5}{\pi \times 269 \times 69}$$

$$= 88{\cdot}26 \text{ N/mm}^2 = 88{\cdot}26 \times 10^6 \text{ N/m}^2$$

$$= 88{\cdot}26 \text{ MN/m}^2$$

Ans. (*ii*)

From $\frac{T}{J} = \frac{\tau}{r} = \frac{G\theta}{l}$ $\qquad \theta = \frac{\tau l}{rG}$

$$\theta = \frac{88{\cdot}26 \times 5 \times 10^3}{6{\cdot}5 \times 103 \times 10^3}$$

$$\begin{aligned} \text{Energy stored} &= \text{work done to twist spring} \\ &= \tfrac{1}{2} T\,\theta \\ &= \tfrac{1}{2} \times 24{\cdot}75 \times 0{\cdot}6594 \\ &= 8{\cdot}16 \text{ J} \end{aligned}$$

Ans. (*iii*)

39. When coming to rest, linear velocity of mass changes from 0·6 m/s to zero, kinetic energy is therefore given out equal to $\frac{1}{2}mv^2$

This energy is absorbed by the cable in the form of elastic strain energy (resilience),

$$\text{Resilience} = \frac{\sigma^2 v}{2\,E}$$

$$\text{Energy taken up by cable} = \text{Energy given out by mass}$$

$$\frac{\sigma^2 v}{2E} = \frac{mv^2}{2}$$

$$\frac{\sigma^2 \times 12 \times 645 \times 10^{-6}}{2 \times 207 \times 10^9} = \frac{1000 \times 0{\cdot}6^2}{2}$$

$$\sigma^2 = \frac{2 \times 207 \times 10^9 \times 1000 \times 0{\cdot}6^2}{12 \times 645 \times 10^{-6} \times 2}$$

$$\sigma = \sqrt{\frac{207 \times 0{\cdot}6^2 \times 2^{18}}{12 \times 645}}$$

$$= 9{\cdot}812 \times 10^7 \text{ N/m}^2$$

$$= 98{\cdot}12 \text{ MN/m}^2$$

Ans. (*i*)

$$\text{Strain} = \frac{\text{extension}}{\text{length}} = \frac{\text{stress}}{E}$$

$$\text{Extension} = \frac{\text{stress} \times \text{length}}{E}$$

$$= \frac{9{\cdot}812 \times 10^7 \times 12}{207 \times 10^9}$$

$$= 5{\cdot}689 \times 10^{-3} \text{ m} = 5{\cdot}689 \text{ mm}$$

Ans. *(ii)*

40. Depth inside $= 160 - (2 \times 20) = 120$ mm

Width inside $= 120 - (2 \times 10) = 100$ mm

Mass of beam $=$ volume $\times$ density

$$= (0{\cdot}16 \times 0{\cdot}12 - 0{\cdot}12 \times 0{\cdot}1) \times 2 \times 7{\cdot}86 \times 10^3$$

$$= 113{\cdot}2 \text{ kg}$$

Weight of beam $= 113{\cdot}2 \times 9{\cdot}81$ N

$$I = \frac{BD^2 - bd^2}{12}$$

$$= \frac{0{\cdot}12 \times 0{\cdot}16^3 - 0{\cdot}1 \times 0{\cdot}12^3}{12}$$

$$= 2{\cdot}656 \times 10^{-5} \text{ m}^4$$

Let m = mass concentrated on free end of beam

$$M_{max} \text{ at wall} = 113{\cdot}2 \times 9{\cdot}81 \times 1 + m \times 9{\cdot}81 \times 2$$

$$= 9{\cdot}81\,(113{\cdot}2 + 2m)$$

$$\frac{M}{I} = \frac{\sigma}{y}$$

$$\frac{9{\cdot}81\,(113{\cdot}2 + 2m)}{2{\cdot}656 \times 10^{-5}} = \frac{40 \times 10^6}{0{\cdot}08}$$

$$113{\cdot}2 + 2m = \frac{40 \times 10^6 \times 2{\cdot}656}{9{\cdot}81 \times 0{\cdot}08 \times 10^5}$$

$$2m = 1354 - 113{\cdot}2$$

$$m = 620{\cdot}4 \text{ kg}$$

Ans.

41.

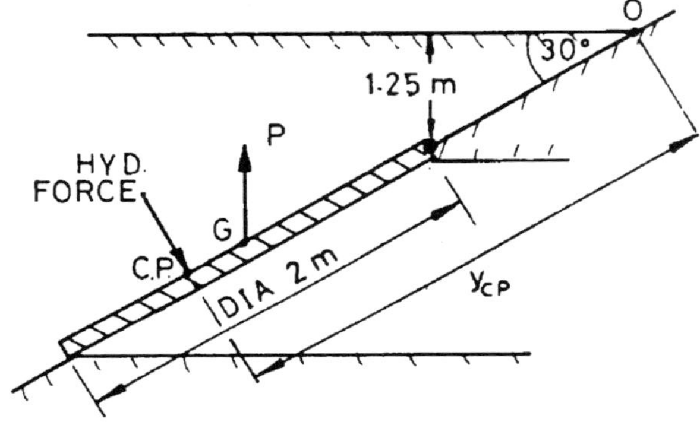

(a) Hydrostatic load $= HA\rho g$

$$= (1{\cdot}25 + 1 \times \sin 30°) \times \frac{\pi}{4} \times 2^2 \times 1000 \times 9{\cdot}81$$

$$= 1{\cdot}75 \times \frac{\pi}{4} \times 4 \times 1000 \times 9{\cdot}81$$

$$= 53{\cdot}93 \text{ kN.}$$ Ans. (a)

(b)

$$y_{cp} = \frac{I_G}{A \times \dfrac{H}{\sin\theta}} + \frac{H}{\sin\theta}$$

$$\frac{\dfrac{\pi \times 2^4}{64}}{\dfrac{\pi D^2}{4} + \dfrac{1{\cdot}75}{\sin 30°}} + \frac{1{\cdot}75}{\sin 30°}$$

$$= 3{\cdot}571 \text{ m}$$

i.e. The *vertical* depth to c.o.p.

$$= 3{\cdot}571 \times \sin 30°$$

$$= 1{\cdot}785 \text{ m}$$ Ans. (b)

(c) Let P = vertical force to open the flap valve.

Take moments about the hinge.

$$P \times 1 \cos 30° = 53{\cdot}93 \times \left(3{\cdot}571 - \frac{1{\cdot}25}{\sin 30°} \right)$$

Force required $= 66{\cdot}7$ kN Ans. (c)

42. Let cross-sectional area of the steel $= A$

then cross-sectional area of the brass $= 2^2 \times A = 4A$

Let modulus of elasticity of brass $= E$

then modulus of elasticity of steel $= 2E$

As bar remains horizontal, extension of each wire must be equal, therefore their strains are equal.

strain in steel = strain in brass

$$\frac{\text{stress in steel}}{E \text{ for steel}} = \frac{\text{stress in brass}}{E \text{ for brass}}$$

$$\frac{\text{load}_S}{\text{area}_S \times E_S} = \frac{\text{load}_B}{\text{area}_B \times E_B}$$

$$\frac{\text{load carried by steel}}{\text{load carried by brass}} = \frac{\text{area}_S \times E_B}{\text{area}_B \times E_B}$$

$$= \frac{A \times 2E}{4\,A \times E} = \tfrac{1}{2}$$

Steel wire carries one-third of mass
Brass wire carries two-thirds of mass } Ans. (*ii*)

The mass must be hung on the bar at one-third of its length from the brass wire, that is, at 0·5 m from the brass wire attachment. Ans.(*i*)

43. In the hose, power loss $= 0{\cdot}5$ kW. (*i*)

In the nozzle, power loss $= 0{\cdot}12 \times$ K.E. gain/s

$= 0{\cdot}12 \times \frac{1}{2}\,\dot{m}\,(v_2^2 - v_1^2)$

$=$ where $\dot{m} =$ mass flow rate

$$= 0{\cdot}12 \times \tfrac{1}{2} \times 8 \times (37^2 - 7^2)$$

$$= 0{\cdot}6336 \text{ kW.} \qquad (ii)$$

Manometric head supplied by pump = suction head + discharge head + velocity head

$$= z_S + z_D + \frac{v_2^{\,2}}{2g}$$

$$= 0 + 20 + \frac{37^2}{2g}$$

$$= 89{\cdot}77 \text{ m}$$

Power to supply this head $= \dot{m}gh$

$$= 8 \times 9{\cdot}81 \times 89{\cdot}77$$

$$= 7{\cdot}046 \text{ kW.} \qquad (iii)$$

From (*i*), (*ii*) and (*iii*)

Total output power $= 0{\cdot}5 + 0{\cdot}6336 + 7{\cdot}046$

$$= 8{\cdot}18 \text{ kW}$$

Pump driving power $= \dfrac{8{\cdot}18}{0{\cdot}8}$

$$= 10{\cdot}22 \text{ kW} \qquad \text{Ans.}$$

44. When flow rate is 3340 m³/h and output signal is $10{\cdot}3 \times 10^4$ N/m²

$$Q = ACd^2 \left(\frac{h}{\rho}\right)^{\frac{1}{2}}$$

$$3340 = 125{\cdot}2 \times 10^{-4} \times 0{\cdot}596 \times 180^2 \times \left(\frac{h}{\rho}\right)^{\frac{1}{2}}$$

$$\left(\frac{h}{1{\cdot}22}\right)^{\frac{1}{2}} = \frac{3340 \times 10^4}{125{\cdot}2 \times 0{\cdot}596 \times 180^2} = 13{\cdot}81$$

$$h = 13{\cdot}81^2 \times 1{\cdot}22 = 232{\cdot}7 \text{ mm water}$$

Putting this value into the second expression to find the value of the constant k,

$$S = (2{\cdot}1 \times 10^4) + kh$$

$$10{\cdot}3 \times 10^4 = (2{\cdot}1 \times 10^4) + k \times 232{\cdot}7$$

$$k = \frac{(10{\cdot}3 - 2{\cdot}1) \times 10^4}{232{\cdot}7} = 352{\cdot}4$$

When output signal is $4{\cdot}2 \times 10^4$ N/m^2, to find the value of h,

$$S = (2{\cdot}1 \times 10^4) + kh$$

$$4{\cdot}2 \times 10^4 = (2{\cdot}1 \times 10^4) + 352{\cdot}4 \times h$$

$$h = \frac{(4{\cdot}2 - 2{\cdot}1) \times 10^4}{352{\cdot}4} = 59{\cdot}6 \text{ mm}$$

To find Q when $h = 59{\cdot}6$

$$Q = ACd^2 \left(\frac{h}{\rho}\right)^{\frac{1}{2}}$$

$$= 125{\cdot}2 \times 10^{-4} \times 0{\cdot}596 \times 180^2 \times \left(\frac{59{\cdot}6}{1{\cdot}22}\right)^{\frac{1}{2}}$$

$$= 1690 \text{ m}^3\text{/h}$$ Ans.

45. For a body moving with S.H.M.,

$$v = \omega\sqrt{r^2 - x^2}$$

$$a = \omega^2 x$$

Crank moves one revolution for each double stroke

$$\omega = \frac{70 \times 2\pi}{60}$$

$$= 7{\cdot}331 \text{ rad/s}$$

$$x = \tfrac{1}{2} \text{ stroke} - 15 \text{ mm}$$

$$= 100 - 15 = 85 \text{ mm} = 0{\cdot}085$$

$$a = \omega^2 x$$

$$= 7{\cdot}331^2 \times 0{\cdot}085 = 4{\cdot}568 \text{ m/s}^2$$

$$F = ma$$

$$= 30 \times 4{\cdot}568 = 137 \text{ N}$$

$$\text{Weight lifted} = mg$$

$$= 30 \times 9{\cdot}81 = 294{\cdot}3 \text{ N}$$

$$\text{Discharge force} = \text{press} \times \text{piston area}$$

$$= 80 \times 10^3 \times 0{\cdot}7854 \times 0{\cdot}15^2$$

$$= 1414 \text{ N}$$

$$\text{Total force} = \text{weight + accel. force + disch. force}$$

$$= 294{\cdot}3 + 137 + 1414$$

$$= 1845{\cdot}3 \text{ N or } 1{\cdot}8453 \text{ kN} \qquad \text{Ans. } (i)$$

$$\text{v} = \omega \sqrt{r^2 - x^2}$$

$$= 7{\cdot}331 \sqrt{0{\cdot}1^2 - 0{\cdot}085^2}$$

$$= 0{\cdot}3862 \text{ m/s}$$

$$\rho = F\text{v}$$

$$= 1845{\cdot}3 \times 0{\cdot}3862$$

$$= 712{\cdot}7 \text{ W} \qquad \text{Ans. } (ii)$$

46.

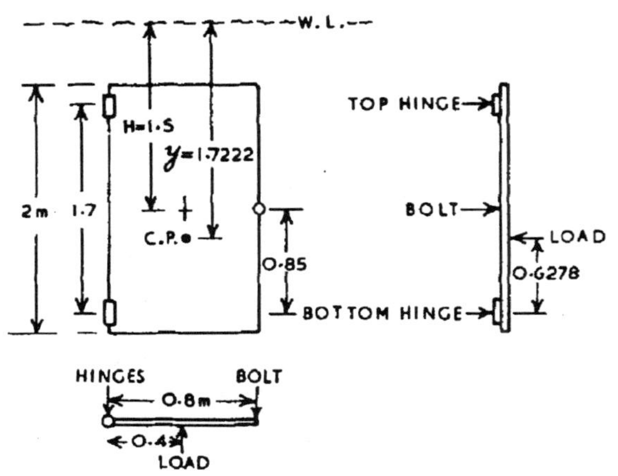

$$\text{Load on immersed area} = HA\rho g$$

$$H = 2{\cdot}5 - \tfrac{1}{2} \times 2 = 1{\cdot}5 \text{ m}$$

$$A = 2 \times 0{\cdot}8 = 1{\cdot}6 \text{ m}^2$$

$$\text{Load on door} = 1{\cdot}5 \times 1{\cdot}6 \times 1024 \times 9{\cdot}81$$

$$= 2{\cdot}411 \times 10^4 \text{ N} = 24{\cdot}11 \text{ kN}$$

From first principles:

Centre of pressure below water level $= y_{cp}$

$$y_{cp} = \frac{\text{2nd moment of area}}{\text{1st moment of area}}$$

$$I_{CG} \text{ of rectangle} = \frac{BD^3}{12}$$

By theorem of parallel axis, second moment of area about water level,

$$I_{WL} = I_{CG} + AH^2 = \frac{BD^3}{12} + BDH^2$$

First moment of area about water level $= BD \times H$

$$y_{cp} = \frac{BD^3}{12 \times BDH} + \frac{BD^2}{BDH}$$

$$= \frac{D^2}{12H} + H = \frac{2^2}{12 \times 1{\cdot}5} + 1{\cdot}5$$

$$= 0{\cdot}2222 + 1{\cdot}5 = 1{\cdot}7222 \text{ m}$$

which is $0{\cdot}85 - 0{\cdot}2222 = 0{\cdot}6278$ m above bottom hinge

Referring to plan view, moments about hinges,

$$\text{Load on door} \times 0{\cdot}4 = \text{Force on bolt} \times 0{\cdot}8$$

$$\text{Force on bolt} = \frac{24{\cdot}11 \times 0{\cdot}4}{0{\cdot}8} = 12{\cdot}05 \text{ kN} \qquad \text{Ans.}$$

Referring to end elevation, moments about bottom hinge,

$$\text{Force on top hinge} \times 1{\cdot}7 + 12{\cdot}05 \times 0{\cdot}85 = 24{\cdot}11 \times 0{\cdot}6278$$

$$\text{Top hinge force} \times 1{\cdot}7 = 15{\cdot}14 - 10{\cdot}24$$

$$\text{Top hinge force} = 2{\cdot}882 \text{ kN} \qquad \text{Ans.}$$

$$\text{Forces to right} = \text{Forces to left}$$

$$\text{Bottom hinge force} + 2{\cdot}882 + 12{\cdot}05 = 24{\cdot}11$$

$$\text{Bottom hinge force} = 9{\cdot}178 \text{ kN} \qquad \text{Ans.}$$

47. Cross-sectional area of 2 steel wires

$= 2 \times 0{\cdot}7854 \times 2{\cdot}8^2 = 12{\cdot}32 \text{ mm}^2 = 12{\cdot}32 \times 10^{-6} \text{ m}^2$

Cross-sectional area of 6 bronze wires

$= 6 \times 0{\cdot}7854 \times 2{\cdot}5^2 = 29{\cdot}45 \text{ mm}^2 = 29{\cdot}45 \times 10^{-6} \text{ m}^2$

To find the separate loads carried by the steel wires and the bronze wires and then to find the stresses in them:

$$\text{Strain in steel} = \text{Strain in bronze}$$

$$\frac{\text{stress}_S}{E_S} = \frac{\text{stress}_B}{E_B}$$

$$\frac{\text{load}_S}{\text{area}_S \times E_S} = \frac{\text{load}_B}{\text{area}_B \times E_B}$$

$$\text{load}_S = \text{load}_B \times \frac{12{\cdot}32 \times 10^{-6} \times 210 \times 10^9}{29{\cdot}45 \times 10^{-6} \times 84 \times 10^9}$$

$$\text{load}_S = 1{\cdot}046 \times \text{load}_B \qquad \dots \dots \dots \dots (i)$$

$$\text{Also, load}_S = \text{total load} - \text{load}_B$$

$$= 4{\cdot}5 \times 10^3 - \text{load}_B \qquad \dots \dots \dots \dots (ii)$$

Equating (*i*) and (*ii*),

$$1{\cdot}046 \times \text{load}_B = 4500 - \text{load}_B$$

$$2{\cdot}046 \times \text{load}_B = 4500$$

$$\text{load}_B = 2199 \text{ N}$$

From (*ii*),

$$\text{load}_S = 4500 - 2199 = 2301 \text{ N}$$

$$\text{stress}_S = \frac{\text{load}_S}{\text{area}_S} = \frac{2301}{12{\cdot}32 \times 10^{-6}}$$

$$= 1{\cdot}868 \times 10^8 \text{ N/m}^2 \text{ or } 186{\cdot}8 \text{ MN/m}^2$$

Ans. (*ia*)

$$\text{stress}_B = \frac{\text{load}_B}{\text{area}_B} = \frac{2199}{29{\cdot}45 \times 10^{-6}}$$

$$= 7{\cdot}467 \times 10^7 \text{ N/m}^2 \text{ or } 74{\cdot}67 \text{ MN/m}^2$$

Ans. (*ib*)

Equivalent modulus of elasticity for the composite rope:
From

$$E = \frac{\text{stress}}{\text{strain}} = \frac{\text{load}}{\text{area} \times \text{strain}}$$

$$\text{Equivalent } E = \frac{\text{total load}}{\text{total area}} \times \frac{1}{\text{strain}}$$

The strain can be obtained from either of the values for steel or bronze because their strains are equal, thus,

$$\text{strain} = \frac{\text{stress}}{E}$$

$$= \frac{186{\cdot}8 \times 10^6}{210 \times 10^9} \quad \text{or} \quad \frac{74{\cdot}67 \times 10^6}{84 \times 10^9}$$

$$\text{Equivalent } E = \frac{4{\cdot}5 \times 10^3 \times 210 \times 10^9}{(12{\cdot}32 + 29{\cdot}45) \times 10^{-6} \times 10^{-6} \times 186{\cdot}8 \times 10^6}$$

$$= \frac{4{\cdot}5 \times 210}{41{\cdot}77 \times 186{\cdot}8} \times 10^{12}$$

$$= 1{\cdot}211 \times 10^{11} \text{ N/m}^2 \text{ or } 121{\cdot}1 \text{ GN/m}^2$$ Ans. (*ii*)

Alternatively, the equivalent E could be found by the following expression as explained in Chapter 9:

$$\text{Equivalent } E = \frac{A_A\, E_A + A_B\, E_B}{A_A + A_B}$$

48. Referring to Fig., the distances from the hinge of the tie-rod connections to the beam are in ratio 5:1 therefore, since the beam remains rigid, elongation of rod B is 5 times elongation of rod A.

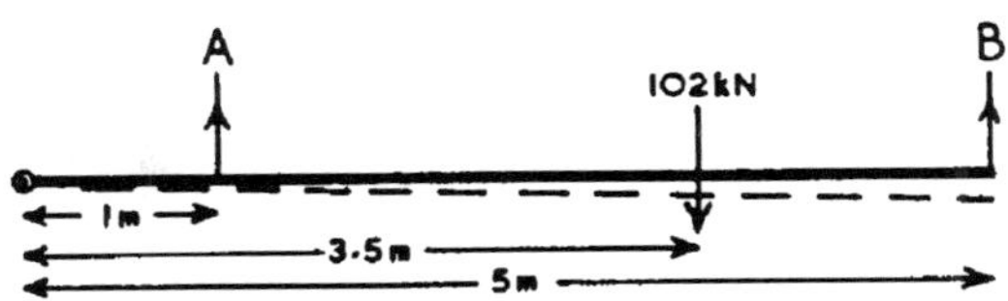

$$\text{Strain} = \frac{\text{stress}}{E} \quad \text{and} \quad \text{Strain} = \frac{\text{extension}}{\text{length}}$$

$$\text{extension} = \frac{\text{stress} \times \text{length}}{E} = \frac{\text{load} \times \text{length}}{\text{area} \times E}$$

$$\text{extension of B} = (\text{extension of A}) \times 5$$

$$\frac{\text{load}_B \times 6}{0{\cdot}7854 \times 0{\cdot}02^2 \times E_B} = \frac{\text{load}_A \times 3 \times 5}{0{\cdot}7854 \times 0{\cdot}01^2 \times E_A}$$

$$\text{load}_B = \frac{\text{load}_A \times 3 \times 5 \times 0{\cdot}02^2}{6 \times 0{\cdot}01^2}$$

$$\text{load}_B = 10 \times \text{load}_A \qquad \ldots \ldots \ldots (i)$$

Moments about hinge:

$$\text{load}_A \times 1 + \text{load}_B \times 5 = 102 \times 3{\cdot}5$$

$$\text{load}_A = 357 - 5 \times \text{load}_B \qquad \ldots \ldots \ldots \ldots (ii)$$

Substituting for load_B from (*i*) into (*ii*),

$$\text{load}_A = 357 - 5 \times 10 \times \text{load}_A$$

$$51 \times \text{load}_A = 357$$

$$\text{load}_A = 7 \text{ kN}$$

$$\text{load}_B = 10 \times \text{load}_A = 70 \text{ kN}$$

$$\text{Stress}_A = \frac{\text{load}}{\text{area}} = \frac{7 \times 10^3}{0{\cdot}7854 \times 0{\cdot}01^2}$$

$$= 8{\cdot}913 \times 10^7 \text{ N/m}^2 \text{ or } 89{\cdot}13 \text{ MN/m}^2$$

Ans. (*ia*)

$$\text{stress}_B = \frac{70 \times 10^3}{0.7854 \times 0.02^2}$$

$$= 2.228 \times 10^8 \text{ N/m}^2 \text{ or } 222.8 \text{ MN/m}^2$$

Ans. (*ib*)

Maximum deflection

$$= \text{extension of rod B}$$

$$= \frac{\text{stress} \times \text{length}}{E}$$

$$= \frac{2.228 \times 10^8 \times 6}{206 \times 10^9}$$

$$= 6.489 \times 10^{-3} \text{ m or } 6.489 \text{ mm}$$

Ans. (*ii*)

49. Height of exit vertically above inlet

$$= 500 \sin 45° = 353.5 \text{ mm} = 0.3535 \text{ m}$$

$$z_1 + \frac{p_1}{w} + \frac{v_1^2}{2g} = z_2 + \frac{p_2}{w} + \frac{v_2^2}{2g}$$

Taking inlet as datum level, $z_1 = 0$, $z_2 = 0.3535$ m

Pressure at exit, p_2 = nil

$$\rho = 10^3 \text{ kg/m}^3$$

$$w = 10^3 \times 9.81 \text{ N/m}^3$$

$$0 + \frac{140 \times 10^3}{10^3 \times 9.81} + \frac{v_1^2}{2 \times 9.81} = 0.3535 + 0 + \frac{v_2^2}{2 \times 9.81}$$

$$280 + v_1^2 = 6.936 + v_2^2$$

$$v_2^2 = v_1^2 + 273.064 \qquad \dots \dots (i)$$

Since volume flow rate remains constant:

$$v_1 \times a_1 = v_2 \times a_2$$

$$v_2 = v_1 \times \frac{75^2}{25^2} = 9v_1 \qquad \dots \dots (ii)$$

Substituting value of v_2 from (*ii*) into (*i*),

$$81v_1^2 = v_1^2 + 273{\cdot}064$$

$$v_1 = \sqrt{\frac{273{\cdot}064}{80}} = 1{\cdot}848 \text{ m/s}$$

$$v_2 = 9 \times 1{\cdot}848 = 16{\cdot}63 \text{ m/s} \quad \text{Ans. } (a)$$

$$\dot{V} = a\text{v}$$

$$= 0{\cdot}7854 \times 25^2 \times 10^{-6} \times 16{\cdot}63$$

$$= 8{\cdot}163 \times 10^{-3} \text{ m}^3\text{/s}$$

$$\dot{m} = \dot{V}\rho$$

$$= 8{\cdot}163 \times 10^{-3} \times 10^3$$

$$= 8{\cdot}163 \text{ kg/s} \quad \text{Ans. } (b)$$

50.

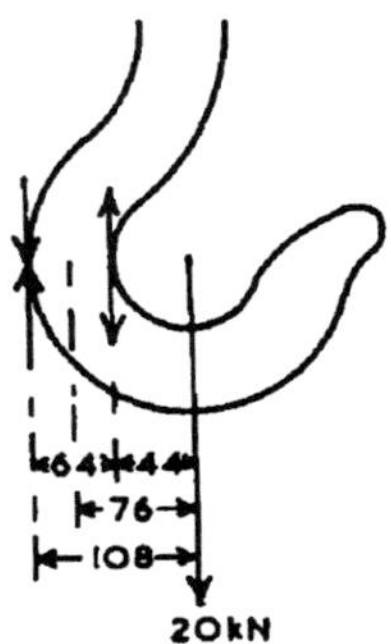

$$\text{Area of elliptical section} = 0{\cdot}7854\, bd$$

$$= 0{\cdot}7854 \times 38 \times 64 = 1910 \text{ mm}^2$$

$$\text{Direct stress} = \frac{\text{load}}{\text{area}}$$

$$= \frac{20 \times 10^3}{1910 \times 10^{-6}}$$

$$= 1{\cdot}047 \times 10^7 \text{ N/m}^2 = 10{\cdot}47 \text{ MN/m}^2$$

From $\dfrac{M}{I} = \dfrac{\sigma}{y}$ Bending stress $\sigma = \dfrac{My}{I}$

$$M = \text{load} \times \text{eccentricity}$$

$$= 20 \times 10^3 \times 76 \times 10^{-3} = 1520 \text{ N m}$$

$$I = \frac{\pi}{64} bd^3 = \frac{\pi}{64} \times 38 \times 64^3 \times 10^{-12} \text{ m}^4$$

$$= 4{\cdot}89 \times 10^{-7} \text{ m}^4$$

$$y = \tfrac{1}{2} \times 64 \times 10^{-3} = 32 \times 10^{-3} \text{ m}$$

$$\sigma = \frac{My}{I}$$

$$= \frac{1520 \times 32 \times 10^{-3}}{4{\cdot}89 \times 10^{-7}}$$

$$= 9{\cdot}947 \times 10^7 \text{ N/m}^2 = 99{\cdot}47 \text{ MN/m}^2$$

The stress due to bending is tensile at the inner edge and compressive at the outer edge

Stress at inner edge = 99·47 tensile + 10·47 tensile

Max. tensile stress = 109·94 MN/m² Ans. (*i*)

Stress at outer edge = 99·47 compressive – 10·47 tensile

Max. compressive stress = 89 MN/m² Ans. (*ii*)

51. Total torque = torque in steel + torque in bronze

$$6000 \text{ N m} = T_S + T_B$$

$$T_B = 6000 - T_S \qquad \ldots \ldots \ldots \ldots (i)$$

Angle of twist in each bar is the same:

From $\dfrac{T}{J} = \dfrac{G\theta}{l}$ $\theta = \dfrac{Tl}{GJ}$

$$\theta \text{ for steel} = \theta \text{ for bronze}$$

$$\frac{Tl}{GJ}\text{ for steel} = \frac{Tl}{GJ}\text{ for bronze}$$

$$\frac{T_S \times 1{\cdot}5 \times 32 \times 10^{12}}{84 \times 10^9 \times \pi \times 50^4} = \frac{T_B \times 2{\cdot}5 \times 32 \times 10^{12}}{42 \times 10^9 \times \pi \times 75^4}$$

$$T_B = \frac{1{\cdot}5 \times 42 \times 75^4}{2{\cdot}5 \times 84 \times 50^4} \times T_S$$

$$T_B = 1{\cdot}519\, T_S \qquad \ldots\ \ldots\ \ldots\ (ii)$$

Equating (*i*) and (*ii*),

$$1{\cdot}519\, T_S = 6000 - T_S$$

$$2{\cdot}519\, T_S = 6000$$

$$\text{Torque in steel} = 2382\text{ N m} \qquad \text{Ans. }(a)(i)$$

From (*i*)

$$\text{Torque in bronze} = 6000 - 2382 = 3618\text{ N m} \qquad \text{Ans. }(a)(ii)$$

From $\frac{T}{J} = \frac{\tau}{r}$ $\qquad T = \frac{\pi}{16} d^3\tau$ $\qquad \tau = \frac{16T}{\pi d^2}$

$$\text{Stress in steel, } \tau = \frac{16 \times 2382 \times 10^9}{\pi \times 50^3}$$

$$= 9{\cdot}704 \times 10^7\text{ N/m}^2\text{ or }97{\cdot}04\text{ MN/m}^2$$

Ans. (*b*)(*i*)

Stress in bronze

$$= \frac{16 \times 3618 \times 10^3}{\pi \times 75^3}$$

$$= 4{\cdot}367 \times 10^7\text{ N/m}^2\text{ or }43{\cdot}67\text{ MN/m}^2$$

Ans (b)(ii)

Steel and bronze bars twist an equal amount, inserting values for the steel bar:

$$\frac{\tau}{r} = \frac{G\theta}{l} \qquad \theta = \frac{l}{Gr}$$

$$\theta = \frac{9{\cdot}704 \times 10^7 \times 1{\cdot}5}{84 \times 10^9 \times 25 \times 10^{-3}} \times \frac{360}{2\pi}$$

$$= 3{\cdot}97^\circ \qquad \text{Ans. }(c)$$

52.

$$p = h_1 - \frac{xh_2}{A} - \frac{x^2h_3}{A}$$

When $x = 0,\ p = 20{\cdot}7,$

$$20{\cdot}7 = h_1 - 0 - 0 \qquad h_1 = 20{\cdot}7 \quad \ldots\ \ldots\ (i)$$

When $x = 6,\ p = 48$

$$48 = 20{\cdot}7 - \frac{6h_2}{6540} - \frac{36h_3}{6540}$$

$$27{\cdot}3 \times 6540 = -6h_2 - 36h_3$$

$$h_2 = -6h_3 - 29760 \quad \ldots\ \ldots\ (ii)$$

When $x = 25,\ p = 103{\cdot}5,$

$$103{\cdot}5 = 20{\cdot}7 - \frac{25h_2}{6540} - \frac{625h_3}{6540}$$

$$82{\cdot}8 \times 6540 = 25h_2 - 625h_3$$

$$h_2 = -25h_3 - 21660 \quad \ldots\ \ldots\ (iii)$$

Equating (*ii*) and (*iii*),

$$-6h_3 - 29760 = -25h_3 - 21660$$

$$19h_3 = 8100$$

$$h_3 = 426{\cdot}3$$

Substituting for h_3 into (*ii*),

$$h_2 = -6 \times 426{\cdot}3 - 29760$$

$$= -32318$$

When $p = 76$ to find x,

$$p = h_2 - \frac{xh_2}{A} - \frac{x^2\,h_3}{A}$$

$$76 = 20{\cdot}7 - \frac{x \times (-32318)}{6540} - \frac{x^2 \times 426{\cdot}3}{6540}$$

$$55{\cdot}3 \times 6540 = 32318x - 426{\cdot}3x^2$$

$$x^2 - 75{\cdot}81x + 848{\cdot}4 = 0$$

$$x = \frac{75{\cdot}81 \pm \sqrt{5747 - 3394}}{2}$$

$$= \frac{75{\cdot}81 \pm 48{\cdot}51}{2} = \frac{124{\cdot}32}{2} \text{ or } \frac{27{\cdot}3}{2}$$

$$= 62{\cdot}16 \text{ or } 13{\cdot}65$$

$$\text{Valve lift} = 13{\cdot}65 \text{ mm}$$ Ans.

53. $$\dot{V} = \frac{\dot{m}}{\rho}$$

$$= \frac{4{\cdot}5 \times 10^3}{60 \times 1024} = 0{\cdot}07324 \text{ m}^3/\text{s}$$

$$\text{Area of flow} = \frac{\text{volume flow}}{\text{radial velocity}}$$

$$= \frac{0{\cdot}07324}{3} = 0{\cdot}02441 \text{ m}^2$$

$$\text{Width} = \frac{\text{area}}{\text{circumference}}$$

$$\text{Width at entrance} = \frac{0{\cdot}02441}{\pi \times 0{\cdot}1}$$

$$= 0{\cdot}07769 \text{ m} = 77{\cdot}69 \text{ mm}$$ Ans. (*i*)(*a*)

$$\text{Width at exit} = \frac{0{\cdot}02441}{\pi \times 0{\cdot}3} \text{ m}$$

$$\text{or } 77{\cdot}69 \div 3 = 25{\cdot}9 \text{ mm}$$ Ans. (*i*)(*b*).

Referring to inlet velocity diagram of Fig. 224, Chapter 13:

Linear velocity of vanes at entrance $= s_1$

$$s_1 = \text{circumference} \times \text{rev/s}$$

$$= \pi \times 0{\cdot}1 \times 10 = 3{\cdot}142 \text{ m/s}$$

$$\tan \theta_1 = \frac{v_1}{s_1} = \frac{3}{3{\cdot}142} = 0{\cdot}9548$$

Entrance angle of vanes = 43°40′ Ans. (*ii*)

Referring to outlet velocity diagram of Fig. 224:

Linear velocity of vanes at exit = s_2

$$s_2 = \pi \times 0{\cdot}3 \times 10 = 9{\cdot}426 \text{ m/s}$$

$$(s_2 - v_W) = \frac{v_1}{\tan \theta_2} = \frac{3}{\tan 30^\circ} = 5{\cdot}196 \text{ m/s}$$

$$v_W = s_2 - (s_2 - v_W)$$

$$= 9{\cdot}426 - 5{\cdot}196 = 4{\cdot}23 \text{ m/s}$$

Absolute velocity of water leaving vanes = v_2

$$v_2^2 = v_W^2 + v_1^2$$

$$= 4{\cdot}23^2 + 3^2 = 26{\cdot}89$$

$$\text{Kinetic energy} = \tfrac{1}{2} m v_2^2$$

$$K.E. \text{ per kg} = \tfrac{1}{2} \times 1 \times 26{\cdot}89$$

$$= 13{\cdot}45 \text{ J/Kg}$$ Ans. (*iii*)

54. See Chapter 11 for derivation of formulae

Total load on spring = area of valve × pressure

$$= 0{\cdot}7854 \times 0{\cdot}15^2 \times 70 \times 10^3$$

$$= 1{\cdot}237 \times 10^3 \text{ N}$$

$$\text{Deflection } \delta = 210 - 170 = 40 \text{ mm}$$

$$\delta = \frac{8WD^3N}{Gd^4}$$

Working in mm.

$$G = 90 \text{ GN/m}^2 = 90 \text{ kN/mm}^2$$

$$D^3 = \frac{\delta \times G \times d^4}{8 \times W \times N}$$

$$= \frac{40 \times 90 \times 10^3 \times 18^4}{8 \times 1{\cdot}237 \times 10^3 \times 8}$$

$$D = \sqrt[3]{4{\cdot}774 \times 10^6} = 168{\cdot}4 \text{ mm} \qquad \text{Ans. } (i)$$

$$\text{Stress in wire} = \frac{8WD}{\pi d^2}$$

$$= \frac{8 \times 1{\cdot}237 \times 10^3 \times 168{\cdot}4}{\pi \times 18^3}$$

$$= 90{\cdot}94 \text{ N/mm}^2 = 90{\cdot}94 \text{ MN/m}^2 \qquad \text{Ans. } (ii)$$

55.

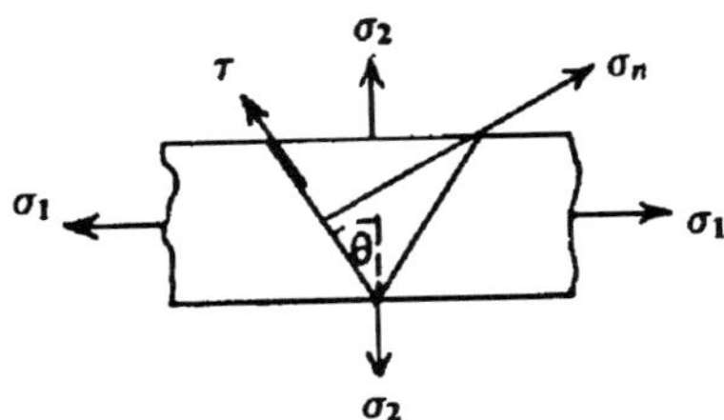

Let σ_1 = stress on circumferential section

$$= \frac{pd}{4t}$$

$$= \frac{3{\cdot}5 \times 10^6 \times 1{\cdot}2}{4 \times 0{\cdot}05} = 21 \times 10^6 \text{ N/m}^2$$

Let σ_2 = stress on longitudinal section

$$= \frac{pd}{2t} = 42 \times 10^6 \text{ N/m}^2$$

Normal tensile stress on seam = σ_n

$$\sigma_n = \sigma_1 \cos^2\theta + \sigma_2 \sin^2\theta$$

$$= 21 \times 10^6 \times \cos^2 30° + 42 \times 10^6 \times \sin^2 30°$$

$$= 15{\cdot}75 \times 10^6 + 10{\cdot}5 \times 10^6$$

$$= 26{\cdot}25 \times 10^6 \text{ N/m}^2 \text{ or } 26{\cdot}25 \text{ MN/m}^2 \qquad \text{Ans. } (i)$$

Shear stress on seam $= \tau$

$$\tau = \tfrac{1}{2}\sin 2\theta\,(\sigma_2 - \sigma_1)$$

$$= 0{\cdot}5 \times \sin 60° \times (42 - 21) \times 10^6$$

$$= 9{\cdot}093 \times 10^6 \text{ N/m}^2 \text{ or } 9{\cdot}093 \text{ MN/m}^2 \quad \text{Ans. } (ii)$$

56. Coefficient of contraction of area

$$= \frac{\text{area of contracted section}}{\text{area of orifice}}$$

$$= \frac{0{\cdot}7854 \times 20^2}{0{\cdot}7854 \times 25^2} = 0{\cdot}64$$

$$\text{Theo. velocity} = \sqrt{2gh}$$

$$= \sqrt{2 \times 9{\cdot}81 \times 1{\cdot}22} = 4{\cdot}893 \text{ m/s}$$

$$\text{Theo discharge} = \text{theo. velocity} \times \text{area of orifice}$$

$$= 4{\cdot}893 \times 0{\cdot}7854 \times 0{\cdot}025^2$$

$$= 2{\cdot}401 \times 10^{-3} \text{ m}^3\text{/s}$$

$$\text{Actual discharge} = \frac{89{\cdot}7 \times 10^{-3}}{60} = 1{\cdot}495 \times 10^{-3} \text{ m}^3\text{/s}$$

$$\text{Coeff. of discharge} = \frac{\text{actual discharge}}{\text{theoretical discharge}}$$

$$= \frac{1{\cdot}495 \times 10^{-3}}{2{\cdot}401 \times 10^{-3}} = 0{\cdot}6226$$

$$\text{Coeff. of discharge} = \text{coeff. of vel.} \times \text{coeff. of area}$$

$$\text{Coeff. of velocity} = \frac{0{\cdot}6226}{0{\cdot}64} = 0{\cdot}9728$$

Coefficient of velocity, area and discharge are, respectively:
0·9728, 0·64, and 0·6226 Ans. (*i*)

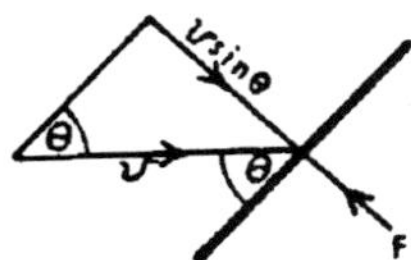

Mass of 89·7 l of fresh water = 89·7 kg

Force = change of momentum per second

= mass per second × change of velocity

Loss of momentum per second in direction normal to the plane

= mass per second × v sin θ

$$\text{Force on plate} = \frac{89{\cdot}7}{60} \times 4{\cdot}893 \times 0{\cdot}9728 \times 0{\cdot}7071$$

= 5·032 N Ans. (*ii*)

57. Centrifugal force = $m\omega^2 r$

Since all masses rotate at same angular velocity, each force can be represented by mr.

A = 2·8 × 0·125 = 0·35

B = 3·5 × 0·1 = 0·35

C = 4·2 × 0·075 = 0·315

D = 4·8 × 0·125 = 0·6

E = m × 0·2 = 0·2m

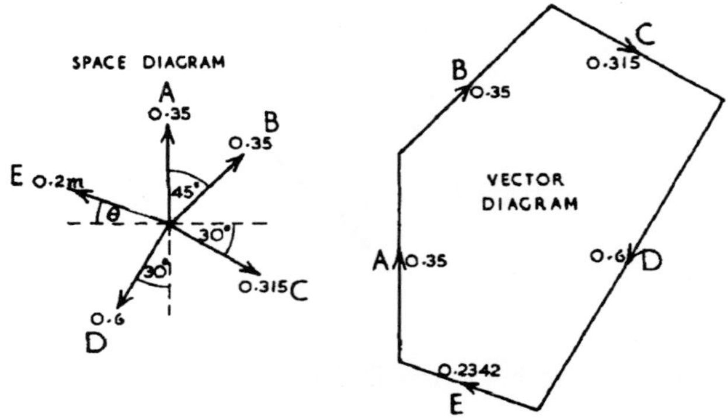

Resolving into vertical and horizontal forces:

A = 0·35 up

B = 0·35 cos 45° = 0·2475 up, 0·35 sin 45° = 0·2475 right

C = 0·315 sin 30° = 0·1575 down, 0·315 cos 30° = 0·2728 right

D = 0·6 cos 30° = 0·5196 down, 0·6 sin 30° = 0·3 left

Vertical forces, 0·35 + 0·2475 = 0·5975 up

0·1575 + 0·5196 = 0·6771 down

Net vertical force = 0·6771 – 0·5975 = 0·0796 down

Horizontal forces, 0·2475 + 0·2728 = 0·5203 right

Net horizontal force = 0·5203 – 0·3 = 0·2203 right

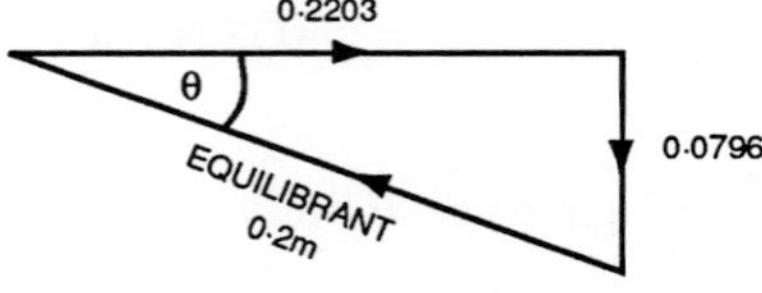

$$\tan\theta = \frac{0{\cdot}0796}{0{\cdot}2203} = 0{\cdot}3613$$

$$\theta = 19°52' \qquad 270° + 19°\,52' = 289°52'$$

$$\text{Equilibrant} = 0{\cdot}2m = \frac{0{\cdot}0796}{\sin 19°\,52'} = 0{\cdot}2342$$

$$m = \frac{0{\cdot}2342}{0{\cdot}2} = 1{\cdot}71$$

Equilibrant E = 1·171 kg at 289° 52' Ans

58. Let W = load, and x = eccentricity

Direct stress

$$= \frac{\text{load}}{\text{area}} = \frac{4W}{\pi D^2}$$

Bending stress

$$= \frac{32M}{\pi D^3} = \frac{32Wx}{\pi D^3}$$

Let compressive stress be positive and tensile stress negative:
Compressive stress on one side,

$$= \frac{4W}{\pi \times 0{\cdot}12^2} + \frac{32Wx}{\pi \times 0{\cdot}12^3} = 15 \times 10^6 \quad \ldots (i)$$

Compressive stress on other side,

$$= \frac{4W}{\pi \times 0{\cdot}12^2} - \frac{32Wx}{\pi \times 0{\cdot}12^3} = -3 \times 10^6 \quad \ldots (ii)$$

Adding (*i*) and (*ii*),

$$\frac{2 \times 4 \times W}{\pi \times 0{\cdot}12^2} = 12 \times 10^6$$

$$W = \frac{12 \times 10^6 \times \pi \times 0{\cdot}12^2}{2 \times 4}$$

$$= 6{\cdot}787 \times 10^4 \text{N} \qquad \text{Ans. } (ia)$$

$$\text{Mass} = \frac{6{\cdot}787 \times 10^4}{9{\cdot}81} = 6{\cdot}918 \times 10^3 \text{ kg}$$

$$= 6{\cdot}918 \text{ t} \qquad \text{Ans. } (ib)$$

Dividing (*i*) by $\frac{4W}{\pi \times 0{\cdot}12^2}$ and inserting value of W,

$$1 + \frac{8x}{0{\cdot}12} = \frac{15 \times 10^6 \times \pi\, 0{\cdot}12^2 \times 2 \times 4}{4 \times 12 \times 10^6 \times \pi \times 0{\cdot}12^2}$$

$$\frac{8x}{0 \cdot 12} = 2 \cdot 5 - 1$$

$$x = \frac{1 \cdot 5 \times 0 \cdot 12}{8}$$

$$= 0 \cdot 0225 \text{ m or } 22 \cdot 5 \text{ mm}$$ Ans. (*ii*)

59. (a) The applied torque is transmitted partly by the solid shaft and partly by the hollow shaft.

i.e. $T_A + T_B = 6 \times 10^3$ Nm. (i)

Each shaft must be subject to the same angle of twist, at the section where they are coupled together.

i.e. $\theta_A = \theta_B$

From the torsion equation,

$$\theta = \frac{Tl}{JG}$$

$$\frac{T_A l_A}{J_A G_A} = \frac{T_B l_B}{J_B G_B}$$

$$\frac{T_A \times 2 \cdot 5}{\frac{\pi}{32} \times 0 \cdot 075^4} = \frac{T_B \times 5}{\frac{\pi}{32} \times (0 \cdot 1^4 - 0 \cdot 05^4)}$$

$T_A = 0 \cdot 675\, T_B$ (ii)

Substituting $T_A = 0 \cdot 675\, T_B$ in equation (i):

$$0 \cdot 675\, T_B + T_B = 6 \times 10^3$$

$$1 \cdot 675\, T_B = 6 \times 10^3$$

$$T_B = 3 \cdot 58 \text{ kNm}$$

Substituting in

equation (i) $T_A = 6 - 3 \cdot 58$

$$T_A = 2 \cdot 42 \text{ kNm.}$$

Torque in the hollow and solid shafts = 3·58 kNm and 2·42 kNm.

Ans. (a)

(b) Considering the solid shaft,

$$\theta_A = \frac{T_A l_A}{J_A G_A}$$

$$= \frac{2{\cdot}42 \times 10^3 \times 2{\cdot}5}{\frac{\pi}{32} \times 0{\cdot}075^4 \times 83 \times 10^9}$$

$$= 0{\cdot}0235 \text{ rad}$$

$$= \frac{2\pi}{360} \times 0{\cdot}0235 \text{ deg}$$

Angular movement of coupling = 1·35° Ans. (b)

60.

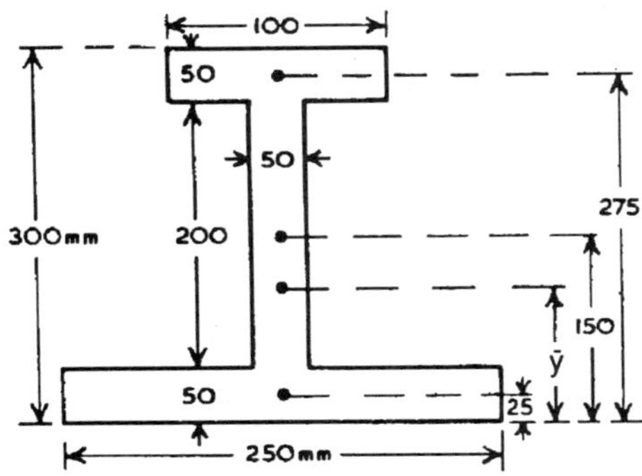

The neutral axis passes through the centre of gravity of the section, let this by at $\bar{y}$ from the base, working throughout in metres:

$$\bar{y} = \frac{\sum \text{moments of areas}}{\sum \text{areas}}$$

$$= \frac{0{\cdot}05 \times 0{\cdot}1 \times 0{\cdot}275 + 0{\cdot}05 \times 0{\cdot}2 \times 0{\cdot}15 + 0{\cdot}05 \times 0{\cdot}25 \times 0{\cdot}025}{0{\cdot}05 \times 0{\cdot}1 + 0{\cdot}05 \times 0{\cdot}2 + 0{\cdot}05 \times 0{\cdot}25}$$

$$\bar{y} = \frac{0{\cdot}0275 + 0{\cdot}03 + 0{\cdot}00625}{0{\cdot}55} = 0{\cdot}1159 \text{ m}$$

For maximum tensile stress in outer fibres at bottom:

Let total load including weight of beam $= W$

Maximum bending moment (at centre) $= \frac{WL}{8}$

$$\frac{M}{I} = \frac{\sigma}{\bar{y}}$$

$$\frac{W \times 3}{8 \times 2{\cdot}785 \times 10^{-4}} = \frac{15{\cdot}4 \times 10^6}{0{\cdot}1159}$$

$$W = \frac{8 \times 2{\cdot}785 \times 15{\cdot}4 \times 10^2}{3 \times 0{\cdot}1159}$$

$$= 9{\cdot}868 \times 10^4 \text{ N} = 98{\cdot}68 \text{ kN}$$

Weight of beam $=$ area $\times$ length $\times$ density $\times g$

$= 0{\cdot}05\,(0{\cdot}1 + 0{\cdot}2 + 0{\cdot}25) \times 3 \times 7210 \times 9{\cdot}81$

$= 5835$ N $= 5{\cdot}835$ kN

Load to be carried $=$ total load $-$ weight of beam

$= 98{\cdot}68 - 5{\cdot}835$

$= 92{\cdot}845$ kN

or, $92{\cdot}845 \div 3 = 30{\cdot}95$ kN per metre run Ans. (*i*)

From neutral axis to outer fibres of top flange

$= 0{\cdot}3 - 0{\cdot}1159 = 0{\cdot}1841$ m

Compressive stress at top

$$= 15{\cdot}4 \times \frac{0{\cdot}1841}{0{\cdot}1159}$$

$= 24{\cdot}46$ MN/m^2 Ans. (*ii*)

61.

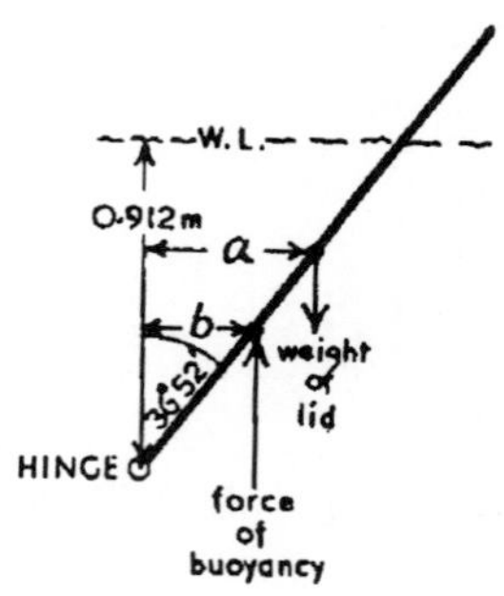

$$\text{Let } \rho = \text{density of lid material}$$
$$\text{Mass of lid} = \text{volume} \times \text{density}$$
$$= 1{\cdot}52 \times 1{\cdot}52 \times 0{\cdot}025 \times \rho$$
$$= 0{\cdot}05776\,\rho$$
$$\text{Weight} = mg$$
$$\text{C.G. of lid from hinge} = \tfrac{1}{2} \times 1{\cdot}52 = 0{\cdot}76 \text{ m}$$

Perpendicular distance of weight from hinge

$$a = 0{\cdot}76 \times \sin 36° \, 52' = 0{\cdot}4559 \text{ m}$$

Length of lid immersed from hinge to water level

$$= \frac{0{\cdot}912}{\cos 36° \, 52'} = 1{\cdot}14 \text{ m}$$

C.G. of displaced water through which force of buoyancy acts, from hinge $= \frac{1}{2} \times 1{\cdot}14 = 0{\cdot}57$ m

Perpendicular distance,

$$b = 0{\cdot}57 \times \sin 36° \, 52' = 0{\cdot}3419 \text{ m}$$
$$\text{Mass of water displaced} = 1{\cdot}14 \times 1{\cdot}52 \times 0{\cdot}025 \times 1024$$
$$= 44{\cdot}36 \text{ kg}$$
$$\text{Force of buoyancy} = mg$$

Moments about hinge

$$\text{Weight of lid} \times a = \text{Force of buoyancy} \times b$$
$$0{\cdot}05776 \times \rho \times 0{\cdot}4559 = 44{\cdot}36 \times 0{\cdot}3419$$
$$\rho = 576 \text{ kg/m}^2$$

$$\text{relative density} = \frac{\text{density of wood}}{\text{density of fresh water}}$$

$$= \frac{576}{1000} = 0{\cdot}576 \qquad \text{Ans.}$$

62.

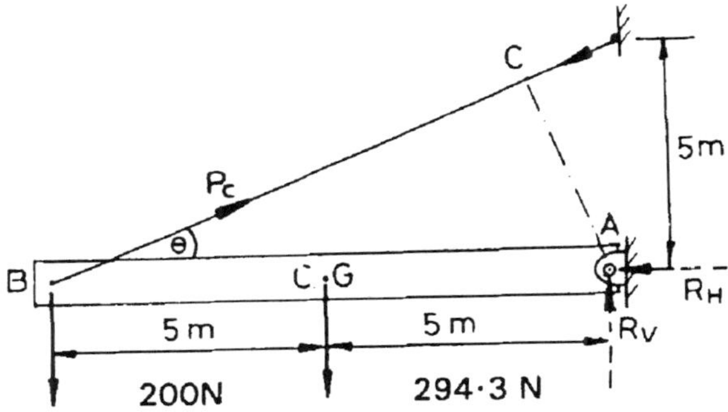

$$\text{The weight of the beam} = 3 \times 10\ 9{\cdot}81$$
$$= 294{\cdot}3 \text{ N}$$

$$\text{Tan } \theta = \frac{5}{10}$$

$$\text{Angle } \theta = 26{\cdot}5°$$

Let P_C = force in the cable

Considering the equilibrium of the beam AB, taking moments about A,

$P_C \times$ perpendicular

$$\text{distance } AC = (294{\cdot}3 \times 5) + (200 \times 10)$$
$$P_C \times 10 \sin 26{\cdot}5° = 3471{\cdot}5$$

Force in the cable = 778 N at 26·5° above horizontal.

Ans.

Resolve force P_C into horizontal and vertical components:

$$\text{Horizontal component} = 778 \times \cos 26{\cdot}5°$$
$$= 696{\cdot}2 \text{ N (to right)}$$
$$\text{Vertical component} = 778 \times \sin 26{\cdot}5°$$
$$= 347{\cdot}1 \text{ N (upward)}$$

Let R_H = horizontal reaction force at pin

R_V = vertical reaction force at pin

For equilibrium, equating vertical forces:

$$R_V + 347{\cdot}1 = 200 + 294{\cdot}3$$
$$\therefore\ R_V = 147{\cdot}2 \text{ N}$$

and, equating

horizontal forces, $R_H = 696{\cdot}2$ N

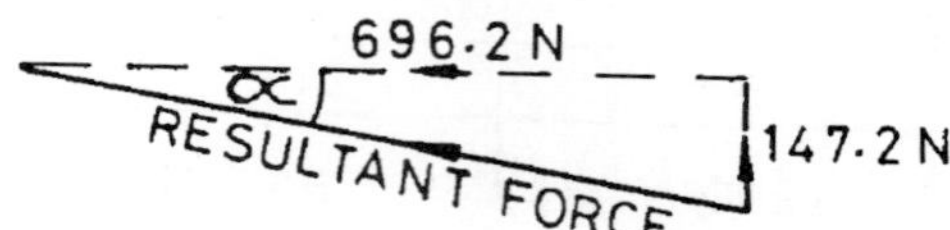

$$\text{Resultant reaction force} = \sqrt{696{\cdot}2^2 + 147{\cdot}2^2}$$

$$= 711{\cdot}6\ \text{N}$$

$$\text{Tan}\ \alpha = \frac{147{\cdot}2}{696{\cdot}2}$$

$$\text{angle}\ \alpha = 11{\cdot}9°$$

Reaction force in the pin = 711·6 N at 11·9° below horizontal.
Ans.

63. $$H = 0{\cdot}15 + 0{\cdot}45 = 0{\cdot}6\ \text{m}$$

$$\text{Length AB} = \frac{0{\cdot}9}{\sin 60°} = 1{\cdot}039\ \text{m}$$

$$\text{Hyd. load on gate} = HA\rho g$$

$$= 0{\cdot}6 \times 1{\cdot}039 \times 1{\cdot}2 \times 1000 \times 9{\cdot}81$$

$$= 7{\cdot}339\ \text{kN}$$

$$y_{cp} = \frac{\text{2nd mom. of area about surface}}{\text{1st mom. of area about surface}}$$

$$= \frac{I_G}{AH/\sin\theta} + \frac{H}{\sin\theta}$$

$$= \frac{1{\cdot}2 \times 1{\cdot}039^2 \times \sin 60°}{12 \times 1{\cdot}2 \times 1{\cdot}039 \times 0{\cdot}6} + \frac{0{\cdot}6}{\sin 60°}$$

$$= 0{\cdot}125 + 0{\cdot}6928$$

$$= 0{\cdot}822\ \text{m from surface at 0}$$

$$\text{Distance to C.o.}P.\ \text{from hinge} = 0{\cdot}822 - \frac{0{\cdot}15}{\sin 60°}$$

$$= 0{\cdot}6488 \text{ m}$$

Let x = distance from hinge to position of added mass

Take moments about the hinge:

$$(7339 \times 0{\cdot}6488) = \left(1600 \times 9{\cdot}81 \times \frac{1{\cdot}039}{2} \times \cos 60°\right) + (2000 \times x \times \cos 60°)$$

$$4761{\cdot}5 = 4077 + 1000x$$

$$x = 0{\cdot}685 \text{ m}$$

Position of 2 kN weight = 685 mm from hinge. Ans.

64. Camshaft speed = 5 rev/s

time for 1 rev. = $\frac{1}{5}$ s

$$\text{Time for 1 osc. of valve} = \frac{1}{5} \times \frac{95°}{360°}$$

$$t = 0{\cdot}05278 \text{ s}$$

$$\omega = \frac{2\pi}{0{\cdot}05278}$$

$$\omega = 119 \text{ rad/s}$$

v_{MAX} (at mid travel) $= \omega r$

$= 119 \times 0{\cdot}025$

Maximum velocity = 2·975 m/s Ans. (a)

a_{MAX} (at ends) $= \omega^2 r$

$= 119^2 \times 0{\cdot}025$

Maximum acceleration = 354 m/s^2 Ans. (b)

Top: resultant force = Spring force + $mg - ma$

$= 1000 + (1{\cdot}1 \times 9{\cdot}81) - (1{\cdot}1 \times 354)$

$= 621{\cdot}4$ N

Bottom: resultant force = Spring force + $mg + ma$

$= 150 + (1{\cdot}1 \times 9{\cdot}81) + 1{\cdot}1 \times 354)$

$= 550{\cdot}2$ N Ans. (c)

65.

$$\text{Extension } x = \frac{Fl}{AE}$$

$$1{\cdot}67 \times 10^{-3} = \frac{81 \times 10^3 \times 3}{\frac{\pi}{4} \times 0{\cdot}032^2 \times E}$$

Modules of Elasticity = 180·9 GN/m² Ans. (a)

When mass is dropped:

$$\text{Inst. extension } x = \frac{\sigma l}{E}$$

$$3{\cdot}34 \times 10^{-3} = \frac{\sigma \times 3}{180{\cdot}9 \times 10^9}$$

$$\text{Inst. stress } \sigma = 201{\cdot}4 \text{ MN/m}^2$$

$$P.E. \text{ lost by mass} = S.E. \text{ gained by rod}$$

$$m \times g \times (h + x) = \frac{\sigma^2}{2E} \times \text{volume}$$

$$m \times g \times (22{\cdot}34 \times 10^{-3}) = \frac{(201{\cdot}4 \times 10^6)^2}{2 \times 180{\cdot}9 \times 109} \times \frac{\pi}{4} \times 0{\cdot}032^2 \times 3$$

Magnitude of mass = 1234·3 kg Ans. (b)

66. Weight of ramp = 2000 × 9·81
= 19620 N

For one cable, taking moments about the hinge at 0:

$$P \times 4 \sin 30° = \frac{19620}{2} \times 2{\cdot}5 \cos 10°$$

Minimum force in a cable = 12·076 kN Ans. (a)

For equilibrium the three forces must be concurrent:

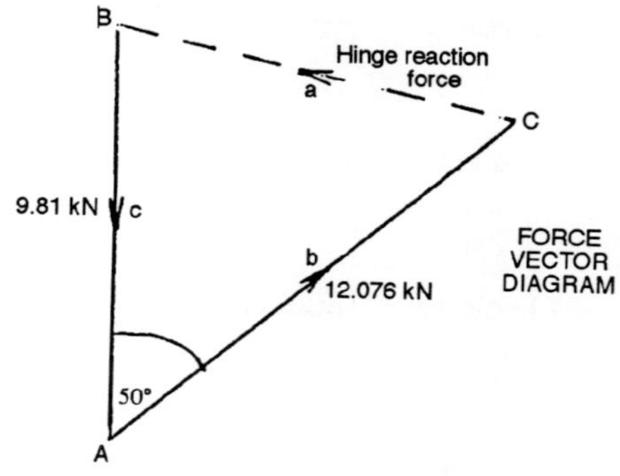

$$a^2 = 12{\cdot}076^2 + 9{\cdot}81^2 - 2 \times 12{\cdot}076 \times 9{\cdot}81 \times \cos 50°$$
$$= 145{\cdot}8 + 96{\cdot}24 - 152{\cdot}3$$
$$a = 9{\cdot}473 \text{ kN}$$

$$\text{Total hinge force} = 2 \times 9{\cdot}473 = 18{\cdot}95 \text{ kN}$$

$$\frac{9{\cdot}473}{\sin 50°} = \frac{12{\cdot}076}{\sin B}$$

$$\text{Angle B} = 77{\cdot}6°$$

Reaction at hinge= 18·95 kN at 77·6° to vertical Ans. (b)

67. $$\text{Depth to C.o.P.} = \frac{I_G}{AH} + H$$

$$= \frac{I_G}{\frac{\pi}{4} D^2 \times H} + H$$

$$\text{Distance from hinge to C.o.P.} = \frac{I_G}{\frac{\pi}{4} D^2 \times H} + H - H$$

$$= \frac{I_G}{\frac{\pi}{4} D^2 \times H}$$

$$\text{Hydrostatic load} = HA\rho g$$

$$= H \times \frac{\pi}{4} D_2^2 \times \rho \times g$$

Moment of force about hinge $= \dfrac{I_G}{\frac{\pi}{4} D^2 \times H} \times H \times \dfrac{\pi}{4} D^2 \times \rho \times g$

Torque required $= I_G \, \rho \, g$

Torque is independent of head H Ans.

68. Mean radius of pads $= \dfrac{r_1 + r_2}{2}$

$$\text{Friction force} = \mu R_N$$
$$= \mu \times P$$
$$\text{Friction torque} = \text{friction force} \times \text{mean radius}$$

Torque to overcome friction $= \dfrac{r_1 + r_2}{2} \times \mu \times P$ Ans. (a)

$$\text{Friction torque} = 0{\cdot}02 \times 160 \times 10^3 \times \frac{0{\cdot}15 + 0{\cdot}3}{2}$$
$$= 720 \text{ Nm}$$
$$\text{Power} = T\omega$$
$$= 720 \times 2\pi \times 2{\cdot}5$$
$$= 11310 \; W$$

$$\%\text{ Power loss} = \frac{11310}{2 \times 10^6} \times 100$$

$= 0{\cdot}565$ Ans. (b)

69. Container: take moments about T_1:

$$T_2 \times 8 = 392{\cdot}4 \times 6$$
$$T_2 = 294{\cdot}3 \text{ kN}$$
$$T_1 = 392{\cdot}4 - 294{\cdot}3$$
$$= 98{\cdot}1 \text{ kN}$$

Beam: take moments about R_1

$$(25{\cdot}5 \times 3) + (294{\cdot}3 \times 6) = (R_2 \times 8) + (2 \times 98{\cdot}1)$$

$$R_2 = 205{\cdot}8 \text{ kN}$$
$$\text{and } R_1 = 392{\cdot}4 + 25{\cdot}5 - 205{\cdot}8$$
$$R_1 = 212{\cdot}1 \text{ kN}$$

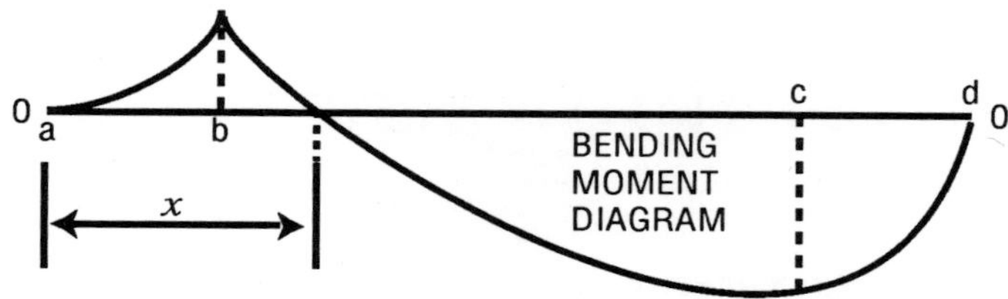

$$M_b = (98{\cdot}1 \times 2) + (2{\cdot}55 \times 2 \times 1) = 201{\cdot}3 \text{ kNm (hog)}$$
$$M_C = -(205{\cdot}8 \times 2) + (2{\cdot}55 \times 2 \times 1) = -406{\cdot}5 \text{ kNm (sag)}$$

Point of contraflexure is at x (where M is zero; slope changes)

Ans. (a)

$$O = (98{\cdot}1 \times x) + (2{\cdot}55 \times x \times \frac{x}{2}) - 212{\cdot}1 \times (x - 2)$$

$$O = 98{\cdot}1x + 1{\cdot}275x^2 - 212{\cdot}1x + 424{\cdot}2$$
$$x^2 - 89{\cdot}4x + 332{\cdot}7 = O$$

$$x = \frac{+89{\cdot}4 \pm \sqrt{89{\cdot}4^2 - (4 \times 1 \times 332{\cdot}7)}}{2}$$

$$x = 3{\cdot}891 \text{ (other answer inadmissible)}$$

Point of contraflexure = 3·891 m from L.H. end of beam

Ans. (b)

70. Zero flow: $O = d(1{\cdot}1d - 40) \times \sqrt{h\rho}$

$$O = 1{\cdot}1d - 40$$

Orifice diameter = 36·36 mm Ans. (a)(*i*)

Max. flow: $30\,000 = d\,(1{\cdot}1d - 40) \times \sqrt{4 \times 920}$

$$30\,000 = (1{\cdot}1d^2 - 40\,d) \times 60{\cdot}66$$
$$1{\cdot}1d^2 - 40d - 494{\cdot}5 = 0$$

$$d = \frac{40 \pm \sqrt{40^2 - (4 \times 1{\cdot}1 \times -494{\cdot}5)}}{2 \times 1{\cdot}1}$$

$$d = \frac{40 + 61{\cdot}45}{2{\cdot}2}$$

Orifice diameter $= 46{\cdot}11$ mm Ans. (a)(*ii*)

At half way: $d = \dfrac{36{\cdot}36 + 46{\cdot}11}{2}$

$$d = 41{\cdot}24 \text{ mm}$$

$$\dot{m} = 41{\cdot}24\,(1{\cdot}1 \times 41{\cdot}24 - 40) \times 60{\cdot}66$$

Mass flow rate $= 13429$ kg/h Ans. (b)

71. Pressure due to liquid head $= 816 - 770$

$= 46 \text{ kN/m}^2$

$$p = \rho g h$$

$$h = \frac{46 \times 10^3}{518 \times 9{\cdot}81}$$

Depth of liquid $= 9{\cdot}052$ m Ans. (a)

Volume of liquid $= \dfrac{\pi h^2}{3}(3r - h)$

$$= \frac{\pi \times 9{\cdot}052^2}{3}(15 - 9{\cdot}052)$$

$= 510{\cdot}4 \text{ m}^3$

Mass of liquid $= V \times \rho$

$= 510{\cdot}4 \times 518$ kg

$= 264{\cdot}4$ t

Weight of liquid $=$ mg

$= 264{\cdot}4 \times 10^3 \times 9{\cdot}81$

$= 2594$ kN Ans. (b)

72. At exit:

Linear velocity of vanes $= \pi \times 0{\cdot}3 \times \dfrac{900}{60}$

$$= 14{\cdot}137 \text{ m/s}$$

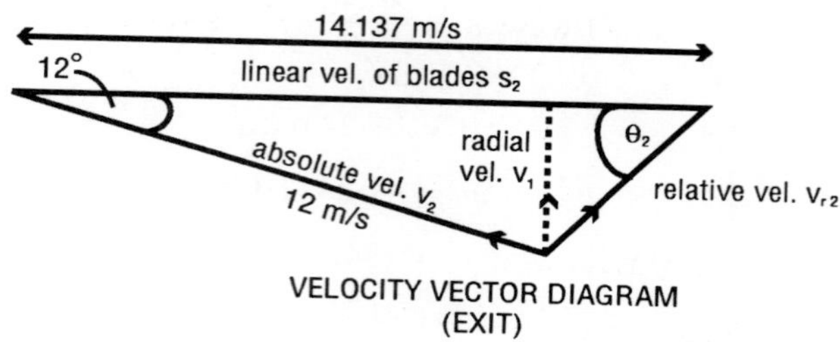

VELOCITY VECTOR DIAGRAM (EXIT)

$$(\text{Rel. velocity})^2 = 12^2 + 14{\cdot}137^2 - (2 \times 12 \times 14{\cdot}137 \times \cos 12°)$$

$$= 11{\cdot}981$$

$$\text{Relative velocity} = 3{\cdot}461 \text{ m/s}$$

$$\frac{3{\cdot}461}{\sin 12°} = \frac{12}{\sin \theta_2}$$

Vane exit angle $\theta_2 = 46{\cdot}14°$ Ans. (a)

$$\text{Radial velocity} = 12 \times \sin 12° = 2{\cdot}495 \text{ m/s}$$

$$\dot{V} = \text{exit flow area} \times \text{exit radial velocity.}$$

$$= \pi \times 0{\cdot}3 \times 0{\cdot}03 \times 2{\cdot}495$$

$$= 0{\cdot}07055 \text{ m}^3\text{/s}$$

$$\dot{m} = 0{\cdot}07055 \times 1025$$

Mass flow rate $= 72{\cdot}31$ kg/s Ans. (b)

$$\dot{V} = \text{inlet width} \times \text{circ.} \times \text{radial vel.}$$

$$0{\cdot}07055 = b_1 \times \pi \times 0{\cdot}1 \times 2$$

Inlet width $= 0{\cdot}1123$ m $= 112{\cdot}3$ mm Ans. (c)

73.

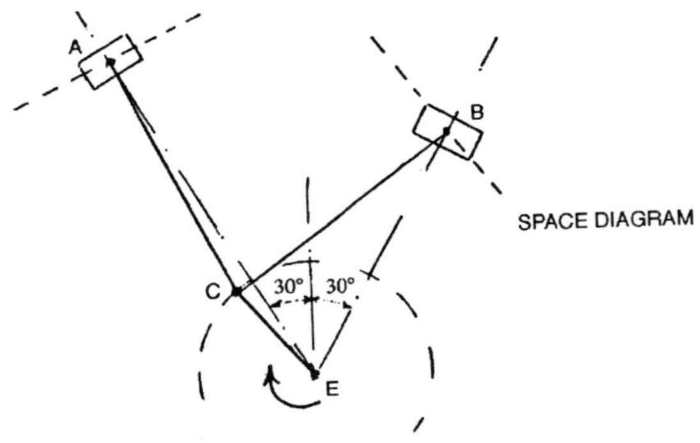

$$\text{Crankpin: } \omega = \frac{2\pi \times 600}{60}$$

$$= 62{\cdot}832 \text{ rad/s}$$

$$v = 62{\cdot}832 \times 0{\cdot}3$$

$$= 18{\cdot}85 \text{ m/s}$$

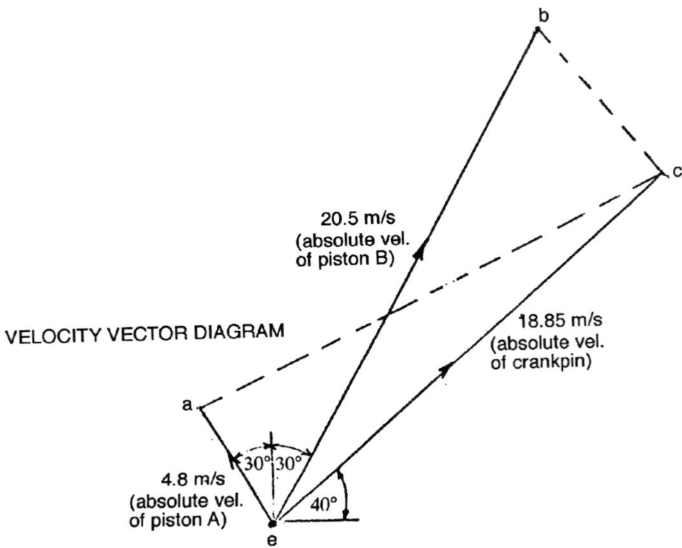

Piston A vel. = 4·8 m/s
Piston B vel. = 20·5 m/s } Ans.

74. Force due to pressure head:

$$p = \rho g h$$
$$= 860 \times 9{\cdot}81 \times 20$$
$$= 168{\cdot}7 \text{ kN/m}^2$$

Force in pipe $= 168{\cdot}7 \times 10^3 \times \frac{\pi}{4} \times 0{\cdot}6^2$

$= 47{\cdot}7$ kN

From the force vector diagram:

Resultant force $= 47{\cdot}7 \div \cos 45°$

$= 67{\cdot}5$ kN (*i*)

Force due to change of velocity:

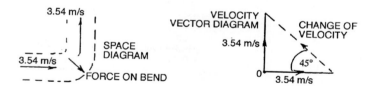

$$\text{Flow velocity} = \frac{1}{0.7854 \times 0{\cdot}6^2}$$
$$= 3{\cdot}54 \text{ m/s}$$

From the velocity vector diagram:

Change of velocity $= 3{\cdot}54 \div \cos 45°$

$= 5{\cdot}01$ m/s

$$\text{Force on bend} = \dot{m} \times \text{change of velocity}$$
$$= 1 \times 860 \times 5{\cdot}01$$
$$= 4{\cdot}31 \text{ kN} \qquad (ii)$$

From (*i*) and (*ii*)
total force on bend $= 71{\cdot}81$ kN Ans.
(acting outward at 45°)

75.

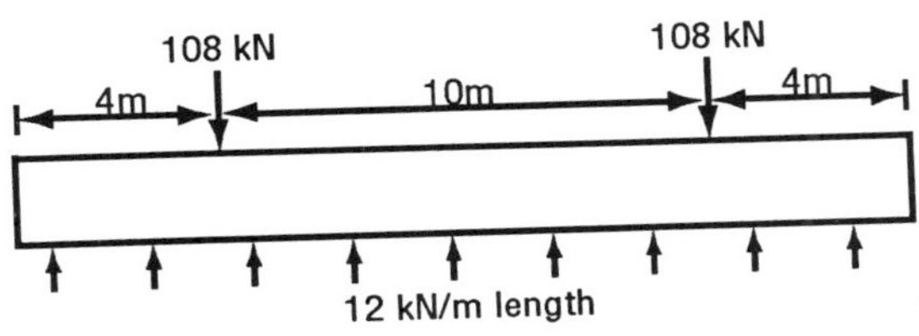

By Archimedes, upthrust force = 216 kN = 12 kN/m.

$$\text{M at ends} = 0$$
$$\text{M at loads} = 12 \times 4 \times \frac{4}{2} = 96 \text{ kNm}$$
$$\text{M at mid-length} = (12 \times 9 \times \frac{9}{2}) - (108 \times 5) = -54 \text{ kNm}$$

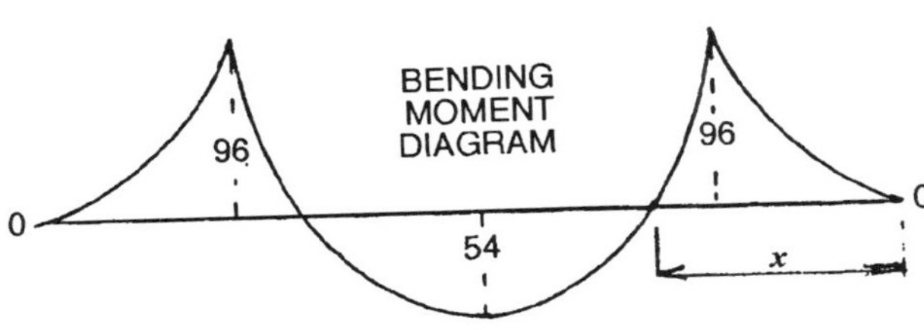

Ans. (a)

Let x = distance from end to point of contraflexure

$$M = (12 \times x \times \frac{x}{2}) - 108\,(x-4)$$
$$O = 6x^2 - 108x + 432$$
$$O = x^2 - 18x + 72$$
$$O = (x-6)\,(x-12)$$
$$x = 6 \text{ and } x = 12$$

Points of contraflexure at 6 m from each end. Ans. (b)

$$\text{Mass of water displaced} = \frac{216 \times 10^3}{8} = 22018 \text{ kg}$$

$$\text{Volume displaced} = \frac{22018}{10^3} = 22{\cdot}018 \text{ m}^3$$

$$\text{Increase of draught} = \frac{22{\cdot}018}{3 \times 18}$$

$$= 0{\cdot}4078 \text{ m}$$ Ans. (c)

76. 4·5m side: hyd. load $= HA\rho g$

$$= 2{\cdot}25 \times 3 \times 4{\cdot}5 \times 10^3 \times 9{\cdot}81$$

$$= 298 \text{ kN}$$

1·8 m side: hyd. load $= 0{\cdot}9 \times 3 \times 1{\cdot}8 \times 10^3 \times 9{\cdot}81$

$$= 47{\cdot}7 \text{ kN}$$

Resultant hyd. load $= 298 - 47{\cdot}7$

$$= 250{\cdot}3 \text{ kN}$$ Ans. (a)

C.o.P. at $\frac{1}{3}$ depth from bottom on each side

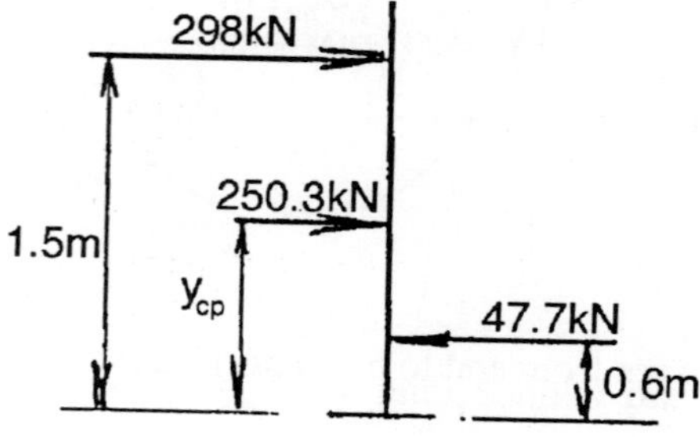

Take moments about the bottom at 0:

$$250{\cdot}3 \times y_{cp} = (298 \times 1{\cdot}5) - (47{\cdot}7 \times 0{\cdot}6)$$

Resultant C.o.P. $= 1{\cdot}672$ m up from bottom Ans. (b)

Take moments about the hinge:

for equilibrium $R \times 3 \sin 20° = 250{\cdot}3 \times 1{\cdot}5$

Reaction force $R = 365{\cdot}9$ kN Ans. (c)

77. Frequency is the number of oscillations made per second by the body.

Periodic-time is the time taken by the body to make one oscillation. Ans. (a)

For one spring:

Additional load = 6000 kg

$$\text{Spring stiffness (s)} = \frac{6000 \times 9{\cdot}81}{0{\cdot}075}$$

$= 784{\cdot}8$ kN/m Ans. (b)

$$\text{Load on one spring} = \frac{5000}{4}$$

$= 1250$ kg

$$\text{Frequency (f)} = \tfrac{1}{2}\pi\sqrt{\frac{k}{\text{mass}}}$$

$$= \tfrac{1}{2}\pi\sqrt{\frac{784{\cdot}8 \times 10^3}{1250}}$$

$= 3{\cdot}988$ osc/s Ans. (c)(*i*)

$$\text{Load on one spring} = \frac{29\,000}{4}$$

$= 7250$ kg

$$\text{Frequency (f)} = \tfrac{1}{2}\pi\sqrt{\frac{784{\cdot}8 \times 10^3}{7250}}$$

$= 1{\cdot}656$ osc/s Ans. (c) (*ii*)

78. At engagement: $C.F. = \text{spring force}$

$$m\omega^2 r = \text{spring stiffness} \times \text{deflection}$$

$$0{\cdot}75 \times \omega^2 \times 0{\cdot}15 = 100 \times 10^3 \times 10 \times 10^{-3}$$

$$\therefore \omega = 94{\cdot}28 \text{ rad/s}$$

$$\text{Speed} = 900{\cdot}3 \text{ rev/min}$$ Ans.(a)

At 1500 rev/min: $$C.F. = 0{\cdot}75 \times \left(\frac{2\pi \times 1500}{60}\right)^2 \times 0{\cdot}15$$

$$= 2776 \text{ N}$$

$$\text{Force between shoe and rim} = C.F. - \text{spring force}$$

$$= 2776 - 1000$$

$$= 1776 \text{ N}$$ Ans. (b)

$$\text{For one shoe friction force} = \mu R_N$$

$$\text{torque} = \mu R_N \times \text{radius of rim}$$

for four shoes $$T = 4 \times 0{\cdot}3 \times 1776 \times 0{\cdot}18$$

$$\text{Torque} = 383{\cdot}6 \text{ Nm}$$ Ans. (c)

79. Mass flow rate = 7500 kg/h in each branch

At the fixed orifice: $$7500 = 30\sqrt{h \times 900}$$

$$\left(\frac{7500}{30}\right)^2 = h \times 900$$

$$h = 69{\cdot}44 \text{ mm}$$

At the variable orifice: $$7500 = C \times 69{\cdot}44^{1{\cdot}3}$$

$$C = 30{\cdot}27$$ Ans. (a)

Mass flow rate = 12 000 kg/h at fixed orifice

= 3000 kg/h at variable orifice

For the fixed orifice; $$12000 = 30\sqrt{h \times 900}$$

$$h = 177{\cdot}8 \text{ mm}$$

At the variable orifice: $$3000 = C \times 177{\cdot}8^{1{\cdot}3}$$

$$C = 3{\cdot}566$$ Ans. (b)

80. Water side: hyd. load $= HA\rho g$

$$= 2{\cdot}4 \times \frac{\pi}{4} \times 1{\cdot}2^2 \times 1025 \times 9{\cdot}81$$

$$= 27{\cdot}29 \text{ kN}$$

Air side: load $= 12 \times 10^3 \times \frac{\pi}{4} \times 1{\cdot}2^2$

$$= 13{\cdot}57 \text{ kN}$$

Resultant load $= 27{\cdot}29 - 13{\cdot}57$

$= 13{\cdot}72$ kN (on water side) Ans. (a)

Water side: $y_{cp} = \dfrac{\text{2nd mom. of area about surface}}{\text{1st mom. of area about surface}}$

$$= \frac{I_G + AH^2}{AH}$$

$$= \frac{I_G}{AH} + H$$

$$= \frac{\frac{\pi D^4}{64}}{\frac{\pi}{4} D^2 \times H} + H$$

$$= \frac{1{\cdot}2^2}{16 \times 2{\cdot}4} + 2{\cdot}4$$

$= 2{\cdot}437$ m below surface

Air side: force acts at centroid of area.

Taking moments about the water surface:

$(13{\cdot}72 \times y_{cp}) = (27{\cdot}29 \times 2{\cdot}437) - (13{\cdot}57 \times 2{\cdot}4)$

$y_{cp} = 2{\cdot}475$ m below surface (resultant; vertically)

C.o.P. = 75 mm below door centre Ans. (b)

INDEX

L

M

N

O

P

R

REED'S MARINE ENGINEERING SERIES

Vol. 1 MATHEMATICS
Vol. 2 APPLIED MECHANICS
Vol. 3 APPLIED HEAT
Vol. 4 NAVAL ARCHITECTURE
Vol. 5 SHIP CONSTRUCTION
Vol. 6 BASIC ELECTROTECHNOLOGY
Vol. 7 ADVANCED ELECTROTECHNOLOGY
Vol. 8 GENERAL ENGINEERING KNOWLEDGE
Vol. 9 STEAM ENGINEERING KNOWLEDGE
Vol. 10 INSTRUMENTATION AND CONTROL SYSTEMS
Vol. 11 ENGINEERING DRAWING
Vol. 12 MOTOR ENGINEERING KNOWLEDGE

REED'S ENGINEERING KNOWLEDGE FOR DECK OFFICERS
REED'S MATHS TABLES AND ENGINEERING FORMULAE
REED'S MARINE DISTANCE TABLES
REED'S OCEAN NAVIGATOR
REED'S SEXTANT SIMPLIFIED
REED'S SKIPPERS HANDBOOK
REED'S COMMERCIAL SALVAGE PRACTICE
REED'S MARITIME METEOROLOGY
SEA TRANSPORT – OPERATION AND ECONOMICS

These books are obtainable from all good Nautical Booksellers or direct from:

THOMAS REED PUBLICATIONS
The Barn, Ford Farm
Bradford Leigh
Bradford-on-Avon
Wiltshire BA15 2RP
United Kingdom

Tel: 01225 868821
Fax: 01225 868831

Email: tugsrus@abreed.demon.co.uk